Living Together

The Biology of Animal Parasitism

Living Together
The Biology of Animal Parasitism

William Trager

The Rockefeller University
New York, New York

PLENUM PRESS • NEW YORK AND LONDON

Library of Congress Cataloging in Publication Data

Trager, William, 1910–
Living together.

Includes bibliographical references and index.
1. Host-parasite relationships. I. Title. II. Title: Biology of animal parasitism.
QL757.T67 1986 591.5′249 86-15097
ISBN 0-306-42310-3

A Division of Plenum Publishing Corporation
233 Spring Street, New York, N.Y. 10013

Printed in the United States of America

To Ida

Foreword

William Trager has been an avid student of parasites for over 50 years at the Rockefeller University. Around the turn of this century, parasitology enjoyed a certain vogue, inspired by colonial responsibilities of the technically advanced countries, and by the exciting etiological and therapeutic discoveries of Ross, Manson, Ehrlich, and others. For some decades, the Western hemisphere's interest in animal parasites has been eclipsed by concern for bacteria and viruses as agents of transmissible disease. Only very recently, initiatives like the Tropical Disease Research programs of WHO–World Bank–UNDP, and the Great Neglected Disease networks of the Rockefeller and MacArthur Foundations have begun to compensate for the neglect of these problems by United States federal health research agencies. Throughout that period, however, the Rockefeller Institute (later University) has given high priority to the challenges of parasitism, corresponding during a formidable period with Dr. Trager's own career.

The present work then, is a distillation of the insight collected by our principal doyen of parasite biology, informed but by no means confined to his own research. It is addressed to the reader of broad biological interest and training, not to the specialist. The disarmingly unpretentious style makes the work readily accessible to college undergraduates or even to gifted high school students; but do not be deceived thereby, as it has an enormous range of factual information and theoretical insight, familiar to few, but potentially important to most biologists. This was a shrewd and well-contrived choice, and I am sure the book will add much to the current momentum of interest in the field.

Trager's work is organized by themes of biological interest, not by the taxonomy of the parasites or of the disease syndromes of the hosts; but these are not neglected where pertinent. As the title implies it concerns the biology of parasitism, not just a survey of parasites. The writing is therefore in the tradition of Theobald Smith, McFarlane Burnet, and Rene Dubos whose writings have been so important in bringing bacteria and viruses, and their parasitic behavior, into the mainstream framework of evolutionary biology. Like these forerunners, Trager focuses on the developmental, biochemical, and genetic adaptations by which the parasite exploits its special ecological niche, and by which the host seeks to retain its own Malthusian fitness in the face of that challenge. Extraordinarily, he is able to unite a half-century of experience with the latest findings and perspectives of molecular biology, which is, of course, bringing this field of study into a revolutionary new phase.

What impressed me, and what will enrich a generation of new molecular parasitology entrants seeking key research problems, is the range of fabulous stories in this book. It is a veritable Arabian Nights of narrative, not of the human imagination, but of Nature's, in the exposition of phenomena of adaptation and specificity. On every page, the author exhibits his profound awareness of the conundrums they pose for physico–chemical and developmental–genetic principles still to be elaborated—to explain specificity for hosts and organs, tropisms, response to host rhythms, and endocrinology; the morphogenetic cycling of vegetative/reproductive phases, and the questions these raise for the differential control of gene expression. Nothing in the biology of the parasitic relationship escapes notice, be it the nutritional requirements of the parasite, the molecular genetics of the kinetoplasts, the mechanisms of pathogenesis, the host defenses, or the rationale and means of chemotherapy. Amusing and challenging are the reports of ways in which parasites alter host behavior and even growth towards the ends of the parasite.

I am tempted to borrow his examples; but that would be transparent and redundant plagiarism—the reader has but to turn to random pages, or scan the logically organized Contents. The work is also enriched by a systematic set of life cycle diagrams, indispensable for an overview of parasite natural history. For writing so easily digested, it is also fully documented in the bibliography following each chapter. It should be said that parasite here is meant to embrace animal parasites of other animal species, though the principles will be of great pertinence to parasitism by fungi, bacteria, and even viruses.

Many young scientists will, I hope, read this work: There are enough research challenges to keep them all fully occupied in an area which is as rich with human needs as it is with challenges to biological imagination. Others will find great stimulation and enjoyment, and a small lament that we do not have multiple lifetimes to enjoy and observe what the next decades of research will bring to the field, which Trager sings of so eloquently.

Joshua Lederberg
President
The Rockefeller University
New York, New York

Acknowledgments

Much of this book was written during several summers at the Marine Biological Laboratory at Woods Hole, Massachusetts. I have been very fortunate in having available to me the resources of two outstanding biological libraries: one at the Marine Biological Laboratory and the other at the Rockefeller University, my scientific home for over 50 years. I have also been fortunate in having a dedicated staff. In particular I want to thank Mr. James Stanorski for accurate and rapid preparation of the manuscript and Mr. Erminio Gubert for skillful mounting of the illustrations. My laboratory work, meanwhile, was kept going through the outstanding ability of Mrs. Marika Tershakovec, research technician, and the devoted help of Mrs. Cora Fields.

I am indebted to the following fellow scientists who kindly reviewed particular sections of the book and gave me the benefit of their comments and suggestions. Professor P.A. D'Alesandro, School of Public Health, Columbia University; Professor Joel E. Cohen, Populations Laboratory, the Rockefeller University; Professor G.A.M. Cross, Laboratory of Molecular Parasitology, the Rockefeller University; Professor D. Despommier, School of Public Health, Columbia University; Professor H.N. Lanners, Tulane University Medical School; Professor F. von Lichtenberg, Harvard University Medical School; Professor A.J. MacInnis, Department of Zoology, University of California, Los Angeles; Professor M. Müller, Laboratory of Biochemical Cytology, the Rockefeller University; Professor N. Noguiera, New York University Medical School; Professor Margaret Perkins, Laboratory of Biochemical Cytology, the Rockefeller University; Dr. A. Sher, Laboratory of Parasitic Diseases, National Institute of Allergy and Infectious Diseases; Professor L. Simpson, Biology Department, University of California, Los Angeles; Professor M.J. Ulmer, Iowa State University and University of Bridgeport, Connecticut; Professor C.C. Wang, Department of Pharmaceutical Chemistry, University of California, San Francisco; and Professor L.P. Weiss, School of Veterinary Medicine, University of Pennsylvania.

I am especially grateful to Dr. Joshua Lederberg, President of the Rockefeller University, for having been so kind as to write the Foreword.

William Trager

Contents

CHAPTER 1

Introduction

Parasitism involves an intimate association between two different kinds of organisms. One of these, the host, provides food and shelter for the other, the parasite. The host may or may not be injured by the parasite. It may soon expel the parasite, or it may harbor it for many years. Since the parasite cannot exist in nature without its host, it is not to the parasite's advantage to destroy its host. At least it must not destroy it until it is ready to move to another. Some hosts are benefited by certain parasites and some are actually dependent on their parasites, a special type of association called *mutualism*. Such mutualistic or symbiotic associations may have been at the origin of chloroplasts and mitochondria, and so at the basis of most eukaryotic cells.

Throughout the living world, from prokaryotes to man, parasitic associations are very common. There is no organism (except for the viruses) that does not have its parasites. Furthermore, all the major taxonomic groups include organisms that are parasitic. To study parasites as organisms in their own right is relatively simple and straightforward. But to study the interrelations between the parasite and its host, i.e., to study parasitism, requires all the disciplines of biology from ecology to biophysics. It is this approach, study of the physiology, biochemistry, and cell biology of host–parasite relationships, that will constitute the main body of future work in parasitology, and it is this approach that will be followed in this book.

I begin with a discussion of the establishment of infection. How do parasites get from one host to another? How do they recognize and enter appropriate hosts and then find their way to particular organs and cells? Equally important are the factors in the host permitting it to accept the parasite. We must remember that all organisms are well equipped with mechanisms for the rejection of foreign structures, living as well as dead. This discussion will lead in a logical way to consideration of what occurs at the parasite–host interface, the roles of surfaces and membranes in sheltering and nourishing the parasite. It is here that uptake of nutrients occurs. I then treat the nutritional requirements of the parasites for growth and differentiation, with some emphasis on their cultivation *in vitro*. This is followed by discussions of the energy metabolism of parasites, their genetics, and the newly burgeoning field of molecular parasitology.

The emphasis then shifts somewhat as I proceed to discuss modification of the host by a parasite and host reaction to it. This is a major aspect of great practical importance; it includes the nature of parasitic disease and of innate and acquired immunity. After general discussions of these subjects, I consider in detail certain specific parasites that provide particularly instructive examples of host–parasite interactions. Possibilities for vaccines are treated here. Mutualism or symbiosis may be viewed as an ultimate evolutionary development from parasitic associations and is briefly considered. The two closing chapters deal with the two methods that have been mainly relied on in the past for treatment and control of parasitic diseases: chemotherapy and the ecological approaches, such as sanitation and vector control.

The various phenomena of parasitism are discussed in relation to specific examples. Most of these are drawn from among the protozoan and helminthic parasites of humans and other vertebrates for two reasons: (1) They are of medical or economic importance; (2) as a consequence, they have been more studied and more is known about them. Since there can be no meaningful discussion of these organisms without reference to their life cycles, a group of life cycle diagrams is included in this introductory chapter. These illustrate the developmental cycles of 11 representative protozoan parasites, of 11 helminthic parasites, and of 3 different arthropods that are involved as vectors. These diagrams are referred to throughout the book. Even a superficial examination reveals interesting parallelisms and differences among them in the various ways in which different parasites have solved the problem of getting from one host to another. Also evident in many of the life cycles are the marked morphological divergences of different stages, so marked that it is not surprising that some were considered separate species before their developmental relationships had been discovered. Even as recently as 1984, what had been considered two distinct unrelated species classified in two separate classes, one a parasite of trout and the cause of whirling disease, the other parasitic in annelid worms, were shown to be two stages in the life cycle of the same organism, each infective only to the other host. There may well be many more such species.

It is the aim of this book not only to present what is known about parasitism, but also to reveal the great gaps in our knowledge and the fascinating work yet to be done. This book is not intended as a substitute for books on parasitic diseases or on classical parasitology with emphasis on morphology and taxonomy. These books will continue to be indispensable to the student of parasitism, and a brief list of some of these general texts is presented at the end of this chapter. Also listed there are a number of more specialized books of value with regard to particular groups of parasites.

At the end of each of the subsequent chapters, a bibliography is provided dealing with the subjects of the chapter. This list consists in part of review articles, in part of a few specific older papers that I consider important to mention, in part of papers from which tables or figures have been taken, and finally of very recent papers not yet mentioned in reviews. These and the reviews should serve to lead the interested reader into the detailed research now in progress. The reviews of course refer to many workers not cited here. To them, and especially to those who may find their work mentioned without any reference at all, I extend my apologies. Perhaps they can take some satisfaction from the thought that Peyton Rous offered to a group of young researchers having lunch with him some time in the 1950s. One of these young men was complaining that his work was discussed in a paper without any reference to him or to his publication. Dr. Rous said: "That should make you feel very good. It shows your work has become part of the main body of science."

Diagrams

The following diagrams (with the exception of Diagram XIV) were prepared by artist John Smith in accordance with instructions from the author.

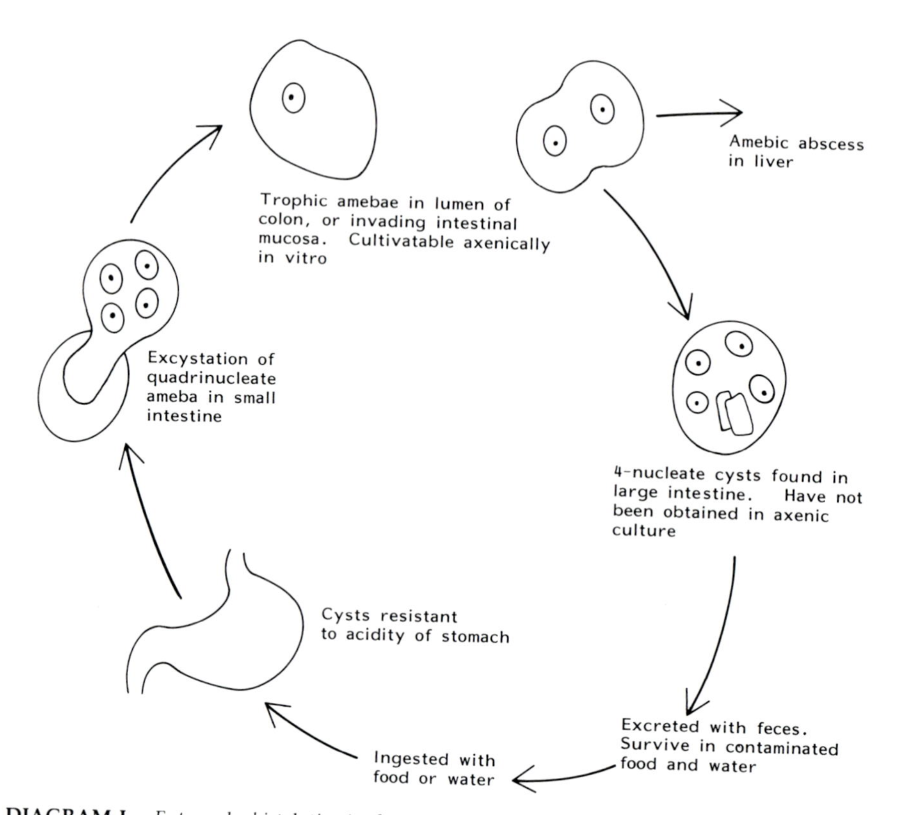

DIAGRAM I. *Entamoeba histolytica* (order Amoebida), cause of amebic dysentery. Amebic abscess of the liver is a common complication. This cycle is typical of many kinds of parasites, especially of the alimentary tract. It involves differentiation of a resistant form (here a quadrinucleate cyst) able to survive in the external environment until ingested by a suitable host.

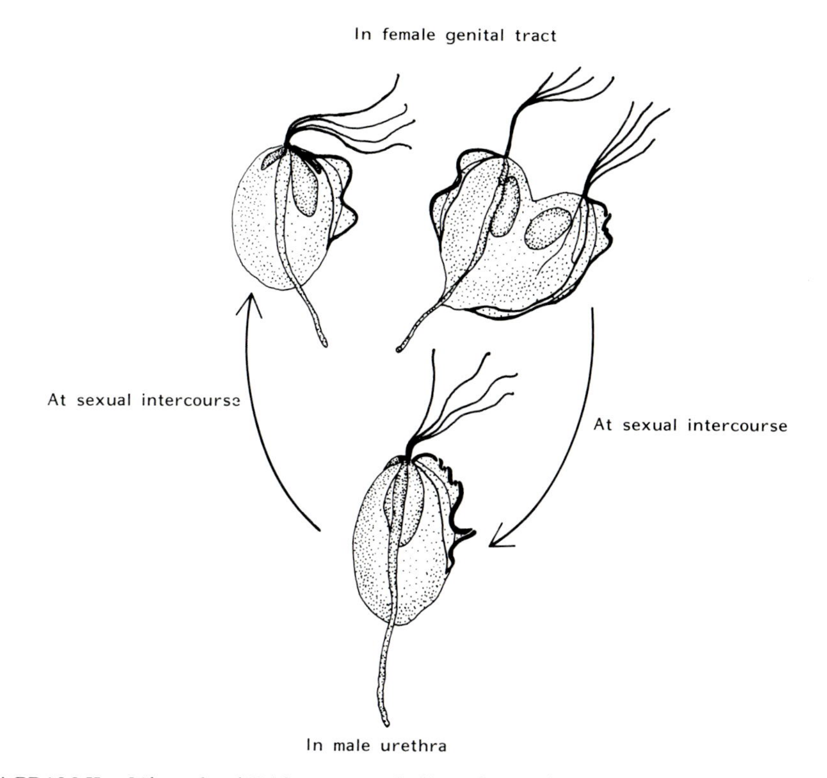

DIAGRAM II. Life cycle of *Trichomonas vaginalis* (order Trichomonadida), a cause of vaginitis in women.

This is an example of the simplest type of cycle in which asexually propagating forms are transferred directly from host to host, in this case by sexual intercourse. No special forms for transmission exist.

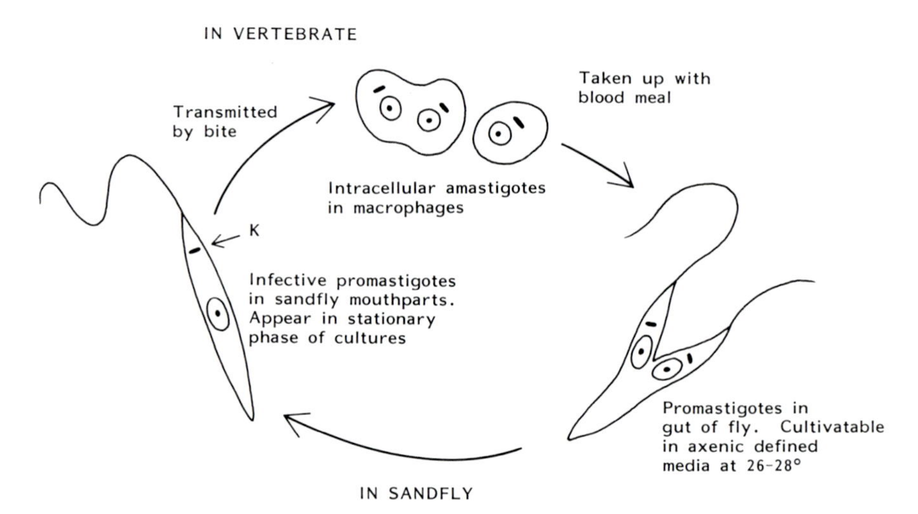

DIAGRAM III. *Leishmania donovani* and other species of *Leishmania* (order Kinetoplastida). The cycle requires two hosts, a vertebrate in which the organisms multiply intracellularly, usually in macrophages, and a sand fly (genera *Phlebotomus, Lutzomyia*) in which they develop in the alimentary tract. *L. donovani* is the agent of visceral leishmaniasis (kala-azar). Other species cause a variety of other human diseases, such as mucocutaneous leishmaniasis (espundia) and dermal leishmaniasis (Oriental sore). All of these are zoonoses. (K: kinetoplast, a region of specialized DNA within the mitochondrion.)

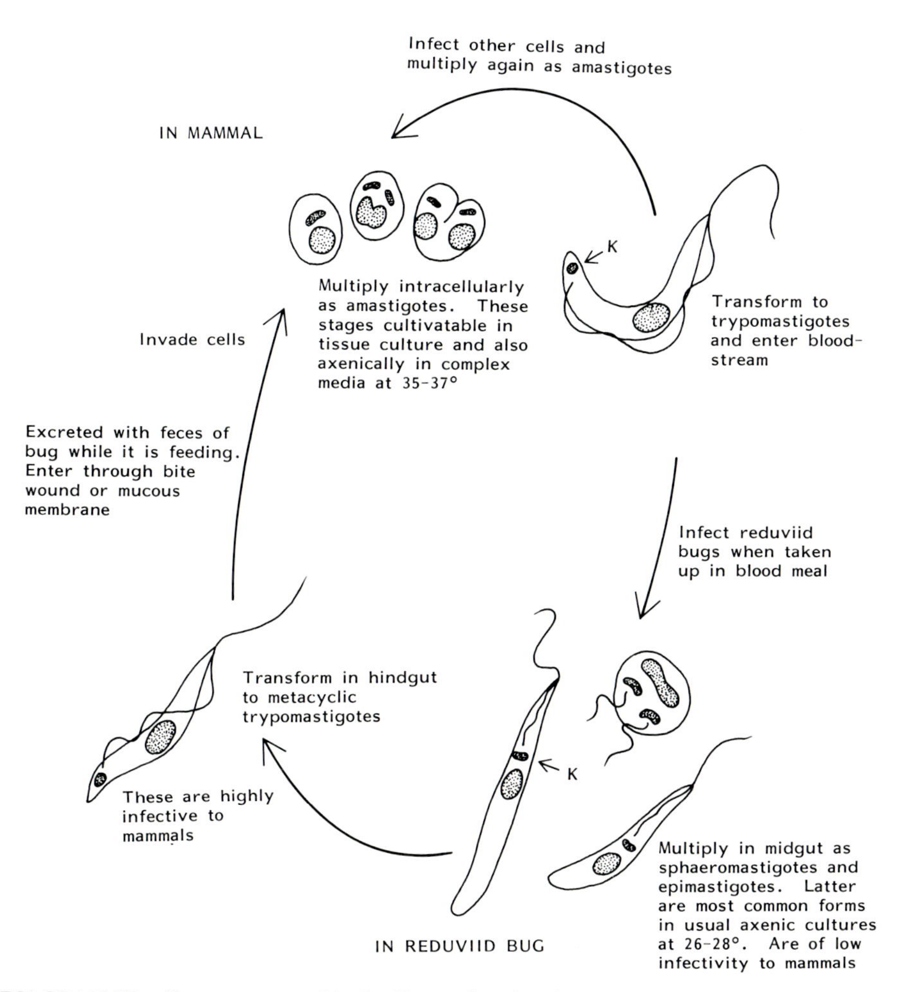

DIAGRAM IV. *Trypanosoma cruzi* (order Kinetoplastida), the agent of Chagas' disease in South and Central America. This is also a zoonosis prevalent among wild mammals and transmitted by species of bugs of the family Reduviidae. Certain species live in houses, particularly in thatch roofs and in cracks in mud walls. Chagas' disease is a principal cause of heart disease in relatively young people.

Note the changes in morphology involving loss and gain of the external flagellum and the position of the kinetoplast, anterior to the nucleus in epimastigote forms, and posterior to the nucleus in trypomastigotes.

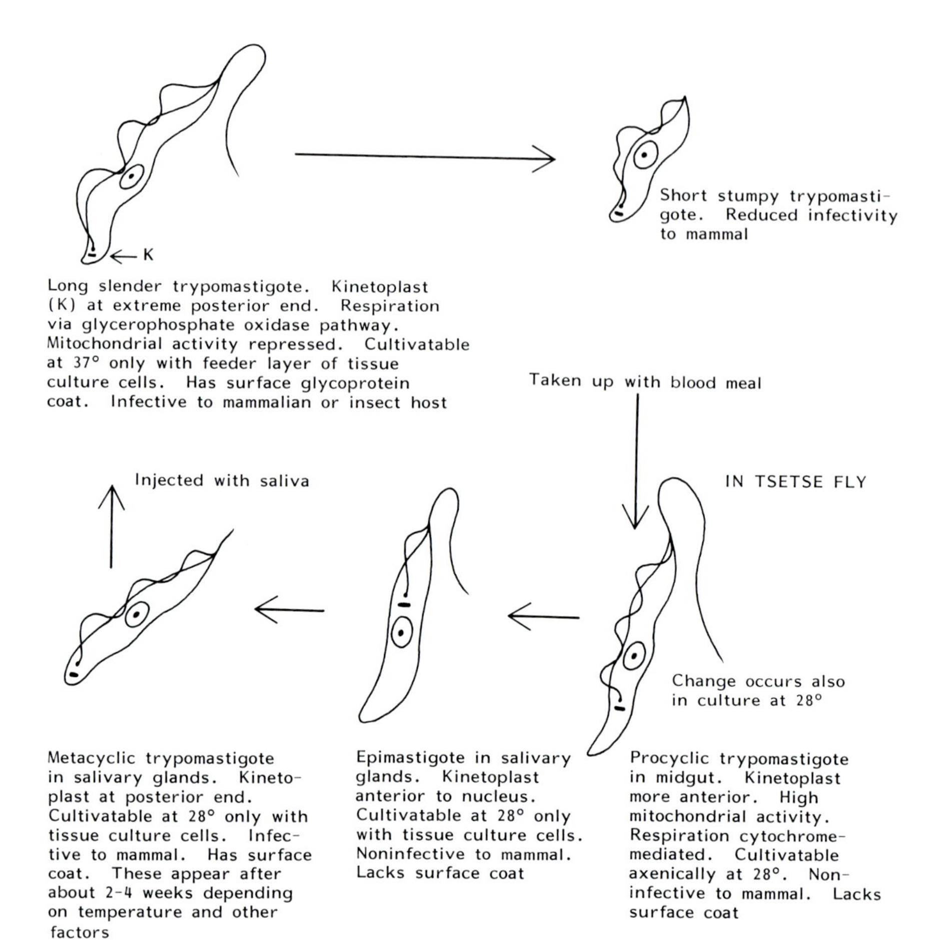

DIAGRAM V. *Trypanosoma brucei brucei* (order Kinetoplastida), the cause of nagana of cattle. Two other subspecies, *T. b. gambiense* and *T. b. rhodesiense,* are the agents of two forms of African sleeping sickness. *T. b. rhodesiense* is a zoonosis, various African game animals serving as reservoir hosts. Tsetse flies of the genus *Glossina* are the sole vectors.

As in *T. cruzi,* there are changes in the position of the kinetoplast, here reflecting changes in mitochondrial function. Multiplication by binary fission occurs in the blood and lymph, and sometimes in the central nervous system of the mammalian host, and in the midgut and salivary glands of the tsetse fly.

T. congolense and *T. vivax,* the agents of major disease of cattle in Africa, have similar cycles of development in *Glossina.* In *T. congolense* the metacyclic forms develop in the proboscis rather than the salivary glands. In *T. vivax* the entire developmental cycle occurs in the proboscis and is much shorter (about 10 days).

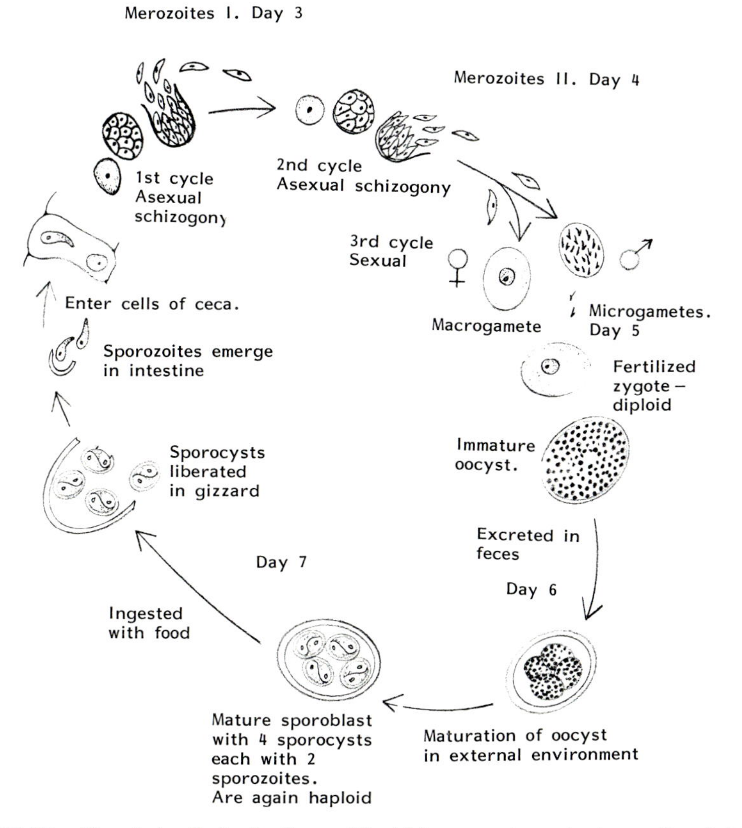

DIAGRAM VI. *Eimeria tenella* (order Eucoccidiorida), a very important parasite of chickens. This life cycle is typical of coccidian parasites of domestic animals. Both the asexual cycles and the sexual cycle occur within a single host. Transmission is effected by very resistant oocysts.

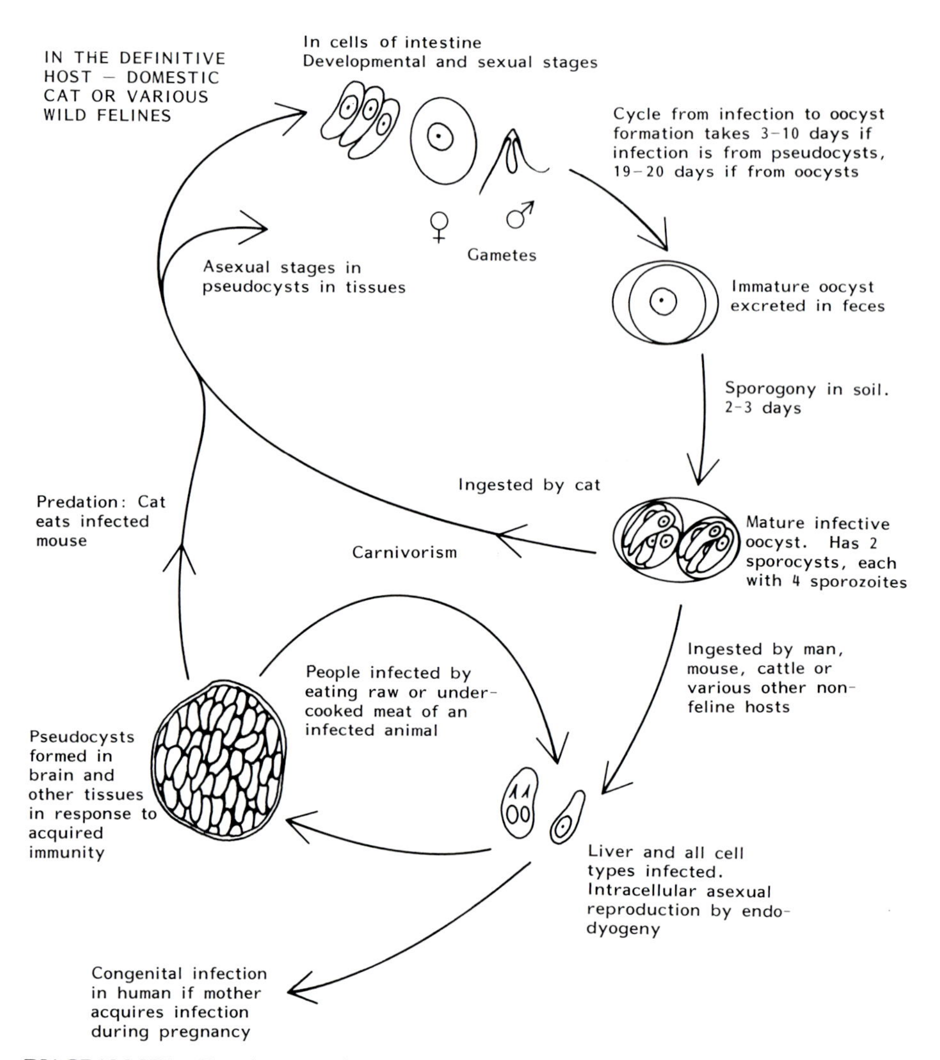

DIAGRAM VII. *Toxoplasma gondii* (order Eucoccidiorida), a widespread parasite of humans and many other animals. Only felines, however, can serve as definitive hosts and in them the developmental cycle is essentially the same as that of typical coccidia like *Eimeria* (Diagram VI). Superimposed on this cycle is an additional wholly asexual cycle of multiplication and development that occurs in rodents and many other vertebrates including domestic animals and people. Some of these then serve as intermediate hosts; felines that prey upon them become infected upon eating the infected tissues. Eating of infected meat, raw or undercooked, is the principal route of infection of people, though they may also be infected via ingestion of oocysts excreted by cats.

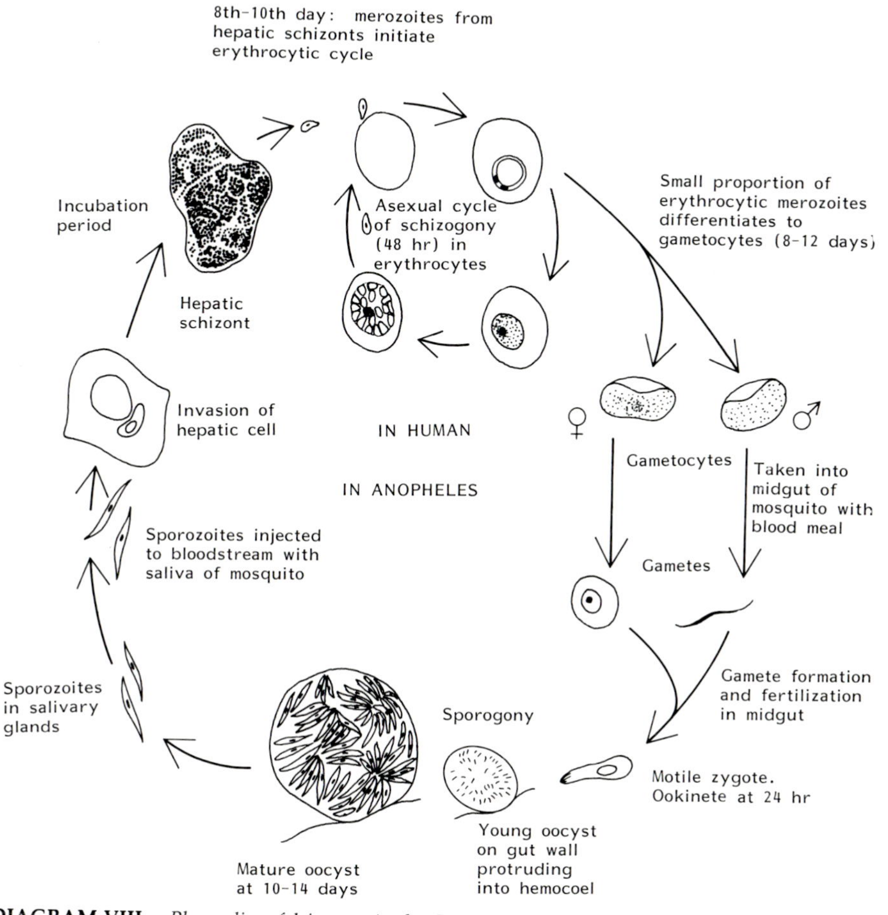

DIAGRAM VIII. *Plasmodium falciparum* (order Eucoccidiorida), the parasite causing malignant tertian malaria. This cycle is typical of all other species of malarial parasites of primates. The cycle of malarial parasites of rodents is also very similar. Among avian malarial parasites, however, the preerythrocytic cycle occurs in reticuloendothelial cells rather than in hepatic cells and merozoites formed in this cycle can either repeat it or initiate the erythrocytic cycle.

Note that the sexual and asexual cycles now occur in two different hosts, the sexual cycle in a mosquito and the asexual in a vertebrate.

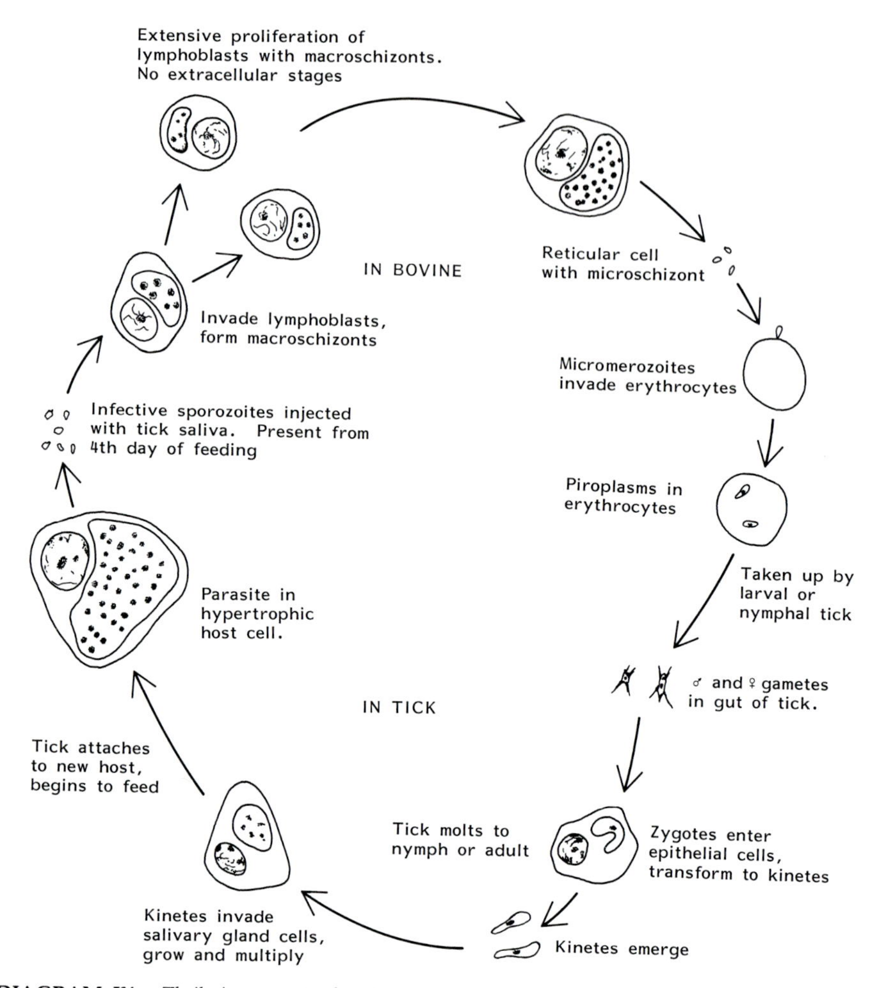

DIAGRAM IX. *Theileria parva* (order Piroplasmorida), cause of East Coast fever, a serious disease of cattle in East Africa. Note that transmission by the tick *Rhipicephalus appendiculatus* is trans-stadial. This is a necessary consequence of the fact that the ixodid ticks attach and feed only once in each of their three stages, larva, nymph, and adult. Infection acquired by a larval tick is transmitted when it feeds again as a nymph. Infection acquired by a nymph is transmitted when it feeds again as an adult. Portions of the cycle in the tick are not fully known. Likewise it is not clear whether the piroplasms in erythrocytes multiply or serve only to initiate infection in the tick.

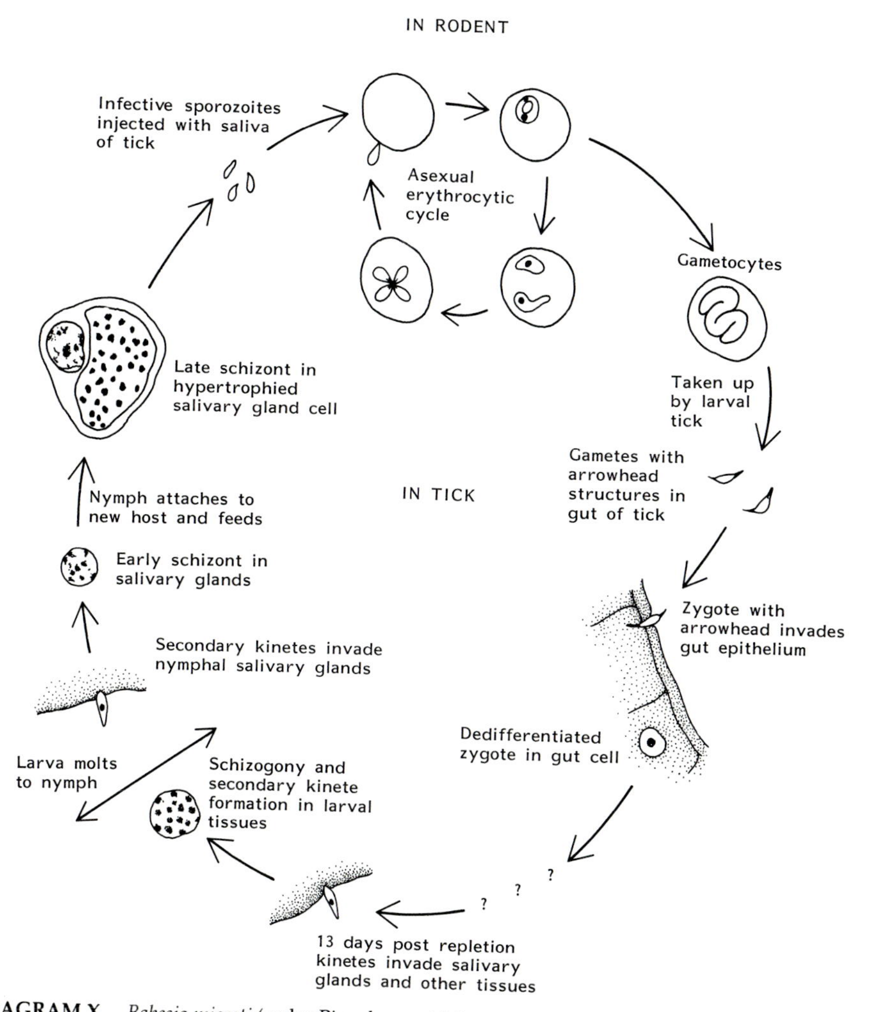

DIAGRAM X. *Babesia microti* (order Piroplasmorida), a parasite of rodents that can nonetheless infect humans. Such infection, transmitted by the bite of an infected nymphal *Ixodes dammini*, provides another example of a zoonosis. As in *Theileria*, transmission among the natural rodent hosts (usually the vole *Microtus agrestis*) is trans-stadial. The larval tick acquires the infection and transmits it when, as a nymph, it feeds again on another host. The adult ticks of this species feed on deer and are apparently not involved in the usual cycle of transmission.

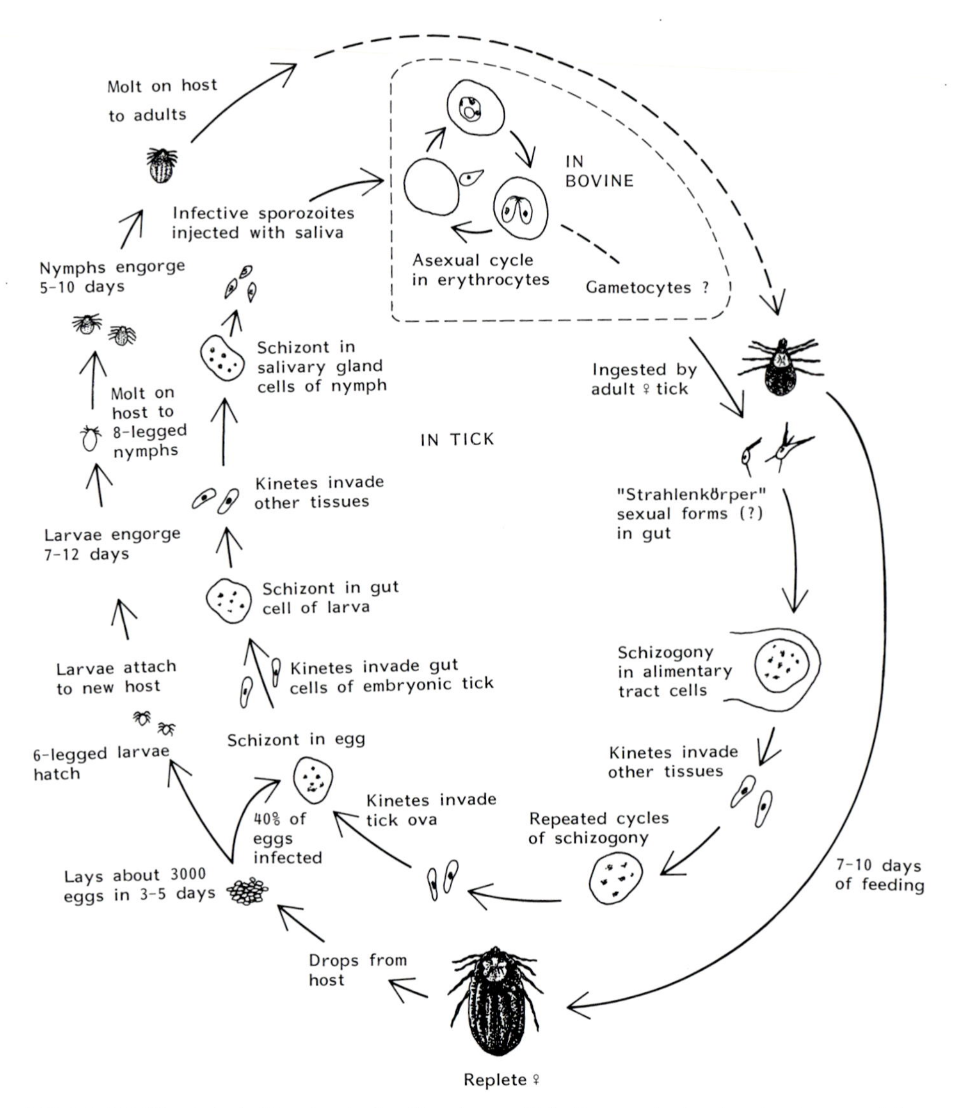

DIAGRAM XI. *Babesia bigemina* (order Piroplasmorida), the agent of red water fever or tick fever of cattle and the vector *Boophilus microplus* (order Acarina). This was the first infective agent shown to be transmitted by an arthropod. After Theobald Smith and F. L. Kilbourne had demonstrated tick transmission, the disease was controlled in the United States by acaricidal treatment of cattle to free them of the vector ticks. Such control, however, has not succeeded elsewhere and the disease continues to be of immense importance in Mexico, Central and South America, southern Europe, and Africa.

In this diagram are shown in two concentric cycles both the life cycle of the vector and that of the parasite. Note that only a small part of the *Babesia* life cycle occurs in the bovine as a cycle of reproduction in the erythrocytes. The major portion of the parasite's development takes place during the life cycle of the tick. *Boophilus* is a one-host tick so that transmission must be transovarial. Larval ticks having attached to a cow engorge and molt on the host, engorge again as nymphs, molt again on the host and feed again, dropping off only as replete adults. As with all ixodid ticks, the adult males take only a relatively small amount of blood.

If a female *Boophilus* feeds on an infected cow, many of her tissues, including the ova, become populated with schizonts of *Babesia*. When the replete female drops from the host and begins

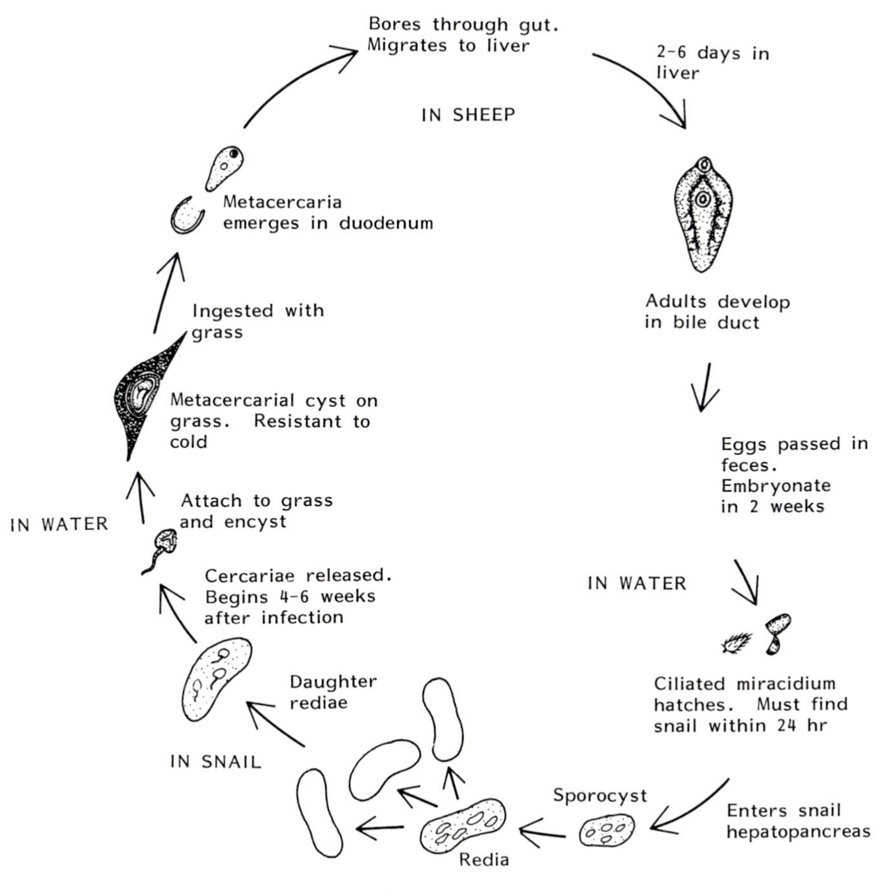

DIAGRAM XII. *Fasciola hepatica* (a digenean trematode), the liver fluke of sheep and other ruminants. Note that the life cycle involves not only two hosts, a mammal and a snail, but also two free-living stages in water than effect the transfer from one kind of host to the other. Of great importance is the extensive reproduction by polyembryony that takes place in the snail.

---

forming eggs, about 40% of these are infected with the parasite. The larvae that hatch from these infected eggs subsequently have infective sporozoites of *Babesia* in their saliva and transmit the infection to their new host when they feed as nymphs. Infected male as well as female nymphs transmit the disease but adult males play no role in its further spread. A new generation of ticks is in each case infected transovarially from an adult female that ingests infected bovine blood. Thus, a single such infected female will produce about 1200 infected larvae, each potentially able to transmit the infection to a bovine. This provides for extensive propagation and dispersion of the protozoan parasite.

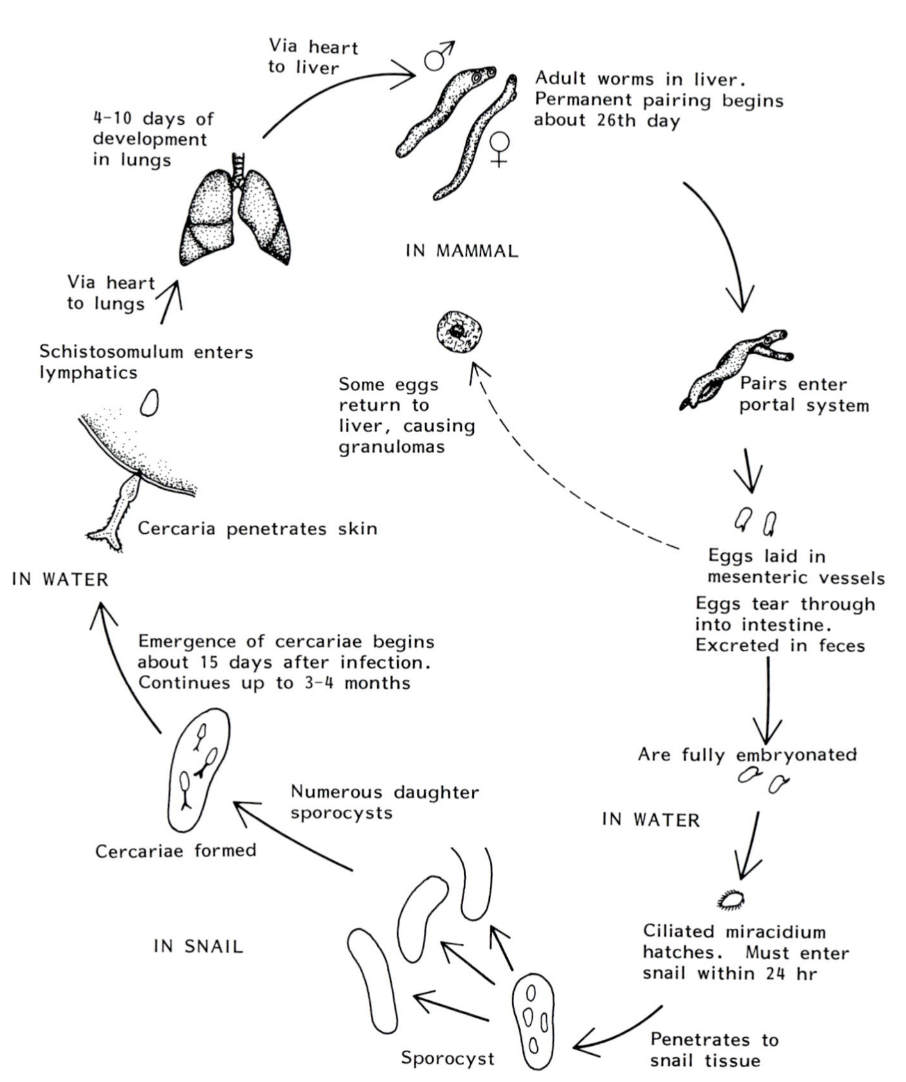

DIAGRAM XIII. *Schistosoma mansoni* (a digenean trematode), the blood fluke of humans. This is one of the three species of *Schistosoma* that cause schistosomiasis, a widespread disease affecting 200 to 300 million people. *S. mansoni* occurs in Africa, South America, and the Caribbean. Its snail hosts are in the genus *Biomphalaria* and reservoir hosts other than humans are of minor significance. *S. japonicum* occurs mainly in China, Malaysia, and the Philippines, and many kinds of mammals, including cattle, pigs, monkeys, and rats, are important reservoir hosts. Its snail hosts are in the genus *Oncomelania*. Finally, *S. haematobium*, which lives in the vessels of the bladder where it causes an inflammation, is found in Africa, the Middle East, and parts of India. It is transmitted through snails of the genus *Bulinus*. As for *S. mansoni*, reservoir hosts are of no significance.

Note the similarity of the life cycle to that of the liver fluke. The schistosomes, however, are not hermaphroditic; the sexes are separate but remain in a permanent state of copulation. Whereas the cercaria of *Fasciola* forms an encysted metacercaria that is eaten by the mammalian host, that of *Schistosoma* enters its mammalian host by penetration through the skin.

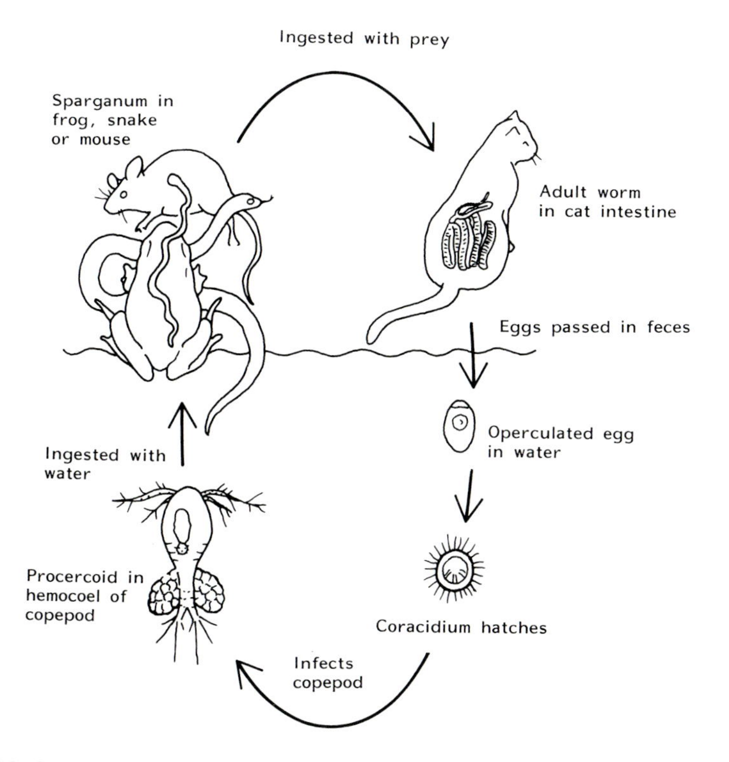

DIAGRAM XIV. *Spirometra* spp. (a cestode, suborder Pseudophyllidea). The sparganum is a larval stage that can develop in people to produce a pathological condition known as sparganosis. The usual hosts of the sparganum stage are small vertebrates that are preyed on by carnivores such as cats and dogs, within which the adult tapeworm develops.

The sparganum is the source of a remarkable growth factor (see Chapter 12). (From Mueller, 1974.)

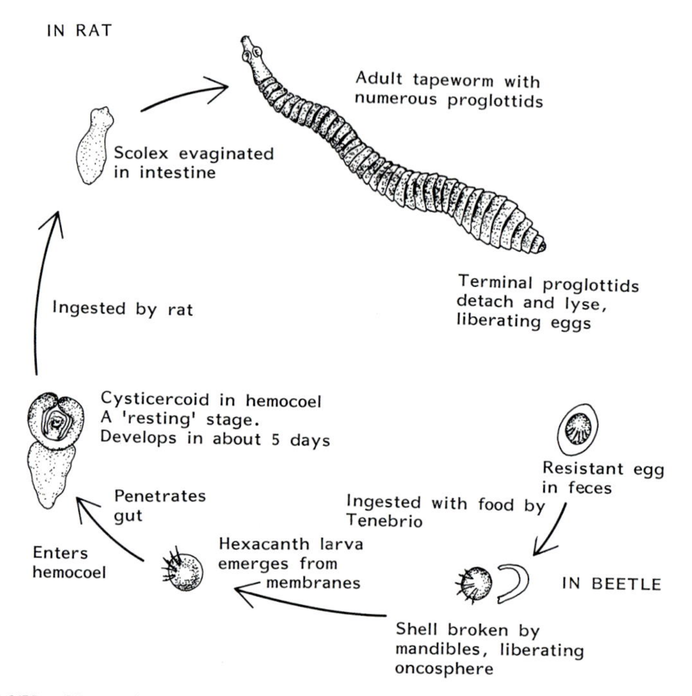

DIAGRAM XV. *Hymenolepis diminuta* (suborder Cyclophyllidea), the rat tapeworm. This is a favorite laboratory organism. Its two-host life cycle is typical of many tapeworms. Note that it has only one multiplicative phase, which occurs in the definitive host, the rat.

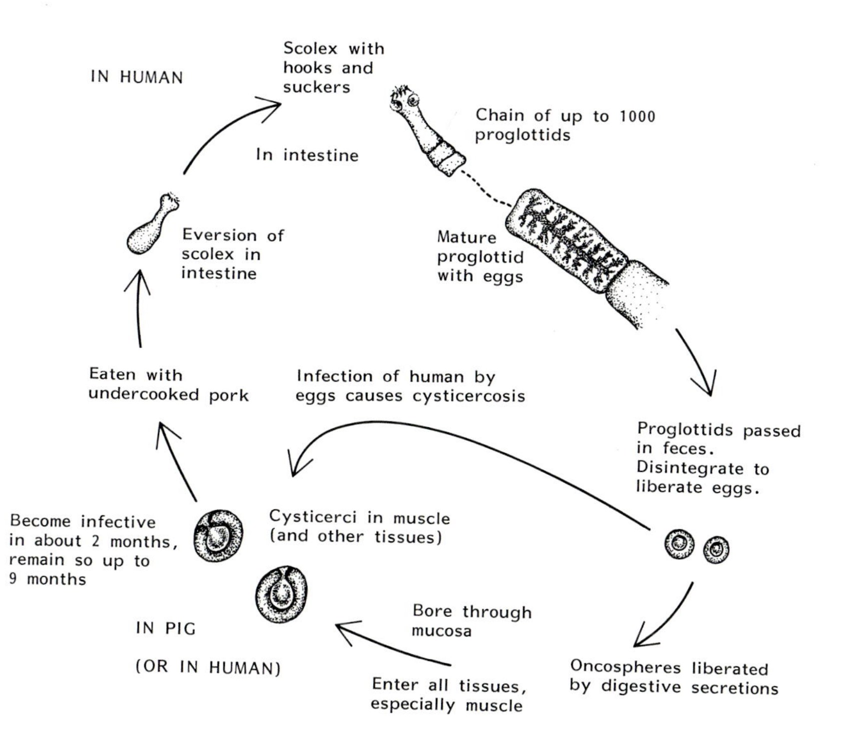

DIAGRAM XVI. *Taenia solium* (suborder Cyclophyllidea), the pork tapeworm of humans. Unlike the situation with the closely related beef tapeworm *T. saginata,* the eggs of *T. solium* are infective to humans, in whom they develop into the larval cysticercal stage normally found in pigs. Cysticercosis may be a serious disease; it occurs under unsanitary conditions. Humans can serve as both the definitive and the intermediate host.

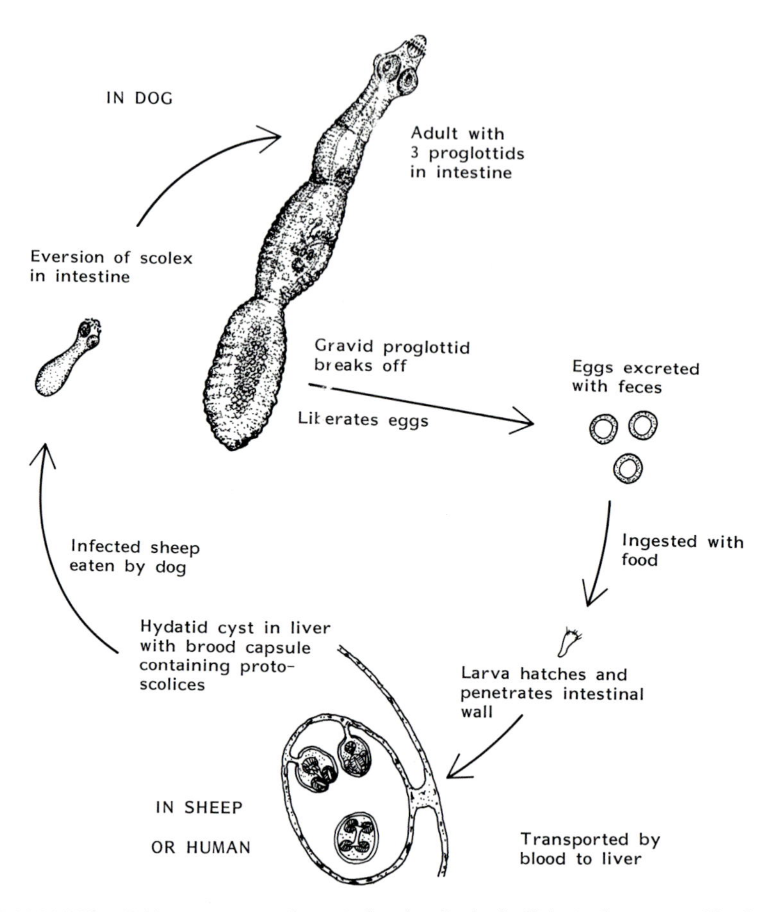

DIAGRAM XVII. *Echinococcus granulosus* (suborder Cyclophyllidea), the cause of hydatid disease. The adult tapeworm in the definitive host, ordinarily a dog, is small with only three or four proglottids. In the intermediate host, however, a single larva forms a cyst within which develop brood capsules each forming numerous protoscolices. Such a cyst is usually 5–10 cm in diameter and holds several liters of liquid containing a million or more scolices. In this species a tremendous reproductive potential is achieved by this polyembryony, rather than by the formation of millions of eggs as in *Taenia solium*. Reproduction occurs in both hosts.

Hydatidosis (echinococcosis) is a serious disease both of people and of sheep. Like cysticercosis, it is caused by the larval rather than the adult worms.

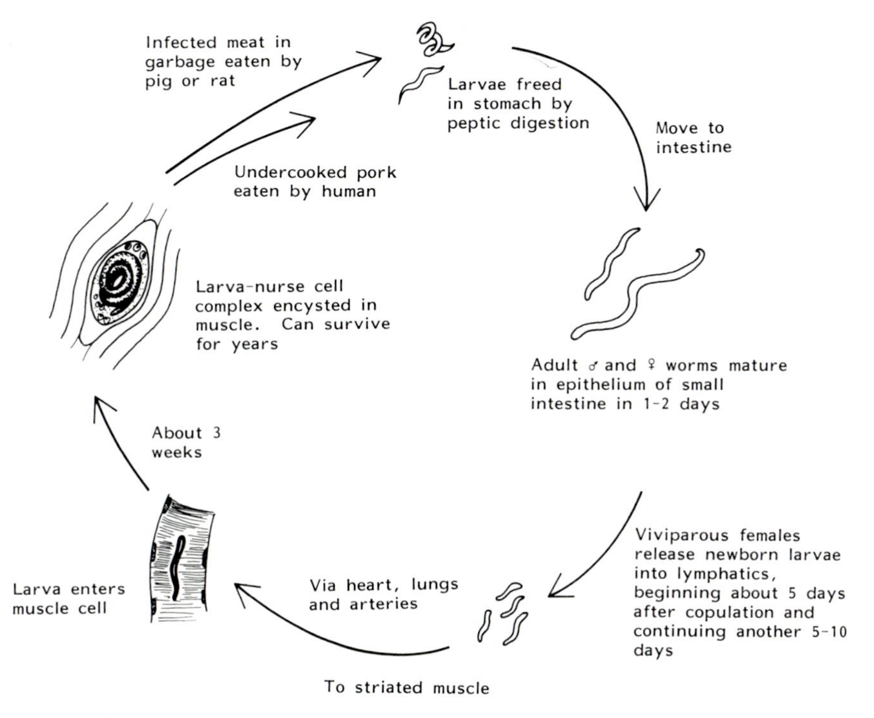

DIAGRAM XVIII. *Trichinella spiralis* (order Trichinellida). As with the tapeworms, this nematode depends on predation and the carnivorous habit for its propagation.

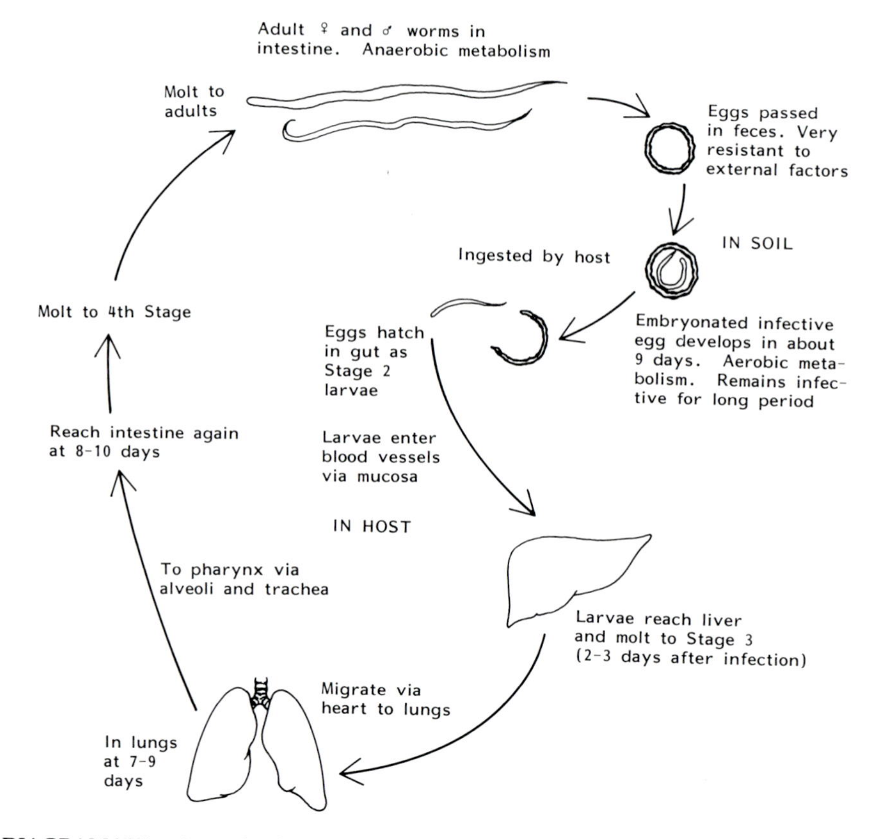

DIAGRAM XIX. *Ascaris lumbricoides* (order Ascarida). This large roundworm is one of the most common of the intestinal nematodes of humans. It is noteworthy that, although the eggs hatch in the alimentary tract, the larvae must migrate to the liver and lungs to undergo their early development, before returning to the intestine where they mature.

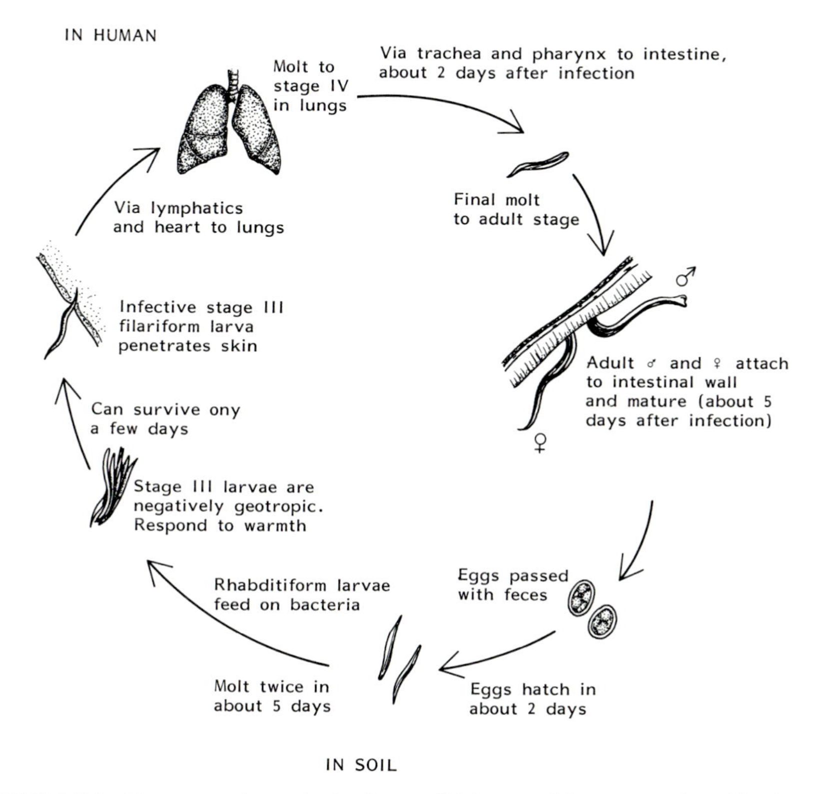

DIAGRAM XX. *Necator americanus* (order Strongylida), one of the two species of hookworms parasitizing humans (the other being *Ancylostoma duodenale*). Both species are widespread and responsible for extensive morbidity, particularly anemia.

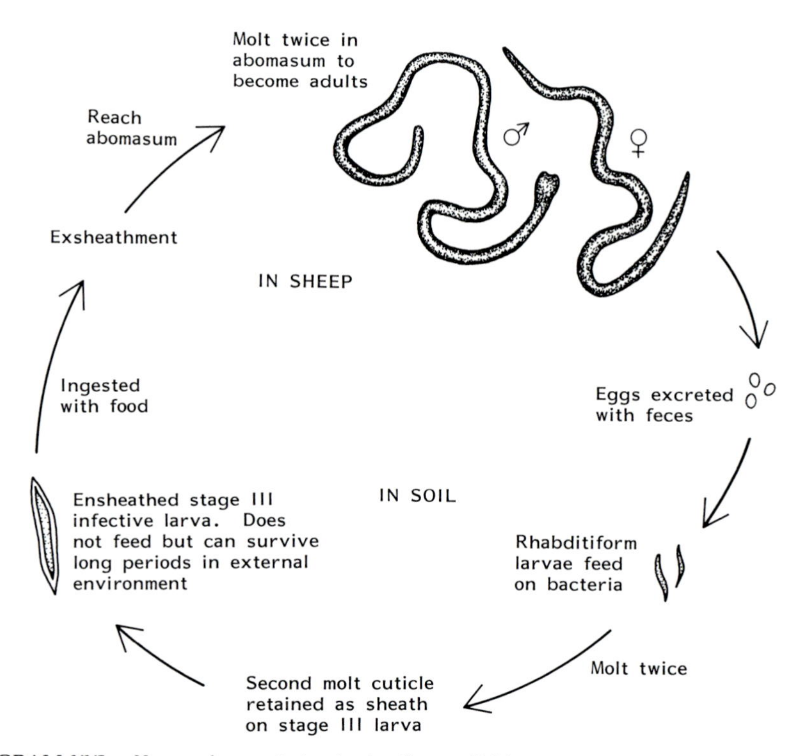

DIAGRAM XXI. *Haemonchus contortus* (order Strongylida), an important pathogenic parasite of sheep. This nematode is typical of an array of gastrointestinal parasites of ruminants, all of economic significance.

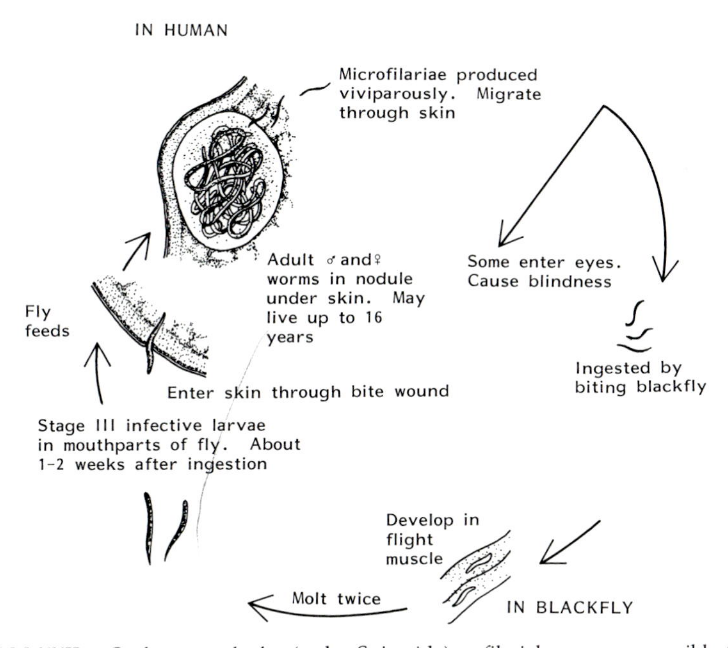

DIAGRAM XXII. *Onchocerca volvulus* (order Spirurida), a filarial worm responsible for "river blindness," particularly in West Africa and highlands of Central America. The parasite occurs in the vicinity of fast-flowing streams that are the habitat of the vector, biting flies of the genus *Simulium*.

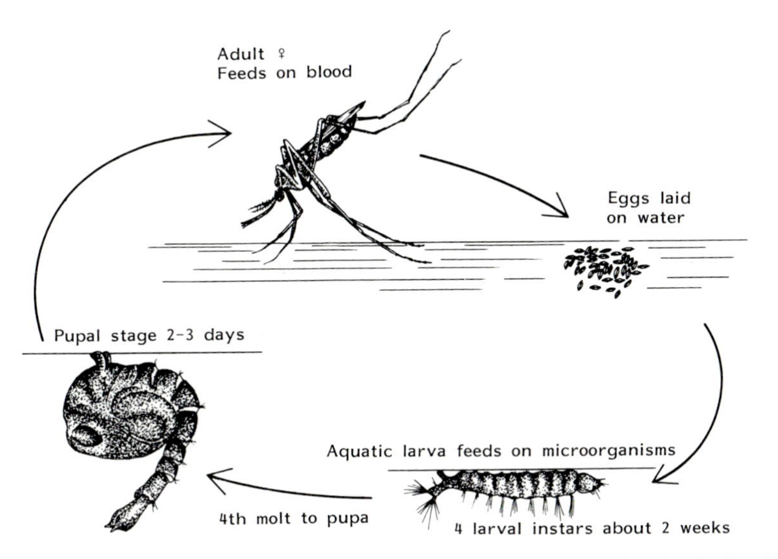

DIAGRAM XXIII. A mosquito (*Anopheles* sp.) (order Diptera). Only the females feed on blood. Both sexes feed on plant juices, but a blood meal is essential for maturation of the eggs. Note that the larvae feed on microorganisms.

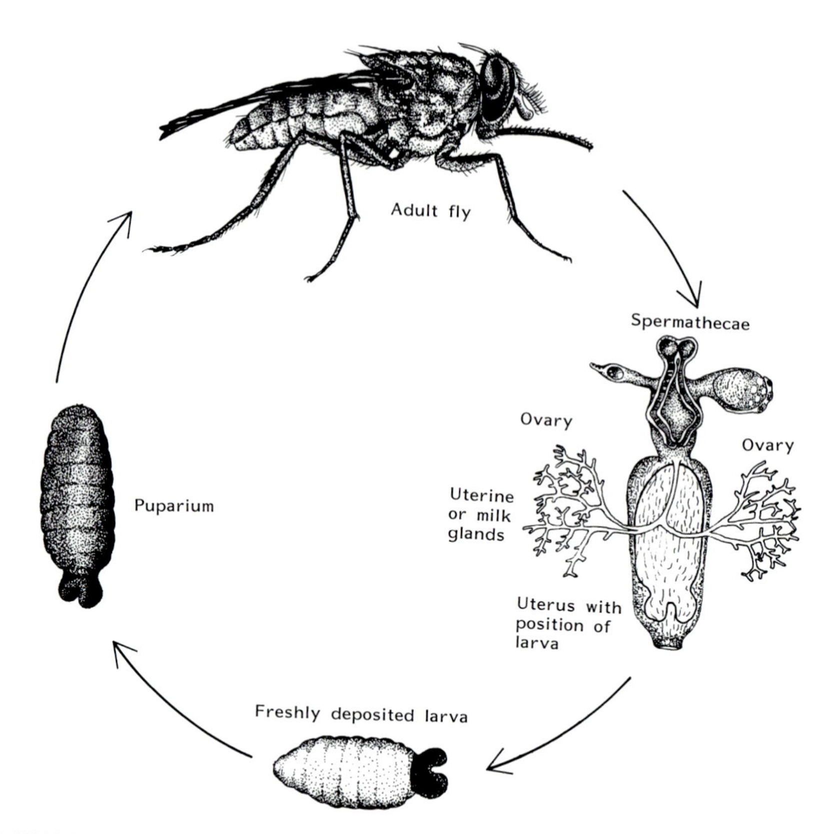

DIAGRAM XXIV. *Glossina* sp. (order Diptera), the tsetse fly. Both sexes feed only on blood. Since larval development takes place entirely with the uterus of the female, the whole life cycle depends on blood. The larva is nourished by secretions from the uterine glands. A single egg matures alternately in the right and left ovaries and is fertilized by sperm from the spermathecae as it enters the uterus, a specialized portion of the oviduct. Larval development within the uterus takes about 10 days. Under optimal conditions of feeding and temperature, within 1 day after birth of one larva the next egg enters the uterus and the next larva begins development. Hence, a female produces six to ten larvae in a life time of 3 months.

The importance of tsetse flies in the ecology and history of Africa cannot be overemphasized; they serve as the sole vectors of the trypanosomes causing African sleeping sickness and several major diseases of cattle and other domestic animals.

(The diagram of the female reproductive tract is copied from Fig. 89 in Buxton, P. A., 1955, The Natural History of Tsetse Flies, London School of Hygiene and Tropical Medicine Memoir No. 10, H. K. Lewis & Co., London.)

Bibliography

Beaver, P. C., Jung, R. C., and Cupp, E. W., 1984, *Clinical Parasitology* (9th edition), Lea & Febiger, Philadelphia.

Brown, H. W., and Neva, F. A., 1980, *Basic Clinical Parasitology* (5th edition), Appleton–Century–Crofts, New York.

Dogiel, V. A., 1966, *General Parasitology* (revised and enlarged by Y. I. Polyansky and E. M. Kheisin; translated by Z. Kabata), Academic Press, New York.

Garnham, P. C. C., 1966, *Malaria Parasites and Other Haemosporidia,* Blackwell, Oxford.

Hoare, C. A., 1972, *The Trypanosomes of Mammals,* Blackwell, Oxford.

Katz, M., Despommier, D. D., and Gwadz, R. W., 1982, *Parasitic Diseases,* Springer-Verlag, Berlin.

Kreier, J. P. (ed.), 1977, *Parasitic Protozoa,* Volumes 1–4, Academic Press, New York.

Levine, N. D., 1968, *Nematode Parasites of Domestic Animals and of Man,* Burgess, Minneapolis, Minn.

Long, P. L. (ed.), 1982, *The Biology of the Coccidia,* University Park Press, Baltimore.

Lumsden, W. H. R., and Evans, D. A. (eds.), 1976, *Biology of the Kinetoplastida,* Volumes 1 and 2, Academic Press, New York.

Mueller, J. F., 1974, The biology of Spirometra *J. Parasitol.* **60:**3–14.

Schmidt, G. D., and Roberts, L. S., 1981, *Foundations of Parasitology,* Mosby, St. Louis.

Smith, T., and Kilbourne, F. L., 1893, Investigations into the Nature, Causation and Prevention of Texas or Southern Cattle Fever, U.S. Bureau of Animal Industry, Bulletin No. 1, Government Printing Office.

Smyth, J. D., 1976, *Introduction to Animal Parasitology* (2nd edition), Wiley, New York.

Strickland, G. T., 1984, *Hunter's Tropical Medicine* (6th edition), Saunders, Philadelphia.

Wolf, K., and Markiw, M. E., 1984, Biology contravenes taxonomy in the Myxozoa: New discoveries show alternation of invertebrate and vertebrate hosts, *Science* **225:**1449–1452.

CHAPTER 2

The Establishment of Infection

Passive Entry into the Host Organism

Many different kinds of parasites passively gain entry into a host. They may be ingested by the host or injected into the host by a feeding vector, such as a bloodsucking insect, or they may be transmitted transovarially or during sexual union of the hosts. In this last situation there is probably no requirement for any special differentiated form of the parasite adapted for transmission. For example, the flagellate *Trichomonas vaginalis,* which may cause severe disease in women, can infect the male genital tract during coitus, and in the same way can be transmitted back from an infected man to a woman (Diagram II). With all the other situations, however, forms exist specially adapted for transmission from one host to another.

Resistant and Dormant Forms

Frequently these are dormant forms capable of prolonged survival in the external environment, such as the resistant cysts of the dysentery ameba *Entamoeba histolytica* or of that more subtle intestinal parasite *Giardia lamblia,* or the very resistant embryonated eggs of the large intestinal worm *Ascaris lumbricoides.* Dormant forms may also exist in the tissues of one host awaiting ingestion by a second, often predatory host. Typical of such forms are the encysted muscle larvae of the nematode *Trichinella spiralis,* which causes trichinosis in humans when eaten in undercooked pork (Diagram XVIII), and the encysted stage of the beef or pork tapeworm, called a cysticercus (Diagram XVI).

All of these dormant stages, whether resistant cysts or eggs ingested accidentally from the outside environment or resting larvae eaten with food, have to be specifically activated to begin their developmental cycle within their new host. It is at this level that host–parasite specificity first plays a role. Since all the parasites we have given as examples enter the human alimentary tract (or that of related mammals), it is not surprising that their activation depends on conditions to which they are exposed as they pass through the stomach and into the small intestine.

The eggs of *Ascaris* (Diagram XIX), passed with the feces of the host, undergo embryonation and a first larval molt all within the tough resistant chitinous shell. These processes (at 30°C) take about 4 and 7–10 days, respectively, so that infective larvae are present within something over 2 weeks. They may survive in the external environment in this stage for long periods, weeks or even months depending on conditions. Hatching of such eggs will occur *in vitro* if they are exposed to all of the following stimuli: (1) temperature of 38°C; (2) brief exposure to slight acidity; (3) continuous exposure for 2–3 hr to a solution of a reducing agent (e.g., 0.1 M cysteine HC1) with 0.1 M $NaHCO_3$ in a gas phase of 5% CO_2, 95% N_2. The further addition of a surface-active agent, such as Tween 80, or the bile salt taurocholate, increases the percent hatching, especially of aged infective eggs. It will be seen at once that these are precisely the conditions found in the intestine of many mammals. In this first interaction with a new host, *Ascaris* would be expected to show little host specificity. The complete hatching stimulus probably first induces an increase in permeability of the vitelline membrane surrounding the fully formed second-instar larva. The larva is activated to produce a chitinase that forms a hole in the chitinous egg shell, and through this hatching occurs.

Other kinds of nematode parasites are ingested not as resistant eggs but as resistant free-living larval forms. These have a tough external sheath that is shed when they establish infection. This exsheathment is induced by much the same factors that induce hatching of *Ascaris* eggs. *Haemonchus contortus,* a parasite of sheep (Diagram XXI), will exsheathe when subjected to a reducing environment (oxidation reduction potential −210 to −260 mV) with high carbonic acid (gas phase of 55–39% CO_2), conditions that exist in the rumen of the sheep. The resting infective larval stages of other types of helminthic parasites are similarly activated in the alimentary tract of the definitive host. The encysted metacercaria of the sheep liver fluke, the trematode *Fasciola hepatica* (Diagram XII), if first gently ground between two glass plates, could then be activated by incubation for 1 hr in a suitable medium with a gas phase of 60% CO_2, 40% N_2. Following activation, excystment would occur in response to the further addition of 10% sheep bile or an equivalent amount of taurocholic acid. In some species of trematodes, the metacercaria is encysted in a fish and excysts when ingested by another predaceous fish. Thus, the metacercariae of *Bucephaloides* encyst in whiting and excyst when eaten by an anglerfish (*Lophius*). Excystation *in vitro* depends on the sequential action of (1) peptic digestion for 1 hr at pH 1.5–2.0, which removes the capsule of host tissue surrounding the larval worm; (2) a rise in pH to 7.2, which triggers excystment in 5 min. A very similar sequence of events occurs when the encysted tissue larvae of *Trichinella* enter a new host except that here a temperature of 37°C is essential for activation. With some kinds of larval tapeworms, such as the cysticercus of *Taenia taeniaformis,* the cyst is digested and the worm released by pepsin HC1 treatment alone. With others, however, a high percentage of excystation occurs only if the pepsin–low pH treatment is followed by exposure to a mixture of bile salts and trypsin.

A similar range of responses is seen among protozoa that enter a new host as resistant cysts ingested with food or water. The cysts of *Entamoeba histolytica* (Diagram I) excyst if placed in an appropriate nutrient medium at 37°C. But the cysts of *Giardia* must first be subjected at 37°C to a pH of 2, and excystation then follows when the cysts are placed, still at 37°C, in culture medium at pH 6.8 (Fig. 2.1). These conditions obviously correspond to passage through the acid stomach, which in some way (probably by increasing the permeability) activates the excystation process, which then occurs at the neutral pH of the duodenum, the natural habitat of *Giardia*.

The Coccidia are a group of parasitic protozoa that form extremely resistant cysts. These not only can survive dormant for long periods in ordinary environments but are not affected by prolonged immersion in dilute sulfuric acid or potassium dichromate solution. The cysts (called oocysts) each contain two smaller cysts, the sporocysts. Each sporocyst contains two or four sporozoites, the products of reduction division of a zygote formed when male and female gametes unite (Diagrams VI, VII). The sporozoites are the infective forms. The sporocysts containing them must first be liberated from the oocysts, and then the sporozoites from the sporocyst. Two distinct stimuli are required. The first, for coccidial parasites of poultry, is mechanical. It occurs in nature in the gizzard of the chicken and can be simulated by grinding, as in a tissue grinder. In parasites of mammals, this first stimulus can be effected *in vitro* by conditioning in CO_2 at 37–40°C. In either case a postulated enzyme system of the coccidian is activated. The second stimulus is supplied by exposure to bile plus trypsin at 37–40°C. Bile salts vary in effectiveness, sodium taurodeoxycholate being by far the most effective for *Eimeria tenella*. A 5% solution of this bile salt wih 0.25% of trypsin 1:250 gave 90% excystation of sporozoites in 90 min at 40°C. Sporocysts of this and many other related species have a specialized structure, the Stieda body, like a plug at one end. This dissolves under the influence of the bile salts and trypsin, leaving an opening through which the sporozoites emerge (Fig. 2.2). Other species, even in the same genera, lack a Stieda body. In these the sporocyst wall has a thick inner layer consisting of four curved plates connected by a thinner layer of material. The latter material is removed by the action of bile salts and trypsin. As the four plates are separated in this way, they curl up, freeing the sporozoites. This is the mechanism found in the sporocysts of *Toxoplasma gondii* (Fig. 2.3), a ubiquitous parasite that infects many people asymptomatically but causes severe disease under certain conditions.

It is worthy of emphasis that oocysts that require mechanical grinding for activation will not excyst and hence will not produce infection in a host that does not have a gizzard. Similar mechanical features play a role in the host specificity of the oncospheres of tapeworms, the only stage in the tapeworm life cycle exposed to the environment. With those species (e.g., *Hy-*

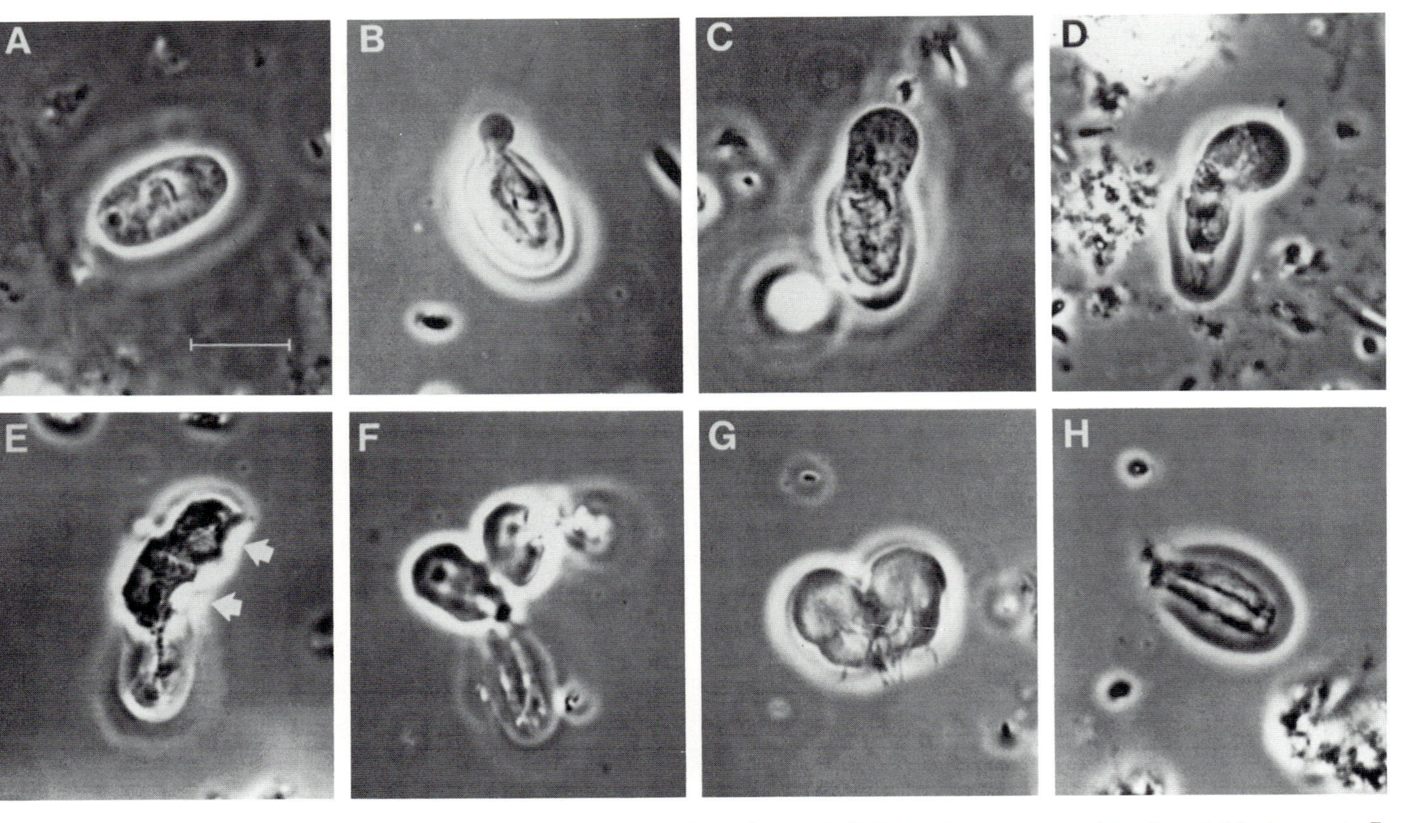

FIGURE 2.1. Photomicrographs showing excystation of *Giardia*. (A) Typical cyst. (B–F) Stages in emergence of trophozoite(s). Arrows in E indicate central adhesive disks of daughter trophozoites. (G) Excystation completed. Division of daughter trophozoites continuing. (H) Empty cyst. Scale bar = 10 μm. (From Bingham and Meyer, 1979. Original prints courtesy of Dr. E. A. Meyer.)

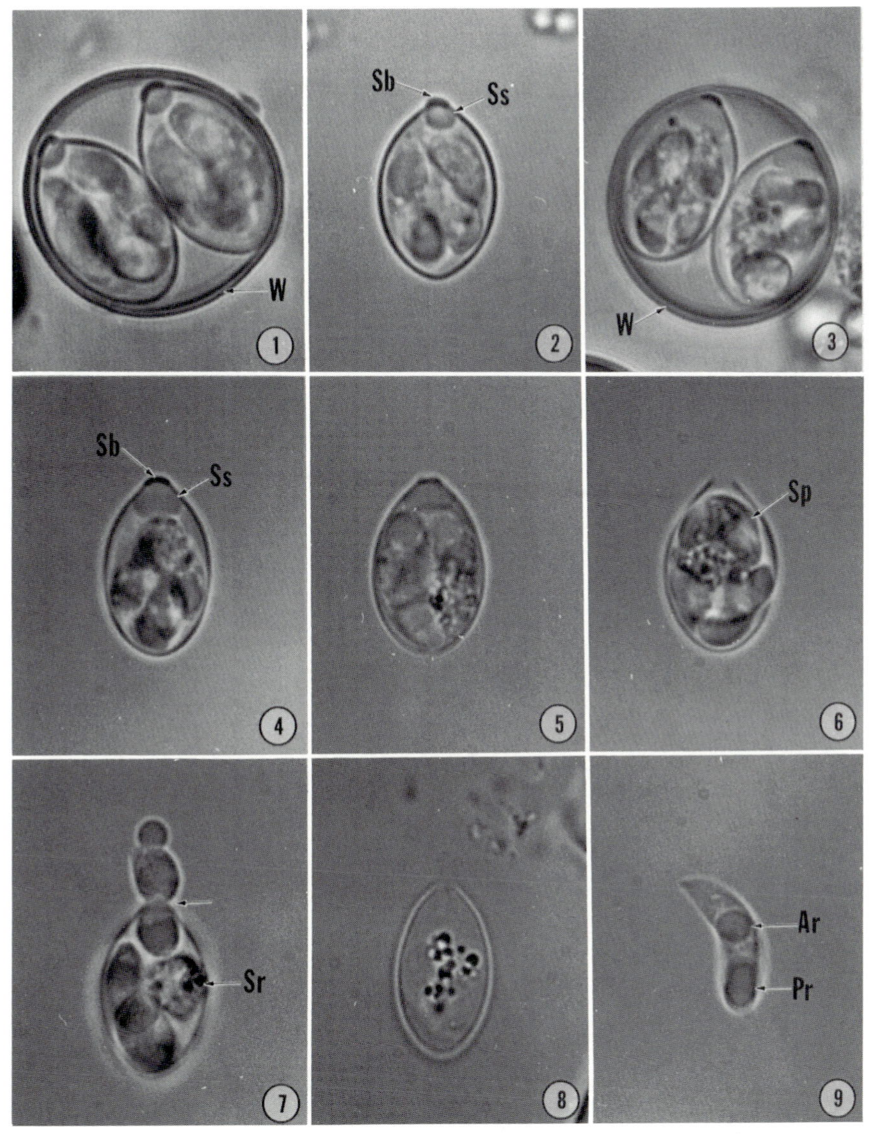

FIGURE 2.2. Photomicrographs showing oocyst structure and emergence of sporozoites in coccidia having sporocysts with a Stieda body. (1,2) *Isospora canaria,* (3–9) *I. serini,* both from the canary. (1,3) Oocysts with wall (W) and two sporocysts within. (2,4) Sporocysts with Stieda (Sb) and substiedal body (Ss). (5) Sporocyst after addition of trypsin–sodium taurocholate mixture. Note that Stieda body has almost disappeared and substiedal body has become more transparent. (6) Sporocyst after further action of the excysting fluid. Substiedal body has disappeared and sporozoites (Sp) may be seen within. (7) A sporozoite emerging. Note constriction (arrow) at point of exit from sporocyst and compact sporocyst residuum (Sr.). (8) Intact empty sporocyst. (9) Free sporozoite. Note anterior (Ar) and posterior (Pr) refractile bodies. All × 2120.

For these observations, oocysts were placed in the excysting fluid under a coverslip and were ruptured by friction of the coverslip to liberate the sporocysts.

(From Speer and Duszynski, 1975. Original prints courtesy of Dr. D. W. Duzynski.)

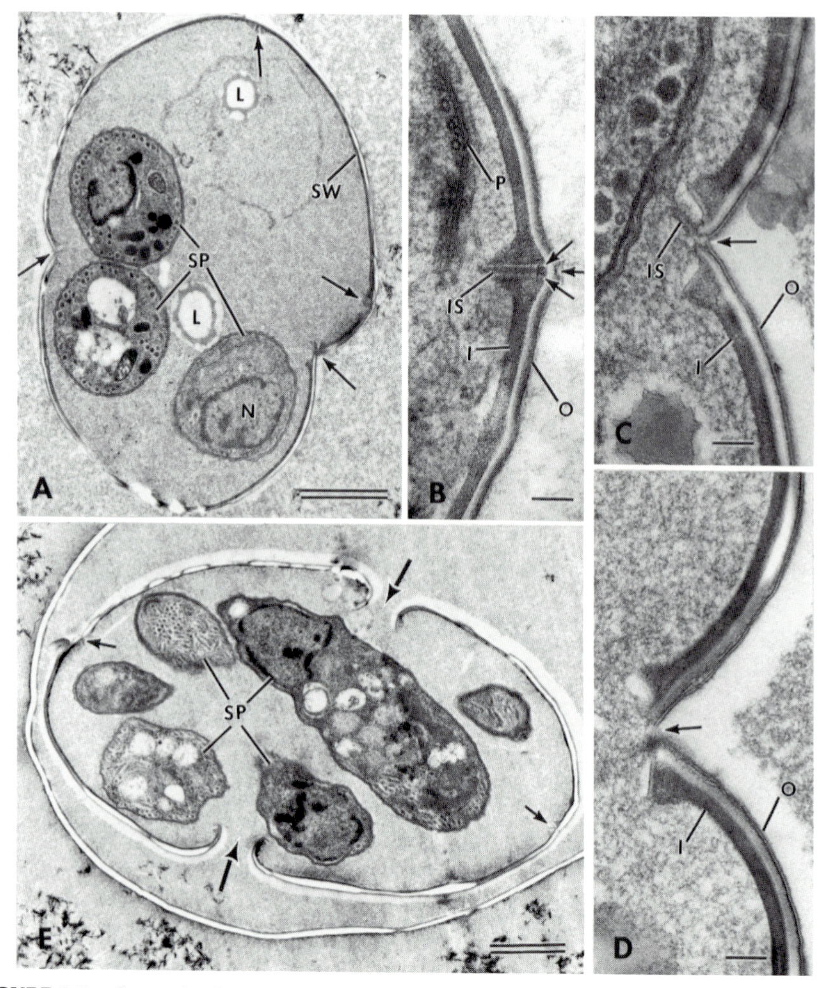

FIGURE 2.3. Stages in the excystation of sporozoites of *Toxoplasma gondii,* as seen in transmission electron micrographs of sections through the sporocysts. The sporocysts were liberated from the oocysts by grinding to rupture the oocyst wall. They were then placed in an excysting medium of 0.25% (w/v) trypsin and 0.75% sodium taurocholate and incubated at 37°C. Samples were removed after 10, 30, and 60 min and processed for electron microscopy. A double scale bar represents 1 μm; a single scale bar represents 100 nm. Abbreviations: I, inner layer of sporocyst wall; IS, interposing strip; L, lipid; N, nucleus; O, outer layer of sporocyst wall; P, plasmalemma; SP, sporozoite; SW, sporocyst wall.

(A) Sporocyst at early stage of excystation. The junctions of the plates forming the inner layer of the sporocyst are shown at arrows. Note the beginning of an inward curling at the two lower junctions.

(B) A cross section showing the structure of the junction between plates of the inner layer prior to treatment with excysting medium. The interposing strip and thin band of osmiophilic material (arrows) joining the plates to the interposing strip can be seen. Note the depression in the outer layer (arrowhead) directly over the junction.

(C) A cross section showing the separation of the plates of the inner layer of the sporocyst wall. Note that the interposing strip has remained attached to the margin of one plate and that the outer layer has bent inward directly above the break (arrow).

menolepis) that develop in an arthropod host, mastication by the host to break the egg shell is an essential first step. Only then will host proteolytic enzymes be effective in digesting the inner envelopes to liberate the activated larva. On the other hand, in the taeniid cestodes, where the egg infects a vertebrate intermediate host (as in *Taenia solium,* Diagram XVI), the poorly developed egg shell is stripped off during passage in the feces, so that the oncosphere is protected only by the embryophore. The embryophore consists of hexagonal blocks of a keratinlike protein held together by a cementing substance. This latter is evidently dissolved, probably by the sequential action of pepsin and bile plus trypsin, permitting the activated oncosphere to liberate itself.

It is interesting to note that even from the first passive contact with the environment of the host, the parasite must have "reactive sites" capable of being appropriately influenced. Thus, with the embryophore of the *Taenia* egg as with the sporocyst of *Toxoplasma* or *Eimeria,* there are special structures, visible with electron microscopy, that are dissolved by the host bile and proteolytic enzymes, permitting escape of activated forms within. More complex and less well understood mechanisms then come into play as these activated infective forms seek special loci within the host where their further development occurs. These will be considered in Chapter 3. Here the discussion of modes of entry into the host must be continued.

Active Forms Ingested or Injected by a Vector

This situation occurs typically with parasites that enter the blood of vertebrates and are transmitted from one to another by bloodsucking invertebrates, especially arthropods. Although such transmission may occasionally occur by direct contamination of the insect's mouthparts, which then serve merely as a "flying pin," it is far more usual for specially differentiated stages of the life cycle to initiate both the infection in the vector and that in the vertebrate. The trypanosomes of human sleeping sickness and the malaria parasites among the protozoa, and the filarial worms among the helminths, provide the best examples (Diagrams V, VIII, XXII).

The trypanosomes of human sleeping sickness (*Trypanosoma gambiense, T. rhodesiense*) and the very similar organisms that cause nagana of cattle (*T. brucei*) parasitize the blood and tissue spaces of their hosts. In an established infection, there occur in the blood stream elongate and relatively short stumpy

---

(D) A similar section to C but at a late stage in excystation. The interposing strip has disappeared. A break is present in the outer layer (arrow) directly above the point of separation of the plates of the inner layer.

(E) A longitudinal section through a sporocyst in a late stage of excystation. Two of the junctions between plates of the inner layer are still intact (small arrows) while at the other two junctions the plates have curled inward to produce an opening in the sporocyst wall (large arrows).

(From Ferguson *et al.*, 1979. Original prints courtesy of Dr. D. J. P. Ferguson.)

forms and forms intermediate between these. When a tsetse fly (of the genus *Glossina*) takes a blood meal from an infected host, it is the stumpy forms that are preadapted for survival and multiplication in the midgut of the fly. Here they undergo further change and establish a population of procyclic forms. These are noninfective to a vertebrate host. Some of them, however, find their way to the salivary glands of the fly where further morphological and physiological changes occur that culminate in a population of metacyclic trypanosomes in the salivary glands and salivary ducts. These forms, which first appear about 2–3 weeks after initial infection of the fly, are preadapted for life in the vertebrate host. When injected with the saliva of the fly into a susceptible host, they proceed to multiply within it.

With malarial parasites, the developmental cycles are even more complex. Here the erythrocytic parasites that multiply asexually within the red cells of a person and that are responsible for the disease, are all killed and digested when taken into the stomach of an anopheline mosquito able to act as a vector. Usually, however, the blood of an infected person also contains a small proportion of parasites that are differentiated into male and female gametocytes. These cannot develop further within the vertebrate, but if ingested by a mosquito, they produce male and female gametes. These unite to form motile ookinetes that enter the stomach wall of the mosquito, grow, and multiply to form numerous minute elongate cells, the sporozoites. The sporozoites enter the salivary glands of the mosquito. When injected with the mosquito's saliva ino a susceptible host, they initiate the new infection. What makes certain erythrocytic parasites differentiate into gametocytes and what the factors are triggering the further changes in the mosquitoes are all fascinating problems in cell and developmental biology to which I will return later (Chapters 8, 19).

The filarial worms, like the protozoa of malaria, are transmitted by blood-sucking insects within which they undergo a developmental cycle. Male and female worms of *Onchocerca volvulus* (Diagram XXII) living in the tissues produce thousands of young. These microfilariae, when they invade the eyes, cause "river blindness," a disease occurring in West Africa around fast-flowing streams. The disease is associated with fast-flowing streams because these provide the habitat for the vector insect, the blackfly *Simulium damnosum*. Microfilariae in the skin develop further only when ingested by such a blackfly with its blood meal. They then undergo a cycle in the muscles of the fly, which, after 1 month, produces third-stage larvae able to infect a person. When the fly feeds, these escape from the mouthparts and through their own activity enter this new host, probably through the wound made by the fly. Here we have a situation where the forms that infect the fly (the microfilariae) are passively ingested with the blood meal, but the later developmental forms infective to the vertebrate are brought to their host by the vector but enter through their own motility. The vector through its sensory physiology finds the host; the worms need only have appropriate responses to enter once deposited on it.

Active Entry

Many kinds of parasites, especially among the helminths (and the arthropods), must find their hosts through their own activity. The hookworms and the schistosomes of humans provide excellent examples. Eggs of the human hookworms (*Necator, Ancylostoma*) (Diagram XX) are passed with the feces. Here they hatch and go through a free-living larval development that ends in infective third-stage larvae. These are very like the resistant ensheathed larvae of the sheep stomach worm *Haemonchus*. Whereas *Haemonchus* larvae are ingested by the sheep with its food, those of *Necator* persist in the moist soil that has been contaminated with feces from infected people. When a barefoot person walks over such soil, the *Necator* larvae respond quickly and penetrate the skin to initiate infection.

Active finding and penetration of the host are particularly well seen in the schistosome parasites, three species of which cause the the three important and widespread types of schistosomiasis in humans. The schistosome life cycle (Diagram XIII) involves a snail and a vertebrate host and both are found and entered through the activities of two different aquatic stages highly specialized for this purpose. The first of these stages, the miracidium, hatches from eggs laid by parent worms in the human host after these eggs reach water with the feces or urine of the host. These active ciliated larvae, capable of only a few hours of life, swim about apparently in a random way until they come close to a snail. Some substance emanating from the snail, shown to be present in snail-conditioned water, increases the extent of miracidial turning, in this way increasing the likelihood that the miracidium will make contact with the snail. There is also evidence from Y-tube experiments that miracidia will follow a gradient of snail-conditioned water. The reaction is a chemokinesis rather than a chemotaxis. The actual substances involved seem to be relatively nonspecific, heat-stable, and of low molecular weight. Several defined substances have been shown to stimulate miracidial activity but the most effective seems to be glutathione. At only 0.1 μmole/ml (hence a concentration of 0.1 mM), the infection rate of snails (*Biomphalaria glabrata*) with miracidia of *Schistosoma mansoni* was raised to 100%. Furthermore, it was of interest that the glutathione content of snails fell quickly after infection, reaching in 1 day to only half its preinfection level.

The cycle of development and asexual reproduction initiated by the miracidium in the snail yields again a free-living form, the cercaria. This escapes from the snail tissue into the water where it can live for several days until it is stimulated to attach to and penetrate the skin of a human host (or other suitable vertebrate). The skin lipids of the host provide the stimulus. In a study of the stimulation of penetration by cercariae of *S. mansoni* into agar, it was found that unsaturated fatty acids were particularly effective. Their effectiveness was increased by the number of double bonds at the *cis* position and not too close to the nonpolar end of the molecule. Of 230 related compounds tested, the most active were precisely among those found in human skin, which, unlike skin of many other mammals, has a high content of free fatty acids. It is remarkable that these compounds effect physiological transformations so that a cercaria, quite able to live in water, becomes osmotically sensitive and dies in water. In nature it would have penetrated into the skin of its host where permeability would be to its advantage. Lipids stimulate the free-living cercaria both to penetrate and to transform into the parasitic schistosomule. For penetration, special glands (the preacetabular glands) secrete enzymes, among which a gelatinase plays an active role. The transformation from free-living cercaria to parasitic schistosome is a rapid and dramatic one and involves all aspects of the physiology of the organism. These will be considered in detail in Chapter 21.

In all of these early interactions the behavior of the host as well as of the parasite plays an essential role. Vegetarian hosts are not likely to become infected with *Trichinella.* Hosts that do not enter fresh water are not likely to acquire schistosomes. Host behavior may actually be modified in such a way as to facilitate transmission of a parasite (see Chapter 12). Wherever active stages of a parasite seek out and invade a host, the parasite is richly equipped with a variety of sense organs. The miracidia and cercariae of trematodes are sensitive to light. These and nematode larvae all have surface structures clearly designed as tactile and chemosensory organelles. The sensory physiology of parasites is a difficult but rapidly growing field. For the nematodes, a useful model system is provided by the free-living worm *Caenorhabditis elegans* for which very detailed information is now available, including complete charts of innervation and of differentiation.

Bibliography

Bazzicalupo, P., 1983, *Caenorhabditis elegans:* A model system for the study of nematodes, in: *Molecular Biology of Parasites* (J. Guardiola, L. Luzzatto, and W. Trager, eds.), Raven Press, New York, pp. 73–92.

Bingham, A. K., and Meyer, E. A., 1979, *Giardia* excystation can be induced *in vitro* in acidic solutions, *Nature* **277**:301–302.

Bingham, A. K., Jarroll, E. L., Jr., and Meyer, E. A., 1979, *Giardia* sp: Physical factors of excystation *in vitro* and excystation vs. eosin exclusion as determinants of viability, *Exp. Parasitol.* **47**:284–291.

Campbell, W. C. (ed.), 1983, *Trichinella and Trichinosis,* Plenum Press, New York.

Canning, E. U., and Wright, C. A. (eds.), 1972, *Behavioural Aspects of Parasite Transmission,* Academic Press, New York.

Chernin, E., 1970, Behavioral responses of miracidia of *Schistosoma mansoni* and other trematodes to substances emitted by snails, *J. Parasitol.* **56**:287–296.

Christie, E., Pappas, P. W., and Dubey, V. P., 1978, Ultrastructure of excystment of *Toxoplasma gondii* oocysts, *J. Protozool.* **25**:438–443.

Doran, D. J., and Farr, M. M., 1962, Excystation of the poultry coccidium *Eimeria acervulina, J. Protozool.* **9**:154–161.

Erlandsen, S. L., and Meyer, E. A. (eds.), 1984, *Giardia and Giardiasis: Biology, Pathogenesis, and Epidemiology,* Plenum Press, New York.

Fairbairn, D., 1960, Physiological aspects of egg hatching and larval exsheathment in nematodes, in: *Host Influence on Parasite Physiology* (L. A. Stauber, ed.), Rutgers University Press, New Brunswick, N.J., pp. 50–64.

Fairbairn, D., 1961, The *in vitro* hatching of *Ascaris lumbricoides* eggs, *Can. J. Zool.* **39:**153–162.

Ferguson, D. J. P., Birch-Andersen, A., Siim, J. C., and Hutchison, W. M., 1979, An ultrastructural study on the excystation of the sporozoites of *Toxoplasma gondii*, *Acta Pathol. Microbiol. Scand. Sect. B* **87:**277–283.

Hass, W., and Schmitt, R., 1982, Characterization of chemical stimuli for the penetration of *Schistosoma mansoni* cercariae. I. Effective substances, host specificity. II. Conditions and mode of action, *Z. Parasitenkd.* **66:**293–307, 309–319.

Jensen, J. B., Nyberg, P. A., Burton, S. D., and Jolley, W. R., 1976, The effects of selected gasses on excystation of coccidian oocysts, *J. Parasitol* **62:**195–198.

Johnston, B. R., and Holton, D. W., 1981, Excystation *in vitro* of *Bucephaloides gracilescens* metacercaria (Trematode: Bucephalidae), *Z. Parasitenkd* **65:**71–78.

Kirschner, K., and Bacha, W. J., Jr., 1980, Excystment of *Himasthla quissetensis* (Trematode: Echinostomatidae) metacercariae *in vitro*, *J. Parasitol.* **66:**263–267.

Lethbridge, R. C., 1980, The biology of the oncosphere of cyclophyllidian cestodes: Review, *Helminthol. Abstr.* **49:**59–72.

MacInnis, A. J., Bethel, W. M., and Cornford, E. M., 1974, Identification of chemicals of snail origin that attract *Schistosoma mansoni* miracidia, *Nature* **248:**361–363.

Mettrick, D. F., and Podesta, R. B., 1974, Ecological and physiological aspects of helminth–host interactions in the mammalian gastro-intestinal canal, *Adv. Parasitol.* **12:**183–278.

Miller, T. A., 1979, Hookworm infection in man, *Adv. Parasitol.* **17:**315–384.

Osum-Carillo, A., Mascero-Lazcano, M. C., Guevara-Pozo, D., and Guevara-Benitez, D., 1978, Efecto de la tripsina, pancreatina y bilis canina en la desinvaginacia *in vitro* de *Cysticercus tenuicollis*, *Rev. Iber. Parasitol.* **38:**569–578.

Patton, W. H., and Brigman, G. P., 1979, The use of sodium taurodeoxycholate for excystation of *Eimeria tenella* sporozoites, *J. Parasitol.* **65:**526–530.

Rogers, W. P., 1960, The physiology of infective processes of nematode parasites; the stimulus from the animal host, *Proc. R. Soc. London Ser. B* **152:**367–386.

Rothman, A. H., 1959, Studies on the excystment of tapeworms, *Exp. Parasitol.* **8:**336–364.

Saladin, K. S., 1979, Behavioral parasitology and perspectives on miracidial host-finding, *Z. Parasitenkd* **60:**197–210.

Schiff, C. J., and Kriel, R. L., 1970, A water-soluble product of *Bulinus (Physopsis) globosus* attractive to *Schistosoma haematobium* miracidia, *J. Parasitol.* **56:**281–286.

Speer, C. A., and Duszynski, D. W., 1975, Fine structure of the oocyst walls of *Isospora serini* and *I. canaria* and excystation of *Isospora serini* from the canary, *Serinus canarius* L., *J. Protozool.* **22:**476–481.

Speer, C. A., Marchiondo, A. A., Duszynski, D. W., and File, S. K., 1976, Ultrastructure of the sporocyst wall during excystation of *Isospora endocallimici*, *J. Parasitol.* **62:**984–987.

Stirewalt, M. A., 1974, *Schistosoma mansoni:* Cercaria to schistosomule, *Adv. Parasitol.* **12:**115–182.

Tielens, A. G. M., van der Meer, P., and van den Bergh, S. G., 1981, *Fasciola hepatica:* Simple, large-scale, *in vitro* excystment of metacercariae and subsequent isolation of juvenile liver flukes, *Exp. Parasitol.* **51:**8–12.

CHAPTER 3

Site Selection within the Host: Entry into Specific Organs and Cells

For most parasites, entry into a suitable host is only a first step. They must then find their way to those particular parts of the host within which they can develop. There are many degrees of complexity, some parasites taking a rather direct route, others a roundabout one. Some are highly restricted to a certain location, such as a particular tissue or type of cell, whereas others are more catholic in their preferences. Parasites of the alimentary tract ingested through the mouth as cysts or as eggs or infective larvae, would seem to have the simplest and most direct path. Even here, however, there are preferred sites for localization. Thus, the infective larvae of the nematode parasite of mice, *Nematospiroides dubius*, enter with food, exsheathe in the stomach, and then lodge in the small intestine. Their specific site is in the region where bile enters and it has been shown experimentally that bile activates motility of the larvae and that they will accumulate to a much greater extent in bile than in water. Furthermore, if bile is surgically redirected in a living mouse to other than the normal region of the intestine, the larvae will establish in that region. Thus, bile can act not only in processes of excystation and hatching, but also at the next stage, in site selection. Most stimuli for site selection within the alimentary tract, however, are not well understood. It is becoming clear that even a tapeworm may exhibit a highly complex behavioral pattern.

The common rat tapeworm *Hymenolepis diminuta* shows an ontogenetic shift in site of scolex attachment (anteriad) from the initial location near the midregion of the small intestine to a point in the duodenum by 7–10 days postinfection. Even more interesting is the circadian migration of this worm from a more anterior location at 0800 hr to a more posterior one at 1600 hr in rats allowed to feed in their normal manner mostly at night. This pattern was reversed in rats fed during the day, indicating a relation to the circadian activity of the host. The relationship, however, does not depend directly on quality or quantity of ingested nutrient, though presumably it has an adaptive, probably nutritional significance. A circadian rise in 5-hydroxytryptamine

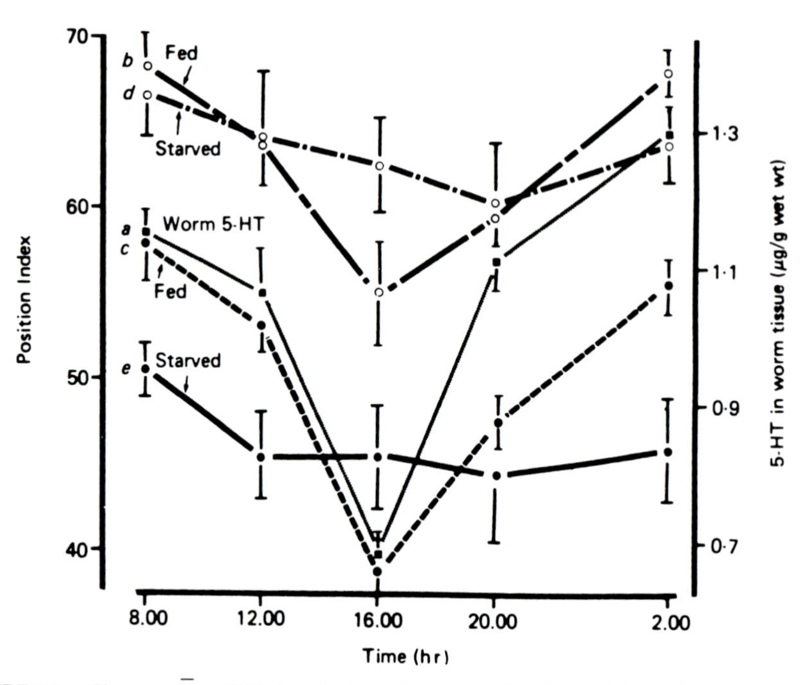

FIGURE 3.1. Changes ($\bar{X}$ ± S.E.) in 5-hydroxytryptamine levels of 16-day-old *Hymenolepis diminuta* (*a*) and worm migratory response expressed as position indices in *ad libitum* fed rats of scolices (*b*) and biomass (*c*). Also shown are the position indices for scolices (*d*) and biomass (*e*) during the second 24-hr period of a 48-hr fast of infected rats, during which only water was available. (From Cho and Mettrick, 1982.)

levels was highly correlated with the anteriad movement of the worms (Fig. 3.1). The responses are also affected by pancreatic, gastric, and biliary secretions as well as to a lesser extent by glucose.

With protozoa also there is evidence for attraction to particular sites within the alimentary tract. For species of the coccidian genus *Eimeria* (of great economic importance in the poultry industry), site localization within the alimentary tract is partly determined by the factors that produce excystation of the sporozoites (see Chapter 2). The sporozoites, however, are motile and capable of traveling 2500 times their own length in 1 hr. Although *Eimeria tenella* sporozoites excyst in the small intestine, they infect only cells of the cecum and are already found there by 1.5 hr after the introduction of oocysts. Even if hatched sporozoites of this species are introduced intravenously, they produce infection in the normal site, the cecum, suggesting a chemotaxis. Similarly, with the species (*E. praecox*) that develops in the cells of the duodenum, sporozoites injected into the cecum or via the cloaca will establish infection only in the duodenum.

Many species of parasites exhibit complex migrations from one part of the host to another. Hookworm larvae, having entered through the skin, rapidly reach the lungs. Here they molt and the stage IV larvae thus formed migrate to the small intestine via the respiratory system, esophagus, and stomach (Diagram XX). It is interesting to note that larval nematodes that enter an abnormal host in which they cannot find an appropriate developmental site may wander aimlessly through the host's body for a long time. For example, if a human ingests eggs of the dog or cat hookworm (*Toxocara canis* or *T. cati*), the eggs hatch and the larvae migrate via the circulation to the lungs. They are, however, unable to penetrate the alveoli and they then wander throughout the body to produce a syndrome called visceral larva migrans. This may cause serious consequences if larvae enter the eye or brain.

The schistosomula of *Schistosoma mansoni* after penetrating the skin get into the lymphatics and veins to be carried to the right heart and then to the lungs (Diagram XIII). The liver is invaded via the blood system and the mature flukes then move from the liver into the vessels of the hepatic portal system. Finally, they migrate against the blood flow into the small mesenteric veins of the intestine. The lung fluke *Paragonimus westermani,* on the other hand, enters its host via the alimentary tract. It passes from the gut into the abdominal cavity and then into the wall of the abdominal cavity. Here it undergoes a period of development, then returns to the abdominal cavity to find its way to the lungs where the adult worm remains. So accurate are its tropisms that an average of 68% of the metacercariae fed to experimental cats succeeded in reaching the lungs.

Parasites dwelling in parts of a host not directly accessible to the external environment have developed a variety of mechanisms for transmission of their offspring to new hosts: they may have encysted quiescent stages that await ingestion of the host by a predator, they may have spined eggs that eventually get excreted in urine or feces, or they may depend on blood- or tissue-feeding ectoparasites. In this last case, as already noted (Chapter 2), specialized forms for infection of the vector occur. The larvae of the filarial nematodes, known as microfilariae, provide an example of special interest that again illustrates complex adaptive behavior. Microfilariae of species transmitted by mosquitoes occur in the circulating blood where they exhibit a circadian periodicity related to the time when their mosquito host does most of its biting. The microfilariae of species of *Onchocerca* transmitted by biting flies of the genus *Simulium* migrate into the skin, where their presence in large numbers causes intense pruritus. What is of special interest here is that some species localize in the skin of the legs, others in the skin of the upper back or neck; these sites are primarily those preferred by the local species of *Simulium* that serves as the vector.

The specificity of such behavior and the reactions that must be involved in such migration are well illustrated by the localization of certain larval trematodes in the lenses of the eyes of their fish hosts. The life cycle of such worms, e.g., *Diplostomum flexicaudum,* resembles that of schistosomes but requires three rather than two hosts for its completion. In the definitive host, a fish-eating bird such as a gull or a heron, the adult fluke lives in the alimentary tract and lays eggs that pass into water with the bird's feces. Miracidia hatch from the eggs and infect snails within which development and multiplication occur to form large numbers of free-swimming cercariae. These do not directly infect the definitive host, as do the schistosomes, but instead penetrate through the skin of certain fish and migrate to the lens. Under normal conditions, no matter where the cercaria enters, it finds its way to either eye and into the lens. If one eye is removed before infection, the cercariae localize in the other. If both eyes are removed, the cercariae appear to be "lost" and fail to develop. If both lenses are removed, some worms reach the eyes but in smaller numbers and again these fail to develop. In the normal situation, development occurs within the lens into a metacercaria, a form representing an intermediate stage toward the adult fluke. This form then remains quiescent until the fish is eaten by an appropriate bird, where maturation to the adult quickly occurs. Cercariae allowed to penetrate into a fish and then, within an hour, removed aseptically and placed *in vitro* with intact lens or even with crushed lens material, developed into morphologically normal metacercariae (but these were not infective to birds). In view of the very high proportion of cercariae that reach the eyes from any part of the fish's body, we must postulate the existence of some substance in the eyes and especially in the lens, to which the cercariae respond. In turn, the response must be by way of appropriate chemosensory organelles. Just as sensory organelles are essential for host location by the free-living infective stages of helminthic parasites, so they also function in the migration of such parasites within the host. The surface of schistosomes well illustrates the numbers and diversity of structures that are probably tactile and chemosensory organelles (Fig. 3.2). Little is known concerning the detailed function of these organelles. How they respond, and what are the substances they recognize that guide them to appropriate locations are problems that could now be approached experimentally. The answers might be of as much significance in animal physiology and in cell biology as in parasitology. They might provide a beginning toward explanation of the migrations I have already mentioned and many others, such as localization of certain parasites in the brain or the movement of sporozoites of malaria from the hemocoel of mosquitoes into the salivary glands, or the affinity of still other parasites for striated muscle.

The large multinucleate cells of striated muscle provide a site for the development of intracellular parasitic metazoa, the larval forms of certain nematodes. For example, when microfilariae of species of *Brugia* are ingested with a blood meal by an appropriate species of mosquito, they shed their sheath and pass through the gut into the hemocoel and then into the flight muscles. The muscle fibers here extend the full length of the muscle and are from 100 μm to 1 mm in diameter. They have numerous peripheral nuclei, many large mitochondria, a well-developed transverse tubular system, and an elaborate network of intracellular tracheoles. The filarial larva tunnels into a myofiber by passing between myofibrils, so that at first there is minimal disruption of the myofiber. The larva occupies its interfibrillar tunnel during the entire period of intracellular development, about a week. Having molted

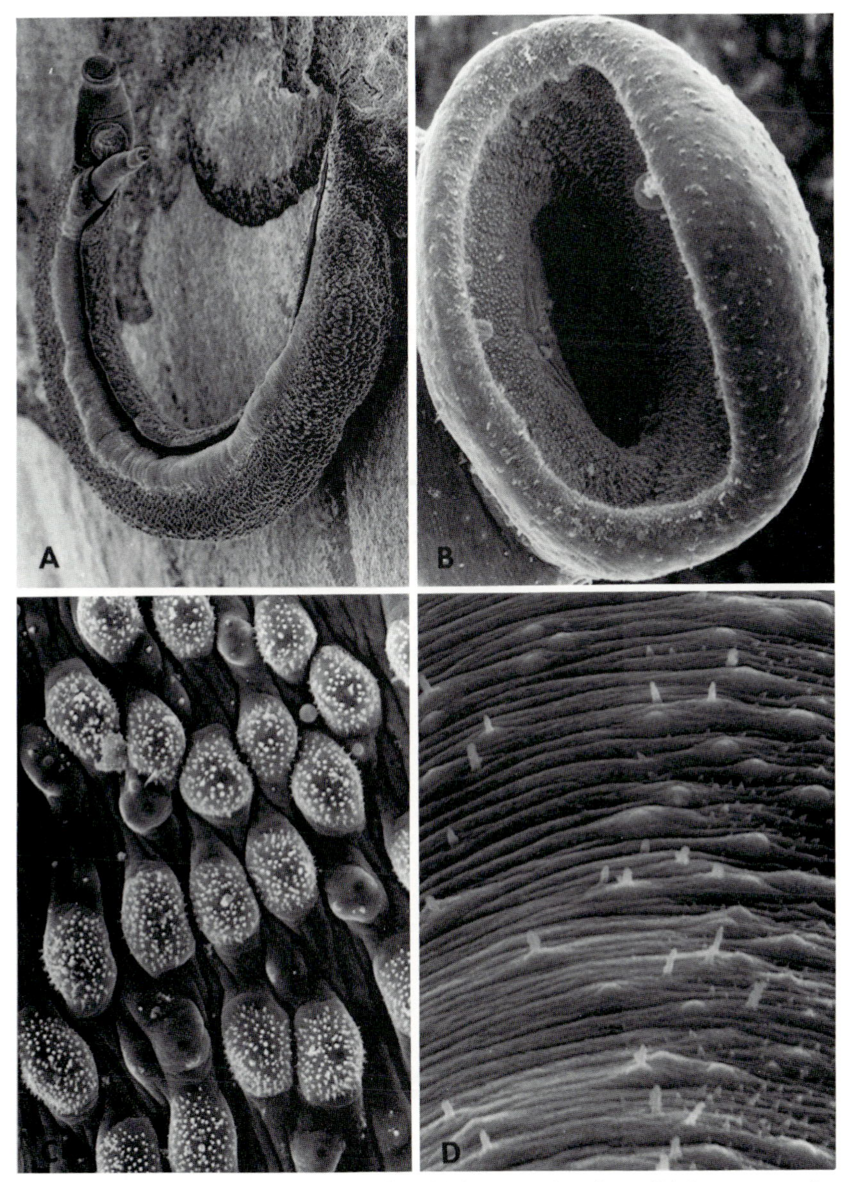

FIGURE 3.2. Scanning electron micrographs of *Schistosoma japonicum*. (A) Low-power view of a pair to show the highly tuberculate surface of the male. (B) Oral sucker of the male. Note the low papules, which may be sensory, on the outer surface. The oral cavity is lined with many small spicules, giving a furry appearance. At the esophageal border a sharp junction to circularly oriented rugae can be seen. Material adherent to the surface of the oral cavity may be from a recent blood meal. × 550. (C) Detail of the dorsal surface at the middle third of the body of a male. Note many tubercles fully invested with setae and some that have no setae but a single central spicule at a nipplelike apex. × 750. (D) Detail of ventral surface of a female where it protrudes from the gynecophoral canal. Note the spicules and fine setae which can interdigitate with larger spikes in the male gynecophoral canal. × 3250. (From Senft and Gibler, 1977. Original prints courtesy of Dr. A. W. Senft.)

meanwhile to the second stage, the larva leaves the flight muscle for the hemocoel, molts again to become a third-stage infective larva that migrates to the proboscis. By the second day after infection of the muscle, the host cell mitochondria accumulate in the spaces at the head and tail of the worm and the host nuclei become hypochromic and enlarged. These changes may be associated with nutrition of the larvae, which do not develop a patent alimentary tract until day 5 after infection. Large numbers of host mitochondria are then found in the gut.

Much more extensive are the changes associated with the intracellular development in mammalian muscle of the larvae of *Trichinella spiralis,* an intestinal worm that depends on predation for transmission to new hosts (Diagram XVIII). The adult female worm lives between the lamina propria and the columnar epithelium of the intestine. Newborn larvae are deposited into the lymphatic spaces of the lamina propria, enter the circulatory system via the superior vena cava, and pass through the heart into the aorta to be distributed via the arterial blood throughout the body. Further development occurs only within myofibers of skeletal muscle. The larva escapes from a capillary into a striated muscle cell. Nothing is known as to the receptors concerned in this behavior, but the process of entry and later development have been well described. The larva becomes aligned with the sarcolemma and then penetrates through it, leaving a ragged hole, into the muscle cell. No host membrane surrounds the larva at this stage. By the third day, muscle fibers immediately around the larva are in disarray and the host cell nuclei have enlarged and moved from the periphery. Later the central part of the muscle becomes converted into a specialized nurse cell. These changes will be considered in more detail in Chapter 5.

Whereas intracellular development in metazoan parasites is limited to muscle cells and relatively few species of nematodes, among the parasitic protozoa there are large groups, including many species, all of which are intracellular in all or some of their developmental stages. Three modes of entry into the host cell are seen: (1) by injection; (2) by phagocytosis by the host cell; (3) by activity of the parasite accompanied by an induced endocytosis by the host cell. Having entered, the parasite may develop directly in the host's cytoplasm, or it may be present in either a phagolysosome or a special parasitophorous vacuole.

Direct injection into the host cell is the mode of entry for the infective stages of all members of the large protozoan group, the Microsporidia, all of which are obligate intracellular parasites in all of their developmental stages. The highly resistant spores of these protozoa contain an intricate mechanism for discharge and injection of the contained sporoplasm into the cytoplasm of a host cell. At one end of the spore lies a tightly coiled tubule, at the other end a complex system of membranes, the polaroplast (Fig. 3.3). When the spore is ingested by an appropriate host, or subjected *in vitro* to the appropriate stimuli, the tubule is everted with explosive rapidity to form a fine filament 50 to 150 μm long and 0.1 μm in diameter. The sporoplasm is forced

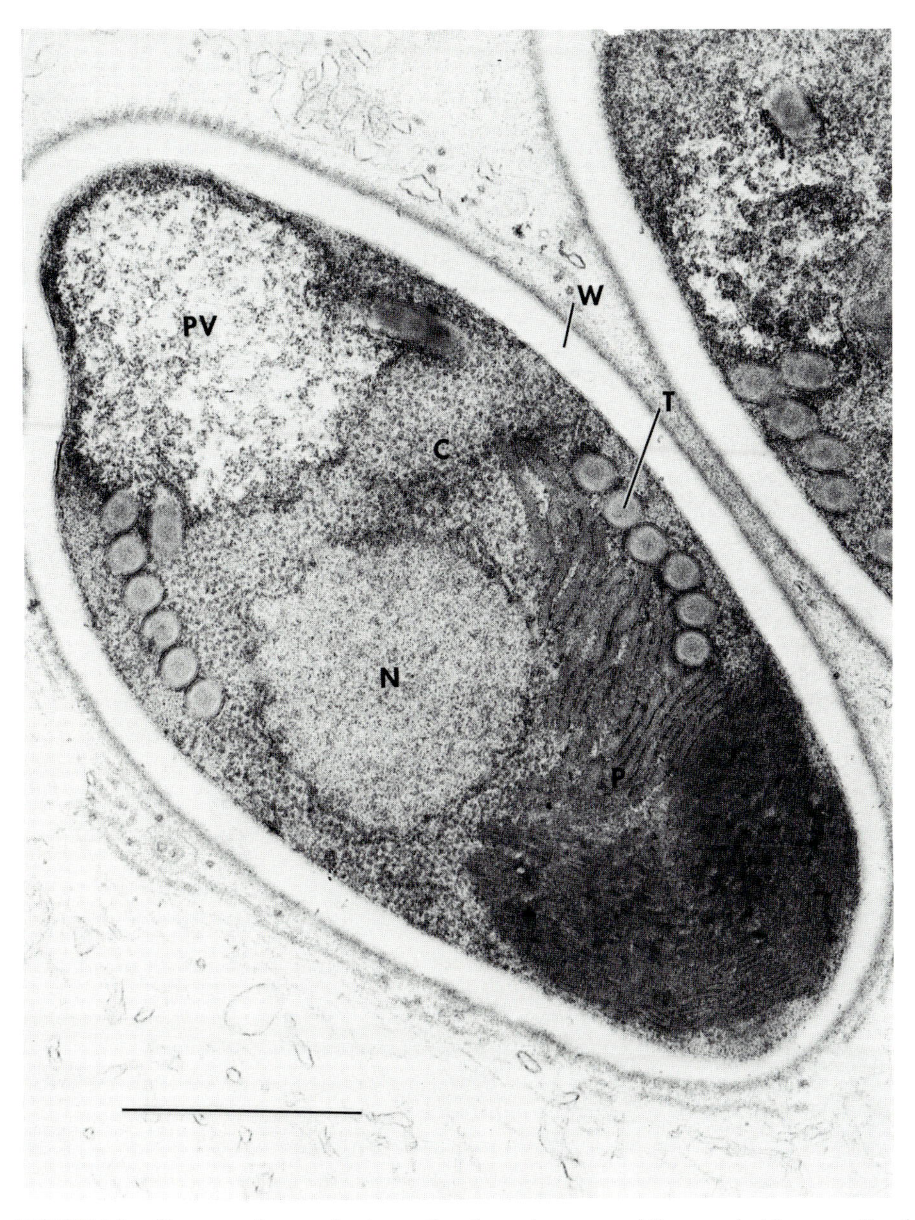

FIGURE 3.3. Electron micrograph of a section through a spore of *Spraguea lophii*, a parasite of the fish *Lophius*, showing polaroplast (P), nucleus (N), ribosome-rich cytoplasm (C), polar tube (T), posterior vacuole (PV), and wall (W). Bar = 0.5 μm. (From Weidner, 1972. Original print courtesy of Dr. E. Weidner.)

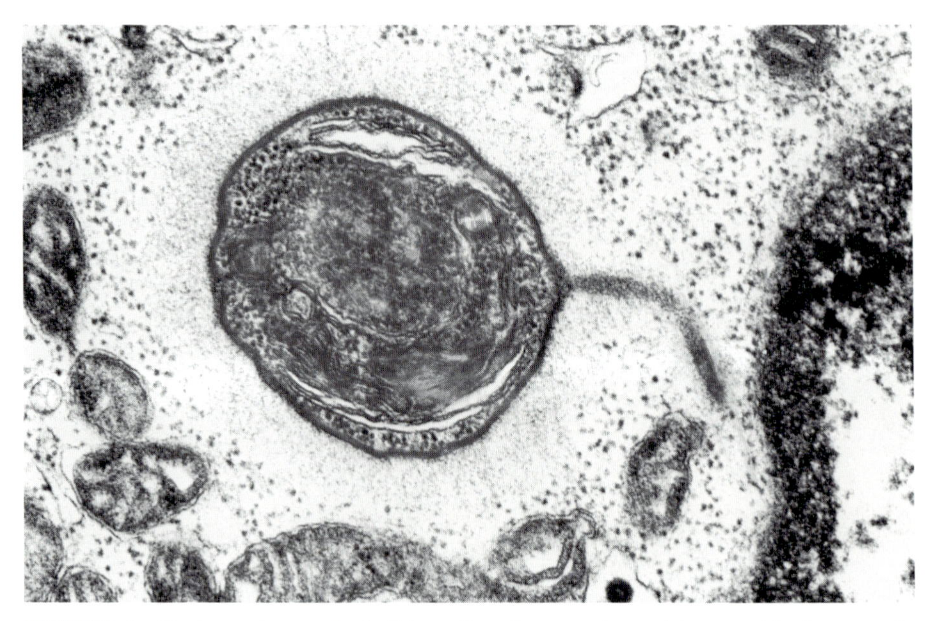

FIGURE 3.4. Electron micrograph of a thin section of a sporoplasm of *Ameson michaelis,* a parasite of the blue crab, 5 min after its injection into an EL-4 cell (a mouse cell line). The polar tube is still attached. (From Weidner, 1972. Original print courtesy of Dr. E. Weidner.)

through this filament to emerge at its distal end. If any nearby cell is struck by the everting tubule, its plasma membrane is penetrated so that the ejected sporoplasm lies within the cell cytoplasm (Fig. 3.4). At the point of entry, cytoplasm from the host cell may spread along the tube (see Fig. 3.6A). The polar filament is so fine and its penetration of the host cell plasma membrane so rapid that it does not disrupt the cell. Even red blood cells were not lysed by penetration and injection of the sporoplasm of *Ameson michaelis.*

Specificity is determined first by the conditions for hatching of the spores, and later by the suitability of the cytoplasm within which the sporoplasm is present. The entry process is entirely mechanical. The stimuli for extrusion are known for only a few species, and only very recently has some light been shed on the physiological mechanism of extrusion. For *Nosema bombycis,* the cause of pebrine disease of silkworms, there must first be exposure to a strongly alkaline medium, as occurs in the gut of the silkworm, followed by exposure to a nutrient medium. Similarly for *A. michaelis* of the blue crab, the spore is "primed" for extrusion by exposure to a high pH. It then extrudes if placed in a tissue culture medium.

Recent work has shown that discharge of spores of *Glugea hertwigi,* a parasite of smelt, depends on calcium displacement from the membranes to the matrix of the polaroplast. The "eversion" of the polar filament, until now difficult to understand, has also been clarified in recent work by Weidner. Again working with spores of *G. hertwigi,* he has shown that the tube assem-

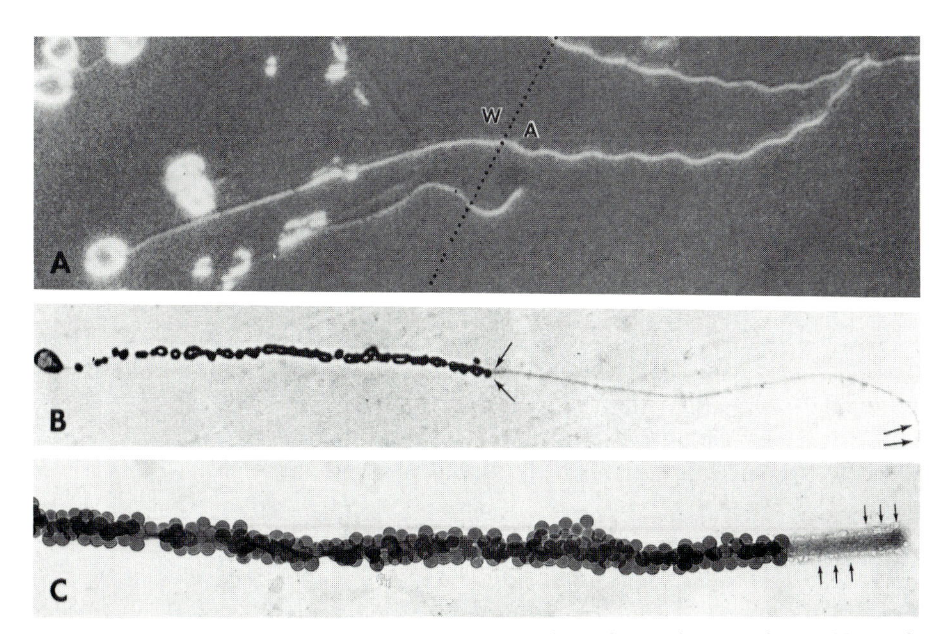

FIGURE 3.5. Discharge of spores of *Glugea hertwigi* from the rainbow smelt. (A) Nomarski interference optics of discharge across water–air (W–A) interface (indicated by dotted line). Note twisting motion of extruding tube on air side. × 3000. (B) Phase optics of discharging polar tube pulse-labeled for 2–3 sec with 0.3-μm latex particles. Arrows indicate end of latex labeling and end of discharged tube. × 1500. (C) Negatively stained discharging tube pulse-labeled for 2–3 sec with 0.1 μm latex particles. Electron micrograph. × 30,000.

For preparations B and C, spores were induced to discharge in small aqueous pools with latex particles on glass slides. Since the extruded tubes attached to the glass, the unbound latex and unhatched spores could be washed away. Note that after 2–3 sec exposure to the latex, the latex was bound only on the basal part of the tube; the distal part was devoid of latex. This shows that the discharging material assembles into a tube at the growing distal end. If the extruding spores were left in the latex pool for 1–5 min, the tubes were completely labeled. (From Weidner, 1982. Original prints courtesy of Dr. E. Weidner.)

bles by a flow of interior material that emerges at the tip and there forms a cylinder (Fig. 3.5). Both calcium and shift of pH are involved in the stabilization of the assembled protein. The polar tube protein is remarkable in being resistant to proteinase K and sodium dodecyl sulfate; it is, however, reduced by mercaptoethanol or dithiothreitol. The polaroplast membranes probably provide the plasma membrane of the sporoplasm, as illustrated diagrammatically in Fig. 3.6B. The old plasma membrane is left behind in the spore.

It is important to realize that a few spores ingested by an appropriate host may multiply to fill most of the tissues, as in fatal pebrine of silkworms. Alternatively, there may be more limited multiplication so that the host does not die. If it is a female, the ovaries and the eggs become infected, providing an additional mode of transmission to the next generation, a fact observed by Pasteur in his classic studies. In either case there must be a stage of the

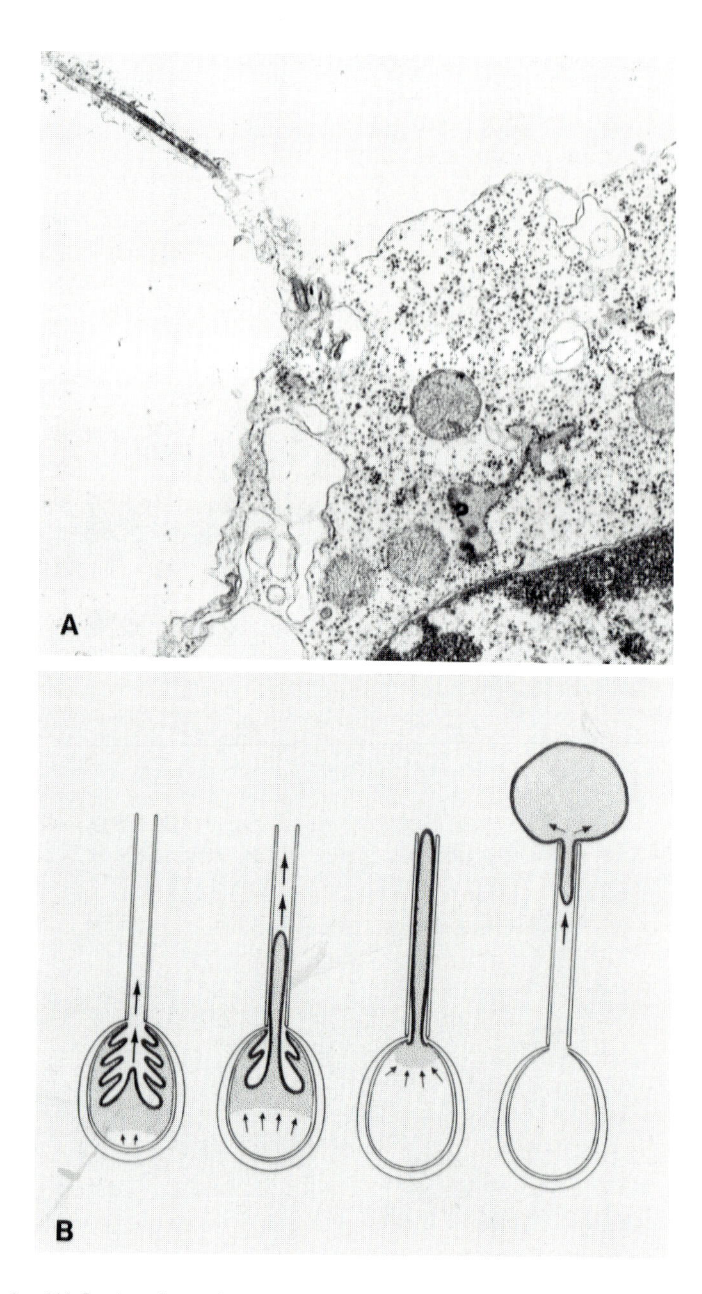

FIGURE 3.6. (A) Section through point of entry of polar tube of *A. michaelis* into an EL-4 cell. (From Weidner, 1972). (B) Diagram to show the origin of the plasma membrane of the sporoplasm from the polaroplast membrane. (From Weidner *et al.*, 1984).

parasite and a mechanism for intracellular infection that provides for this spread within the already infected host and that probably does not involve the spore. Indications for such a stage were obtained in an early experiment in which it was shown that hemolymph taken from a silkworm 2 days after its infection by mouth with spores of *N. bombycis*, and added to a silkworm tissue culture, produced an infection with spore formation. Hemolymph taken 1 day after infection of the silkworm was not infective. The 2-day hemolymph was presumed to contain infective forms responsible for the spread of the infection. The existence and nature of these forms have been clearly demonstrated in cultures of *N. bombycis* in silkworm tissue cultures. There are present characteristic round binucleate forms somewhat larger than the original infective sporoplasm but smaller than developing sporonts. These round forms were observed leaving infected cells and entering other cells. Their mode of entry has not been determined; it probably involves induced endocytosis as seen in the entry of sporozoa into nonphagocytic cells (see later in this chapter).

Uptake of a parasite by phagocytic cells would seem at first glance to be a simple and direct way for a parasite to become intracellular. Actually, for most microbes it is a route to certain death. Phagocytosis does not bring ingested particles into the cytoplasm but rather into a phagosome, a membrane-lined structure that fuses with secondary lysosomes to form a vacuole with low pH into which a battery of hydrolytic enzymes is secreted. Oxidant radicals may also be formed here. The usual result is the killing and digestion of ingested organisms. The polymorphonuclear leukocytes and the macrophages of vertebrates constitute one of the first lines of defense against invading parasites; it is not surprising that few types of parasites occur within them. There is, however, an important group of protozoa that has made the phagolysosome of macophages its home, the leishmanial parasites of humans, rodents, and some other vertebrates.

These organisms, etiological agents of a wide spectrum of human diseases, are transmitted by blood-feeding phlebotomine sand flies. In the alimentary tract of the insect they multiply as elongate flagellated forms, the promastigotes (Diagram III). Promastigotes introduced into a vertebrate host by the bite of an infected fly are ingested by macrophages into phagolysosomes, like any other foreign material. In an appropriate host, however, the infective promastigotes are resistant to the degradative action of the lysosome. The basis for this resistance is the subject of much research and is not known. There are indications that the organisms have specialized surface glycoproteins. In any case, it is certain that the phagolysosome itself is not modified, for other materials ingested with a promastigote into the same vacuole are digested. The promastigotes, however, transform into smaller forms, the amastigotes that lack an external flagellum. These multiply within the cell and may completely fill its cytoplasm. Amastigotes liberated by breakdown of an infected cell (and perhaps also by exocytosis) are then taken up by other macrophages and serve to spread the infection within the host. Specificity is determined presumably first by ligands that bind promastigotes or amastigotes to the surface of the macrophage, and second by the ability to survive in the phagolysosome of a particular type of phagocytic cell. For example, human polymorphonuclear leukocytes show very active formation of H_2O_2 and the anionic free radicals—O_2^-, $OH^{\cdot}$, and $^1O_2$. In these cells both promastigotes and amastigotes of *Leishmania donovani* are killed, whereas they survive and multiply in human monocytes and macophages. The latter cells also form the free radicals but to a lesser extent, a level against which the parasites evidently have an effective protective mechanism.

Most other groups of intracellular protozoa generally infect nonphagocytic cells. If they do enter a phagocyte, they either escape from the phagosome into the cytoplasm, or they somehow prevent fusion of lysosomes with the vacuole containing the parasite. This vacuole may then be called the parasitophorous vacuole. *Trypanosoma cruzi* (Diagram IV), the cause of Chagas' disease in South and Central America, exhibits the first method, escape into the cytoplasm. This flagellate, in the same large taxonomic group as the leishmanias, is transmitted by bloodsucking bugs of the family Reduviidae. In the alimentary tract of the bug the protozoa multiply and differentiate to form metacyclic trypomastigote infective forms. These are excreted as the bug defecates during feeding, and they penetrate the skin. They then enter connective tissue cells, fibroblasts, and macophages, where they multiply as amastigotes directly in the cytoplasm. As the host cell is destroyed, the parasites transform again into motile trypomastigotes; these enter the blood and spread the infection to other cells, in particular to striated muscle cells. In the muscle they multiply as amastigotes in the cytoplasm, with eventual transformation to trypomastigotes that reenter the blood. These bloodstream trypomastigotes serve for the infection of other tissues and, most importantly, for the infection of reduviid bugs that feed on the host. How the trypomastigotes enter a muscle cell is not known; a positive attraction to such cells has, however, been demonstrated. *In vitro* the attractive index for one strain of *T. cruzi* to secondary cultures of bovine embryo skeletal muscle was twice as high as to HeLa cells and ten times as great as to green monkey kidney cells. Killed muscle cells (heated 5 min at 60°C) exerted no attraction whatever. There was no difference in endogenous oxygen consumption among the three types of cells, indicating that pO_2 or pCO_2 gradients were not responsible for the differing degrees of attraction. Further work with such a system might reveal the crucial metabolites released by muscle cells making them more attractive to trypomastigotes of myotropic strains of *T. cruzi*. How subtle the interaction with the host cell may be is indicated by the recent finding that HeLa cells in S phase were about twice as susceptible to *T. cruzi* as those in G1 phase. Perhaps receptors on the cell surface are increased during S phase.

When trypomastigotes of *T. cruzi* entered a macrophage, they were taken in apparently by phagocytosis but after about 1 hr they were found free in the cytoplasm (Fig. 3.7). On the other hand, epimastigote forms from cultures taken up by a macrophage remained in the phagosome where they were killed and digested. Epimastigotes accordingly could not establish infection

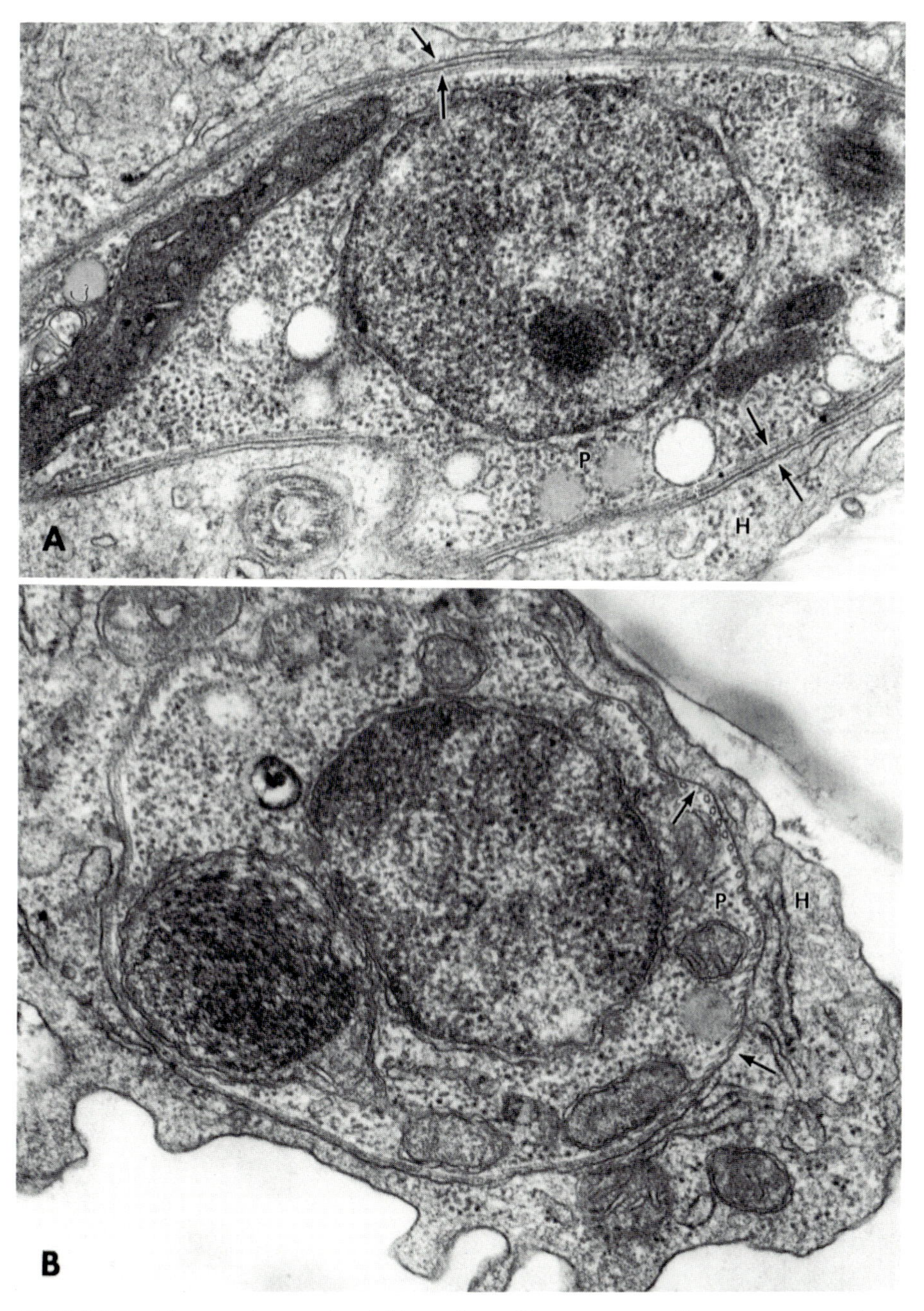

FIGURE 3.7. Electron micrographs of thin sections of *Trypanosoma cruzi* in a mouse macrophage. (A) At 1 hr after entry. Note that two membranes (arrows) separate the cytoplasm of the parasite (P) from that of the host (H). (B) At a later time. There is now only one membrane (arrows), that of the parasite, subtended by microtubules. The parasite no longer lies in a host-membrane-lined vacuole, but directly in the host cytoplasm. × 90,000. (Original prints courtesy of Dr. N. Nogueira. See also Chapter 20, Nogueira and Coura, 1984.)

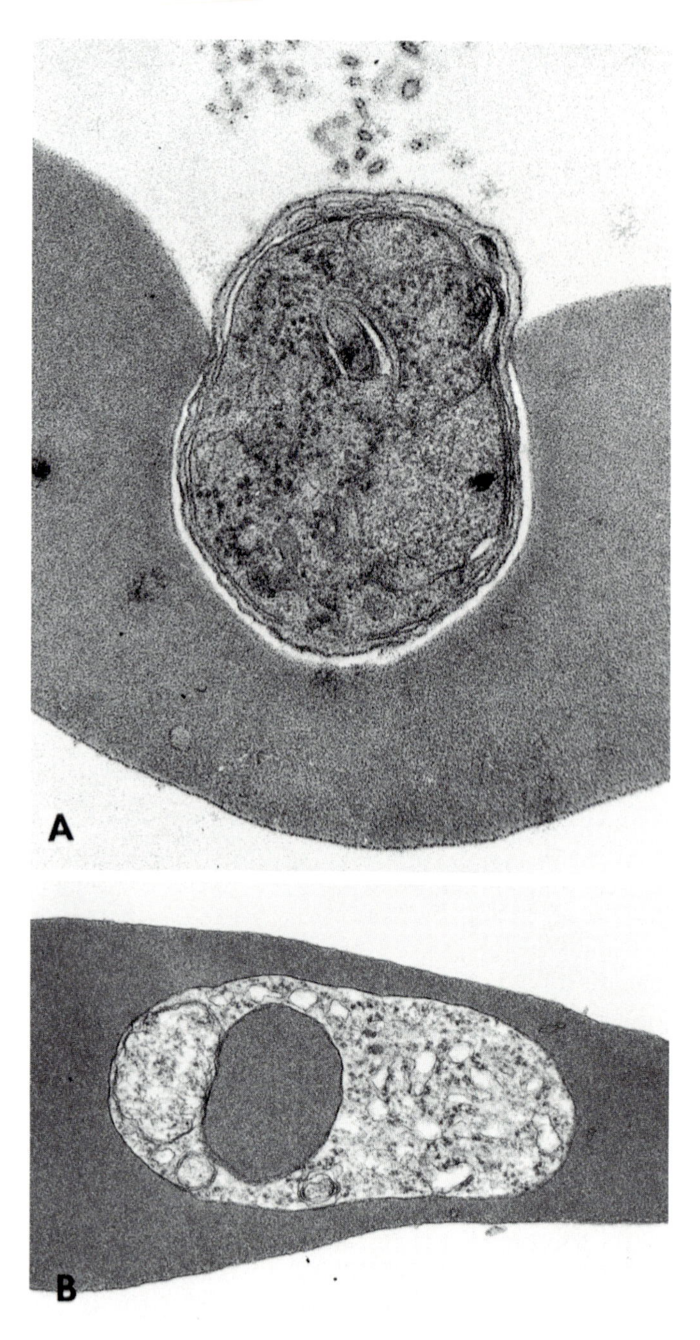

FIGURE 3.8. *Babesia microti.* (A) Entry of merozoite into endocytotic vacuole in hamster erythroycte. Note sloughing of the surface coat of the merozoite and the close junction with the red cell plasma membrane at its invagination. × 52,000. (B) Trophozoite of *B. microti* lying directly in the red cell cytoplasm. × 37,000. (Original prints courtesy of Dr. M. Rudzinska. (A) from Rudzinska, *et al.*, 1976; (B) from Rudzinska and Trager, 1977.)

in macrophages but trypomastigotes could. Likewise, with nonphagocytic cells, epimastigotes could not initiate infection; they had to first change into trypomastigotes. How trypomastigotes in a phagosome of a macrophage escape into the cytoplasm is not known. Most likely, they lyse the phagosomal membrane. The ultrastructure of entry of *T. cruzi* into nonphagocytic cells, such as muscle, has not been studied. A reasonable hypothesis would be that the trypomastigote induces invagination of the cell membrane and then, once inside, lyses the membrane around it to get into the cytoplasm.

Just this type of entry has been observed with the merozoites of *Babesia* (Diagram X) entering an erythrocyte. Species of the genus *Babesia* multiply in the red cells of mammals. They are transmitted by ticks in which they undergo a developmental cycle, and are responsible for important diseases of cattle and other domestic animals. At least two species are known to produce zoonosis in humans: *B. microti*, a parasite of deer mice, and *B. bigemina*, a parasite of cattle. For the rodent parasite *B. rodhaini* it has been shown that entry into a rat erythrocyte requires the C3 component of serum complement and activation of the alternative pathway. Uptake of C3b occurs via a C3b receptor on the parasite. This presumably facilitates interaction with a C3b receptor on the red cell. The process of entry has been seen in transmission electron micrographs prepared from blood of hamsters heavily infected with *B. microti* (Fig. 3.8A).

The invasive form, a merozoite, must attach to the red cell surface by its apical end, the end at which there is a pair of organelles called rhoptries. These organelles are present in the invasive stages of malaria and other sporozoa, and from studies with these other species there is good basis for the belief that they secrete some substance that causes the host cell membrane to invaginate. Such induced endocytosis is the mechanism whereby a merozoite of *Babesia* is taken into an erythrocyte. Although the newly entered merozoite is at first surrounded by the invaginated red cell membrane, the membrane soon disappears (probably in less than 1 hr) and the *Babesia* undergoes its further development directly in the red cell cytoplasm (Fig. 3.8B). It forms two or four daughter parasites; how these merozoites escape from their host cell to infect other cells is not known.

Much more is known about the entry of malarial merozoites into their host erythrocytes. This again occurs by induced endocytosis (Fig. 3.9). Complement is not required, however. The invaginated red cell membrane does not disappear but instead forms the parasitophorous vacuole membrane and becomes closely apposed to the plasma membrane of the developing parasite (Fig. 3.9C). There is increasing evidence for the role of specific receptors in the interaction between a malarial merozoite and its host erythrocyte. This had been forecast in early studies by R. B. McGhee who introduced mammalian erythrocytes into the circulation of chick embryos infected with an avian malaria *Plasmodium lophurae* and observed that certain species, notably mouse erythrocytes, were invaded by this avian parasite. Most interesting was the observation that red cells of newborn rats, but not those of weaned

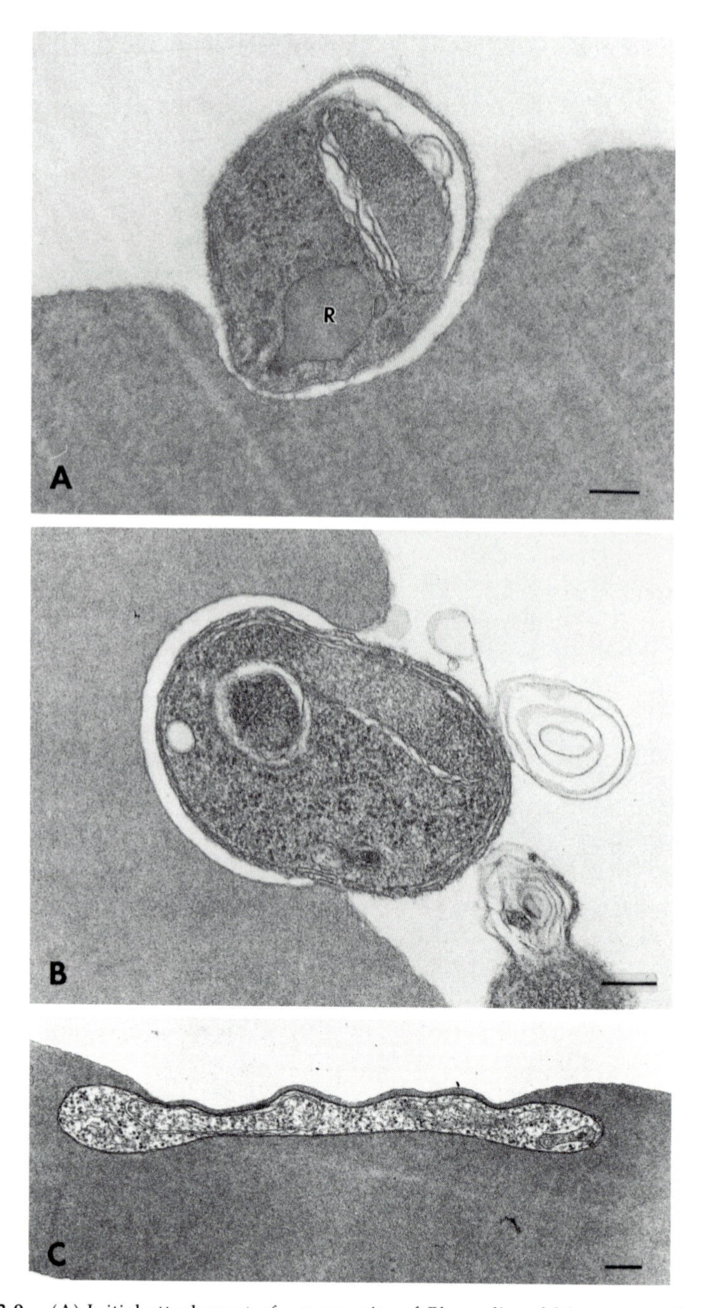

FIGURE 3.9. (A) Initial attachment of a merozoite of *Plasmodium falciparum* to a human erythrocyte. Note the fibrous surface coat surrounding the merozoite and the fibrils extending from its apex to the erythrocyte plasma membrane. R, rhoptry. (B) Merozoite about halfway in. Note the junction at the orifice of the invagination and the lack of surface coat on that portion of the merozoite that has entered the forming parasitophorous vacuole. (C) Section through an appliqué form of *P. falciparum*. This recently entered parasite lies in a membrane-bound vacuole. Two

rats, were susceptible, suggesting loss of a receptor. In studies with *P. knowlesi*, a malarial parasite of monkeys that can also infect humans, it has been found that *in vitro* merozoites will attach only to the red cells of susceptible host species. Furthermore, among human erythrocytes the merozoites of *P. knowlesi* would enter only those positive for the Duffy blood group. They would attach to but would not enter Duffy-negative human red cells. If such Duffy-negative cells were treated with trypsin or neuraminidase, however, they would be invaded. Moreover, erythrocytes of New World monkeys (*Aotus trivirgatus*) were readily invaded, even though they lack the Duffy determinants Fy^a and Fy^b. Hence, it is not likely that the Fy^a and Fy^b antigens are themselves the erythrocyte receptors for invasion. Nevertheless, the findings with *P. knowlesi* and Duffy-negative human erythrocytes have led to an important correlation in human malaria. It was known that *P. vivax* does not occur in West Africa and that many American blacks, most of whose ancestors came originally from West Africa, are refractory to this infection even though fully susceptible to the other three species of human malaria (*P. falciparum, P. malariae,* and *P. ovale*). Since 99% of the population in West Africa is Duffy-negative, it was suggested that the Duffy antigens are essential receptors for the invasion of human erythrocytes by *P. vivax,* and that their absence is responsible for resistance to this parasite. This hypothesis is strongly supported by the following observations. In a survey of villages in Honduras where about half the population was Duffy-negative, *P. falciparum* infections were uniformly distributed throughout, but the 14 *P. vivax* infections were all in Duffy-positive individuals. Among 13 American blacks who had acquired *P. vivax* in Vietnam, all were Duffy-positive, whereas the American black population in general, as determined for blood donors, has only 40–50% Duffy-positive. The probability of this result occurring by chance alone is less than 0.001. Final validation of this hypothesis must await *in vitro* experiments, not easy to do since *P. vivax* cannot be readily grown in culture.

Such *in vitro* experiments with the major human malarial parasite *P. falciparum* show clearly that the Duffy antigen is not involved, and indicate a likely receptor role for the membrane proteins, glycophorin A and glycophorin B. At low concentrations these proteins, which are the major sialoglycoproteins of human red cells, inhibit invasion by merozoites of *P. falciparum in vitro.* Mutant cells deficient in glycophorins are relatively resistant to invasion. Furthermore, the merozoites have been shown to have two surface antigens of 155,000 and 130,000 daltons that bind avidly to human glycophorin (Figs. 3.10, 3.11). The evidence is strong that these two proteins and the glycophorins are the mutually interacting receptors of the merozoite and the

membranes may be clearly seen separating parasite from host, the parasite plasma membrane and the parasitophorous membrane. Bars = 0.2 μm (B from Langreth *et al.*, 1978. Original prints courtesy of Dr. S. G. Langreth.)

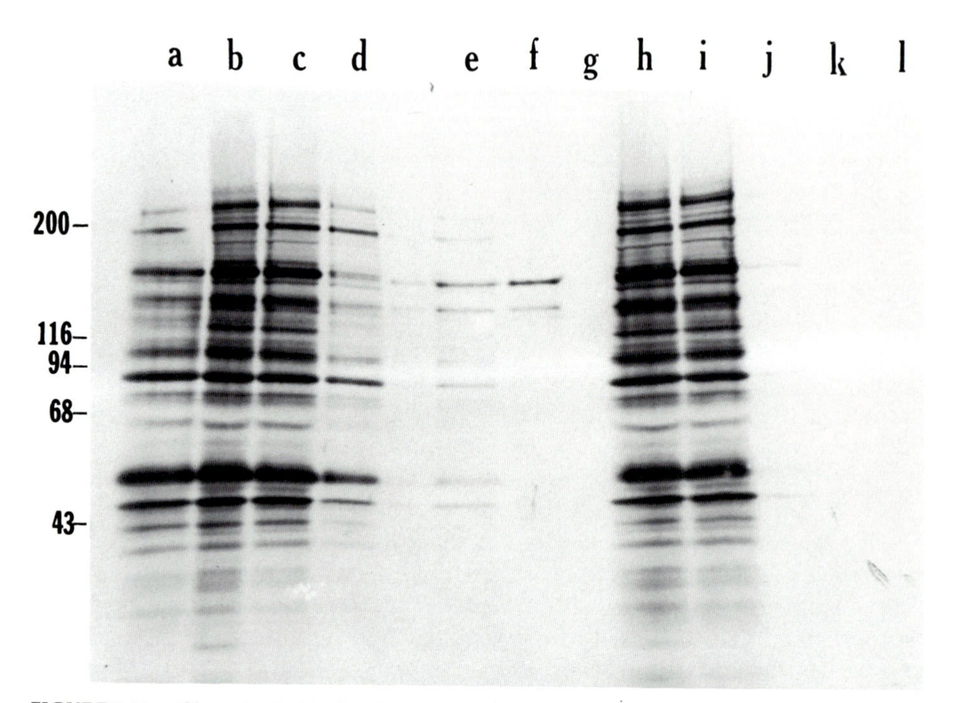

FIGURE 3.10. Glycophorin binding by proteins from merozoites of *P. falciparum*. Extracts were prepared from ^{3}H-glycine-labeled merozoites. Lanes *b, c,* and *h* show this starting material, *c* after 30 min at 4°C. The extracts were mixed with either glycophorin-coupled acrylamide (GA) beads (lanes *a, d–g*) or fetuin-coupled (FA) beads (lanes *i–l*) and the beads were washed and eluted with different solutions. *a:* proteins not binding to GA; *d:* 2 × phosphate-buffered saline (PBS) eluate from GA; *e:* 0.5 M NaCl/PBS eluate from GA; *f:* 2% sodium dodecyl sulfate (SDS)/PBS eluate from GA; *g:* 4% SDS, 100°C eluate; *i:* proteins not binding to FA; *j:* 2 × PBS eluate from FA; *k:* 0.5 M NaCl/PBS eluate from FA; *l:* 2% SDS/PBS eluate from FA. The materials were electrophoresed on a 5–15% polyacrylamide gradient gel. Molecular weight standards indicated at left.

Note that proteins were bound only by the GA beads. Loosely bound proteins were eluted with PBS and several with 0.5 M NaCl (lanes *d* and *e*). The only proteins that remained bound after NaCl elution were a major one of 155,000 daltons and a minor one of 130,000 daltons. Both of these were eluted with 2% SDS (lane *f*). (From Perkins, 1984.)

red cell, respectively, and are responsible for the initial attachment of *P. falciparum* merozoites to human erythrocytes.

Attachment, however, is only the first step in the complex process of entry. A detailed morphological description of this process is now available, and indicates the kinds of biochemical questions to be asked and answered. The apical end of the merozoite, i.e., the end at which open the ductules from the paired organelles or rhoptries (Fig. 3.9A), makes contact with the surface of a susceptible erythrocyte and forms a small depression in the erythrocyte membrane. This part of the membrane becomes thickened and forms a junction with the plasma membrane of the merozoite. As the red cell mem-

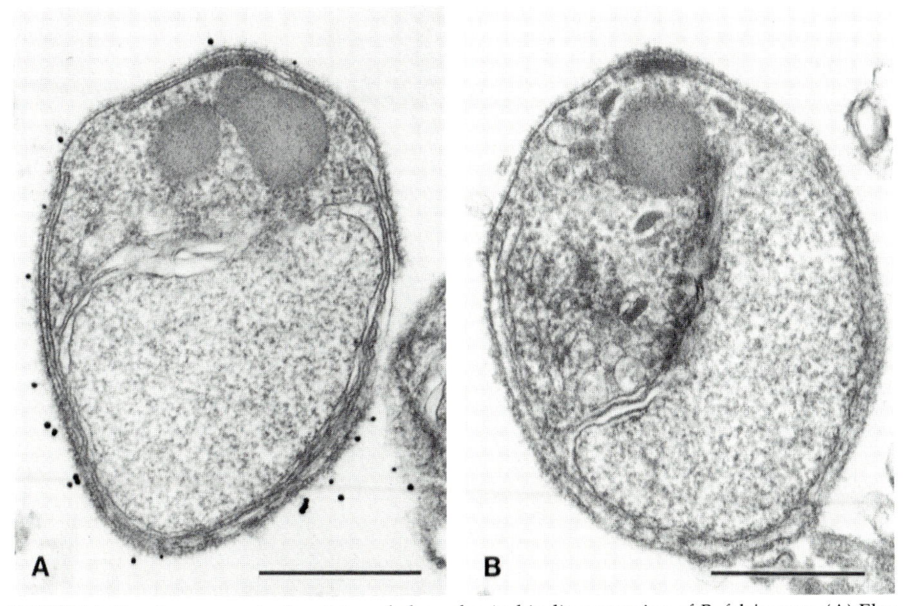

FIGURE 3.11. Immunocytochemistry of glycophorin-binding proteins of *P. falciparum*. (A) Electron micrograph of thin section of merozoites treated with immune rabbit serum (to 155,000-dalton protein prepared from culture supernatant) and protein A–gold. (B) Electron micrograph of thin section of merozoites treated with preimmune rabbit serum and protein A–gold. Bars = 0.5 μm. Gold particles were seen only with the immune serum and only in the surface coat of the merozoite. (From Perkins, 1984. Original prints courtesy of Dr. M. Perkins.)

brane invaginates to receive the parasite, this junction becomes a circumferential zone of attachment between the erythrocyte and the merozoite (Fig. 3.9B). The junction moves along the membranes to maintain its position at the orifice of the invagination. When entry is completed (within 20–30 sec after attachment), the orifice closes behind the parasite to seal the red cell plasma membrane and to complete the parasitophorous vacuole membrane (hereafter called the parasitophorous membrane). As the merozoite is drawn in, its brushy surface coat is lost from the portion inside the vacuole. This facilitates the subsequent close apposition of the parasite membrane and the parasitophorous membrane. This latter, host-derived membrane rapidly becomes altered. It loses its intramembranous particles on the P face and at least two of its enzymes reverse polarity. Red cell membrane ATPase is localized cytochemically at the electron microscope level on the inner surface of the erythrocyte plasma membrane. It would be expected therefore on the outer surface of the invaginated, inside-out, parasitophorous membrane. It is indeed there just after entry, but very soon comes to be on the inner surface (Fig. 3.12A,B). Just the opposite occurs with an NADH oxidase. Located on the outside of the red cell membrane, it would be expected on the inside of the parasitophorous membrane; but it is on the outside. These changes occur

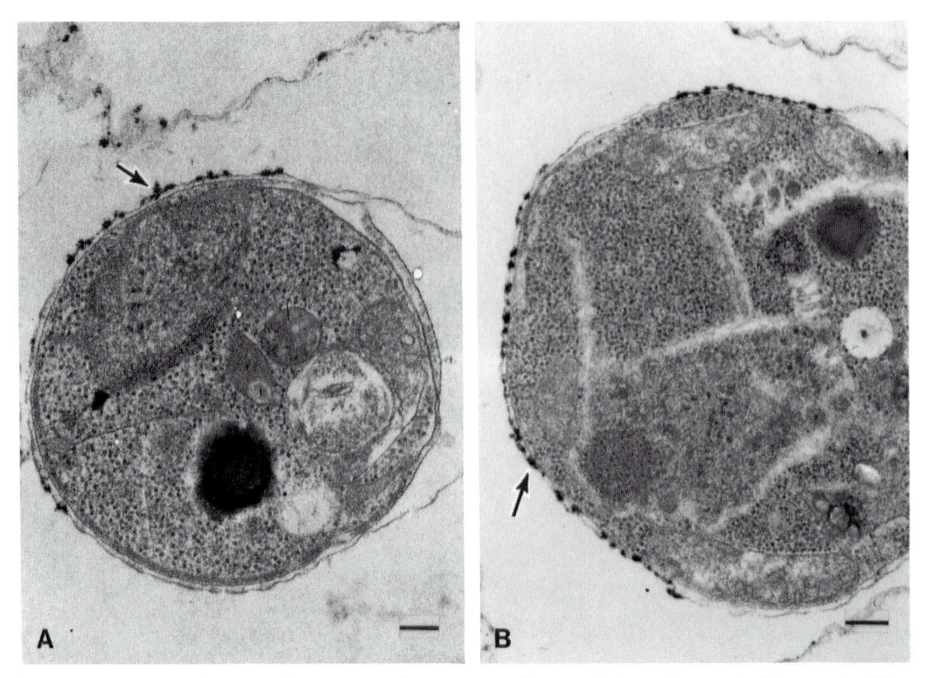

FIGURE 3.12. ATPase on the parasitophorous membrane of the avian malarial parasite *Plasmodium lophurae.* (A) A newly invaded very young trophozoite; (B) an older trophozoite. Bars = 0.2 μm. Parasites were freed with saponin. Note that whereas in A the reaction product is mostly on the outer surface of the parasitophorous membrane (arrow), in B it is entirely on the inner surface (arrow). (From Langreth, 1977. Original prints courtesy of Dr. S. G. Langreth.)

within minutes after entry and continue to the point where the parsitophorous membrane is much more parasite than red cell (see Chapter 5). Merozoites of *P. knowlesi* treated with cytochalasin B attached to susceptible erythrocytes but did not invade. Nevertheless, a small parasitophorous membrane formed in advance of the apical tip of the merozoite, and the P face of this membrane was devoid of intramembranous particles. Cytochalasin B evidently does not interfere with the initial effect of the rhoptry secretion; possibly it prevents movement of the junction and in this way prevents entry.

The *in vitro* system of continuous culture of *P. falciparum* in human erythrocytes is especially suitable for the further study of this process and has already been used to good advantage. If human red cells were carefully hemolyzed by exposure to a hypotonic medium in a dialysis bag and then allowed to reseal, the cells so treated could be tested for their susceptibility to invasion by falciparum merozoites *in vitro*. In such experiments it was shown that Mg-ATP had to be present for invasion of the resealed ghosts to occur, and the nearer the ATP content of the ghosts to that of normal cells, the nearer was the extent of invasion to that of normal cells. We will see in a later chapter that exogenous ATP continues to play a role throughout the intraerythrocytic

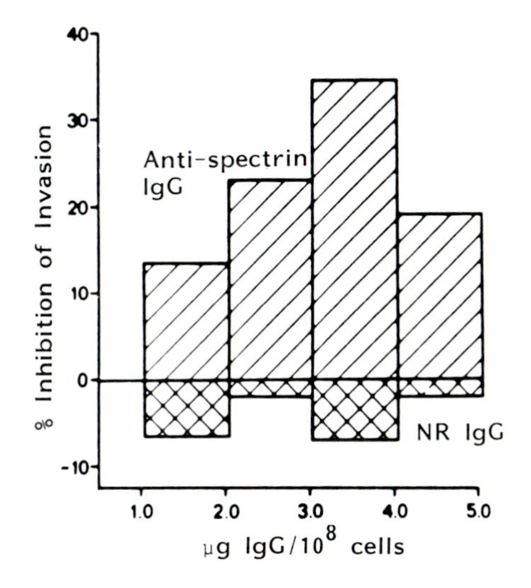

FIGURE 3.13. Effect of antispectrin IgG on invasion of resealed ghosts by *P. falciparum*. Results represent data from four experiments pooled by IgG content. For each group (1.0–1.9, 2.0–2.9, 3.0–3.9, 4.0–4.9 μg IgG/10^8 cells), percent inhibition values were averaged and plotted. Single hatching represents antispectrin IgG; cross-hatching represents normal rabbit IgG. (From Olson and Kilejian, 1982.)

development of malarial parasites. Resealed ghosts have also been used to demonstrate the essential role of spectrin, and hence of the cytoskeleton, in the entry process. Ghosts allowed to reseal in the presence of antispectrin IgG at 3–4 μg/10^8 cells were 35% less susceptible than normal cells (Fig. 3.13). Normal IgG had no effect. Most significantly, the monovalent Fab′ of antispectrin IgG had no effect, showing that cross-linking of the spectrin interfered with its role in the invasion process. Not surprisingly, calcium also is essential for entry.

It seems likely that most of the intracellular sporozoa other than malarial parasites enter their host cells by some process of induced endocytosis resembling the mode of entry of the malarial merozoite into a red cell. Since these other species (and even other stages of the malarial life cycle—see Diagram VIII) parasitize cells other than erythrocytes which have a more active metabolism and which may exhibit phagocytic activity and have a system of effective lysosomal hydrolytic enzymes, the situation becomes somewhat more complicated. Furthermore, these other invasive stages have not only a more extensively developed system of rhoptries and microenemes, all opening at the anterior apical end; they also have additional differentiation of the apical end into a motile conoid, seemingly adapted for penetration into a cell, and they are capable of considerable motility. Detailed observations have been made with *Toxoplasma gondii* and with several other species of coccidia.

For many years *T. gondii* was known only as a ubiquitous intracellular parasite of many kinds of vertebrates including humans. It showed little specificity with regard to either its host organism or the type of host cell in which it would develop. It grew well within different kinds of cells in tissue cultures, but its extracellular survival was very short. The only known method of transmission was through predation. Thus, people became infected by

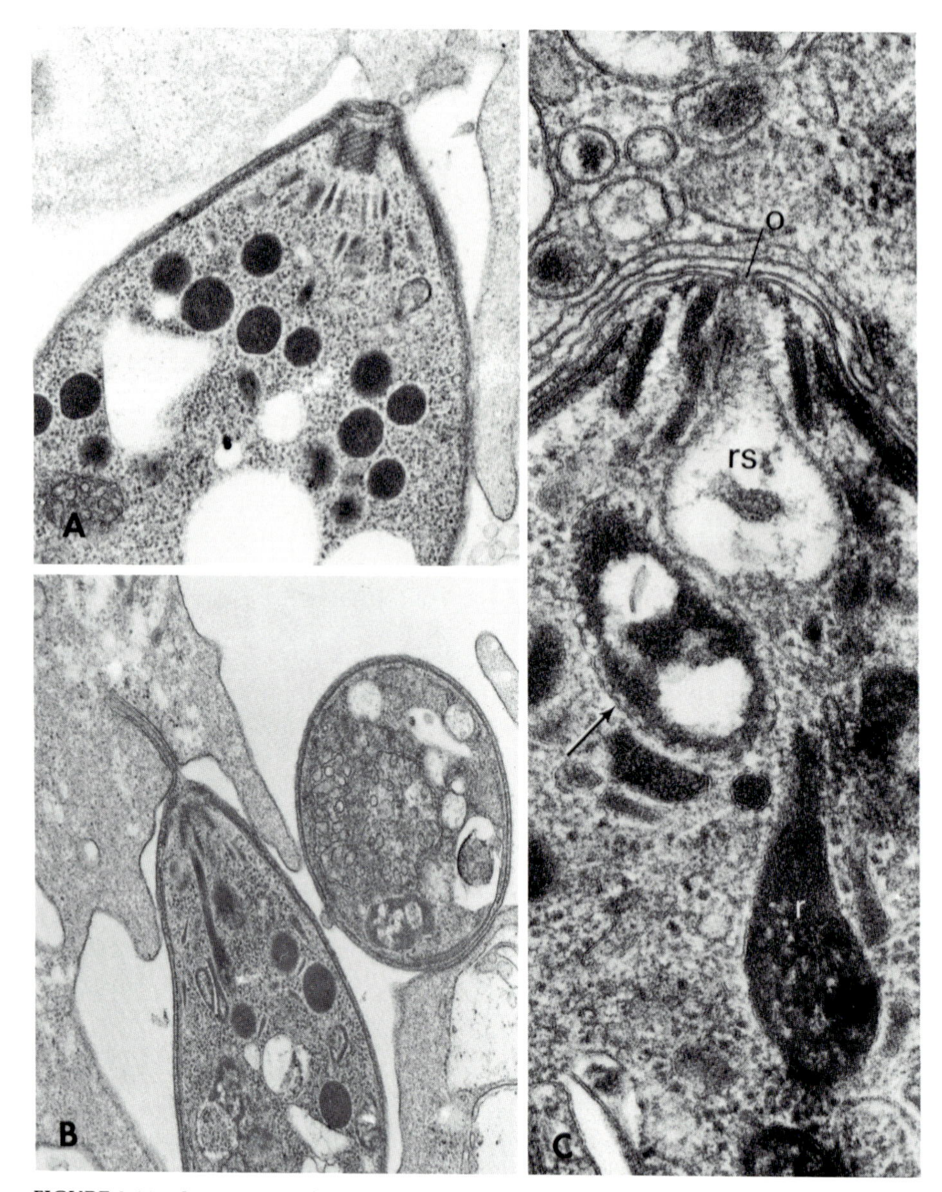

FIGURE 3.14. Interactions of *Toxoplasma gondii* with host cells as seen in electron micrographs of thin sections. (A, B) Early interactions. In A note small portion of host cytoplasm protruding into anterior end of parasite. B shows a structure sometimes seen protruding from the parasite into the cell. It is continuous with the parasite plasmalemma. A, × 25,000; B, × 30,000. (C) Anterior end of *Toxoplasma* within a mouse peritoneal cell 1 min after inoculation. Note the empty rhoptry (rs) and its opening (O). A partially empty rhoptry is seen at the arrow, and below it an apparently unaltered rhoptry (r). × 90,200. (A and B from Aikawa *et al.*, 1977; original prints courtesy of Dr. M. Aikawa. C from Nichols *et al.*, 1983; original print courtesy of Dr. B. A. Nichols.)

eating raw or undercooked meat from an infected animal. Now we know that these stages represent only a small portion of the life cycle (Diagram VII).

Within the intestinal cells of members of the cat family (Felidae) there occurs a typical coccidian cycle of development with formation of male and female gametes, their fusion to form oocysts, and the subsequent development of sporocysts and sporozoites within the oocyst. Excreted oocysts ingested by a rodent or a human initiate the tissue cycle of development, the only cycle previously known. It is of great interest that the sexual cycle is highly specific in its host requirement, only members of the Felidae being suitable. Hence, it is not surprising that little experimental work has been done with the sporozoites of *Toxoplasma,* whereas there is a great bulk of work with the tissue stages, the so-called tachyzoites.

It is clear that these enter host cells typically through an active process involving induced endocytosis. When entry is into a macrophage, this process cannot be easily distinguished from ordinary phagocytosis, but when entry is into nonphagocytic cells, e.g., HeLa cells, it can be seen to involve an active entry by the parasite. By light microscopy, zoites of *Toxoplasma* can be seen moving toward a host cell by flexion. They then insert the conoid and appear to squeeze through a narrow opening. Electron microscopy shows that the plasma membrane of the host cell most probably is not broken but is invaginated to receive the entering zoite (Fig. 3.14). This has been especially nicely demonstrated with entry into red cells, where the picture is much like that we have seen with the malarial merozoite. In some instances the invaginating membrane seems to be partially disrupted and it may be lost after the parasite is interiorized. The parasite then lies briefly directly in the cytoplasm until a new membrane is formed to give rise to the parasitophorous vacuole. This membrane is clearly different from the membrane of a phagosome and secondary lysosomes do not fuse with it. On the other hand, a killed *Toxoplasma* ingested by a phagocytic cell comes to lie in a typical phagosome with which lysosomes fuse, resulting in digestion of the parasite. The invasion of *Toxoplasma* into chick embryo erythrocytes was dependent on an adequate level of ATP, suggesting that, as for malarial merozoites, host ATP is an essential factor.

Among the coccidia other than *Toxoplasma,* most observations have been made with the sporozoites of several species of *Eimeria*. It will be remembered (Chapter 2) that these are the stages that escape from ingested oocysts and initiate infection usually in cells of the alimentary tract of a suitable new host. *In vitro* sporozoites of *Eimeria* will enter and undergo schizogony in tissue culture cells of types other than those to which they are restricted in nature. That these *in vitro* conditions are not altogether adequate is indicated by the fact that the complete developmental cycle (Diagram VI) including growth of merozoites formed from the sporozoites and subsequent gametogony, has been obtained in tissue culture for only one species (*E. tenella* of chickens). With this qualification in mind, the weight of evidence again indicates that sporozoites of *Eimeria* enter host cells by an active process that involves induced endocytosis by the host cell.

Motility of the sporozoites is of prime importance. Chilling of the sporozoites reduced their entry far more than chilling of the host cells, and treatment of the sporozoites with cytochialasin B completely prevented entry. On the other hand, treatment of the host cells with agents that inhibited phagocytosis by blocking glycolysis (e.g., 2-deoxyglucose) or by blocking activity of microtubules (e.g., colchicine) had no effect on entry. Colchicine did not affect the motility of the sporozoites but cytochalasin B (an agent affecting microfilament function) did inhibit it. As with malarial merozoites, the entry is an active process by the parasite but it is not a crude break-in with rupture of the host cell membrane. Rather, the activities of the parasite exert a subtle effect on the host membrane, causing it to interact in a specific way with the parasite membrane to result in engulfment of the parasite. Where the parasitophorous membrane thus formed persists, as in *Plasmodium, Toxoplasma,* and *Eimeria,* it must have special and remarkable properties to provide for passage of essential ions and nutrients to the parasite and for excretion of waste products of the parasite's metabolism. This fascinating membrane will be considered further in Chapter 5.

With *Eimeria,* as with *Toxoplasma, Plasmodium,* and *Babesia,* entry is always at that end of the organism where the rhoptries open. It seems reasonable to suppose that secretions from the rhoptries are responsible for effects on the host cell plasma membrane. Such secretion has been especially well visualized for the zoites of *Toxoplasma* (Fig. 3.14). The rhoptries clearly are partially emptied at the time of entry. Of great interest is the appearance in some rhoptries of tubular structures of the same dimensions and appearance as those seen in the parasitophorous vacuole (Fig. 3.15). These may contribute to the formation of the parasitophorous membrane.

Little is known concerning the nature of the secretion from the rhoptries. A protein has been partially purified from extracts of *Toxoplasma* that greatly enhances penetration of *Toxoplasma* zoites into HeLa cells in culture. Fluorescein-labeled antibody to this protein labels the anterior end of *Toxoplasma,* suggesting that the protein may indeed be in the rhoptries. Autoradiography at the electron microscope level with [^{3}H] histidine indicates a histidine-rich protein in the rhoptries of merozoites of the avian malaria *P. lophurae.* Such a protein has been purified from certain larger granules formed by *P. lophurae* when it develops in duck erythrocytes and has been observed to deform erythrocyte membranes, again suggesting that it might play a role in the entry of merozoites into erythrocytes. Furthermore, polycationic polypeptides, such as polylysine and polyarginine as well as polyhistidine, mimicked the effect of the penetration-enhancing factor for *Toxoplasma.* Basic proteins are present in exceptionally high amount in the rhoptries of *Toxoplasma* as indicated by phosphotungstic acid staining at the ultrastructural level (Fig. 3.16).

In invertebrates especially, intracellular protozoa may first have to penetrate a tough noncellular membrane to gain access to a host cell. For example,

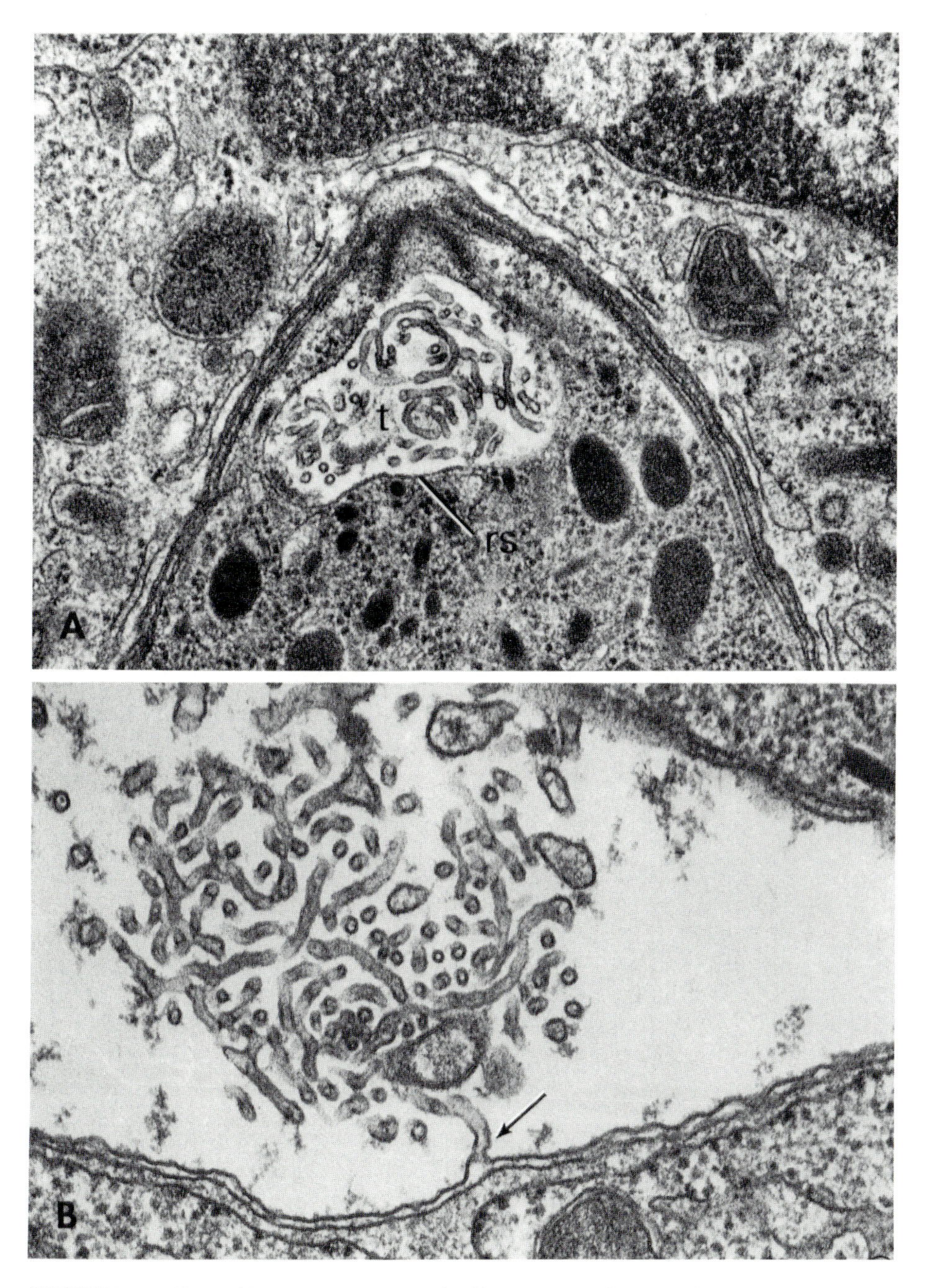

FIGURE 3.15. *T. gondii* in mouse peritoneal cells at 15 min after inoculation. Electron micrographs of thin sections. (A) Rhoptry sac (RS) with tubules (t) resembling those present in the parasitophorous vacuole. × 55,800. (B) Mass of tubules in the parasitophorous vacuole. Note continuity of a tubule with vacuole membrane (arrow). × 74,900. (From Nichols *et al.*, 1983. Original prints courtesy of Dr. B. A. Nichols.)

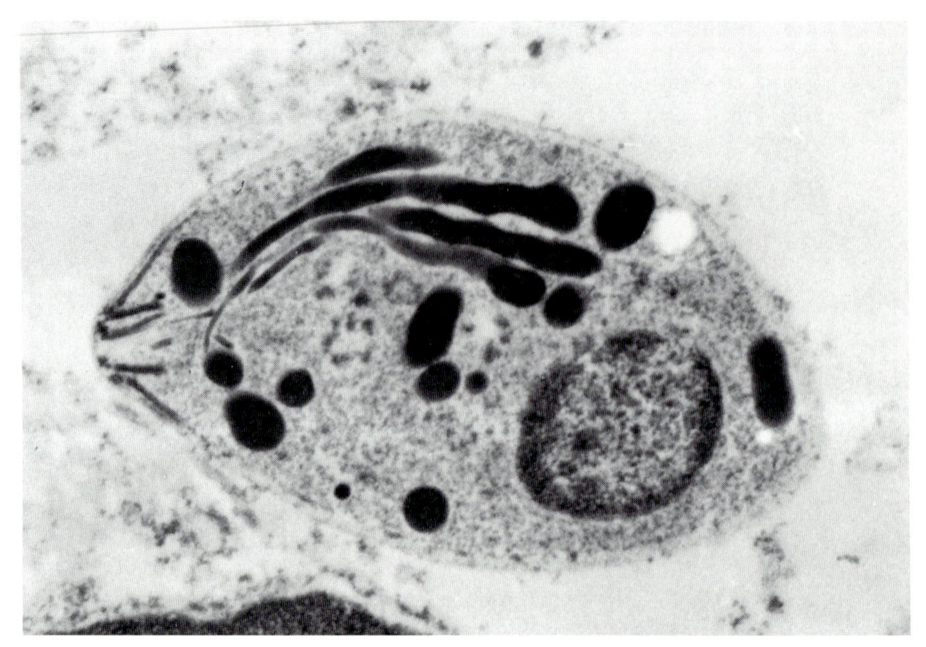

FIGURE 3.16. *T. gondii* tachyzoite. Electron micrograph of thin section treated with phosphotungstic acid to reveal the basic protein of the rhoptry contents. × 30,000. (From De Souza and Sauto-Padrón, 1978. Original print courtesy of Dr. W. De Souza.)

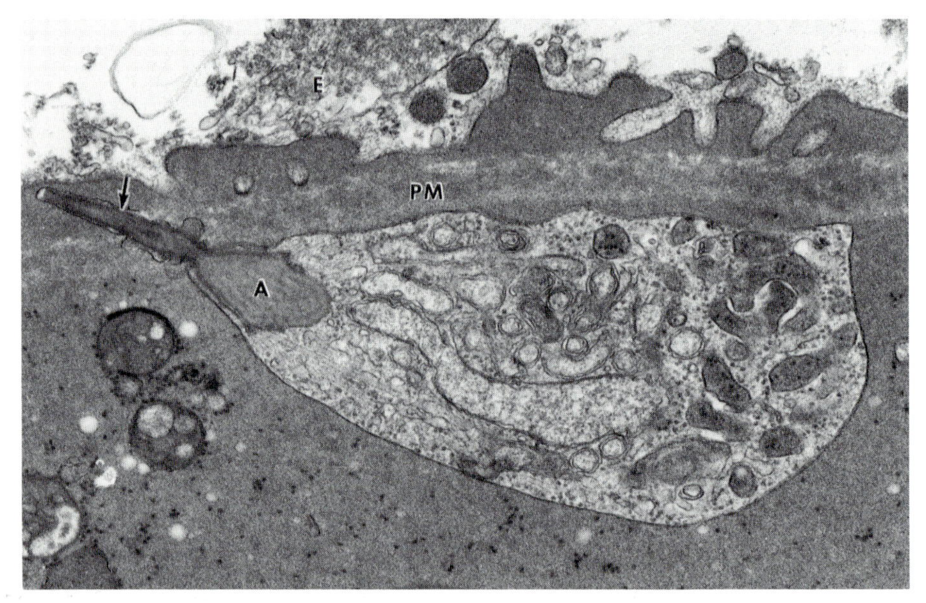

FIGURE 3.17. *Babesia microti* in the vector tick *Ixodes dammini*. Penetration of the peritrophic membrane by the zygote equipped with its arrowhead organelle (A). The arrowhead (at arrow) has just entered the peritrophic membrane (PM). The body of the organism lies in the lumen of the gut. E, epithelial cell. × 37,500. (From Rudzinska *et al.*, 1983. Original print courtesy of Dr. M. A. Rudzinska.)

the sporozoites of *Aggregata eberthi*, a parasite of crabs with a mollusk as alternate host, must penetrate the basal lamina of the crab intestine to reach a connective tissue cell. To do this, a mass of dense granular material appears at the apex of the sporozoite. Probably by enzymatic action, this seems to perforate the basal lamina, forming a narrow orifice through which the sporozoite rapidly squeezes. Even more striking is the so-called arrowhead organelle formed by gametes of*Babesia microti* in the gut of the tick *Ixodes dammini*. This remarkable organelle is clearly specially designed for perforation of the peritrophic membrane that lines the inside of the tick's gut (Fig. 3.17). Parasitism seems to bring out the potentialities of protoplasm even more than other modes of life.

Bibliography

Aikawa, M., and Miller, L. H., 1983, Structural alteration of the erythrocyte membrane during malarial parasite invasion and intraerythrocytic development, *Ciba Found. Symp.* **94**:45–63.

Aikawa, M., Yoshitaka, K., Toshikatsu, A., and Midorikawa, O., 1977, Transmission and scanning electron microscopy of host cell entry by *Toxoplasma gondii*, *Am. J. Pathol.* **87**:285–296.

Aikawa, M., Miller, L. H., Johnson, J., and Rabbege, J., 1978, Erythrocytic entry by malarial parasites: A moving junction between erythrocyte and parasite, *J. Cell Biol.* **77**:72–82.

Bannister, L. H., 1977, Invasion of red cells by *Plasmodium*, in: *Parasite Invasion* (A. E. R. Taylor and R. Mullers, eds.), Blackwell, Oxford, pp. 27–55.

Bannister, L. H., 1979, The interactions of intracellular Protista and their host cells, with special reference to heterotrophic organisms, *Proc. R. Soc. London Ser. B* **204**:141–163.

Chang, K.-P., 1983, Cellular and molecular mechanisms of intracellular symbiosis in leishmaniasis, *Int. Rev. Cytol. Suppl.* **14**:267–305.

Cho, C. H., and Mettrick, D. F., 1982, Circadian variation in the distribution of *Hymenolepis diminuta* (Cestoda) and 5-hydroxtryptamine levels in the gastrointestinal tract of the laboratory rat, *Parasitology* **84**:431–441.

Crompton, D. W. T., 1976, Entry into the host and site selection, in: *Ecological Aspects of Parasitology* (C. R. Kennedy, ed.), North-Holland, Amsterdam, pp. 41–73.

Dawes, B., and Hughes, D. L., 1964, Fascioliasis: The invasive stages of *Fasciola hepatica* in mammalian hosts, *Adv. Parasitol.* **2**:97–168.

De Souza, W., and Sauto-Padrön, T., 1978, Ultrastructural localization of basic proteins on the conoid, rhoptries and micronemes of *Toxoplasma gondii*, *Z. Parasitenkd* **56**:123–129.

Despommier, D., 1976, Musculature, in: *Ecological Aspects of Parasitology* (C. R. Kennedy, ed.), North-Holland, Amsterdam, pp. 269–285.

Dluzewski, A. R., Rangachari, K., Wilson, R. J. M., and Gratzer, W. B., 1983, A cytoplasmic requirement of red cells for invasion by malarial parasites, *Mol. Biochem. Parasitol.* **9**:145–160.

Dvorak, J. A., and Crane, M. St. J., 1981, Vertebrate cell cycle modulates infection by protozoan parasites, *Science* **214**:1034–1036.

Ferguson, M. S., and Hayford, R. A., 1941, The life history and control of an eyefluke, *Prog. Fish Cult.* **54**:1–13.

Friedman, M. J., Fukuda, M., and Laine, R. A., 1985, Evidence for a malarial parasite interaction site on the major transmembrane protein of the human erythrocyte, *Science* **228**:75–77.

Hawking, F., 1975, Circadian and other rhythms of parasites, *Adv. Parasitol.* **13**:123–182.

Holmes, J. C., 1976, Host selection and its consequences, in: *Ecological Aspects of Parasitology* (C. R. Kennedy, ed.), North-Holland, Amsterdam, pp. 21–39.

Jack, R. M., and Ward, P. A., 1980, *Babesia rodhaini* interactions with complement: Relationship to parasitic entry into red cells, *J. Immunol.* **124**:1566–1573.

Kawarabata, T., and Ishihara, R., 1984, Infection and development of *Nosema bombycis* (Microsporida: Protozoa) in a cell line of *Antheraea eucalypti*, *J. Invert. Pathol.* **44**:52–62.

Kennedy, C. R. (ed.), 1976, *Ecological Aspects of Parasitology*, North-Holland, Amsterdam.

Kilejian, A., Liao, T. H., and Trager, W., 1975, On primary structure and biosynthesis of histidine-rich polypeptide from malarial parasite *Plasmodium lophurae*, *Proc. Natl. Acad. Sci. USA* **72:**3057–3059.

Kimata, I., and Tanabe, K., 1982, Invasion by *Toxoplasma gondii* of ATP-depleted and ATP-restored chick embryo erythrocytes, *J. Gen. Microbiol.* **128:**2499–2501.

Langreth, S. G., 1977, Electron microscope cytochemistry of host–parasite membrane interactions in malaria, *Bull. WHO* **55:**171–178.

Langreth, S. G., Nguyen-Dinh, P., and Trager, W., 1978, *Plasmodium falciparum:* Merozoite invasion *in vitro* in the presence of chloroquine, *Exp. Parasitol.* **46:**235–238.

Long, P. L., 1976, Intracellular coccidia, in: *Ecological Aspects of Parasitology* (C. R. Kennedy, ed.), North-Holland, Amsterdam, pp. 409–427.

Lycke, E., Carlberg, K., and Norrby, R., 1975, Interactions between *Toxoplasma gondii* and its host cells: Function of the penetration-enhancing factor of toxoplasma, *Infect. Immun.* **11:**853–861.

McGhee, R. B., 1949, Infection of mammalian erythrocytes by the avian malaria parasite, *Plasmodium lophurae*, *Proc. Soc. Exp. Biol. Med.* **71:**92–93.

MacInnis, A. J., 1976, How parasites find hosts: Some thoughts on the inception of host–parasite integration, in: *Ecological Aspects of Parasitology* (C. R. Kennedy, ed.), North-Holland, Amsterdam, pp. 3–20.

Miller, L. H., McGiniss, M. H., Holland, P. V., and Sigmar, P., 1978, The Duffy blood group phenotype in American blacks infected with *Plasmodium vivax* in Vietnam, *Am. J. Trop. Med. Hyg.* **27:**1069–1072.

Nichols, B. A., Chiappino, M. L., and O'Connor, G. R., 1983, Secretion from the rhoptries of *Toxoplasma gondii* during host-cell invasion, *J. Ultrastruct. Res.* **83:**85–98.

Olson, J. A., and Kilejian, A., 1982, Involvement of spectrin and ATP in infection of resealed erythrocyte ghosts by the human malarial parasite, *Plasmodium falciparum*, *J. Cell Biol.* **95:**757–762.

Pasteur, L., 1922–1939, *Oeuvres réunis par Pasteur Vallery-Radot, Etudes sur la maladie des vers à soie*, Volume 4, Masson, Paris.

Pasvol, G., and Jungery, M., 1983, Glycophorins and red cell invasion by *Plasmodium falciparum*, *Ciba Found. Symp.* **94:**174–195.

Perkins, M. E., 1984, Surface proteins of *Plasmodium falciparum* merozoites binding to the erythrocyte receptor, glycophorin, *J. Exp. Med.* **160:**788–798.

Porchet, E., Richard, A., and Ferreira, E., 1981, Contamination expérimentale *in vivo* et *in vitro* par *Aggregata eberthi:* Étude ultrastructurale, *J. Protozool.* **28:**228–239.

Porchet-Hennere, E., and Nicholas, G., 1983, Are rhoptries of coccidia really extrusomes?, *J. Ultrastruct. Res.* **84:**194–203.

Ravetch, J. V., Kochan, J., and Perkins, M., 1985, Isolation of the gene for a glycophorin-binding protein implicated in erythrocyte invasion by a malaria parasite, *Science* **227:**1593–1597.

Read, C. P., and Kilejian, A. Z., 1969, Circadian migratory behavior of a cestode symbiote in the rat host, *J. Parasitol.* **55:**574–578.

Rose, J. H., 1976, Lungs, in: *Ecological Aspects of Parasitology* (C. R. Kennedy, ed.), North-Holland, Amsterdam, pp. 227–242.

Rudzinska, M. A., 1981, Morphological aspects of host-cell–parasite relationships in babesiosis, in: *Babesiosis* (M. Ristic and J. P. Kreier, eds.), Academic Press, New York, pp. 87–141.

Rudzinska, M. A., and Trager, W., 1977, Formation of merozoites in intraerythrocytic *Babesia microti:* An ultrastructural study, *Can. J. Zool.* **55:**928–938.

Rudzinska, M. A., Trager, W., Lewengrub, S. J., and Gubert, E., 1976, An electron microscopic study of *Babesia microti* invading erythrocytes, *Cell Tissue Res.* **169:**323–334.

Rudzinska, M. A., Lewengrub, S., Spielman, A., and Piesman, J., 1983, Invasion of *Babesia microti* into epithelial cells of the tick gut, *J. Protozool.* **30:**338–346.

Russell, D. G., 1983, Host cell invasion by Apicomplexa: An expression of the parasite's contractile system?, *Parasitology* **87:**199–209.

Schwartzman, J. D., and Pfefferkorn, E. R., 1983, Immunofluorescent localization of myosin at the anterior pole of the coccidian, *Toxoplasma gondii*, *J. Protozool.* **30:**657–661.

Senft, A. W., and Gibler, W. B., 1977, *Schistosoma mansoni* tegumental appendages: Scanning microscopy following thiocarbohydrazideosmium preparation, *Am. J. Trop. Med. Hyg.* **26:**1169–1177.

Smithers, S. R., and Worms, M. J., 1976, Blood fluids—helminths, in: *Ecological Aspects of Parasitology* (C. R. Kennedy, ed.), North-Holland, Amsterdam, pp. 349–369.

Sukhdeo, M. V., and Croll, N. A., 1981, The location of parasites within their hosts: Bile and the site selection behaviour of *Nematospiroides dubius, Int. J. Parasitol.* **11:**157–162.

Sukhdeo, M. V. K., and Mettrick, D. F., 1984, Migrational responses of *Hymenolepis diminuta* to surgical alteration of gastro-intestinal secretions, *Parasitology* **88:**421–430.

Weidner, E., 1972, Ultrastructural study of microsporidian invasion into cells, *Z. Parasitenkd* **40:**227–242.

Weidner, E., 1982, The microsporidian spore invasion tube. III. Tube extrusion and assembly, *J. Cell Biol.* **93:**976–979.

Weidner, E., and Byrd, W., 1982, The microsporidian spore invasion tube. II. Role of calcium in the activation of invasion tube discharge, *J. Cell Biol.* **93:**970–975.

Weidner, E., Byrd, W., Scarborough, A., Pleshinger, J., and Sibley, D., 1984, Microsporidian spore discharge and the transfer of polaroplast organelle membrane into plasma membrane, *J. Protozool.* **31:**195–198.

Werk, R., Dunker, R., and Fischer, S., 1984, Polycationic polypeptides: A possible model for the penetration-enhancing factor in the invasion of host cells by *Toxoplasma gondii, J. Gen. Microbiol.* **130:**927–933.

CHAPTER 4

The Host–Parasite Interface I

In Extracellular Parasites

The functional interface between parasite and host is the site of molecular exchanges between the organisms: nutrients from the host enter the parasite and various products of the parasite metabolism enter the host. This is where the action is. It is also the place where the host can react against the parasite. Detailed studies of this interface have been done with only a relatively few host–parasite associations and we are only beginning to learn of the many complex interactions that take place. The structural nature of the interface varies widely with the type of parasite and the site it occupies in the host. For a parasite like a hookworm with a relatively impermeable cuticle and a complete alimentary tract into which blood and other host materials are ingested, the functional interface involves both the alimentary tract lining and the external surface of the worm. For a cestode, with no alimentary tract, the interface is the surface of the worm; for a protozoan cell, it is the plasma membrane and the lysosomal membrane system. For intracellular parasites, additional membranes derived in part from the host cell are involved. Intimately related to the host–parasite interface are structures used by parasites to maintain an appropriate favorable position: hooks, suckers, attachment mechanisms of different kinds, including even modifications of the host cell induced by certain intracellular parasites. I will consider in detail both the structure and what is known about the function of certain selected host–parasite interfaces, proceeding from parasites of the alimentary tract of vertebrates to those of blood and tissue spaces and finally to intracellular parasites.

Two important pathogenic parasites of the human alimentary tract, *Entamoeba histolytica* and *Giardia lamblia*, provide some sharp contrasts in their mode of association with the host. *E. histolytica* inhabits the large intestine where it creeps about in ameboid fashion through the mucosal lining and over the microvilli of the epithelial cells. Very little is known about its activities in the 80% of infected individuals who show no signs or symptoms (see

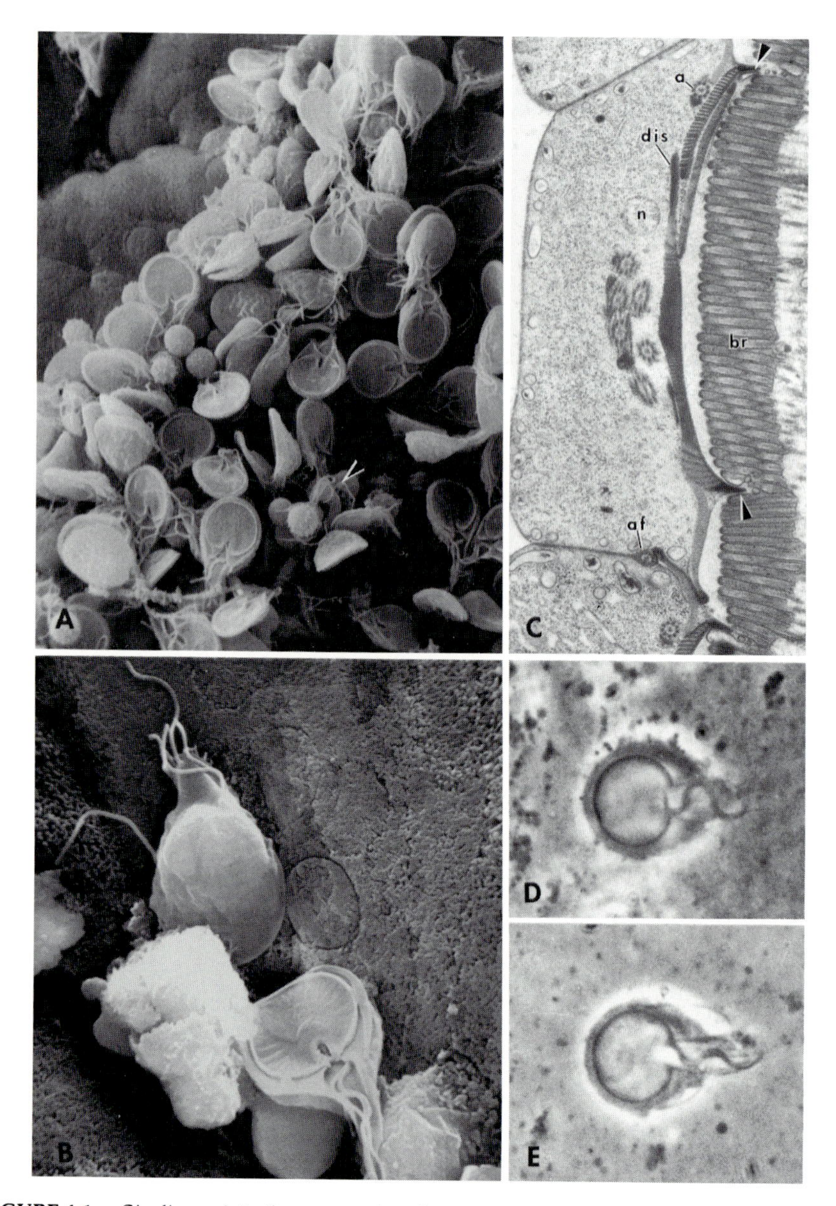

FIGURE 4.1. *Giardia muris* in the mouse. Attachment to intestinal wall. (A) Mass of *Giardia* and lymphocytes at base of villus. The arrowhead points to a single *Hexamita muris,* a common nonpathogenic flagellate, with an adjacent lymphocyte. SEM. × 1958. (B) Lower flagellate shows ventral surface with adhesive disk and cytoplasmic flange. Next to the adherent upper organism is a circular mark made by an adhesive disk. SEM. × 8026. (C) Section through attached trophozoites to show ultrastructure of the ventral surface and its relationship to the epithelium (at right). Note the close lateral packing of the parasites. The central organism is cut through the midregion; the posterior pole of one nucleus is seen (*n*). Three pairs of intracytoplasmic axonemes are seen centrally and one anterior axoneme (*a*) laterally. The second anterior axoneme is seen as a free flagellum (*af*). The roof of the ventral chamber is supported by three component plates

Chapter 22). Presumably, it satisfies its complex nutritional requirements by phagocytosis of particulate matter, including cell debris and bacteria, and by uptake of solutes such as glucose by specific transport mechanisms. When, for reasons not presently understood and that may be largely host-related, the amebae produce disease, they interact with host tissue in a much more intimate fashion. They now come into close contact with epithelial cells and leukocytes and they phagocytize these cells as well as erythrocytes. *In vitro* the amebae exert a cytocidal effect on contact, but whether this also occurs *in vivo* is not known. Clearly the properties of the plasma membrane of *Entamoeba* must be of prime importance. It provides the dynamic interface with the host tissue; its activity is responsible for destruction of host cells and also for interaction with the host's immune system. Detailed studies of the plasma membrane of *E. histolytica* have only recently been initiated. The membrane contains 12 major peptides, all of them glycoproteins, ranging in size from 12,000 to 200,000 daltons. Of interest and possible significance is the finding that the membrane contains a large amount of an unusual phospholipid, ceramide aminoethyl phosphate, that is resistant to phospholipase. A correlation has been observed between pathogenicity of different strains of *E. histolytica* and their agglutinability with concanavalin A. The pathogenic strains are much more readily agglutinated; their surface has both a lower charge (as shown by microelectrophoresis and by failure to bind ferritin) and a high binding capacity for concanavalin A. Exactly how such surface alterations affect the interface with the host in such a way as to produce or permit pathogenic effects remains to be determined (see Chapter 22 for a more detailed discussion).

Giardia lamblia is a flagellate that colonizes the proximal 25% of the small intestine. It maintains its position through its own motility and by attachment to the villi by means of a remarkable adhesive disk (Fig. 4.1). The ventral surface of the somewhat pear-shaped binucleate bilaterally symmetrical organism is modified to form what looks like a suction cup surrounded by a flange. The centrally located posteriorly directed flagella serve to create the suction. The intensity of the suction exerted is attested to by the numerous lesions, having the form of a mirror image of the disk, left on the villous surface of the intestine (Fig. 4.1B). It is not known whether this intimate interface with the host serves only for attachment or whether it has some nutrient function. Endocytosis has been observed on both dorsal and ventral surfaces of the *Giardia* cell. In any case, it seems reasonable to suppose that

of the striated disk (*dis*). The rim of the ventral disk (arrowhead) intercalates with microvilli of the host brush border (*br*), locally distorting it and leaving the circular imprint seen in B. TEM. × 15,000. (D, E) Light micrographs of trophozoites attached to glass to show the wave form of the paired ventral flagella beating within the ventrocaudal groove. These create currents to facilitate attachment. Electronic flash photo. × 2500. (A and B from Owen *et al.*, 1979; original prints coutesy of Dr. R. L. Owen. C–E from Holberton, 1973; original prints courtesy of Dr. D. V. Holberton.)

the coating by the parasites of a large portion of the surface of the small intestine, as occurs in many infections, may well be responsible for the malabsorption and other gastrointestinal symptoms associated with giardiasis. With *Giardia* as with *Entamoeba,* however, many infections are asymptomatic and those with symptoms show a wide range from borderline to severe pathogenicity.

The hookworms, in their adult stage widespread and very important parasites of humans and domestic animals, provide excellent examples of an interface that involves specialized attachment mechanisms and ingestion of host material into a fully developed alimentary tract. Adults of the human hookworms *Ancylostoma duodenale* and *Necator americanus* have a large mouth equipped with characteristic toothed plates adapted for both attachment to and laceration of the intestinal surface of the host (Fig. 4.2). These worms ingest large amounts of blood, much of which may pass through their alimentary tract relatively unaltered. The amount of blood taken by human hookworms is enormous. Estimates vary around 0.1–0.2 ml of blood per worm per day. This means 2–3 ml a day for a light infestation, up to 100–200 ml a day in a heavy infection. It is not surprising that microcytic anemia is a major consequence of hookworm infection. Since the blood takes only 1–2 minutes to pass through the worm from mouth to anus, it may well have functions

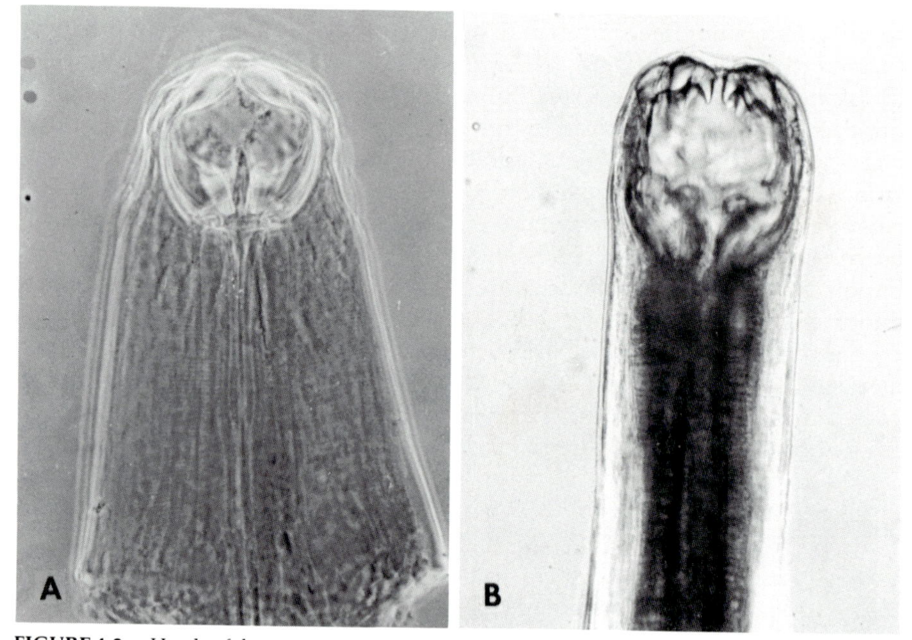

FIGURE 4.2. Heads of the two main species of human hookworm (× 100). (A) *Necator americanus;* (B) *Ancylostoma duodenale.* Note the teeth in upper region of oral cavity. (From Katz *et al.,* 1982. Original prints courtesy of Dr. D. Despommier.)

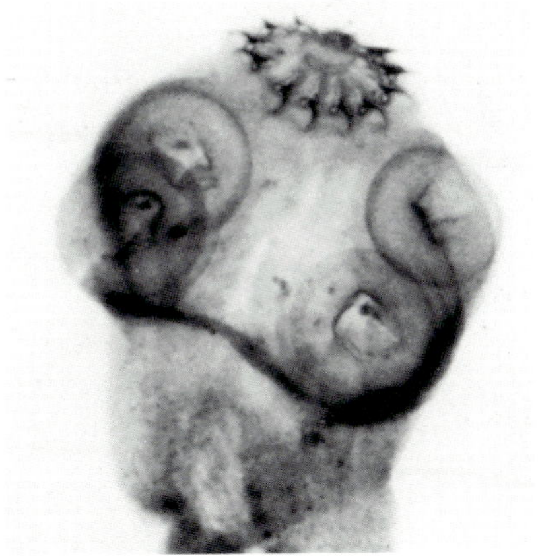

FIGURE 4.3. Scolex or "head" of the port tapeworm *Taenia solium* showing the four suckers and the single row of hooks serving for attachment to the wall of the small intestine. ×33. (From Katz *et al.*, 1982. Original print courtesy of Dr. D. Despommier.)

in addition to strictly nutritional ones. Although these worms live in the alimentary tract of their host, their main absorptive surface is actually in contact with an intermittent flow of blood; perhaps physiologically they are more parasites of the blood than of the alimentary tract.

Adult cestodes, the tapeworms, on the other hand, are perfect examples of true parasites of the alimentary tract itself. They have no digestive canal of their own, but must assimilate across their tegument. They have developed remarkable anchoring structures, a "head" with suckers and hooks of diverse form (very useful to taxonomists as well as for attachment) (Figs. 4.3, 4.4). In back of this structure is a germinative layer that buds off segments, the proglottids, each equipped mainly with reproductive organs for the formation and fertilization of eggs. The interface of each proglottid with the alimentary tract of the host consists of a tegument clearly adapted for absorption. This is formed of a syncytial cytoplasmic layer on the external surface. Cytoplasmic processes pass from this layer through the basal lamina and through the muscle layer to the perikaryon containing the cell nuclei. This external cytoplasmic layer is rich in mitochondria and its surface is vastly increased by numerous digitiform projections, the microtriches (Fig. 4.5). These resemble microvilli of mucosal epithelial cells, including the presence of medullary microfilaments. In addition, however, they have a solid or densely fibrillar distal tip set off from the rest of the shaft by a multilaminate base plate (Fig. 4.6). The whole surface is covered with a characteristic glycocalyx formed by the worm. In *Hymenolepis diminuta*, a small tapeworm of the rat, this shows a turnover time for ^{3}H-galactose-labeled constituents of about 6 hr. This highly specialized surface is the site of uptake by pinocytosis as well as by diffusion and transport, both facilitated and active. The surface has a variety of enzymatic activites, including very active phosphatases. These enable the worm to take up nucleosides and sugars supplied as phosphate esters. Thus, glucose liberated from glucose phosphate at the worm surface does not diffuse

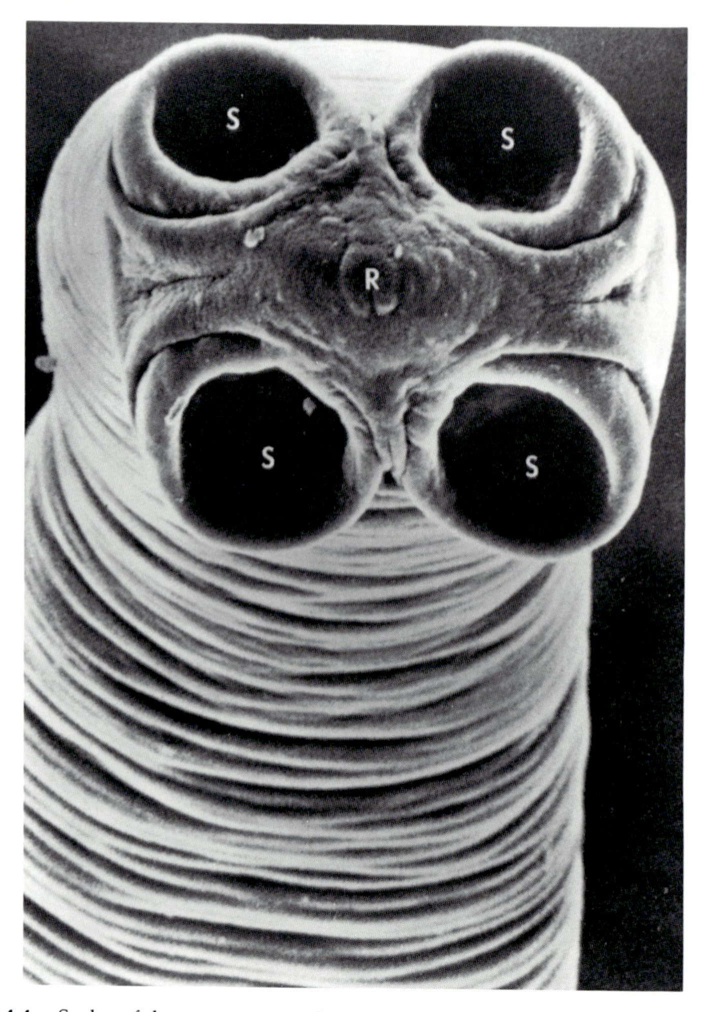

FIGURE 4.4. Scolex of the rat tapeworm *Hymenolepis diminuta* to show apical rostellum (R) and bilateral suckers (S). SEM. × 435. (From Ubelaker *et al.*, 1973. Original print courtesy of Dr. R. D. Lumsden.)

into the medium but is directly absorbed by the worm. Surface lipase and RNase activities have also been demonstrated, as well as certain host enzymes (e.g., α-amylase) adsorbed to the worm surface. Other host enzymes, however (e.g., trypsin and chymotrypsin), are inactivated at the surface. The immediate environment of a tapeworm in the vertebrate intestine is surprisingly uniform in its physiological characteristics, despite the periodic intake and digestion of food by the host, an important fact to which Clark Read first drew attention.

Trematodes, unlike the cestodes, have in their adult stage an alimentary canal with a single opening through which host material may be ingested. Nevertheless, it is clear from the structure of their surface that this must also

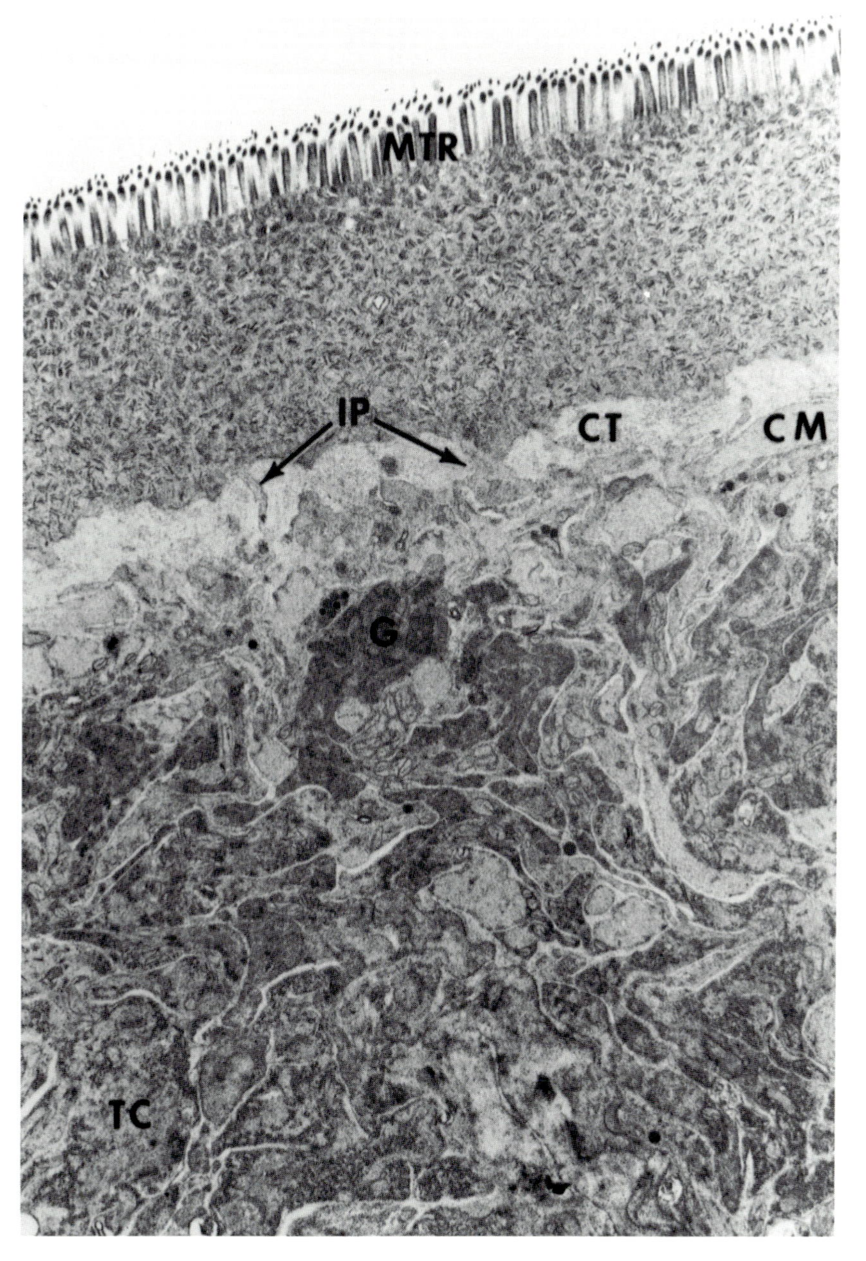

FIGURE 4.5. Tegmentary cortical region of the rat tapeworm *Hymenolepis diminuta*. MTR, microtriches, or brush border extending from the syncytial cytoplasmic layer; TC, basal tegmentary cytons; IP, internuncial processes from tegmentary cytons; CM, circular muscles; CT, connective tissue (basement lamina); G, glycogen. TEM. × 5900. (From Lumsden and Specian, 1980. Original print courtesy of Dr. R. D. Lumsden.)

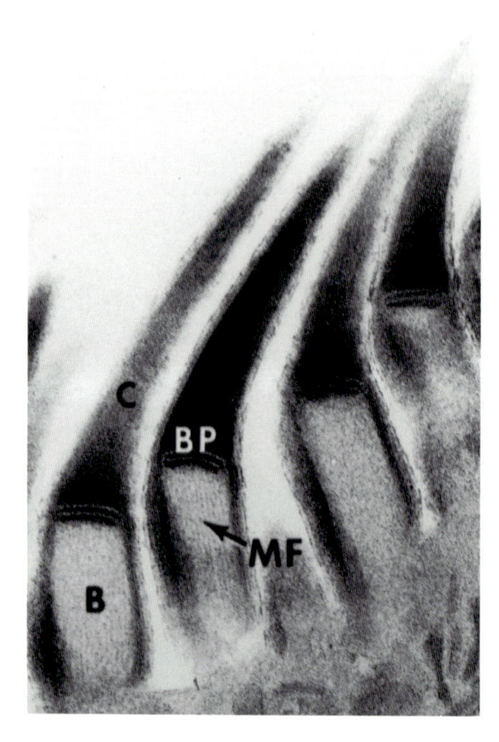

FIGURE 4.6. Tegmental microtriches of *H. diminuta*. C, Electron-opaque cap; BP, multilaminate base plate; B, base; MF, microfilaments within base. TEM. × 71,000. (From Lumsden and Specian, 1980. Original print courtesy of Dr. R. D. Lumsden.)

serve for uptake of nutrients, and, in larval stages, which lack an alimentary canal, this is the only route. The tegument of an adult fluke, e.g., the blood fluke *Schistosoma mansoni* (one of the three species that cause schistosomiasis in humans), resembles in its ultrastructure that of tapeworms except that there are no microtriches. Instead the increase in surface area is provided by extensive invaginations of the plasma membrane (see also Chapter 21). Here again, as in tapeworms, phosphatases have been demonstrated on the outer surface, but it is not known whether they are involved in transport or only as digestive enzymes. The adult schistosome's interface with its host, as it lies in a blood vessel, involves both its surface, active in absorption, and its alimentary tract, in which the digestion of blood occurs.

The basic physiological situation is much the same for the blood forms of the protozoa of African sleeping sickness (*Trypanosoma brucei* subspecies). Since they are single-celled organisms, absorption involves only transport across plasma membrane, and alimentary tract uptake is replaced by endocytosis, a mechanism for the uptake of lipids and macromolecules. Of special importance is a relatively thick surface coat of glycoprotein. This is highly antigenic, but the trypansomes have a large portion of their genome devoted to coding for an extensive repertoire of variant surface glycoproteins, enabling them to evade the hosts' immune response (see Chapter 18). Since so far the blood forms of these trypanosomes can be cultured *in vitro* only with a feeder layer of living host cells (see Chapter 8), the actual interface of these organisms with their host may be more complex than at first appears. At present we do not know whether the live host cells serve to supply some essential but labile low-molecular-weight metabolite or whether they might supply a membrane component. The actively multiplying trypanosomes in a culture are always found intimately associated with the feeder layer of cells. Perhaps this represents an evolutionary step on the way to intracellular parasitism.

Bibliography

Aley, S. B., Scott, W. A., and Cohn, Z. A., 1980, Plasma membrane of *Entamoeba histolytica*, *J. Exp. Med.* **152:**391–404.

Erasmus, D. A., 1977, The host–parasite interface of trematodes, *Adv. Parasitol.* **15:**201–242.

Featherston, D. W., 1972, *Taenia hydatigena*. IV. Ultrastructure study of the tegument, *Z. Parasitenkd.* **38:**214–232.

Griffin, J. L., 1972, Human amebic dysentery: Electron microscopy of *Entamoeba histolytica* contacting, ingesting and digesting inflammatory cells, *Am. J. Trop. Med. Hyg.* **21:**895–906.

Holberton, D. V., 1973, Fine structure of the ventral disk apparatus and the mechanism of attachment in the flagellate *Giardia muris*, *J. Cell Sci.* **13:**11–41.

Katz, M., Despommier, D. D., and Gwadz, R., 1982, *Parasitic Diseases*, Springer-Verlag, New York.

Lumsden, R. D., 1975, Surface ultrastructure and cytochemistry of parasitic helminths, *Exp. Parasitol.* **37:**267–339.

Lumsden, R. D., and Specian, R., 1980, The morphology, histology, and fine structure of the adult stage of the cyclophyllidean tapeworm, in: *Biology of the Tapeworm Hymenolepis diminuta* (H. P. Arai, ed.), Academic Press, New York, pp. 157–280.

Mettrick, D. F., and Podesta, R. B., 1974, Ecological and physiological aspects of helminth–host interactions in the mammalian gastro-intestinal canal, *Adv. Parasitol.* **12:**183–278.

Nickol, B. B. (ed.), 1979, *Host–Parasite Interfaces*, Academic Press, New York.

Owen, R. L., Nemanic, P. C., and Stevens, D. P., 1979, Ultrastructural observations on giardiasis in a murine model, *Gastroenterology* **76:**757–769.

Pappas, P. W., 1980, Enzyme interactions at the host–parasite interface, in: *Cellular Interactions in Symbiosis and Parasitism* (C. B. Cook, P. W. Pappas, and E. D. Rudolph, eds.), Ohio State University Press, Columbus, pp. 145–172.

Smyth, J. D., 1973, Some interface phenomena in parasitic protozoa and platyhelminths, *Can. J. Zool.* **51:**367–377.

Taylor, A. E. R., and Muller, R. (eds.), 1972, *Functional Aspects of Parasite Surfaces*, Blackwell, Oxford.

Ubelaker, J. E., Allison, V. F., and Specian, R. D., 1973, Surface topography of *Hymenolepis diminuta* by scanning electron microscopy, *J. Parasitol.* **59:**667–671.

CHAPTER 5

The Host–Parasite Interface II

In Intracellular Parasites (Protozoa and the Nematode *Trichinella spiralis*)

The association of a eukaryotic protozoan cell as an obligate internal parasite of a eukaryotic host cell provides material for the study of basic problems in cell biology. Furthermore, this widespread phenomenon found with a great many species of protozoan parasites in both vertebrate and invertebrate animals includes a considerable number of pathogenic species of great economic and medical importance. One need only mention as examples the parasites of human malaria, of Chagas' disease, of leishmaniasis and toxoplasmosis, the coccidial parasites of chickens and other domestic animals, the babesias and theilerias of ruminants, and the microsporidian parasites of insects. All are obligate intracellular parasites. Their interface with the host cell depends in part on their mode of entry, already discussed in a previous chapter. There are three kinds of situations: (1) the parasite lies in a phagosome; (2) it lies in a special parasitophorous vacuole lined by the parasitophorous membrane; (3) it lies directly in the host cytoplasm. The last two cases in particular may change with the developmental stage of the parasite.

Those parasites that develop in the phagosome may be considered the most daring; they challenge successfully the very heart of the cell's defensive machinery. Into the phagosome of a macrophage are secreted powerful hydrolytic and oxidative enzymes, serving to kill and digest most living organisms. But protozoa of the genus *Leishmania* (Diagram III) somehow manage not only to survive but to grow and multiply within the phagosome. We know they are in a functional phagosome for other material in the same vacuole can be seen undergoing digestion (Fig. 5.1). How do they escape? There is very recent evidence for glycoproteins on the surface of *Leishmania* that are resistant to the lysosomal enzymes. Perhaps they resemble the proteins in the limiting membrane of the lysosome itself. Acid phosphatase has been demonstrated cytochemically in phagosomes containing leishmanial parasites

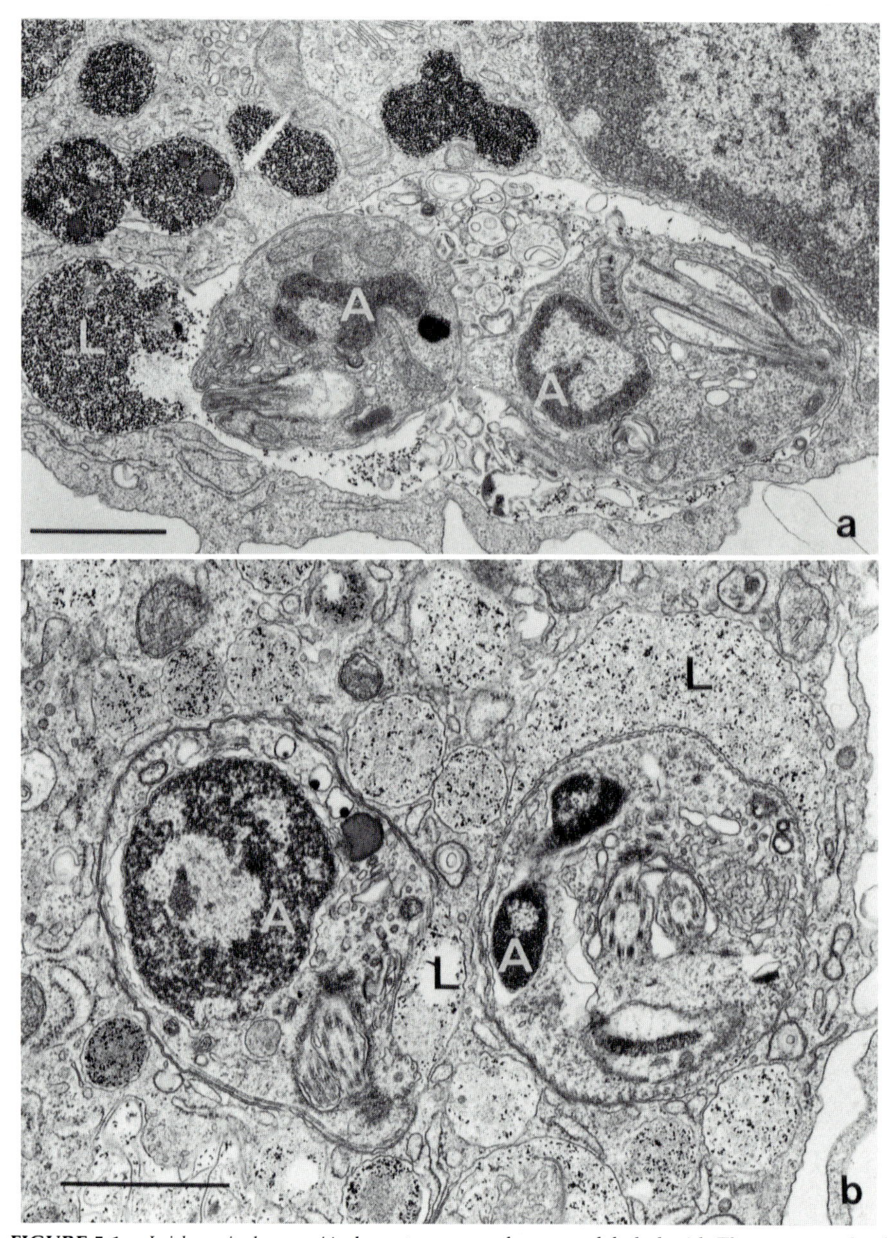

FIGURE 5.1. *Leishmania donovani* in hamster macrophages prelabeled with Thorotrast and subsequently infected *in vitro* with amastigotes. TEM. (a) Fixed 1 hr after infection, showing the fusion of a secondary lysosome (L) and a parasitophorous vacuole containing two amastigotes (A). Note the membrane debris lodged with the parasites in the same vacuole. (b) Fixed 3 days after infection, showing two dividing amastigotes (A) each in a Thorotrast-containing vacuole, i.e., a secondary lysosome (L). Bars = 1 μm. (From Chang and Dwyer, 1976. Original prints courtesy of Dr. K.-P. Chang.)

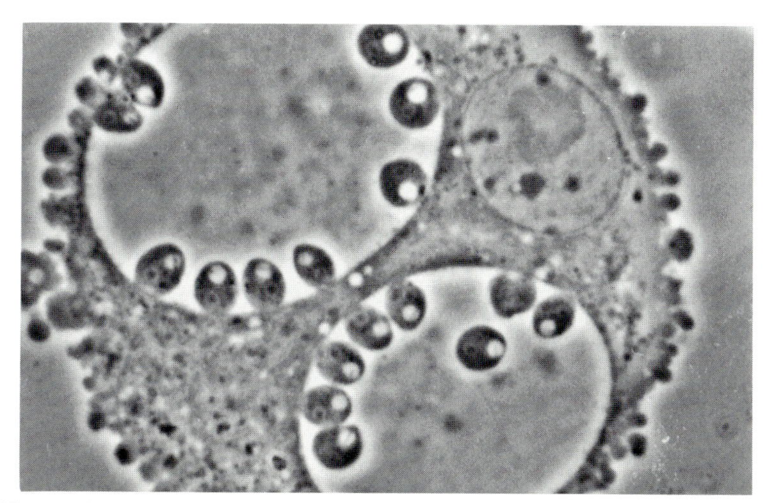

FIGURE 5.2. *L. mexicana* amastigotes growing in a large parasitophorous vacuole in a cell of the transformed mouse cell line J774 GH. ×1000 phase-contrast photomicrograph of living material obtained courtesy of Dr. K.-P. Chang. (From Chang and Hendricks, 1985.)

and has been assumed to be entirely of host origin. That it could also be derived from the parasites is indicated by the recent demonstration of high acid phosphatase activity limited to the external surface of the plasma membrane of promastigotes of *Leishmania donovani.* It is of great interest that leishmanial phosphatase has been found to block superoxide formation in neutrophils and macrophages stimulated by the chemoattractant peptide formyl-methionyl-leucyl-phenylalanine. The phosphatase may dephosphorylate a receptor for the peptide and perhaps its function is to neutralize the oxidant activity of the host cell. It could also serve a nutritive role, like that of the phosphatases in the tegument of tapeworms, by hydrolyzing organic phosphates. The same functions might be served also by the nucleotidases present on the surface. Since the parasite lies in a functioning phagosome, it must have available to it nutrients ingested by the host cell from the extracellular milieu. Little is known about this, since the intracellular amastigote stages of one species of *Leishmania* have only recently been grown extracellularly in culture (Chapter 8). Whether the phagosomal membrane is modified in any way by the leishmania within it is not known. Of much interest are the different behaviors of this membrane depending on the species of *Leishmania.* With *L. mexicana,* a parasite that causes a cutaneous disease, the phagosome becomes greatly enlarged as the parasites multiply within it, and each parasite adheres to the membrane at one end (Fig. 5.2). With *L. donovani,* on the other hand, a parasite causing visceral disease especially in the spleen and liver, each parasite is usually surrounded by its own individual membrane (Fig. 5.3). Here there must be some mechanism whereby the surrounding phagosomal vacuole divides in two in synchrony with the division of the parasite.

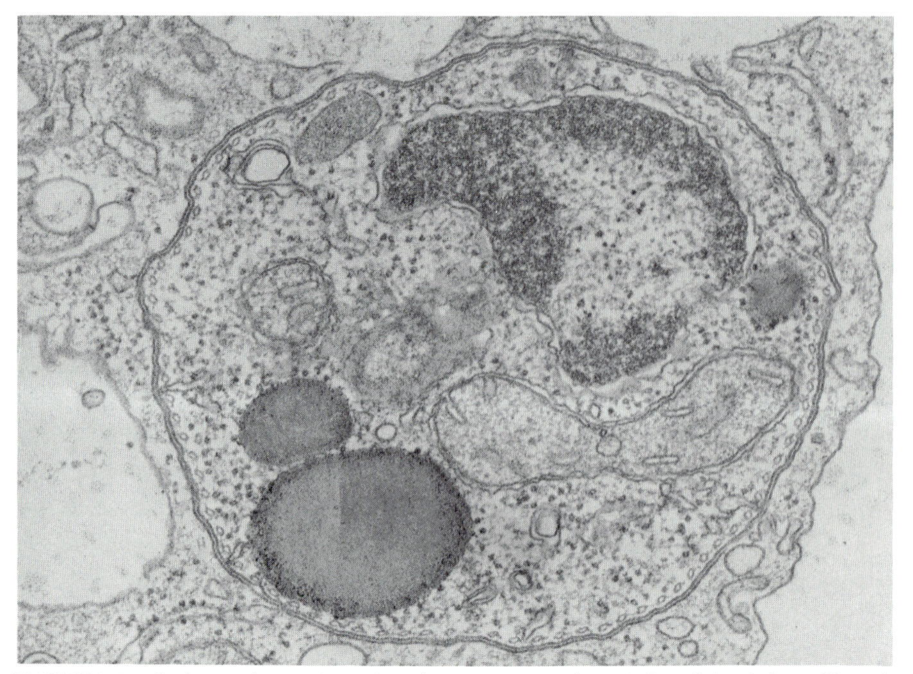

FIGURE 5.3. *L. donovani* amastigote in a hamster macrophage infected for 3 days. Note the close apposition of the parasite plasma membrane and the surrounding parasitophorous membrane. TEM. × 47,000. (From Chang and Dwyer, 1978. Original print courtesy of Dr. K.-P. Chang.)

When a parasite lies in a membrane-bound vacuole with which secondary lysosomes do not fuse, we consider this a parasitophorous vacuole. This is clearly a descriptive term with no implication as to the nature or origin of the membrane. As noted in Chapter 3, *Toxoplasma* and other coccidia such as those of the genus *Eimeria*, and erythrocytic stages of malarial parasites, provide good examples of organisms dwelling within a parasitophorous membrane. This membrane and the contents of the vacuole within it, are the interface between the host cell and the plasma membrane of the parasite. In *Toxoplasma* and *Eimeria* the space between the parasite and the parasitophorous membrane may be filled with vesicles or other membranous material. For some species, e.g., *Eimeria auburnensis* and *E. canadensis* of cattle, a complex system of intravacuolar folds has been demonstrated (Fig. 5.4). The folds consist of two closely apposed unit membranes extending into the vacuole for a maximum depth of about 0.7 μm. They apparently arise from the parasitophorous membrane. Presumably, they consist of host material elaborated in response to stimuli from the parasite, and serve some nutritive function for the parasite. At least two types of endocytosis of vacuolar material by coccidian parasites have been noted in electron micrographs, one involving

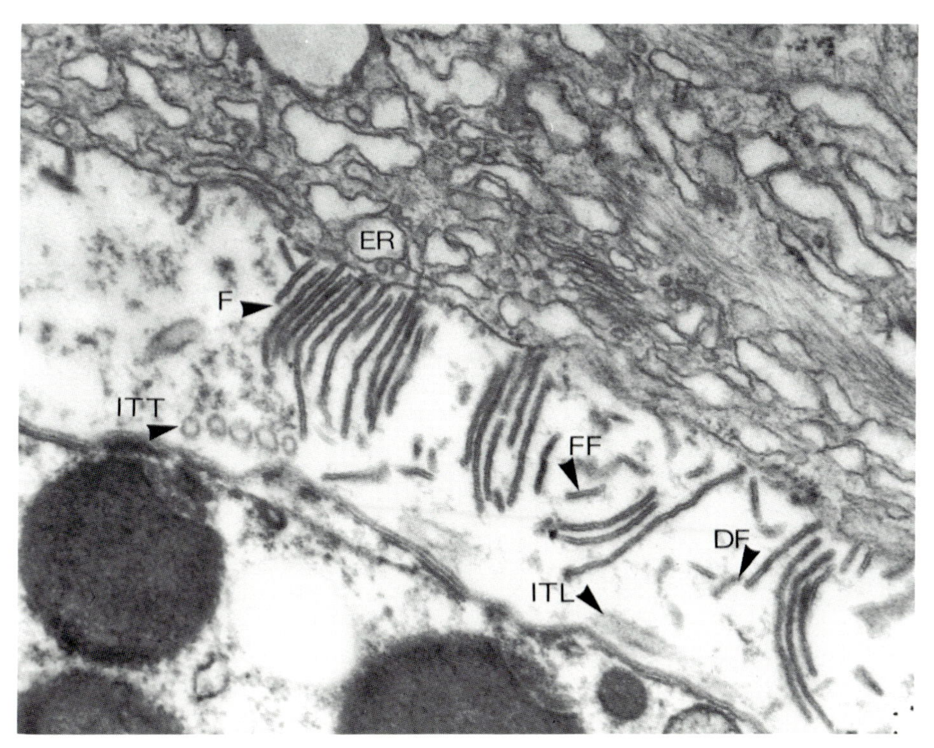

FIGURE 5.4. Peripheral portion of a macrogamete of *Eimeria auburnensis* from an infected calf to show in the parasitophorous vacuole: intravacuolar folds (F), intravacuolar tubules in cross section (ITT) and in longitudinal section (ITL), smooth endoplasmic reticulum of the host cell (ER), disintegrating (DF) and detached (FF) intravacuolar folds. TEM. × 60,000. (From Hammond *et al.*, 1967. Original print courtesy of Dr. E. Scholtyseck.)

relatively shallow pinocytotic invaginations and the other the formation of narrow deep channels. The host cell cytoplasm adjacent to the parasitophorous membrane generally has well-developed endoplasmic reticulum with numerous mitochondria. With some species that form very large schizonts, however, a sharply demarcated region of host cell cytoplasm in the vicinity of the parasitophorous vacuole lacks cytoplasmic organelles except for ribosomes. In addition to the intravacuolar folds, intravacuolar tubules about 0.06 μm in diameter and having periodic peripheral striations are present. At least in the macrogametes of *E. perforans*, these seem to make a structural connection between the parasite and the host cell (Fig. 5.5). Perhaps they serve for direct transport of nutrients. Whether they are formed by the parasite or the host cell is not known.

In *Sarcocystis muris* developing either *in vivo* in cat intestine or *in vitro* in cultures of dog intestinal cells, the invading merozoites have been seen to excrete by exocytosis into the parasitophorous vacuole. The contents of char-

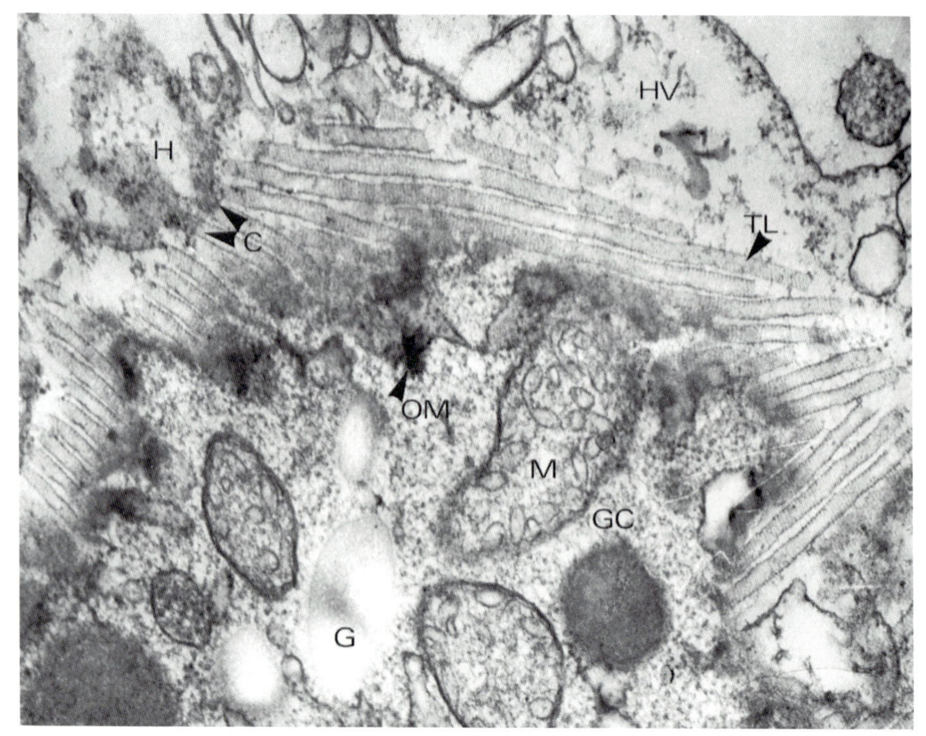

FIGURE 5.5. *Eimeria perforans* in intestinal cell of the European rabbit. Portion of a macrogamete. Microtubules (TL) are shown in longitudinal section making a direct connection (C) with the cytoplasm of the host cell (H). GC, ground cytoplasm of the parasite; HV, vacuole surrounding the parasite. G, glycogen granules; M, mitochondrion; OM, osmiophilic mass. TEM. × 50,000. (From Scholtyseck *et al.*, 1966. Original print courtesy of Dr. E. Scholtyseck.)

acteristic dense granules become more granular and may be seen passing from the parasite into the vacuole (Fig. 5.6). The function of this material is at present entirely conjectural.

Although the parasitophorous membrane of coccidia arises from invagination of the host cell plasmalemma, it does so in response to a stimulus from the invasive form of the parasite and clearly differs from phagosomal membrane. It also seems not to be analogous to the receptosomes, membrane-lined vesicles that are formed from the interaction of specific receptors, as for hormones, and are internalized at the region of bristle-coated pits. The parasitophorous membrane of malarial parasites, derived as it is from red cell plasma membrane, might at first glance appear to offer a less complex material for study. Even here, however, little is known. As we have seen (Chapter 3), very soon after entry of the invasive merozoite, the parasitophorous membrane becomes closely apposed to the plasma membrane of the parasite, it loses its intramembranous particles from its P face, and its enzymes reverse their polarity. In experiments with an avian malaria *Plasmodium lophurae,* viable

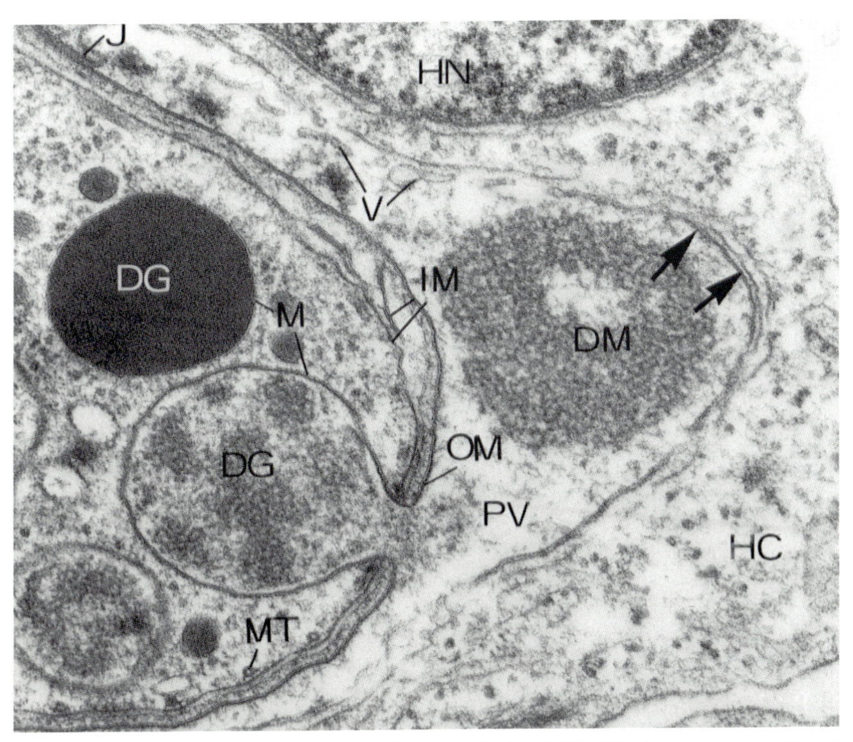

FIGURE 5.6. *Sarcocystis muris* in cultured dog intestinal cells 30–45 min after inoculation. This transmission electron micrograph shows a membrane-bound (M) dense granule (DG) emptying its contents (DM) into the parasitophorous vacuole (PV) by exocytosis. Note the parasite's pellicle with plasmalemma (OM) and inner membrane complex (IM). The parasitophorous vacuole is limited by a membrane (arrows) of the host cell (HC). HN, host cell nucleus. (From Entzeroth, 1984. Original print courtesy of Dr. R. Entzeroth.)

parasites removed from their host erythrocyte were still enclosed in the parasitophorous membrane (Fig. 5.7A,B). Intact viable parasites freed from the parasitophorous membrane have not been demonstrated, except of course for the merozoites which are liberated by the breakdown of both the parasitophorous membrane and the plasma membrane of the host erythrocyte. In the only serious attempt to study both the parasitophorous membrane and the plasma membrane of the parasite within, done again with *P. lophurae,* no significant difference between them could be found. This fits well with the observation that antiserum to the *P. lophurae* parasites agglutinates antigens of the parasitophorous membrane, whereas antiserum to the host duck erythrocytes does not. For the preerythrocytic stages also of *P. berghei,* which develop in hepatic cells, it has been clearly shown that the parasitophorous vacuole membrane, as well as the membrane of the parasite, react with antisera to parasite material; indeed with a monoclonal antibody to the sporozoite surface antigen PB44 (Fig. 5.8). For malarial parasites the evidence so far

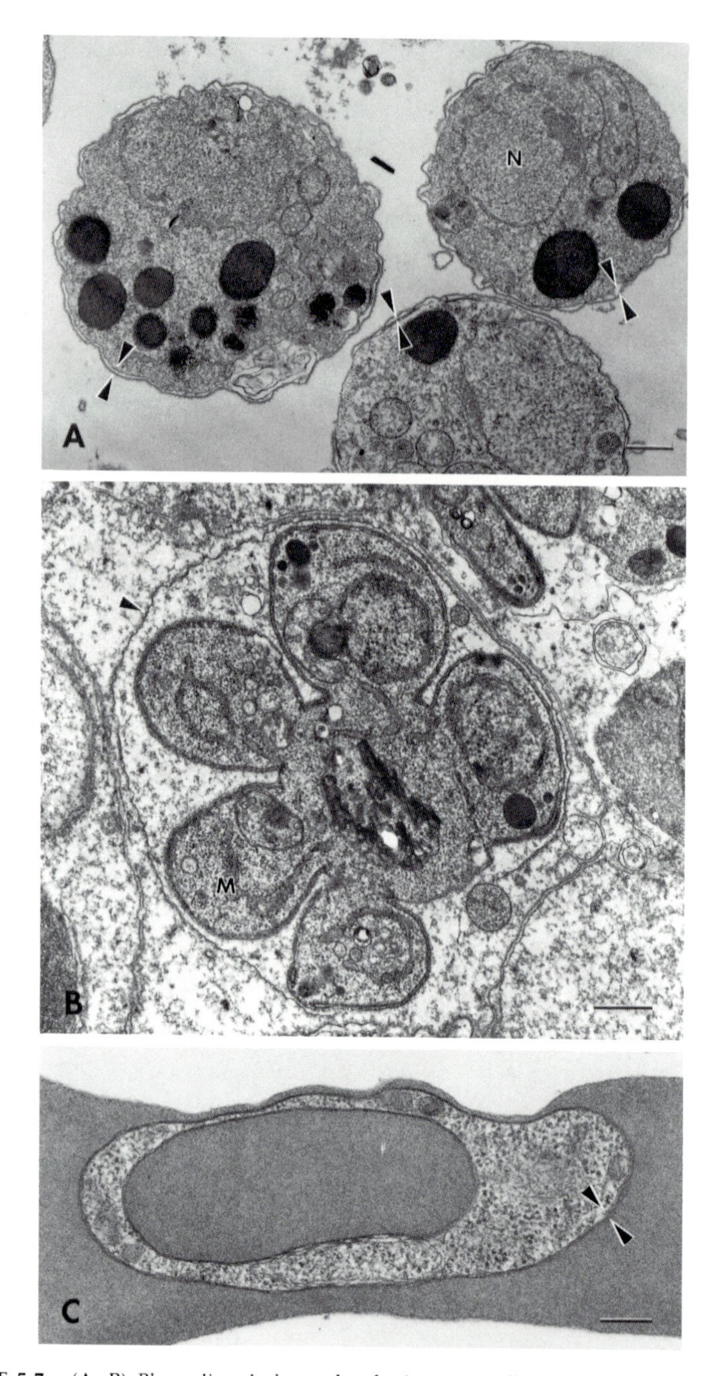

FIGURE 5.7. (A, B) *Plasmodium lophurae* developing extracellularly *in vitro*. (A) Trophozoites after 1 day. Note the two closely apposed membranes (arrowhead pairs) surrounding each parasite; N, nucleus. The dark bodies are granules of the histidine-rich protein. Bar = 0.5 μm.

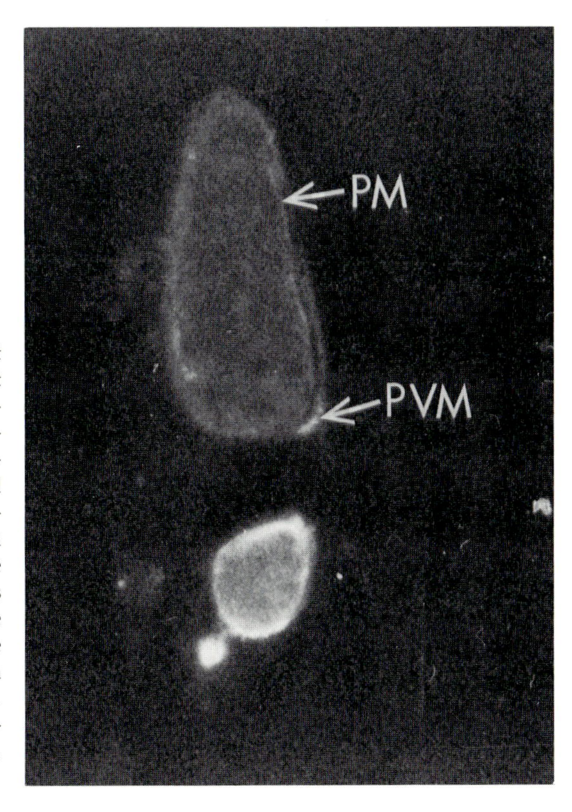

FIGURE 5.8. Immunofluorescent reaction of exoerythrocytic schizont of *Plasmodium berghei* after 48 hr development from sporozoites inoculated to WI 38 cells (human embryonic lung) *in vitro*. A monoclonal antibody (B6) was used, from unpurified ascites fluid, that reacts with a sporozoite surface antigen. Note fluorescence of the parasitophorous membrane (PVM) as well as the plasma membrane of the parasite (PM). Similar reaction was seen with antiserum to the erythrocytic stage. (From Hollingdale *et al.*, 1983. Original print courtesy of Dr. M. R. Hollingdale.)

clearly indicates that the parasitophorous membrane is more parasite than host. The interface (Fig. 5.7C) consists of a parasite plasma membrane closely apposed to a surrounding membrane structurally derived at first from the host cell but rapidly changed to consist largely of parasite material. This membrane grows with the growth of the parasite, as shown by observations on *P. lophurae* trophozoites maintained extracellularly *in vitro*. In these experiments with extracellular *P. lophurae*, an exogenous supply of ATP was an absolute requirement for prolonged viability. Perhaps this is needed to maintain transport functions in the parasitophorous membrane. How transport occurs and how it is integrated with transport across the parasite's own plasma membrane are challenging problems for cell biologists. It is also highly significant that the intracellular parasite's main feeding mechanism involves

(B) A schizont after 2 days. The developing merozoites (M) are enclosed within the parasitophorous membrane (arrowhead). Bar × 0.5 μm. (C) *P. falciparum*. Young ring in a human erythrocyte from *in vitro* culture. Note the two closely apposed membranes (arrowheads) separating the parasite from the host cell cytoplasm. Bar = 0.2 μm. TEM. (Original prints courtesy of Dr. S. G. Langreth.)

further extension of the interface with the host cell by endocytosis. This occurs largely through cytostomes, specialized regions of the surface that appear as thickened rings. Of great interest is the fact that both the plasma membrane of the parasite and the parasitophorous membrane are invaginated through the cytostome in the formation of the food vacuole (see Chapter 6). As digestion of the hemoglobin occurs, one membrane, presumed to be the parasitophorous membrane, disappears.

It is difficult to overemphasize the fluidity and malleability of cellular membranes. They seem to be in constant and rapid cyclic flow connecting the various compartments of the cell. And yet each organelle retains its integrity and its characteristic membrane composition. The repeated membrane fusion events that occur must have an exquisite specificity. And yet the intracellular parasite somehow intercalates or perhaps superimposes its own membrane system on that of the host cell.

Although the interface between malarial parasite and host erythrocyte cytoplasm involves the two membranes, parasite plasma membrane and parasitophorous membrane, the interface with the host as a whole also involves the plasma membrane of the erythrocyte. This membrane is altered by the presence of the parasite within. There are marked changes in permeability, some perhaps adaptive in permitting a greater flow of nutrients to the parasite and of waste products such as lactic acid out of the cell. Small amounts of parasite-derived proteins become inserted into the erythrocyte membrane. In the human malaria *P. falciparum,* certain of these form structures on the surface visible by electron microscopy (see Chapter 13) and even under appropriate conditions by light microscopy. These small knobs furthermore function as organelles of attachment to endothelial cells of the capillaries. In *P. falciparum* infection in humans, only the young rings are found in the circulating blood. At this stage knobs are not yet present. As the parasites develop into trophozoites, after about 16–20 hr of the 48-hr cycle of asexual reproduction (Diagram VIII), the knobs appear and at the same time the infected cells become sequestered in capillaries of organs such as the heart and the brain, where they adhere by the knobs to the endothelial lining. In this way the larger parasites are protected from having to pass through the spleen, the principal organ of defense against parasites of erythrocytes. Thus, through modification of the host cell membrane, the parasite secures itself in a particularly favorable niche.

When a parasite lies directly in the host cell cytoplasm, the interface consists only of its own plasma membrane. This perhaps less complex situation is found with such organisms as *Trypanosoma cruzi,* cause of Chagas' disease, the piroplasms (e.g., *Babesia* and *Theileria*), and the microsporidia. For all of these there is little information other than morphological observations at the ultrastructural level. *T. cruzi* invades a host cell by an endocytotic process induced by the flagellated trypomastigote form. The invaginated host cell membrane rapidly disappears and the parasite then reproduces within the cell as amastigotes. With the consumption of the host cell, these transform again into trypomastigotes. Since all of these stages can now be grown *in vitro* extracellularly, we are in a position for detailed functional analysis of the interface—the properties of the plasma membrane.

Such is not the case for any of the other kinds of protozoa developing directly in the cytoplasm of a host cell. Organisms such as *Babesia* and *Theileria* or any of the microsporidia, not only have not been cultured extracellularly, they have not been maintained viable extracellularly for even relatively short periods of time. As already noted, when a merozoite of *Babesia* invades an erythrocyte, the invaginated red cell membrane disappears probably within about 1 hr after the beginning of entry. Morphological studies during this period would be of great interest and would constitute at least a beginning toward a functional study of this interface. Since some species of *Babesia* have been grown in their host red cells *in vitro* and since infective merozoites can be collected from the cultures, such a study seems quite approachable from a technical standpoint. Once the developing *Babesia* is free in the cytoplasm of its host erythrocyte, its interface with this cytoplasm appears to involve not only its plasma membrane, but also peculiar whorled membranes that extend out from the parasite, and channellike invaginations. Both of these may function in the uptake of nutrients. Experiments with extracellular babesias have shown the uptake of ferritin into long narrow channels formed by pinocytosis of the plasma membrane. The coiled organelle could well be an organ for extracellular digestion (see also Chapter 6).

In the related organisms of the genus *Theileria,* the association with the cytoplasm is so intimate that daughter parasites become attached to host microtubules of the mitotic spindle and in this way are distributed to the two daughter host cells formed at mitotic division. *Theileria parva,* the cause of East Coast fever of cattle in Africa, during a portion of its life cycle infects lymphoblastoid cells and can be grown *in vitro* in cultures of these cells. The presence of the parasites supplies an essential mitogenic stimulus that maintains continued multiplication of the host cells. Like *Babesia* in the red cell, the *Theileria* macroschizont lies directly in the cytoplasm of the lymphoblast. At prophase the infected cell forms annulate lamellae from the Golgi complex. These then contribute, at metaphase, to the formation of microtubules that join the parasites and the host cell centrioles. Other microtubules arise from the host nuclear envelope and join host chromosomes to the centrioles. The daughter parasites are then distributed like chromosomes at mitosis, but they remain in the cytoplasm (see Chapter 13).

Particular species of intracellular protozoa are not necessarily limited to one kind of interface with the host cell; they may change with the stage in the developmental cycle of the parasite. This is well illustrated in the microsporidia. As seen in Chapter 3, the infective stages of microsporidia are injected from a resistant spore by means of an intricate highly specialized apparatus directly into the cytoplasm of a host cell. For many species, initial cycles of multiplication then occur in the cytoplasm to give rise to forms that disseminate the infection to other cells and tissues. When spores are again

FIGURE 5.9. *Trichinella spiralis.* Nurse cell-infective L_1 larva complex. Digested from infected mouse muscle and treated 1 hr at 37°C with 0.25% trypsin at pH 8. Nomarski interference microscopy. Bar = 50 μm. (From Despommier, 1983. Original print courtesy of Dr. D. Despommier.)

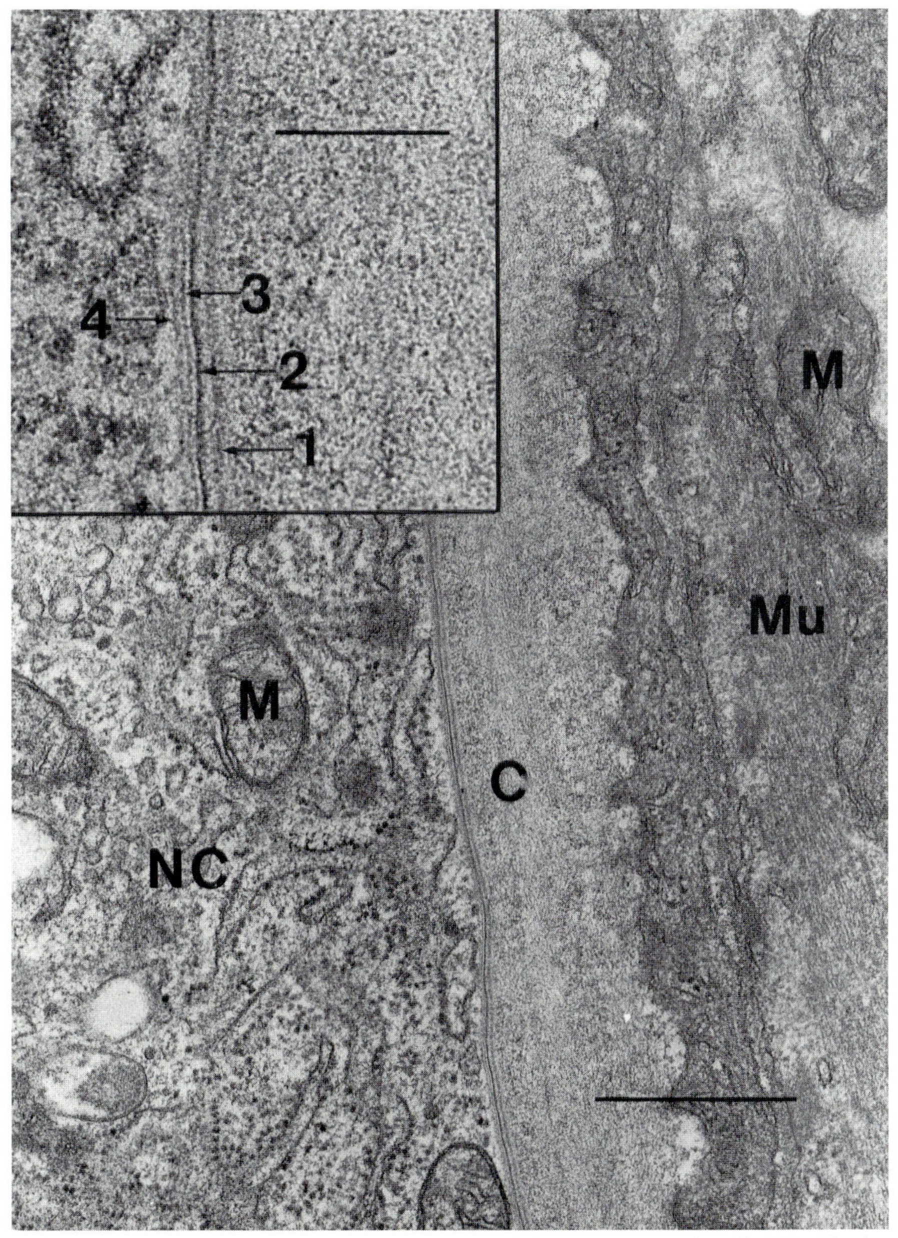

FIGURE 5.10. *T. spiralis*. Interface of a 14-day muscle larva in its muscle cell (NC). The four layers of the interface (1–4) are shown in the inset. Layers 1 and 2 are cuticular. The origins of layers 3 and 4 are not known. M, mitochondrion; C, cuticle; Mu, muscle of larva. Bars: main figure, 2 μm; inset, 0.5 μm. (From Despommier, 1983. Original print courtesy of Dr. D. Despommier.)

being formed, however, the parasites lie in a parasitophorous vacuole limited by a membrane thought to be derived from the rough endoplasmic reticulum of the host cell. It must be emphasized that nothing is really known about the origin of this membrane, let alone its structure and function.

Even more complex is the interface that develops around the one type of intracellular metazoan parasite that has been adequately studied, the muscle-dwelling larva of the nematode *Trichinella spiralis*. A migrating larva of *Trichinella* must enter a striated muscle cell if it is to survive and develop. For the first 3 days after entry, the larva does not grow; during this time, changes in the host myofibers are occurring. There is extensive growth of the T-tubular system. By day 10 after entry, the worm, now growing rapidly, is surrounded by a double membrane provided by the host, which may now be termed a nurse cell (Figs. 5.9, 5.10). This special structure provides the interface through which the larval worm supports its further growth to day 20. Growth then stops, but some exchange with the host must continue, for such quiescent larvae may remain viable and infective up to 30 years. On the other hand, further reactions of the host may bring about calcification of the nurse cell and death of the larva within.

Bibliography

Bannister, L. H., 1979, The interactions of intracellular Protista and their host cells, with special reference to heterotrophic organisms, *Proc. R. Soc. London Ser. B* **204**:141–163.

Chang, K.-P., 1978, *Leishmania* infection of human skin fibroblasts *in vitro:* Absence of phagolysosomal fusion after induced phagocytosis of promastigotes and their intracellular transformation, *Am. J. Trop. Med. Hyg.* **27**:1084–1096.

Chang, K.-P., and Dwyer, D. M., 1976, Multiplication of a human parasite (*Leishmania donovani*) in phagolysosomes of hamster macrophages *in vitro, Science* **193**:678–680.

Chang, K.-P., and Dwyer, D. M., 1978, *Leishmania donovani.* Hamster macrophage interactions *in vitro:* Cell entry, intracellular survival and multiplication of amastigotes, *J. Exp. Med.* **147**:515–530.

Chang, K.-P., and Hendricks, L. D., 1985, Laboratory cultivation and maintenance of *Leishmania,* in: *Leishmaniasis* (K.-P. Chang and R. S. Bray, eds.), Elsevier, Amsterdam, pp. 213–244.

Despommier, D. D., 1976, Musculature, in: *Ecological Aspects of Parasitology* (C. R. Kennedy, ed.), North-Holland, Amsterdam, pp. 269–285.

Despommier, D. D., 1983, Biology, in: *Trichinella and Trichinosis* (W. C. Campbell, ed.), Plenum Press, New York, pp. 75–151.

Dwyer, D. M., and Gottlieb, M., 1983, The surface membrane chemistry of leishmania: Its possible role in parasite sequestration and survival, *J. Cell. Biochem.* **23**:35–45.

Ebert, F., Buse, E., and Mühlpfordt, H., 1979, *In vitro* light and electron microscope studies on different virulent promastigotes of *Leishmania donovani* in hamster peritoneal macrophages, *Z. Parasitenkd.* **59**:31–41.

Entzeroth, R., 1984, Electron microscope study of host–parasite interactions of *Sarcocystis muris* (Protozoa, Coccidia) in tissue culture and *in vivo, Z. Parasitenkd.* **70**:131–134.

Fawcett, D. W., Doxsey, S., Stagg, D. A., and Young, A. S., 1982, The entry of sporozoites of *Theileria parva* into bovine lymphocytes *in vitro:* Electron microscopic observations, *Eur. J. Cell Biol.* **27**:10–21.

Gottlieb, M., and Dwyer, D. M., 1981, Protozoan parasite of humans: Surface membrane with externally disposed acid phosphatase, *Science* **212**:939–941.

Hammond, D. M., Scholtyseck, E., and Chobotar, B., 1967, Fine structures associated with nutrition of the intracellular parasite *Eimeria auburnensis, J. Protozool.* **14:**678–683.

Hollingdale, M. R., Leland, P., Leef, J. L., Leef, M. F., and Beaudoin, R. L., 1983, Serological reactivity of *in vitro* cultured exoerythrocytic stages of *Plasmodium berghei* in indirect immunofluorescent or immunoperoxidase antibody tests, *Am. J. Trop. Med. Hyg.* **32:**24–30.

Ishihara, R., 1969, The life cycle of *Nosema bombycis* as revealed in tissue culture cells of *Bombyx mori, J. Invert. Pathol.* **14:**316–320.

Langreth, S. G., 1976, Feeding mechanisms in extracellular *Babesia microti* and *Plasmodium lophurae, J. Protozool.* **23:**215–223.

Langreth, S. G., and Trager, W., 1973, Fine structure of the malaria parasite *Plasmodium lophurae* developing extracellularly *in vitro, J. Protozool.* **20:**606–613.

Moulder, J. W., 1983, Chlamydial adaptation to intracellular habitats, in: *Microbiology—1983* (D. Schlessinger, ed.), American Society for Microbiology, Washington, D.C., pp. 370–374.

Moulder, J. W., 1985, Comparative biology of intracellular parasitism, *Microbiol. Rev.* **49:**298–337.

Mueller, B. E. G., 1980, *Eimeria canadensis* (Protozoa, Sporozoea): Ultrastructure of the host–parasite interface during development of first-generation schizonts *in vitro, Can. J. Zool.* **58:**2018–2025.

Musisi, F. L., Bird, R. G., Brown, C. G. D., and Smith, M., 1981, The fine structural relationship between *Theileria* schizonts and infected bovine lymphoblasts from culture, *Z. Parasitenkd.* **65:**31–41.

Nogueira, N., and Cohn, Z., 1976, *Trypanosoma cruzi:* Mechanisms of entry and intracellular fate in mammalian cells, *J. Exp. Med.* **143:**1402–1420.

Pastan, I. H., and Willingham, M. C., 1981, Journey to the center of the cell: Role of the receptosome, *Science* **214:**504–509.

Remaley, A. T., Kuhns, D. B., Basford, R. E., Glew, R. H., and Kaplan, S. S., 1984, Leishmanial phosphatase blocks neutrophil O_2^- production, *J. Biol. Chem.* **259:**11173–11175.

Rudzinska, M. A., 1981, Morphological aspects of host-cell–parasite relationships in babesiosis, in: *Babesiosis* (M. Ristic and J. P. Kreier, eds.), Academic Press, New York, pp. 87–141.

Scholtyseck, E., Hammond, D. M., and Ernst, J. V., 1966, Fine structure of the macrogametes of *Eimeria perforans, E. stiedae, E. bovis* and *E. auburnensis, J. Parasitol.* **52:**975–987.

Sherman, I. W., and Jones, L. A., 1979, *Plasmodium lophurae:* Membrane proteins of erythrocyte-free plasmodia and malaria-infected red cells, *J. Protozool.* **26:**489–501.

Steinman, R. M., Mellman, I. S., Muller, W. A., and Cohn, Z. A., 1983, Endocytosis and the recycling of plasma membrane, *J. Cell Biol.* **96:**1–27.

Stewart, G. L., and Giannini, S. H., 1982, *Sarcocystis, Trypanosoma, Toxoplasma, Brugia, Ancylostoma* and *Trichinella* spp: A review of the intracellular parasites of striated muscle, *Exp. Parasitol.* **53:**406–447.

Takahashi, Y., and Sherman, I. W., 1980, *Plasmodium lophurae:* Lectin-mediated agglutination of infected red cells and cytochemical fine structure detection of lectin binding sites on parasite and host cell membranes, *Exp. Parasitol.* **49:**233–247.

Takahashi, Y., Yamada, K., and Sherman, I. W., 1980, *Plasmodium lophurae:* Antibody-induced movement and capping of surface membranes of erythrocyte-free malarial parasites, *Exp. Parasitol.* **50:**201–211.

Trager, W., 1979, Erythrocyte–malaria parasite interactions, *Microbiology* **1979:**120–123.

Trissl, D., Martinez-Palomo, A., Arguello, C., de la Torre, M., and de la Hoz, R., 1977, Surface properties related to concanavalin A-induced agglutination: A comparative study of several Entamoeba strains, *J. Exp. Med.* **145:**652–665.

CHAPTER 6

The Uptake of Nutrients. Digestion

The uptake of nutrients is certainly one of the principal activities that occurs at the host–parasite interface. All of the parasite's nutrients are taken from the host. The modes of uptake basically do not differ from those used by free-living organisms. Metazoan parasites with an alimentary canal ingest food into this with subsequent digestion and absorption by cells of the gut. Ectoparasites, such as the bloodsucking arthropods, are restricted to this type of food ingestion. Endoparasitic metazoa with a gut may in addition take up some nutrients through their external body surface. This is the sole means of uptake for parasites, like tapeworms, that lack an alimentary tract, and for all of the protozoa. Ingestion at the body surface may occur by transport of solutes across the membrane or by endocytosis, either of solid particles (phagocytosis) or of liquids (pinocytosis).

The external body surface of tapeworms, as already noted in Chapter 4, is highly specialized so as to provide a maximal absorptive surface in contact with host fluids. Transport of amino acids and other low-molecular-weight nutrients into cestodes, particularly the rat tapeworm, *Hymenolepis diminuta,* has been extensively studied by Clark Read and his co-workers. They have shown (Table 6.1) that amino acids enter by highly specific transport systems, of which there are at least six: (1) a basic amino acid transport system that also transports histidine; (2) a system for dicarboxylic amino acids with a low affinity for some neutral amino acids; (3–6) at least four neutral amino acid transport systems, one of which also transports aromatic amino acids and histidine. All the systems are specific, one α-amino group being essential for inhibition. The uptake of any single amino acid is influenced by other amino acids in a complex way. It is also affected by the host species (e.g., rats versus hamsters) in which the worms are reared, suggesting it might be very important to study worms reared *in vitro* under more nearly controlled conditions (see Chapter 8). The movement of amino acids in cestodes is not coupled to movement of sodium or an anion so that the driving force for uptake against a gradient is not understood. On the other hand, the active transport of glucose is sodium dependent, as in vertebrate cells. Phlorizin, which acts as a competitive inhibitor of glucose uptake, can be demonstrated (as [^{3}H]phlorizin)

TABLE 6.1. The Amino Acid Transport Systems Identified in the Cestode *Hymenolepis diminuta*[a]

	System					
	Dicarboxylic	Glycine	Serine	Leucine	Phenylalanine	Dibasic
Major amino acids interacting	Aspartic Glutamic Methionine	Glycine Methionine	Serine Alanine Threonine Methionine Valine Proline	Leucine Isoleucine Methionine	Phenylalanine Tyrosine Histidine Methionine	Arginine Lysine Histidine
Other amino acids	Serine Alanine Glycine	Serine Threonine Alanine	Glycine	Glycine Serine Threonine Alanine Valine	Leucine Isoleucine	

[a] Modified from Pappas and Read (1975).

TABLE 6.2. A Summary of the Interactions of Purines and Pyrimidines in *Hymenolepis diminuta*[a,b]

	"Effector"							
Substrate	Thymine	Uracil	5-Bromouracil	5-Aminouracil	6-Methyluracil	Hypoxanthine	Adenine	Purine
Thymine	±	±	±	+	−	−	−	−
Uracil	±	±	±	+	−	−	−	−
5-Bromouracil	+	0	+	+	−	−	−	−
Hypoxanthine	0	−	−	0	−	−	−	−
Adenine	0	−	−	0	−	−	±	0
Guanine	0	−	−	0	−	−	−	0

[a] Modified from Pappas and Read (1975).
[b] A (+) indicates that the "effector" stimulated substrate absorption, while a (−) indicates that the uptake of substrate was inhibited. A "0" indicates that the "effector" had no effect on the uptake of the substrate. A (±) indicates the "effector" either stimulated or inhibited substrate absorption, depending on the "effector"/substrate ratio.

in the tapeworm tegument as well as in the brush border of the host's intestinal mucosa. In either case it indicates sites of glucose uptake.

H. diminuta has mediated systems for uptake of purines, pryimidines, and nucleosides, involving at least three distinct carriers (Table 6.2). There are at least two systems for the uptake of fatty acids, one for short-chain (acetate, butyrate), the other for long-chain fatty acids. With the latter, complex allosteric effects may be involved, since uptake of palmitate was stimulated by laurate (to a maximum of 77% at 2 mM laurate) and also by low concentrations (0.025 M) of oleate, linoleate, and linolenate (but high concentrations were inhibitory). There are also at least two systems for uptake of water-soluble vitamins, one for thiamine and riboflavin, another for nicotinamide and pyridoxine. The tapeworm *Diphyllobothrium latum,* acquired from the eating of undercooked freshwater fish infected with its larval form, accumulates vitamin B_{12} (cobalamin) in such large amounts as to produce a B_{12} deficiency and anemia in its host. It must have a system for active transport of B_{12} and presumably the B_{12} serves some special role in the economy of the worm, but nothing is yet known about either of these interesting and important matters.

The apparent uptake of phosphorylated compounds by *H. diminuta* takes place by the combination of two processes at the surface of the worm. Phosphohydrolases dephosphorylate glucose-6-phosphate and nucleotides in such a way that the glucose and nucleoside formed do not diffuse away but are immediately transported into the worm.

In blood-inhabiting trematodes it is likely that many low-molecular-weight nutrients are taken up through the tegument, just as in cestodes, despite the presence of a cecum. This has been clearly demonstrated for glucose, glycine, and proline in *Schistosoma mansoni.* On the other hand, schistosomes ingest red blood cells into their gut and here the hemoglobin is broken down with the formation of hemin and digestion of the globin by a specific protease. This aspect of the worm's nutrition is not different fundamentally from what occurs in the more highly developed alimentary tract of ectoparasitic blood-sucking insects such as the louse, tsetse fly, or female mosquito. It is interesting to note, however, that in ticks the lysed blood cell material in the gut is endocytosed by the gut cells and then digested intracellularly in food vacuoles, much as in malarial parasites (see below).

The role of endocytosis in the tegumental uptake of nutrients by parasitic helminths has not been adequately defined. In the parasitic protozoa it clearly plays a major role. This is obvious for organisms like *Entamoeba histolytica,* which resembles free-living amebae in its phagocytosis of particles and pinocytosis of solutes. The turnover of membrane as a result of pinocytosis is extensive. In *Acanthamoeba* (a free-living species), roughly ten times the surface area is interiorized during 1 hr of pinocytotic activity (pinocytotic vesicles of average diameter 0.25 μm, cell surface area estimated at 2200 μm^2). In *Entamoeba* the activity is similar and exceeds even that of macrophages. For the latter cells the area of membrane of an average pinocytotic vesicle was 0.27

μm². About 27 μm² of plasma membrane is interiorized per minute; this corresponds to 3% of the cell surface. Thus, the equivalent of the entire cell surface was interiorized every 33 min. But even in forms like the trypanosomes, endocytosis is more important than direct transport across the plasma membrane of the cell. Pinocytosis by trypanosomes was first observed in

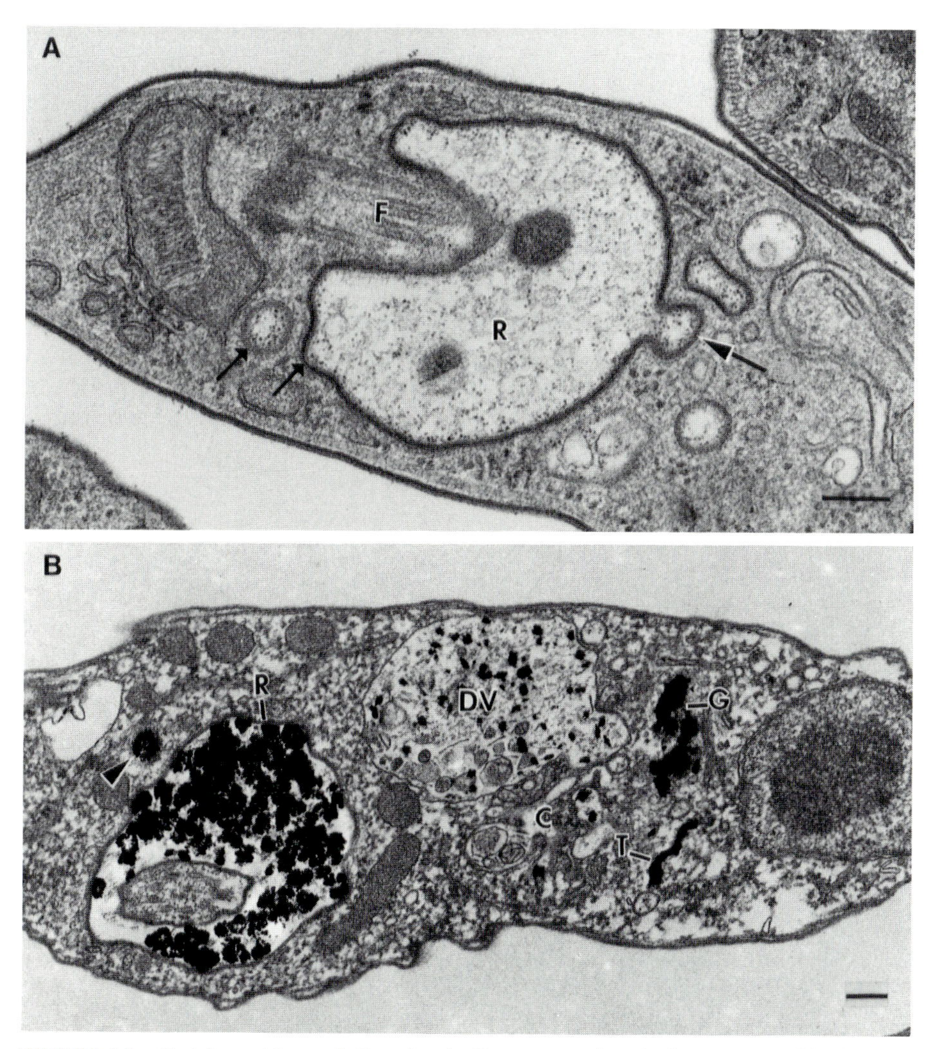

FIGURE 6.1. Protein uptake and digestion in *Trypanosoma brucei*, short stumpy bloodstream forms. (A) Ferritin uptake for 20 min at 25°C. From the flagellar pocket or reservoir (R) the tracer enters large (~110 nm) spiny-coated vesicles (large arrow). The spines on the cytoplasmic surfaces of the vesicles and the flagellar pocket are the same (small arrows). F, flagellum. (B) Acid phosphatase reaction product in the reservoir (R), a large coated vesicle (arrowhead), the digestive vacuole (DV), Golgi cisternae (G), collecting tubules (T), and the collecting system (C). EM. Bars = 0.2 μm. (From Langreth and Balber, 1975. Original prints courtesy of Dr. S. G. Langreth.)

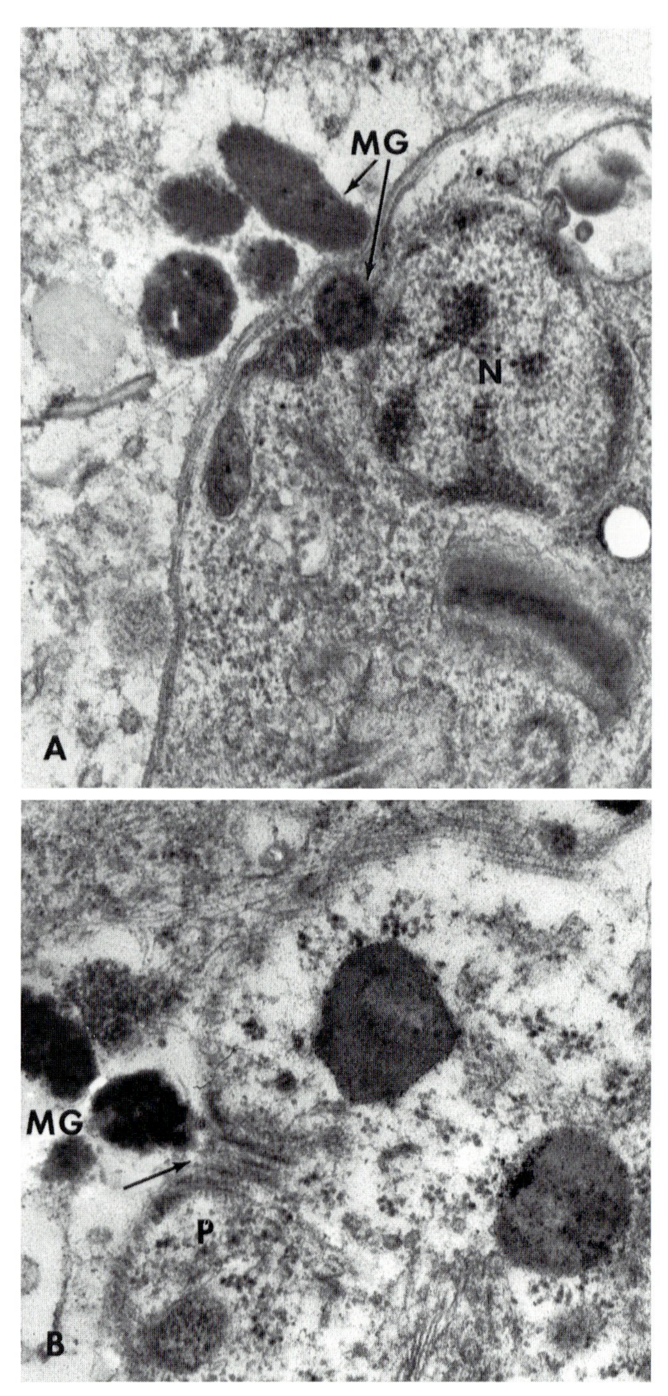

FIGURE 6.2. *T. cruzi* growing in pigment epithelial cells from chick embryo iris in tissue culture, to show uptake of melanin granules by the parasite. (A) Part of parasite with nucleus (N) and

culture forms of *Trypanosoma mega* and in blood forms of *T. brucei*, and has since shown to be a widespread phenomenon. Pinocytotic uptake seems to be more active in the flagellar pocket, which is also the site of secretion of acid phosphatase (Fig. 6.1). Pinocytosis may also occur via a cytopharynx opening through a cytostome in the flagellar pocket or near it. Of special interest is the uptake of host cell material demonstrated for intracellular amastigote forms of *T. cruzi* (Fig. 6.2). By growing the parasites in tissue cultures of the pigmented epithelial cells from the iris of chick embryos, Meyer and De Souza were able to show the phagocytotic ingestion of the melanin granules from the host cytoplasm into the amastigotes. This occurred through a cytostome which could apparently form at any region of the surface of the parasite.

Such intracellular phagocytosis is exactly the principal method for uptake of protein by malarial parasites (as already noted in Chapter 5). Ingestion is mainly through cytostomes, and these can form at any region of the surface of a rapidly growing trophozoite. In *P. falciparum* as many as five have been seen in a single tangential section (Fig. 6.3). It must be remembered that, whereas *T. cruzi* amastigotes lie directly in the host cytoplasm, which they phagocytose through cytostomes, malarial parasites are bounded by a parasitophorous membrane. This has to be ingested along with the host cytoplasm and is subsequently digested in the food vacuole. This turnover processing of the parasitophorous membrane may play a role in the replacement of its originally red cell constituents by materials from the parasite. Endocytosis of material from the parasitophorous vacuole evidently occurs also in coccidia, where we have already noted complex membranous folds extending into the vacuole and tubular structures connecting host and parasite cytoplasm. In malarial parasites, because of the electron density of hemoglobin, digestion of the hemoglobin within the food vacuoles can be seen in electron micrographs, as well as the concomitant formation of the residual pigment hemozoin (mainly hemin). Such grossly visible changes associated with food uptake cannot be seen with the other main group of intraerythrocytic parasites, the babesias. These organisms have a whorled membrane structure extending into the red cell cytoplasm which might be responsible for extracellular (i.e., with reference to the babesial cell) digestion and subsequent uptake through narrow pinocytotic channels (Fig. 6.4) (see also Chapter 5).

The food of parasites of animals is typically rich in protein. Although some parasites may satisfy their nitrogen requirements entirely by the uptake of amino acids from the hosts, it seems more likely that most rely on the uptake and digestion of proteins. Proteases have been demonstrated in nu-

melanin granules (MG) inside and outside parasite. ×45,000. (B) Anterior part of parasite (P) showing cytostome and melanin granule (MG) apparently in act of entering cytostome (arrow). ×52,500. (From Meyer and De Souza, 1973. Original prints courtesy of Dr. H. Meyer.)

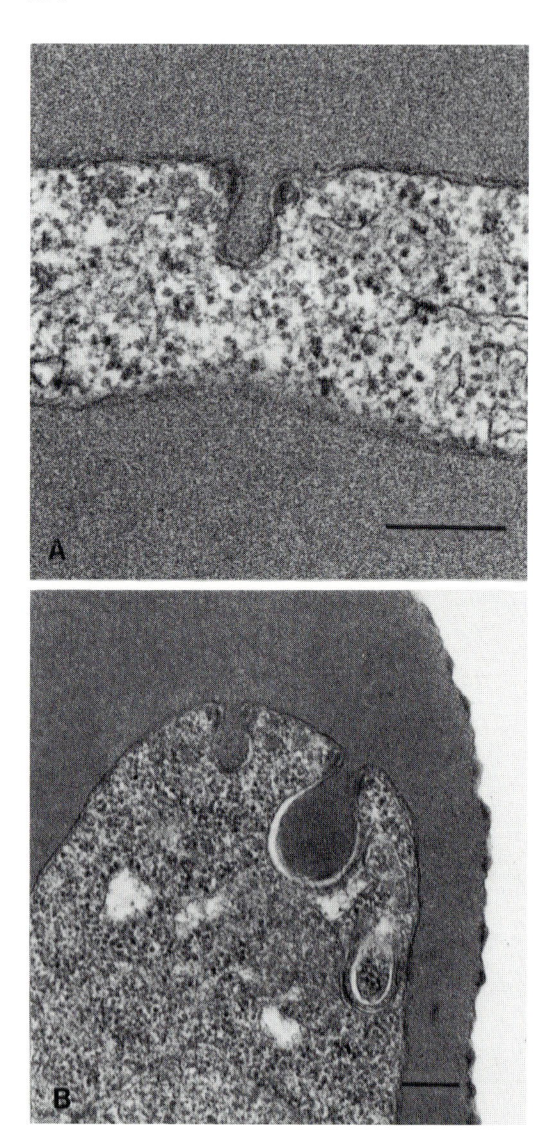

FIGURE 6.3. Ingestion of erythrocyte cytoplasm through cytostomes by *Plasmodium falciparum* in human erythrocytes *in vitro.* (A) In a young ring; (B) in an older trophozoite. Bars = 0.2 μm. Note that the parasitophorous membrane is internalized so that the forming food vacuole has two membranes at first. (B from Langreth *et al.*, 1978. Original prints courtesy of Dr. S. G. Langreth.)

merous species of parasites. Bloodsucking ectoparasitic insects have powerful tryptic enzymes that digest the large blood meal within the lumen of the midgut (tsetse, mosquitoes).

The blood-feeding trematode *Schistosoma mansoni* breaks down hemoglobin to small peptides with the liberation of heme as a black pigment. Since the trematode alimentary tract is a blind one with only one opening that serves both as mouth and anus, the waste products are vomited out. Along with them come digestive enzymes of the worms which then act as antigens to sensitize the host. This has provided the basis for a sensitive diagnostic

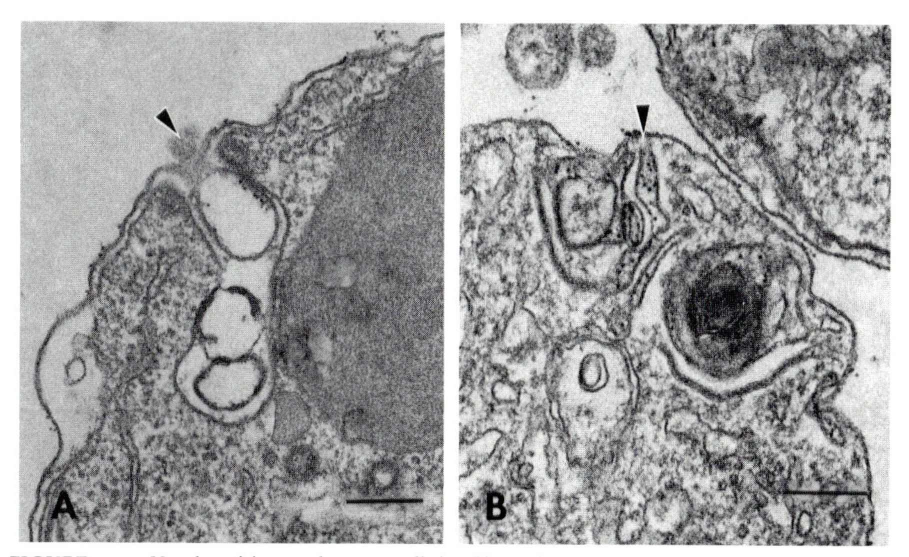

FIGURE 6.4. Uptake of ferritin by extracellular *Plasmodium lophurae* (A) and *Babesia microti* (B). For A the parasites were freed from their host duck erythrocytes by saponin lysis and incubated 30 min at 37°C with ferritin in Trager's buffer. Note the forming food vacuole at the cytostome (arrowhead) with ferritin particles inside the ingested parasitophorous membrane. For B the parasites were freed from hamster erythrocytes by immune lysis and incubated 30 min at 37°C with ferritin in Trager's buffer. Note the ferritin in narrow pinocytotic channels (arrowhead). Bars = 0.2 μm. (A from Langreth, 1976. Original prints courtesy of Dr. S. G. Langreth.)

skin test for schistosomiasis, employing a highly purified preparation of the proteolytic enzyme. Proteolytic enzymes of parasitic protozoa are chiefly thiol-dependent. Such thiol-dependent proteases have been found in trypanosomes, trichomonads, and entamebae. Three kinds of proteases have been found in several species of malarial parasites, including *P. falciparum* (Fig.

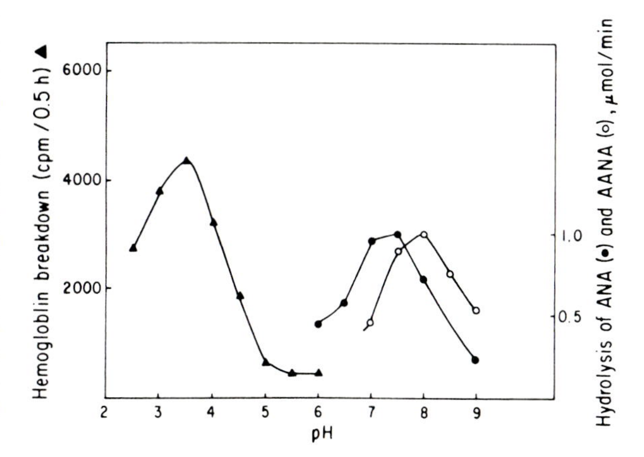

FIGURE 6.5. Effect of pH on the activities of peptidases from *Plasmodium falciparum*. The acid pepidase was measured by the liberation of radioactivity from [^{3}H]acetylhemoglobin. The neutral and alkaline peptidases were measured by converting the change in absorbance per minute of L-alanine-*p*-nitroanilide (ANA) and *N*-acetylalanine-*p*-nitroanilide (AANA) to μmoles *p*-nitroaniline formed per minute from a calibration curve. Buffer systems were acetate, pH 2.5–4.5; phosphate, pH 4.5–7; Tris-HC1, pH 7–9. (From Gyang *et al.*, 1982)

6.5). One is an acid protease resembling mammalian cathepsin D in its low pH optimum (about 3.5 for the enzyme from *P. falciparum*), inhibition by pepstatin, and lack of sensitivity to thiol reagents. This enzyme in *P. falciparum*, however, has a much higher apparent molecular weight (148,000) than the mammalian enzymes. It acts directly on hemoglobin. In view of these properties, it seems likely that this is the enzyme responsible for the digestion of hemoglobin in the food vacuoles of malarial parasites and that these vacuoles resemble mammalian lysosomes. Like mammalian cell lysosomes, they would be expected to concentrate certain antimalarial drugs by proton trapping and hence to play a role in the mode of action of such important antimalarials as chloroquine. The other two proteases are a neutral aminopeptidase and an alkaline endopeptidase. Whether and how these may function, and their subcellular localization, are not known. Proteases may well have various functions other than the digestion of food; they are evidently involved in the entrance of malarial merozoites into red cells (see Chapter 3).

Recent studies with *Leishmania mexicana* have revealed a most interesting situation. Whereas high-molecular-weight (67,000) proteases were present in both the extracellular promastigotes and the intracellular amastigotes, lower-molecular-weight (about 31,000) proteases were present only in the amastigotes. Furthermore, these proteases showed greater substrate specificity and also accounted for the increased proteolytic activity of amastigote extracts. These enzymes evidently are synthesized in the transformation from promastigote to amastigote, and may be assumed to play some specific role in intracellular life.

The possibility that some proteins may be imported as the intact polypeptide directly across the cell membrane has to be kept in mind, especially for intracellular parasites. This is what happens with many proteins imported into mitochondria and transported across one or both of the mitochondrial membranes.

Carbohydrate is available to many parasitic organisms as glucose or other simple sugars, and most have mechanisms for its uptake either by active transport or by facilitated diffusion. In addition, many parasites have hydrolases that make available to them various complex polysaccharides, such as glycogen, starch, and cellulose, and such cellular constituents as blood group factors, sialic acid, and hyaluronic acid. The latter enzymes in particular are probably more concerned with tissue invasion than with nutrition but it would be difficult to separate these activities.

Bibliography

Aley, S. B., Scott, W. A., and Cohn, Z. A., 1980, Plasma membrane of *Entamoeba histolytica*, *J. Exp. Med.* **152:**391–404.

Arme, C., 1976, Feeding, in: *Ecological Aspects of Parasitology* (C. R. Kennedy, ed.), North-Holland, Amsterdam, pp. 75–97.

Charet, P., Slomianny, C., Prensier, G., and Moreau, S., 1984, *Plasmodium* Sp. et dégradation de l'hémoglobine: Mécanisme et relations avec les antimalariques, *Ann. Biol.* **23**:31–60.

Dresden, M. H., 1982, Proteolytic enzymes of *Schistosoma mansoni, Acta Leiden.* **49**:81–99.

Dresden, M. H., Payne, D. C., and Basch, P. F., 1982, Acidic thiol proteinase activity of *Schistosoma mansoni* cultured *in vitro, Mol. Biochem. Parasitol.* **6**:203–208.

Dresden, M. H., Sung, C. K., and Deelder, A. M., 1983, A monoclonal antibody from infected mice to a *Schistosoma mansoni* egg proteinase, *J. Immunol.* **130**:1–3.

Friend, W. B., and Smith, J. J. B., 1972, Feeding stimuli and techniques for studying the feeding of haematophagous arthropods under artificial conditions with special reference to *Rhodnius prolixus,* in: *Insect and Mite Nutrition* (J. G. Rodriguez, ed.), North-Holland/Elsevier, Amsterdam, Holland pp. 241–256.

Gyang, F. N., Poole, B., and Trager, W., 1982, Peptidases from *Plasmodium falciparum in vitro, Mol. Biochem. Parasitol.* **5**:263–273.

Hempelmann, E., and Wilson, R. J. M., 1980, Endopeptidases from *Plasmodium knowlesi, Parasitology* **80**:323–330.

Landsperger, W. J., Stirewalt, M. A., and Dresden, M. H., 1982, Purification and properties of a proteolytic enzyme from the cercariae of the human trematode parasite *Schistosoma mansoni, Biochem. J.* **201**:137–144.

Langreth, S. G., 1976, Feeding mechanisms in extracellular *Babesia microti* and *Plasmodium lophurae, J. Protozool.* **23**:215–223.

Langreth, S. G., and Balber, A. E., 1975, Protein uptake and digestion in bloodstream and culture forms of *Trypanosoma brucei, J. Protozool.* **22**:40–53.

Langreth, S. G., Jensen, J. B., Reese, R. T., and Trager, W., 1978, Fine structure of human malaria, *J. Protozool.* **25**:443–452.

McLaughlin, J., 1982, The hydrolases of the parasitic protozoa, in: *Aspects of Parasitology* (E. Meerovitch, ed.), McGill University, Montreal, pp. 227–262.

Meyer, H., and De Souza, W., 1973, On the fine structure of *Trypanosoma cruzi* in tissue cultures of pigment epithelium from the chick embryo: Uptake of melanin granules by the parasite, *J. Protozool.* **20**:590–593.

North, M. J., 1982, Comparative biochemistry of the proteinases of eucaryotic microorganisms, *Microbiol. Rev.* **46**:308–340.

Pappas, P. W., and Read, C. P., 1975, Membrane transport in helminth parasites: A review, *Exp. Parasitol.* **37**:469–530.

Pappas, P. W., Narcisi, E. M., and Rentko, V., 1983, Alterations in brush border membrane proteins and membrane-bound enzymes of the tapeworm, *Hymenolepis diminuta,* during development in the definitive host, *Mol. Biochem. Parasitol.* **8**:317–323.

Pupkis, M. F., and Coombs, G. H., 1984, Purification and characterization of proteolytic enzymes of *Leishmania mexicana mexicana* amastigotes and promastigotes, *J. Gen. Microbiol.* **130**:2375–2383.

Rudzinska, M. A., and Vickerman, K., 1968, The fine structure, in: *Infectious Blood Diseases of Man and Animals* (D. Weinman and M. Ristic, eds.), Academic Press, New York, pp. 217–306.

Sauer, M. C. V., and Senft, A. W., 1972, Properties of a proteolytic enzyme from *Schistosoma mansoni, Comp. Biochem. Physiol. B* **42**:205–220.

Schatz, G., and Butow, R. A., 1983, How are proteins imported into mitochondria?, *Cell* **32**:316–318.

Senft, A. W., and Maddison, S. E., 1975, Hypersensitivity to parasite proteolytic enzyme in schistosomiasis, *Am. J. Trop. Med. Hyg.* **24**:83–89.

Smithers, S. R., and Worms, M. J., 1976, Blood fluids—Helminths, in: *Ecological Aspects of Parasitology* (C. R. Kennedy, ed.), North-Holland, Amsterdam, pp. 349–369.

Steinman, R. M., Mellman, I. S., Muller, W. A., and Cohn, Z. A., 1983, Endocytosis and the recycling of plasma membrane, *J. Cell Biol.* **96**:1–27.

Von Brand, T., 1973, *Biochemistry of Parasites,* Academic Press, New York.

CHAPTER 7

Nutritional Requirements

The nutritional requirements of organisms are a function of their loss of biosynthetic capabilities. Whereas green plants and some autotrophic bacteria can synthesize all of their complex organic constituents from simple inorganic precursors (as carbon dioxide, water, and mineral salts), many organisms, including prokaryotes as well as eukaryotes, have clearly found an evolutionary advantage in obtaining more or less of their organic constituents preformed in their diet. In this way they dispense with whole arrays of biosynthetic systems, freeing energy for other purposes. Parasitic animals, protozoan as well as metazoan, may be expected to have at least as extensive requirements for such preformed organic molecules as their free-living counterparts. Whether their requirements are any more extensive is an interesting question to which I will return later in this chapter.

In order to establish all of the nutritional requirements of an organism, it has to be grown axenically (without other living organisms) through its complete developmental and reproductive cycle on a diet of known constituents, i.e., in a defined medium. This so far has been done for relatively few free-living animals and for still fewer parasitic ones. Several species of trypanosomatid flagellates, in particular those that live in the alimentary tract of insects and have only the insect as their host, have been grown in various defined media. A single defined medium was later developed by R. F. Steiger and E. Steiger that supports excellent growth of three species of *Leptomonas,* four of *Herpetomonas,* six of *Crithidia,* and one of *Blastocrithidia,* all alimentary tract parasites of insects. In addition, this medium (Table 7.1) supports growth of the promastigotes (insect stages) of *Leishmania tarentolae* (recently shown to be probably synonymous with *Trypanosoma platydactyli*), a parasite of lizards, and *L. donovani* and *L. brasiliensis,* causative agents of the human diseases, visceral and mucocutaneous leishmaniasis, respectively. At least for the promastigotes of *L. donovani,* this medium has been further simplified by the omission of Na-acetate; guanosine; the amino acids arginine, cysteine, cystine, glutamic acid, and serine; the vitamins lipoic acid, menadione, and vitamin A, ascorbic acid, and vitamin B_{12}; and the defatted bovine albumin.

TABLE 7.1. Defined Medium REI for Promastigotes of *Leishmania donovani* and *L. brasiliensis*[a,b]

	Ingredients	Concentration (mg/liter)
A.	NaCl	8,000
	KCl	400
	$MgSO_4 \cdot 7H_2O$	200
	$Na_2HPO_4 \cdot 2H_2O$	60
	KH_2PO_4	60
	$CaCl_2$	70
	Glucose	2,000
	Na-Acetate	600
B.	L-Arginine HCl	200
	L-Cysteine HCl	50
	L-Cystine	50
	L-Glutamic acid	300
	L-Glutamine*	300
	L-Histidine	100
	L-Isoleucine	100
	L-Leucine	300
	L-Lysine HCl	250
	L-Methionine	50
	L-Phenylalanine	100
	L-Proline	300
	DL-Serine	200
	L-Threonine	400
	L-Tryptophan	50
	L-Tyrosine	50
	L-Valine	100
C.	$NaHCO_3$	1,000
	Hepes	14,250
D.	Adenosine	20
	Guanosine	20
E.	D-Biotin	1
	Choline Cl	1
	Folic acid**	11
	i-Inositol	2
	Niacinamide	1
	D-Ca-pantothenate	1
	Pyridoxal HCl	1
	Riboflavin	0.1
	Thiamine HCl	1
F.	Lipoic acid	0.4
	Menadione	0.4
	Vitamin A	0.4
G.	Ascorbic acid	0.2
	Vitamin B_{12}	0.2
	Bovine albumin (defatted)	15
	Hemin**	10
	Phenol red	10

TABLE 7.1. *(continued)*

Ingredients	Concentration (mg/liter)
Redist. H_2O Q.S.	1,000 ml
pH adjusted with 1 N NaOH to 7.3–7.4	

[a] Modified from Steiger and Steiger (1976).
[b] A, B, E from frozen stock solutions: 2×, 5×, and 100× (BME Vitamins, GIBCO), respectively. Fat-soluble vitamins (F) first dissolved in 1 ml EtOH. Hemin and additional folic acid (**) added in 1 N NaOH. L-Glutamine (*) and the other components weighed indirectly. Defatted bovine albumin (Fraction V from Bovine Plasma, METRIX, Armour Pharmaceutical Co., Chicago) prepared according to Cross and Manning (1973, *Parasitology* **67**:315–331). All other reagents (purest grades available) obtained from Sigma Chemical Co. and Calbiochem.

The hemin could be reduced to 2.5 mg/liter and the folic acid to 1 mg/liter. Adenine could be substituted for adenosine but only in the presence of glucose. With adenosine present, glucose could be omitted.

A defined medium of greater complexity that supported growth of procyclic forms of some strains of *Trypanosoma brucei* was modified and somewhat simplified to give a defined medium for epimastigotes of *T. cruzi*. This medium, designated AR-103 (Table 7.2), supported continuous growth of at least five strains of *T. cruzi*. It must be emphasized that for those parasites with two hosts, a vertebrate and an insect vector, only the insect stage has been grown in the defined medium, and that at present we remain ignorant of the nutritional requirements of the vertebrate stages (see Chapter 8).

Comparison of the composition of medium REI (Table 7.1) with media used for axenic culture of free-living protozoa such as *Paramecium*, or for insects such as *Drosophila* and mosquitoes, indicates similar nutritional requirements for all, and indeed most of the requirements are not very different from those of rats. Minimal requirements for amino acids depend in part on other components of the medium. Thus, *Crithidia fasciculata* (an alimentary tract parasite of mosquitoes) could grow with only 11 amino acids (Table 7.3), the 10 essential for rats plus tyrosine. The concentrations of these amino acids, however, could be materially reduced in the presence of six additional nonessential amino acids (Table 7.3). For the promastigotes of *Leishmania tarentolae*, proline was found to play a major role. In defined media, either proline or glucose could be omitted, but not both (Table 7.4). If glucose was omitted, both proline and glutamic acid were required, in addition to the 10 amino acids essential under all conditions. However, if proline was omitted, glucose being present, there was a further requirement for methionine, isoleucine, alanine, and aspartic acid. These flagellates require an exogenous source of purine, adenosine probably being adequate for all of them. A re-

TABLE 7.2. Defined Medium AR-103 for *Trypanosoma cruzi* (pH 7.5)[a]

Compound	g/liter	Compound	mg/liter
Fructose	2.0	Hemin	15.0
β-Na-glycerophosphate·$5H_2O$	20.0	Adenine	50.0
NaCl	4.0	Adenosine	20.0
$Na_3PO_4 \cdot 12H_2O$	5.0	Guanine-HCl	1.5
KCl	0.4	Guanosine	20.5
Na_3-citrate·$2H_2O$	0.6	Hypoxanthine	1.5
Na-acetate·$3H_2O$	0.79	Xanthine	1.5
Na-succinate	0.27	AMP	1.0
L-Alanine	0.285	ATP	50.0
L-Arginine	0.55	Folic acid	30.0
L-Aspartic acid	0.55	D-α-Tocopherol succinate	4.0
L-Asparagine	0.1	DL-α-Lipoic acid	0.1
L-Cysteine	0.1	Menadione	0.45
L-Cystine	0.14	Thiamine	10.0
L-Glutamic acid	1.045	Nicotinamide	10.0
L-Glutamine	0.1	Nicotinic acid	0.125
L-Histidine-HCl	0.26	Ca-Pantothenate	10.0
Glycine	0.35	Pyridoxine-HCl	0.125
L-Isoleucine	0.4	Pryidoxal-HCl	10.0
Glutathione, reduced	0.0025	Inositol	20.0
L-Hydroxyproline	0.05	Riboflavin	10.0
L-Leucine	0.76	Biotin	10.0
L-Lysine-HCl	0.75	Ascorbic acid	0.25
L-Methionine	0.175	*p*-Aminobenzoic acid	0.25
L-Phenylalanine	0.375	Choline Cl	12.0
L-Proline	0.78		
L-Serine	0.285		
L-Threonine	0.27		
L-Tryptophan	0.14		
L-Tyrosine ethylester	0.36		
L-Valine	0.485		

[a]Modified from Azevedo and Roitman (1984).

quirement for each of the water-soluble B vitamins, biotin, folic acid, nicotinamide, pantothenic acid, pyridoxal riboflavin, and thiamine, as well as for choline, has been demonstrated for at least several species and probably applies to all. Menadione, vitamin A, ascorbic acid, and vitamin B_{12} seem not to be required. These parasites have, however, two interesting growth factor requirements not known to be shared with many other kinds of organisms. They require hemin and, in addition to folic acid, a pteridine. The pteridine requirement has been shown for *C. fasciculata* and for *L. tarentolae*, but probably applies to all, since a high level of folate masks the biopterin requirement. Thus, with biopterin at 1.7 ng/ml *L. tarentolae* required folic acid at only 0.34 ng/ml, whereas folic acid had to be present at 1.7 ng/ml in the absence of a pteridine. The minimal hemin requirement has been estimated at only 200

TABLE 7.3. Amino Acid Requirements of *Crithidia fasciculata*[a]

Amino acid	Concentration (μg/ml)	
	In defined medium in absence of any nonessential amino acids	For maximal growth in presence of nonessential amino acids[b]
L-Histidine-HCl	210	80
L-Phenylalanine	500	80
DL-Isoleucine	630	320
DL-Valine	660	200
L-Leucine	970	160
L-Lysine	760	160
L-Arginine-HCl	430	350
L-Tyrosine	200	60
DL-Methionine	340	180
L-Tryptophan	120	24
DL-Threonine	440	80

[a] Data from Kidder and Dutta (1958).
[b] There were present (in μg/ml): DL-alanine, 550; L-aspartic acid, 610; glycine, 50; L-glutamic acid, 1165; L-proline, 770; DL-serine, 440.

ng/ml for *L. tarentolae* but for *L. donovani* and *L. brasiliensis* it is about 1250 ng/ml.

There are, however, three species of trypanosomatid flagellates from insects that have no hemin requirement and can be grown in very simple defined media with perhaps two amino acids and a few vitamins. It has turned out that all three, *Crithidia oncopelti, C. deanei,* and *Blastocrithidia culicis,* harbor intracellular bacteria. When the flagellates were rendered aposymbiotic by appropriate treatment with antibiotics, they had all of the nutritional requirements of related forms not having symbiotic bacteria, and indeed additional ones. Aposymbiotic *C. oncopelti* and *B. culicis* required a factor present in liver and not yet identified. *C. deanei* free from its symbiotes required an unusually high level of nicotinamide, 30 to 50 mg/liter, as compared to the usual level in defined media of 1 to 5 mg/liter. Evidently, the symbiotic bacteria play an essential biosynthetic role for their host cells. Already it has been shown that they supply the enzymes for heme synthesis in *C. oncopelti* and *B. culicis* and those of the urea cycle in *B. culicis* and *C. deanei.* This integration of intracellular microbes into the economy of their host is a subject of major importance to which I will return in Chapter 24.

Other than the trypanosomatids, there is only one group of parasitic protozoa for which nearly defined and defined media have been developed. These are the trichomonad flagellates, a very widespread group of microaerophilic organisms inhabiting the alimentary tract and other cavities of animals. Most important are the reproductive tract parasites *Tritrichomonas foetus* of

TABLE 7.4. Effects of Amino Acid and Glucose Deletion on Growth of *Leishmania tarentolae* in Defined Medium C[a]

Amino acid	Concentration (μg/ml)	Growth[b]		
		In absence of amino acid[c]	In absence of both amino acid and proline[c]	In absence of both amino acid and glucose
L-Histidine	150	−	−	−
DL-Phenylalanine	400	−	−	−
L-Valine	500	−	−	−
L-Leucine	1500	−	−	−
L-Lysine	1250	−	−	−
L-Arginine	300	−	−	−
L-Tyrosine	400	−	−	−
L-Tryptophan	200	−	−	−
L-Threonine	500	−	−	−
DL-Serine	400	−	−	−
L-Proline	500	+		−
L-Glutamic acid	1900	+	−	−
L-Methionine	300	+	−	+
L-Isoleucine	600	+	−	+
L-Alanine	700	+	−	+
L-Aspartic acid	1200	+	−	+
Glycine	100	+	+	+

[a] Data from Krassner and Flory (1971).
[b] + = continuous growth through at least six successive subcultures.
[c] Glucose present at 5 mg/ml.

cattle and *Trichomonas vaginalis*, a cause of serious vaginitis in women. *T. vaginalis* has recently been grown in a newly formulated defined medium (Table 7.5), supplemented (in place of serum) with a mixture of Fraction V bovine albumin with cholesterol and DL-glyceryl-1-palmitate-2-oleate-3-stearate. Furthermore, in experiments with another defined medium where the albumin was supplied as an essentially fatty acid-free preparation, continuous growth, though at a slow rate, was obtained with palmitic, oleic, and stearic acids supplied at final concentrations of 66 μg/ml and cholesterol at 20 μg/ml. The lipid requirement can be satisfied by human low-density lipoprotein, for which the trichomonads have a specific surface receptor. Although the defined media used so far are not minimal, it is clear that *T. vaginalis* requires cholesterol, palmitic, oleic, and stearic acids, a source of purines and pyrimidines in addition to full arrays of amino acids and vitamins. It does not require hemin and probably does not require an unconjugated pteridine. From what is known of the growth in culture of other species of trichomonads, it is likely that their requirements are much like those of *T. vaginalis*. These organisms can use either glucose or maltose as an energy source.

Among the parasitic metazoa, none has been propagated in a defined medium. The closest approach has been with a parasite of insects, *Neoaplectana*

TABLE 7.5. Defined Medium for *Trichomonas vaginalis* (Medium DL 8)[a,b,c]

	Concentration
Salts	
NaCl (mg/liter)	2000
KCl (mg/liter	2000
KH_2PO_4 (mg/liter)	1000
K_2HPO_4 (mg/liter)	500
Ferrous gluconate (Serva) (mg/liter)	55
$MgSO_4 \cdot 7H_2O$ (mg/liter)	750
$CaCl_2 \cdot 6H_2O$ (mg/liter)	350
Trace elements (ml/liter)	5
Amino acids	
Glutamine (mg/liter)	440
MEM amino acid solution 50 × (Gibco) (ml/liter)	30
MEM nonessential amino acid solution 100 × (Gibco) (ml/liter)	15
Nucleic acid precursors (mg/liter)	
Adenine sulfate	70
Guanosine	30
Uridine	50
Cytidine	50
Thymidine	15
Carbohydrates (mg/liter)	
Maltose	5000
DL-α-Glycerophosphate $diNa \cdot 6H_2O$	100
Ascorbic acid	1000
D(+)-Glucosamine HCl	100
D(+)-Galactosamine HCl	50
N-Acetyl-D-glucosamine	10
N-Acetylneuraminic acid (Sigma type VI)	10
Vitamins	
Vitamin concentrate C, 10 ml/liter	
Adjust final pH to 7.1 with 1.0 M NaOH	
Osmolality 220 mOsm	
Vitamin stock solution C 100 × concentrate (mg/100 ml)	
d-Biotin	20
D-Pantothenic acid hemi-calcium salt	100
Nicotinamide	50
Pyridoxamine HCl	50
Riboflavin	10
Thiamine HCl	10
p-Aminobenzoic acid	5
Coenzyme Q_{10} (dissolve in ethanol)	1
Choline Cl	10
i-Inositol	100
Reduced glutathione	100
Folic acid (dissolve with NaOH)	20
Vitamin B_{12}	10
Deoxyribose	50
Dithiothreitol	100
Tween 80	100

[a] Data from Linstead (1981).
[b] This medium must be supplemented with albumin, cholesterol, and palmitic, oleic, and stearic acids.
[c] Storage: Vitamin solution C was adjusted to pH 7.4 and stored frozen and in the dark at −20°C in aliquots of 10 ml. Single-strength medium DL 8 was stored in the dark at −20°C in aliquots of 100 ml.

TABLE 7.6. Defined Medium for *Neoaplectana glaseri*, a Nemotode Parasite of the Japanese Beetle[a]

	mg/liter		mg/liter
Part A	0.11	Part C	1560.0
H_3BO_3	0.44	L-Arginine-HCl	520.
$CoSO_4 \cdot 7H_2O$	0.50	L-Histidine	2080.
$CuSO_4$	4.00	L-Isoleucine	3120.
$MnSO_4 \cdot 4H_2O$	10.00	L-Leucine	3120.
$FeSO_4 \cdot 7H_2O$	22.00	L-Lysine-HCl	2600.
$ZnSO_4 \cdot 7H_2O$	26.00	DL-Methionine	1000.
$CaCl_2$	1000.	DL-Phenylalanine	2600.
$MgSO_4 \cdot 7H_2O$	4000.	DL-Threonine	800.
NaCl	1250.	DL-Tryptophan	2600.
Na_2HPO_4	500.	DL-Valine	
KH_2PO_4	5000.	Part D	700.
Glucose		DL-Alanine	1200.
Part B	0.20	DL-Aspartic acid	1900.
Biotin	1.60	L-Glutamic acid	100.
Folic acid	2.00	Glycine	500.
Pyridoxal	2.00	L-Proline	400.
Pyridoxamine	2.00	DL-Serine	400.
Pyridoxine	2.00	L-Tyrosine	
Riboflavin	2.00	Part E	160.
Thiamine	8.0	Butyric acid	380.
Ca-pantothenate	3.0	Citric acid	110.
para-Aminobenzoic acid	5.0	Fumaric acid	180.
Nicotinamide	3.0	DL-Lactic acid	130.
Inositol	3.0	L-Malic acid	160.
Choline chloride	1.7	Pyruvic acid	110.
Adenine	1.7	Succinic acid	1500.
Guanine	1.7	Taurine	0.2
Xanthine	1.7	Urea	
Uracil	0.5		
Cytidylic acid			

[a] This medium (Jackson, 1962) was based on a defined medium that supported growth of the hemoflagellate *Leishmania tarentolae*. The concentrations of the essential amino acids (Part C) were, however, materially increased and the ingredients of Part E were added. In this medium, third-stage larvae developed into adults and gave low (2×–3×) but consistent levels of reproduction.

glaseri. This nematode, which invades its host as free-living second-instar "dauer larven," then replicates extensively to produce again a large crop of such resistant larvae that leave the now dead and depleted host to search for new ones. It was the first metazoan parasite obtained in continuous culture. In defined medium (Table 7.6) it will give only one reproductive cycle. Not all the ingredients of this complex medium have been investigated, but it has been determined that the vitamins of the B group and the amino acids arginine, histidine, isoleucine, leucine, and lysine are required at the relatively high concentrations used.

The fact that most parasites of animals have been difficult to cultivate *in vitro* apart from their host, and that most of those so cultured required complex nutritive media, suggests that they may have requirements either for special preformed nutrients not known to be required by any free-living organism, or for a very narrowly defined balance of certain of the known essential nutrients, or for both. As we have just seen, with the few species so far grown in defined media, no striking new kinds of growth factors have as yet been uncovered. Hemin is known to be required by certain free-living invertebrates, cholesterol is essential for insects, pteridine has been found essential for the nonparasitic ciliate *Paramecium aurelia*.

Information about nutritional requirements can also be obtained in ways not necessarily dependent on cultivation or on culture in a defined medium. If the biosynthesis of a nutrient by the parasite without its host can be demonstrated or if the enzymes of a biosynthetic pathway can be shown to be present in the parasite, then the substance synthesized can be ruled out as a nutritional requirement. If the biosynthetic pathways are absent, then the substance is probably required as a nutrient. For example, malarial parasites have the enzymes for pyrimidine synthesis but not for purine synthesis. A preformed purine is required, just as has been shown by direct nutritional experiments for hemoflagellates. That the purine is probably utilized as the free base (rather than as a nucleoside) was shown for another type of intracellular parasite (*Toxoplasma gondii*) by using host cells incapable of incorporating hypoxanthine or guanine. E. Pfefferkorn used Lesch-Nyhan cells as the hosts for *Toxoplasma* and showed that labeled hypoxanthine and guanine in the medium were directly incorporated into the intracellular parasites. This illustrates yet another approach to studying the nutritional needs of obligate parasites—the use of hosts lacking certain biosynthetic pathways.

The requirement for a growth factor may be so large or so unusual that it becomes apparent even if the organism cannot be grown in a defined medium, or sometimes even if it has not been grown *in vitro* at all. A particularly interesting example is provided by the vitamin B_{12} requirement of certain helminths. This was first noted with several cestodes as a high uptake and concentration of B_{12}. It is this excessive uptake of B_{12} by the fish tapeworm *Diphyllobothrium latum* that is responsible for the macrocytic anemia that develops in many individuals harboring this tapeworm. The anemia resembles that of pernicious anemia, where there is a defect in absorption of B_{12}. Here the worm (which is acquired from eating raw infected freshwater fish) steals the B_{12} from the host's alimentary tract. B_{12} is also concentrated in specific cells of various species of nematodes, including the large *Ascaris suum*. In this worm it has been shown to be present as a typical cobamide coenzyme. B_{12} coenzyme is required in certain reactions of importance in the metabolism of *Ascaris*, as in propionate formation (see Chapter 9). But why it should exist in such exceptionally high concentrations, enough to give a pinkish color, in certain cells of some species of helminths, remains totally unknown.

The question as to whether the parasitic mode of life might entail special nutrient requirements seems especially relevant with regard to intracellular parasites. The chlamydiae and certain rickettsiae, prokaryotic intracellular parasites, make use of the host cell's energy-supplying system, the ATP. Among the intracellular parasitic protozoa, one species, the bird malaria *Plasmodium lophurae,* can be kept developing extracellularly *in vitro* for a short period in a complex medium containing a duck erythrocyte extract. Under these conditions, the survival and development of the parasites required external sources of both ATP and coenzyme A. These substances therefore may be regarded as additional nutrients that these parasites are able to obtain as a result of their intracellular position. Pantothenic acid is among the growth factors essential for the development of malarial parasites within their host erythrocytes, and indeed even in an intact host. Thus, either pantothenic acid deficiency or antagonists of pantothenic acid will inhibit growth of the parasites. In the extracellular condition, however, *P. lophurae* requires the complete CoA molecule; it lacks the enzymes for the biosynthesis of CoA from pantothenate.

Whether other intracellular parasitic protozoa similarly lack this biosynthetic pathway, or others of equal importance, remains for future work to determine. It could well be that they have the genetic information for these pathways but fail to express them during certain stages of the life cycle.

In general, present knowledge permits us only to say that parasitic animals have at least the same nutrient requirements as are known for some of the major groups of free-living animals, mammals, birds, insects, and certain protozoa. These requirements would include essential minerals including trace elements, an energy source (e.g., glucose), certain essential amino acids, vitamins of the water-soluble B group, a few fatty acids and cholesterol, and other factors such as a purine source and hemin. Since, however, only a relatively few species of parasites have been grown apart from a living host through their complete life cycle, and since these require complex media and even living host cells, there remains the strong possibility that some parasites must obtain from their host unusual growth factors. These could be in the nature of coenzymes, readily available to intracellular parasites, or hormone-like factors that trigger differentiation in successive stages of the life cycle.

Bibliography

Azevedo, H. P., and Roitman, I., 1984, Cultivation of *Trypanosoma cruzi* in defined media, in: *Genes and Antigens of Parasites: A Laboratory Manual* (2nd edition) (C. M. Morel, ed.), UNDP/World Bank/WHO Special Programme for Research and Training in Tropical Diseases, Fundaçao Oswaldo Cruz, Rio de Janeiro, Brazil.

Jackson, G. J., 1962, The parasitic nematode, *Neoaplectana glaseri,* in axenic culture. II. Initial results with defined media, *Exp. Parasitol.* **12:**25–32.

Kidder, G. W., and Dutta, B. M., 1958, The growth and nutrition of *Crithidia fasciculata, J. Gen. Microbiol.* **18:**621–638.

Krassner, S. M., and Flory, B., 1971, Essential amino acids in the culture of *Leishmania tarentolae, J. Parasitol.* **57:**917–920.

Linstead, D., 1981, New defined and semi-defined media for cultivation of the flagellate *Trichomonas vaginalis*, *Parasitology* **83:**125–137.
Nyberg, W., 1952, Microbiological investigations on antipernicious anemia factors in the fish tapeworm, *Acta Med. Scand. Suppl.* **271:**1–71.
Oya, H., and Weinstein, P. P., 1975, Demonstration of cobamide coenzyme in *Ascaris suum*, *Comp. Biochem. Physiol. B* **50:**435–442.
Peterson, K. M., and Alderete, J. F., 1984, *Trichomonas vaginalis* is dependent on uptake and degradation of human low density lipoproteins, *J. Exp. Med.* **160:**1261–1272.
Pfefferkorn, E. R., and Pfefferkorn, L. C., 1977, *Toxoplasma gondii:* Specific labeling of nucleic acids of intracellular parasites in Lesch-Nyhan cells, *Exp. Parasitol.* **41:**95–104.
Pollack, S., and Fleming, J., 1984, *Plasmodium falciparum* takes up iron from transferrin, *Br. J. Haematol.* **58:**289–293.
Sherman, I. W., 1979, Biochemistry of *Plasmodium* (malarial parasites), *Microbiol. Rev.* **43:**453–495.
Shorb, M. S., 1964, The physiology of trichomonads, in: *Biochemistry and Physiology of Protozoa*, Volume 3 (S. H. Hutner, ed.), Academic Press, New York, pp. 383–457.
Steiger, R. F., and Black, C. D. V., 1980, Simplified defined media for cultivating *Leishmania donovani* promastigotes—Short communication, *Acta Trop.* **37:**195–198.
Steiger, R. F., and Steiger, E., 1976, A defined medium for cultivating *Leishmania donovani* and *L. braziliensis*, *J. Parasitol.* **62:**1010–1011.
Trager, W., 1983, *In vitro* growth of parasites, in: *Molecular Biology of Parasites* (J. Guardiola, L. Luzzatto, and W. Trager, eds.), Raven Press, New York, pp. 39–51.
Wallbanks, K. R., Maazoun, R., Canning, E. U., and Rioux, J. A., 1985, The identity of *Leishmania tarentolae* Wenyon 1921, *Parasitology* **90:**67–78.

CHAPTER 8

Cultivation of Parasites *in Vitro* with Special Reference to Differentiation in the Life Cycle

As discussed in Chapter 2, most parasites have forms specially adapted for transmission from one host to another. What controls differentiation of these forms is poorly understood. One approach to study of this important subject is to cultivate the organism *in vitro* and attempt to duplicate the developmental cycle. This has generally proved difficult to do.

A simple illustration is provided by the intestinal parasites *Entamoeba histolytica* and *Giardia lamblia*. In nature both of these organisms depend on resistant cysts, capable of survival in the external environment, for the infection of new hosts. Both organisms have been grown axenically *in vitro* in complex media but encystation has never been obtained. Both form cysts in infected hosts and *E. histolytica* will also form cysts in cultures grown with bacteria and provided with rice starch. The special factors responsible for differentiation of the resistant cysts are unknown. However, for another species of *Entamoeba, E. invadens* parasitic in reptiles, axenic culture has given us some insight as to conditions needed for encystation.

When trophozoites of *E. invadens* growing axenically in Diamond's TP-S medium (Table 8.1) were transferred to a medium containing only phosphate buffer, trypticase, yeast extract, and dialyzed serum, mass encystation occurred within about 30 hr, 70 to 90% of the organisms encysting. The best encystation occurred if the amebae were transferred during the most active logarithmic phase of their vegetative growth. The results suggest that lowered osmotic pressure may be one of the stimuli to encystation *in vitro*.

The situation becomes more complex with parasitic protozoa having two different kinds of hosts, each serving for infection of the other. The leishmanias, trypanosomes, and malarial parasites provide good examples which I will discuss in detail.

In the life cycle of the leishmanial parasites (Diagram III) there are two forms: (1) the elongate, usually flagellated promastigotes found in the gut

TABLE 8.1. Media for Axenic Cultivation of *Entamoeba histolytica* and Related Amebae[a]

I. Diamond's TP-S-1 medium

A. Composition of the nutrient broth TP

Trypticase (BBL)	1.00 g
Panmede liver digest	2.00 g
Glucose	0.50 g
L-Cysteine-HCl	0.10 g
Ascorbic acid	0.02 g
NaCl	0.50 g
KH_2PO_4	0.06 g
K_2HPO_4	0.10 g
H_2O (glass-distilled) to make	87.5 ml
pH adjusted to 7.0 with 1 N NaOH	

B. To the autoclaved TP broth are added aseptically 10 ml of horse serum and 2.5 ml of Vitamin mix 107 (see I.C).

C. Composition of Vitamin mix 107

	mg/liter	Final concn (mg/100 ml medium)
Niacin	16.7	0.04
p-Aminobenzoic acid	33.3	0.08
Niacinamide	16.7	0.04
Pyridoxine-HCl	16.7	0.04
Pyridoxal-HCl	16.7	0.04
Thiamine HCl	6.7	0.017
Ca-pantothenate	6.7	0.017
i-Inositol	33.3	0.08
Choline Cl	333.0	0.80
Riboflavin	6.7	0.017
Biotin	6.7	0.017
Folic acid	6.7	0.017
Vitamin D_2 (calciferol)	66.7	0.17
Vitamin A (crystalline alcohol)	66.7	0.17
Vitamin K (menadione)	12.0	0.03
Vitamin E (α-tocopherol acetate)	6.7	0.017

II. Diamond's TY-I-S-33 medium. Because the Panmede liver digest has given irregular results, it has been replaced in a new and improved medium of the following composition:

A. Composition of the nutrient broth TYI

Trypticase (BBL)	2 g
Yeast extract (BBL)	1 g
Glucose	1 g
NaCl	200 mg

TABLE 8.1. *(continued)*

K_2HPO_4	100 mg
KH_2PO_4	60 mg
L-Cysteine-HCl	100 mg
L-Ascorbic acid	20 mg
Ferric ammonium citrate	2.28 mg
H_2O (glass distilled)	87 ml
pH adjusted to 6.8 with 1 N NaOH	

B. To the autoclaved TYI broth (87 ml) are added 3 ml of vitamin–Tween 80 mixture (see II.C) and 10 ml of bovine serum.

C. Vitamin–Tween 80 mixture. This consists of 1000 ml of Vitamin mix 107 (see I.C) plus the following:

12 ml vitamin B_{12} solution (40 mg/100 ml H_2O)
4 ml DL-6,8-thioctic acid solution (100 mg in 100 ml absolute ethanol)
4 ml Tween 80 solution (50 g in absolute ethanol to 100 ml)
180 ml glass-distilled water

[a] For the details of preparation of these media it is essential to consult the original papers. See Diamond *et al.* (1978).

and mouthparts of the vector sand fly; (2) the smaller, more rounded amastigotes lacking an external flagellum, that multiply intracellularly in the vertebrate host. The transformation of *Leishmania mexicana,* a dermatropic species, from amastigote to promastigote, will occur *in vitro* at 26°C in a suitable medium and requires about 48 hr. Not only are glucose and certain amino acids essential but also the nonesterified fatty acids and possibly other lipids as present in fetal calf serum. This transformation, in which a 3- to 4-μm amastigote changes into a 12- to 16-μm promastigote, proceeds via three fairly well-defined intermediate stages. It requires energy, protein synthesis, and in the third intermediate stage a cell division. It is relevant to note that flagellum development by promastigotes is affected by levels of certain growth factors. Thus, promastigotes of *L. tarentolae* in a defined medium with a suboptimal level of choline multiplied as well as with higher choline but they became aflagellate.

The reverse change from promastigote to amastigote likewise involves changes in protein synthesis, e.g., decrease in synthesis of tubulins and increase in other proteins. Until very recently this change could not be observed except by allowing promastigotes to infect living host cells, notably macrophages, in tissue culture at temperatures of 33–35°C. Such temperatures are lethal to promastigotes unless they have been gradually acclimatized to them, in which case they continue to grow as promastigotes—they do not transform to amastigotes. In 1984, A. A. Pan reported a new medium (Table 8.2) in which promastigotes of *L. mexicana* at 33 or 35°C change rapidly into amas-

TABLE 8.2. Medium for Growth of Amastigotes of *Leishmania* sp. at 33 or 35°C (JH-30)[a,b]

Basal solution (see A below)		92.1 ml
Hepes 1.0 M	Sterilized by filtration	1.0
$NaHCO_3$ 4.2%	Autoclaved 20 min at 20 lb	3.0
Redistilled water	Autoclaved 20 min at 20 lb	39.5
	Total	135.6

A. Composition of the Basal solution. Mixed aseptically.

Trypticase 5%	Autoclaved 20 min at 20 lb	10.0
Methylcellulose 4%	Autoclaved 20 min at 20 lb	5.0
M 199 (10 × concentrated)		10.0
Glucose 8%	Sterilized by filtration	2.5
Vitamin mix (see B below)	Sterilized by filtration	2.0
Folic acid 10 mg/100 ml (dissolved in 0.008% NaOH)	Sterilized by filtration	5.0
Biotin 4 mg/100 ml (dissolved in 0.083% HCl)	Sterilized by filtration	2.0
Vitamin B_{12} 1 mg/100 ml	Sterilized by filtration	2.0
Sodium pyruvate 0.1 M	Sterilized by filtration	3.0
Penicillin 100,000 U/ml	Sterilized by filtration	0.1
Hemin 50 mg/100 ml (dissolved in 0.04% NaOH)	Sterilized by filtration	5.0
Nucleotide mix (see C below)	Sterilized by filtration	10.0
Fetal bovine serum (56°C, 30 min)		33.5
L-Glutamine 3%	Sterilized by filtration	2.0
	Total	92.1 ml

B. Composition of the Vitamin mix

Redistilled water	100 ml
p-Aminobenzoic acid	30 mg
D-Ca-pantothenate	40
Choline	30
i-Inositol	30
Nicotinamide	50
Nicotinic acid	20
Pyridoxal-HCl	20
Pyridoxine-HCl	20
Pyridoxamine-2HCl	20
Riboflavin-5-phosphate-Na-$2H_2O$	4
Thiamine-HCl	20

C. Composition of the Nucleotide mix

Redistilled water	100 ml
ATP	200 mg
ADP	100
AMP	100
Glutathione (reduced)	20
L-Cysteine-HCl	20
Ascorbic acid	20

[a] From Pan (1984).
[b] The pH of this medium is 7.2.

tigotes and thereafter grow continuously as amastigotes (Fig. 8.1). Within 96 hr, after start of the culture with promastigotes, 72% of the organisms were already amastigotes, and by the next 96 hr, in first subculture, 99% were amastigotes. The amastigotes in these cultures had an average size of 3.5 ± 0.2 μm as compared to 3.2 ± 0.5 μm for amastigotes from an infected hamster footpad. They resembled amastigotes in all other properties so far tested, including notably their high infectivity to macrophages. In the absence of either the Vitamin mix or the Nucleotide mix (Table 8.2 B and C, respectively), growth was much reduced, suggesting that the amounts of vitamins and nucleotides supplied by the M 199 tissue culture medium are inadequate. M 199 supplies B vitamins in the general range of about 0.02 to 0.05 mg/liter, whereas the final mixture shown in Table 8.2 would supply them at around 4 mg/liter. Similarly where M 199 furnishes ATP at 1.0 mg/liter, the JH-30 medium supplies it at 150 mg/liter. It will be of great interest for future work to determine whether growth of amastigotes requires simply a higher level of a purine source, which could be supplied as AMP or adenosine, or there is a need for ATP as is the case with the erythrocytic stages of malaria when maintained extracellularly.

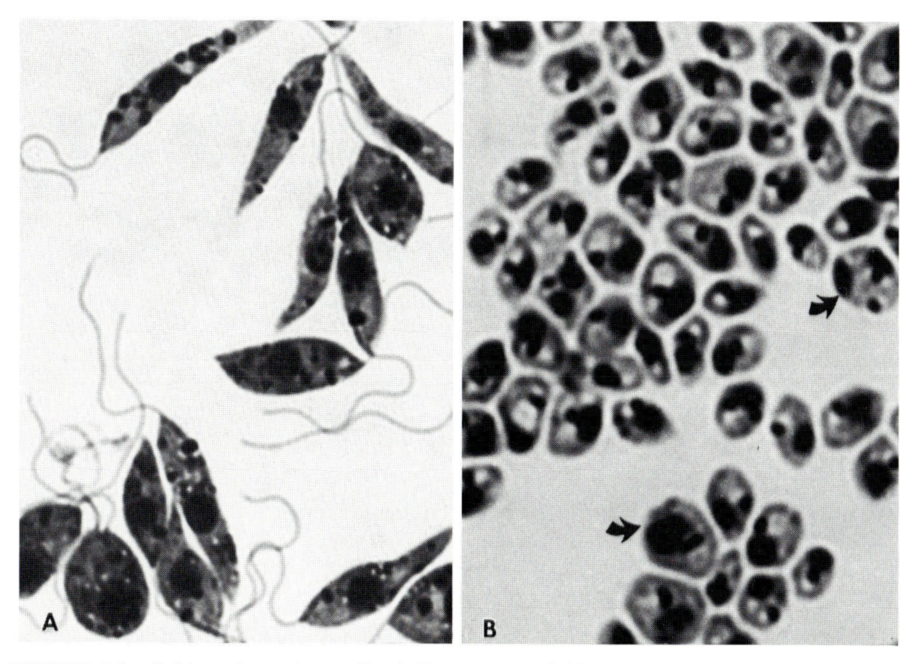

FIGURE 8.1. *Leishmania mexicana pifanoi* Giemsa-stained films. (A) Promastigotes in medium JH-31 at 26°C. ×2300. (B) Amastigotes in medium JH-30 at 33°C. Arrows indicate dividing amastigotes with two nuclei and two kinetoplasts. ×2100. (From Pan, 1984. Original prints courtesy of Dr. A. A. Pan.)

Somewhat analogous results have been obtained with the related hemoflagellate *Trypanosoma cruzi* (Diagram IV). This organism, the cause of Chagas' disease in South and Central America, is transmitted by bloodsucking bugs of the family Triatomidae and has a more complex developmental cycle than that of *Leishmania*. The forms multiplying in the alimentary tract of the bug, chiefly epimastigotes, are easily grown in axenic culture at 28°C, but they are of low infectivity to the vertebrate host. Infective metacyclic trypomastigotes appear in the hindgut of the bug and have been obtained in axenic culture at 26°C under certain conditions. The amastigotes and bloodstream trypomastigotes, forms that in nature develop intracellularly in the vertebrate host, can be easily grown in tissue cultures at 35–37°C. Here they multiply intracellularly as amastigotes until the cell is filled, when they transform to trypomastigotes. This development at 37°C has also been obtained under axenic conditions in two different types of culture media. One medium, used by S. C.-T. Pan, is defined except for the 10% of fetal bovine serum, and in this medium a supplement containing ATP to give a final concentration of 1 mg/100 ml appeared essential. The other medium, used by F. Villalta and F. Kierszenbaum, contained tryptose phosphate broth as well as fetal bovine serum, each at 10%. Both media supported the extracellular growth at 37°C of amastigotes of *T. cruzi* that could not be distinguished from amastigotes found in cells in tissue culture or in a host animal (Fig. 8.2).

With the African trypanosomes, where tsetse flies serves as the vector host (Diagram V), the situation is still more complex. Here again the forms that occur in the midgut of the fly, relatively elongate trypomastigotes called procyclic forms, are easily grown axenically in complex media at temperatures around 25°C. Such cultures are completely noninfective to the vertebrate host; they do not contain the morphologically different forms, metacyclic trypomastigotes, found in the salivary glands or proboscis of tsetse flies able to transmit the infection. These infective forms have, however, been produced *in vitro* in the presence of appropriate kinds and amounts of surviving tissue of tsetse flies.

Recently, the entire life cycle of the important pathogenic trypanosome of cattle, *T. vivax*, has been obtained in culture in the presence of a feeder layer of vertebrate tissue cells with incubation at 25 and 37°C for the fly and vertebrate cycles, respectively. Enough is now known to permit us to say we are in a position to elucidate the precise factors responsible for the morphological and physiological differentiations that transform a noninfective flagellate unable even to survive at the body temperature of a mammal into an infective, pathogenic organism. The bloodstream forms themselves have not yet been cultivated axenically. They have been grown only in the presence of a feeder layer of living host cells. Although the trypanosomes are not intracellular and show no structural attachment to the vertebrate cells, the latter must be actively metabolizing and the trypanosomes must be able to be in direct contact with them. Growth does not occur if the cells are separated by a micropore membrane. If the factors concerned are diffusible ones, they

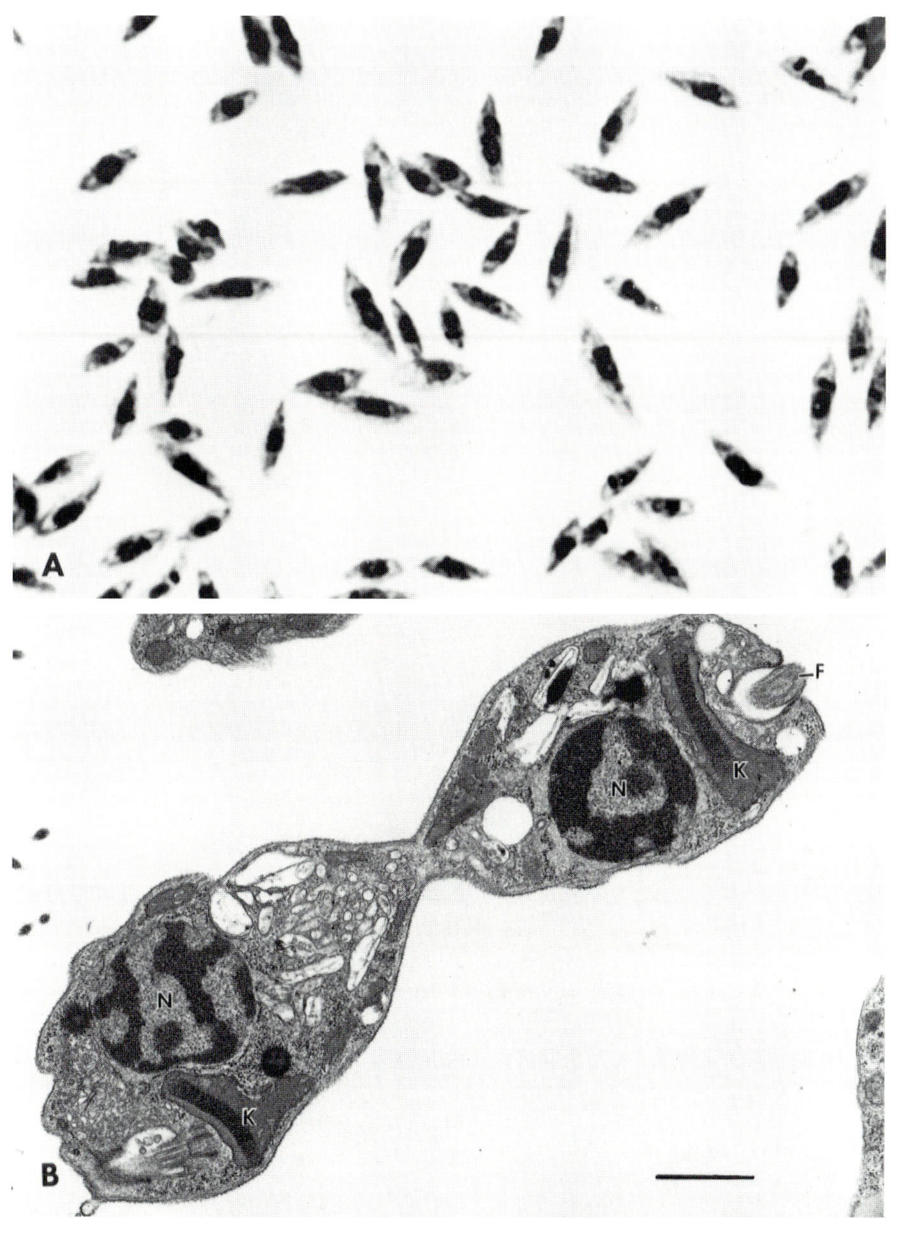

FIGURE 8.2. *Trypanosoma cruzi* (Tulahuen strain) amastigotes from axenic culture in ML-15 HA medium. (A) Giemsa-stained film. ×1200. (B) EM. Amastigote in final phase of division. F, flagellar base; K, kinetoplast; N, nucleus, Bar = 1 μm. (From Villalta and Kierszenbaum, 1982. Original prints courtesy of Dr. F. Kierszenbaum.)

must be highly labile. That this is indeed the case is strongly indicated by recent work from the laboratory of Dr. G. A. M. Cross. The bloodstream forms of *T. brucei* could be grown axenically if the cultures were supplied at regular intervals with appropriate concentrations of L-cysteine. The optimal concentration depended somewhat on cell density and was close to physiological levels in serum. Concentrations of cysteine higher than 2×10^{-4} M (24 mg/liter) were acutely toxic to trypanosome populations of 3×10^7/ml. Cystine could replace cysteine only if accompanied by low concentrations of 2-mercaptoethanol. It may well be that cysteine is the only highly labile metabolite that is essential to the bloodstream trypanosomes and that is provided by the feeder layer.

It is noteworthy that cysteine also serves as a signal for transformation to the tissue phase of the fungal parasite *Histoplasma capsulatum*. This organism exists in a free-living mycelial form in soil. It will change to the yeast phase that occurs in host tissue in response to both elevated temperature and exogenous cysteine. The requirement for cysteine ensures that the transformation in nature will occur only within a host. This requirement also renders the organism sensitive to the sulfhydryl inhibitor *p*-chloromercuriphenylsulfonic acid. This compound was found to prevent the mycelial-to-yeast transition; it had no effect on the reverse transition or on the growth of either form.

Malarial parasites have a life cycle even more complicated than that of trypanosomes and not yet fully reproduced in culture (Diagram VIII). In their vertebrate hosts (lizards, birds, or mammals), the principal propagative cycle, and the one responsible for the human disease malaria, occurs intracellularly within the red blood cells. This is a cycle of asexual reproduction in which a small invasive form, a merozoite, enters an erythrocyte, grows rapidly within it, and then undergoes schizogony (see Chapter 3) to produce 8 to 24 daughter merozoites (depending on the species), each potentially able to invade another erythrocyte and reproduce the cycle. Each cycle takes 24, 48, or 72 hr, again depending on the species. This portion of the life cycle clearly is capable of indefinite propagation as long as the parasites are supplied with appropriate red cells and other environmental conditions. Only relatively recently was this cycle duplicated *in vitro*. The first successful cultures were with the major human malarial parasite *Plasmodium falciparum* (Figs. 8.3, 8.4). The erythrocytic stages of a number of other species have since been grown by the same methods. It is important to note that the parasites are being cultured in suspensions of their living host cells, not axenically. For reasons that are not known, certain of the erythrocytic parasites develop into male and female gametocytes, stages that initiate the developmental cycle in the mosquito. These stages appear in the cultures, and are infective to mosquitoes. Much more work is needed before we will have some idea as to the factors that trigger gametocyte development and those essential for the fertility of the gametes. Probably intrinsic genetic factors as well as environmental ones are involved. Conditions somewhat unfavorable to the asexual development favor

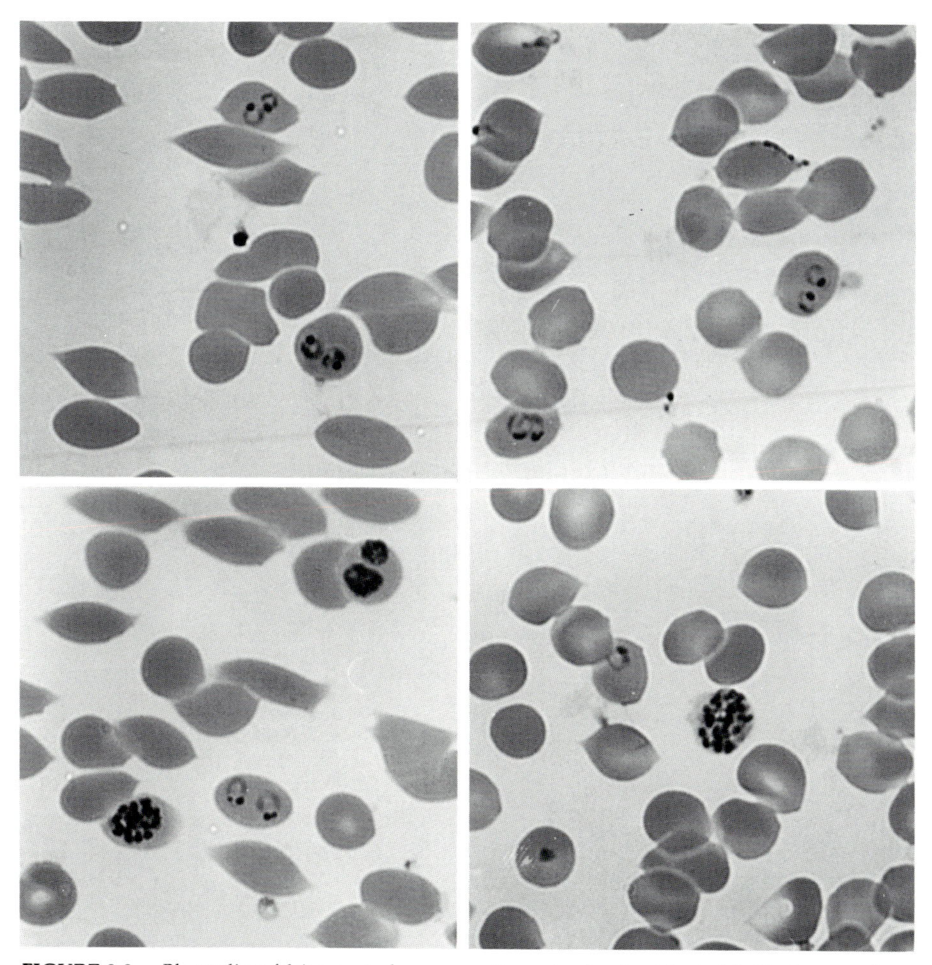

FIGURE 8.3. *Plasmodium falciparum* in human erythrocytes *in vitro*. Typical fields from a culture with 7% parasitemia of line FCR-3 after 14 months of continuous growth *in vitro*. Note rings, trophozoites, late schizonts, some double infections, and, toward the upper right corner, a cell with two appliqué forms. Giemsa-stained films. ×1200. (From Trager, 1978.)

gametocyte formation, but just how this is brought about remains to be discovered. It has long been known that some species of malarial parasites of experimental animals stop forming gametocytes if passaged for long periods by blood inoculation. Similarly, in *in vitro* culture there may well be a selection away from parasites able to form fertile gametocytes. Very similar situations exist with various species of coccidia. The asexual forms of *Eimeria* or *Toxoplasma* are readily propagated in tissue cultures of appropriate host cells, but the sexual forms and the development of oocysts in tissue culture have so far been obtained for only a few species of coccidia. When sporozoites from oocysts developed *in vivo* were used as the inoculum in primary chicken

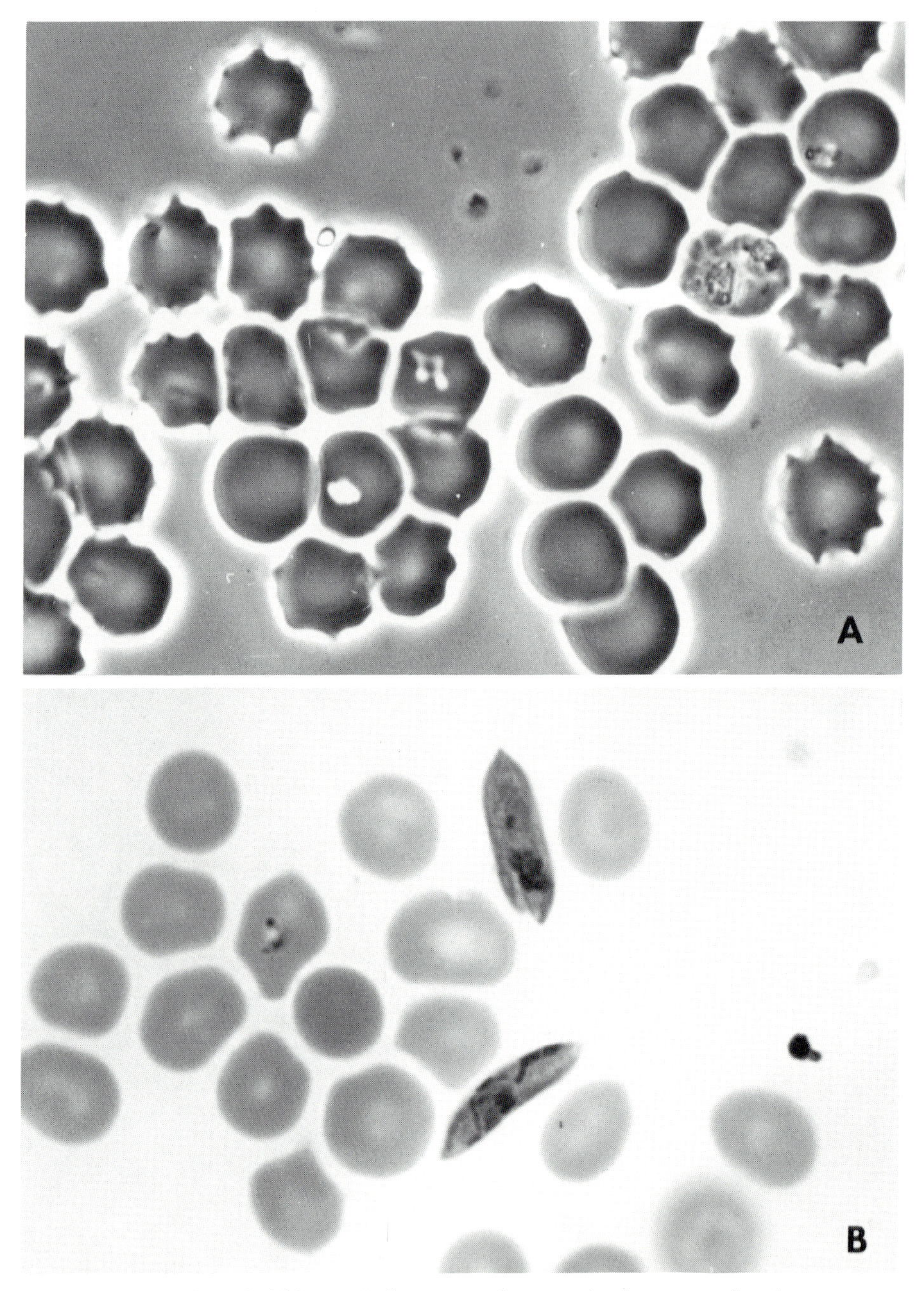

FIGURE 8.4. (A) Live *P. falciparum* in human erythrocytes *in vitro* as seen by phase contrast (× 2000). Note the highly ameboid late ring at the center, several bright trophozoites and a schizont with forming merozoites at the right. (From Trager, 1978.) (B) *P. falciparum* stage IV gametocytes from a culture of a gametocyte-forming clone (HB-3). Giemsa-stained film. ×2000. (Photomicrograph by H. N. Lanners.)

kidney cultures, only *Eimeria tenella* showed oocyst formation. Several other species of *Eimeria* would produce oocysts if the cultures were begun with merozoites taken from an intact host. The complete cycle of development to sporulated infective oocysts has been obtained *in vitro* for the small coccidian parasite *Cryptosporidium*, a species that has recently become prominent as a

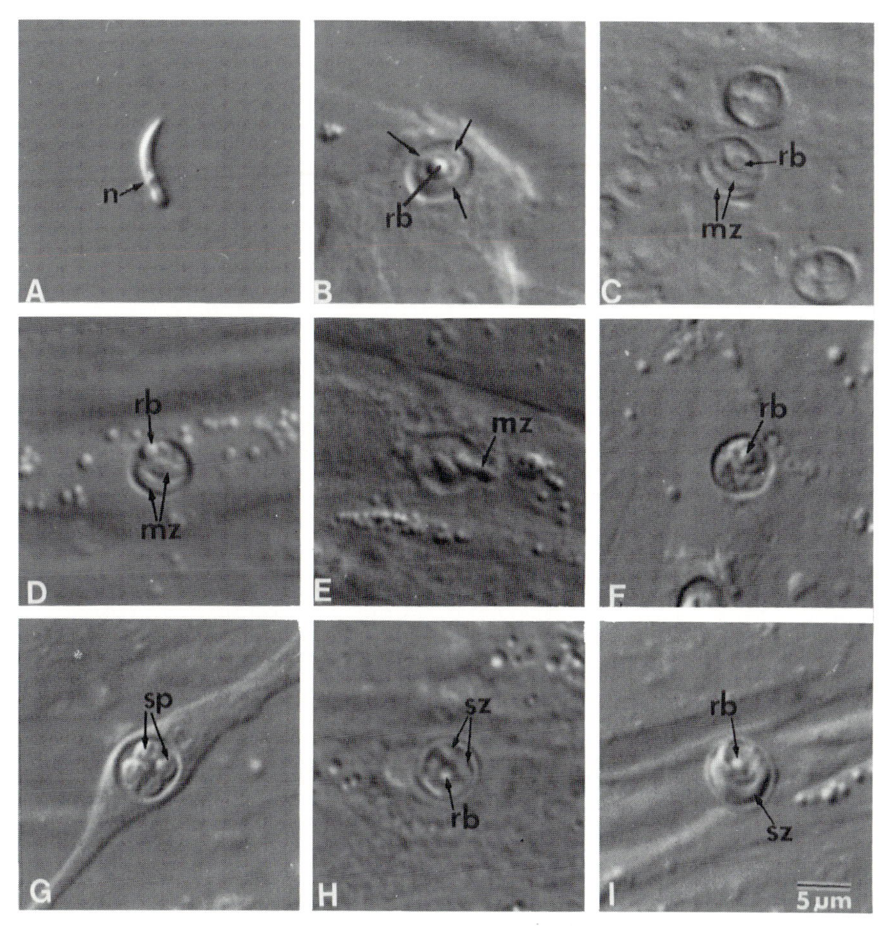

FIGURE 8.5. *Cryptosporidium* in human fetal lung cell cultures inoculated with sporozoites. Nomarski interference contrast photomicrographs of living cells. (A) Sporozoite free in medium 4 hr after inoculation. (B) Immature type I meront with refractile body (rb) and peripheral nuclei (arrows) 12 hr after inoculation. (C) Mature type I meront with merozoites (mz) and refractile body (rb). (D) Mature type II meront with merozoites (mz) and refractile body (rb). (E) Type II merozoite (mz) in process of leaving the parasitophorous vacuole of the host cell. (F) Macrogamete with refractile body (rb) and granular cytoplasm. (G) Oocyst undergoing sporogony in parasitophorous vacuole of host cell. Four sporoblasts are present at this stage. (H) Optical cross section of a fully sporulated oocyst in host cell. Note sporozoites (sz) and refractile body (rb). (I) Side view of fully sporulated oocyst in host cell. Note sporozoite (sz) and refractile body (rb). (From Current and Haynes, 1984. Original print courtesy of Dr. W. L. Current.)

serious pathogen in immunosuppressed people (Fig. 8.5). These difficulties in obtaining sexual development *in vitro* are not so surprising when we realize that even *in vivo* the sexual forms are more fastidious. Thus, whereas the asexual forms of *Toxoplasma* will grow in a wide variety of vertebrate hosts and tissues, the sexual forms develop only in the intestine of felines.

The cycle of malaria in the vector mosquito is a self-limited one that begins with zygote formation by male and female gametes and ends with the formation of sporozoites. Although this development is not intracellular, only limited success has so far attended attempts at its duplication *in vitro.* The third developmental cycle of malarial parasites, the preerythrocytic cycle, begins when infective sporozoites are introduced into a vertebrate host by the bite of an infected mosquito. At this point a sharp distinction has to be made between parasites of birds and those of mammals. The avian malarial parasites invade cells of the reticuloendothelial system, whereas the mammalian ones are restricted to hepatic cells of the liver. Furthermore, most of the avian parasites can undergo repeated cycles of exoerythrocytic development in reticuloendothelial cells, in addition to giving rise to merozoites that initiate the erythrocytic cycle. In keeping with this fact, exoerythrocytic forms of several avian malarias have been continuously propagated in tissue cultures of fibroblastlike cells (Fig. 8.6). The preerythrocytic stages of mammalian malarial parasites apparently undergo within hepatic cells only a single developmental cycle that forms many merozoites that invade erythrocytes. This cycle was first obtained in tissue culture with rodent malarial parasites by M. Hollingdale (see Fig. 5.8) using either hepatic cells or a human lung cell line as hosts for the sporozoites. In recent work, primary cultures of human hepatocytes, infected with sporozoites of either *P. falciparum* or *P. vivax,* have supported development of the preerythrocytic schizonts through to merozoites that were infective to human erythrocytes (Fig. 8.7).

Thus, the complete life cycle of malarial parasites has not yet been propagated *in vitro.* The erythrocytic stages of several species, notably *P. falciparum* and some other primate malarias, can be maintained in continuous culture. Furthermore, differentiation of fertile gametes can be induced. With the sporogonic cycle that follows, only fragmentary development *in vitro* has been obtained. Sporozoites of avian malaria have been grown to exoerythrocytic stages, which in turn can be propagated continuously *in vitro.* The sporozoites of two species of rodent malaria and of *P. falciparum* and *P. vivax* have been grown through the preerythrocytic cycle with formation of infectious merozoites. It is clear that even now we already have material of remarkable interest available for exploitation by the techniques of molecular biology. With the avian species, for example, one could compare exoerythrocytic and erythrocytic stages, with special regard to the differentiation of one into the other. One might ask: what are the factors determining whether a merozoite invades a reticuloendothelial cell or an erythrocyte? With continuous cultures of erythrocytic stages it should be possible to determine both intrinsic and extrinsic factors essential for gametocyte formation. Clones of *P. falciparum* are now

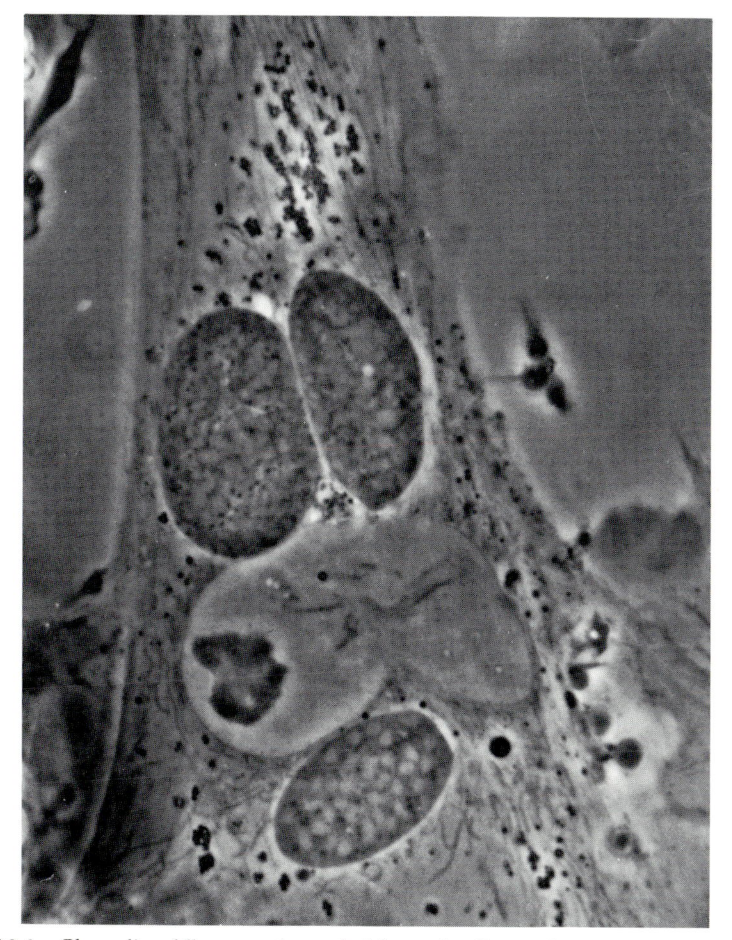

FIGURE 8.6. *Plasmodium fallax,* an avian malarial parasite. Exoerythrocytic stages in continuous culture in chick embryo cells. Three schizonts may be seen, one below and two above the host cell nucleus. ×1000. (From Trager, 1960. Print courtesy of Dr. C. G. Huff. See Huff, 1964.)

available some of which form gametocytes whereas others do not. Possible differences in the DNA of such clones might be of significance (see Chapter 10).

The *in vitro* propagation of helminthic parasites presents problems analogous to those encountered with cultivation of protozoan parasites having complex life cycles. Many nematode parasites have a period of free-living larval development in which they live in feces or soil and feed principally on bacteria. Not surprisingly, these larval stages have been cultured axenically under conditions similar to those used for axenic culture of other free-living invertebrate bacteria-feeders, such as the larvae of mosquitoes and fruit flies.

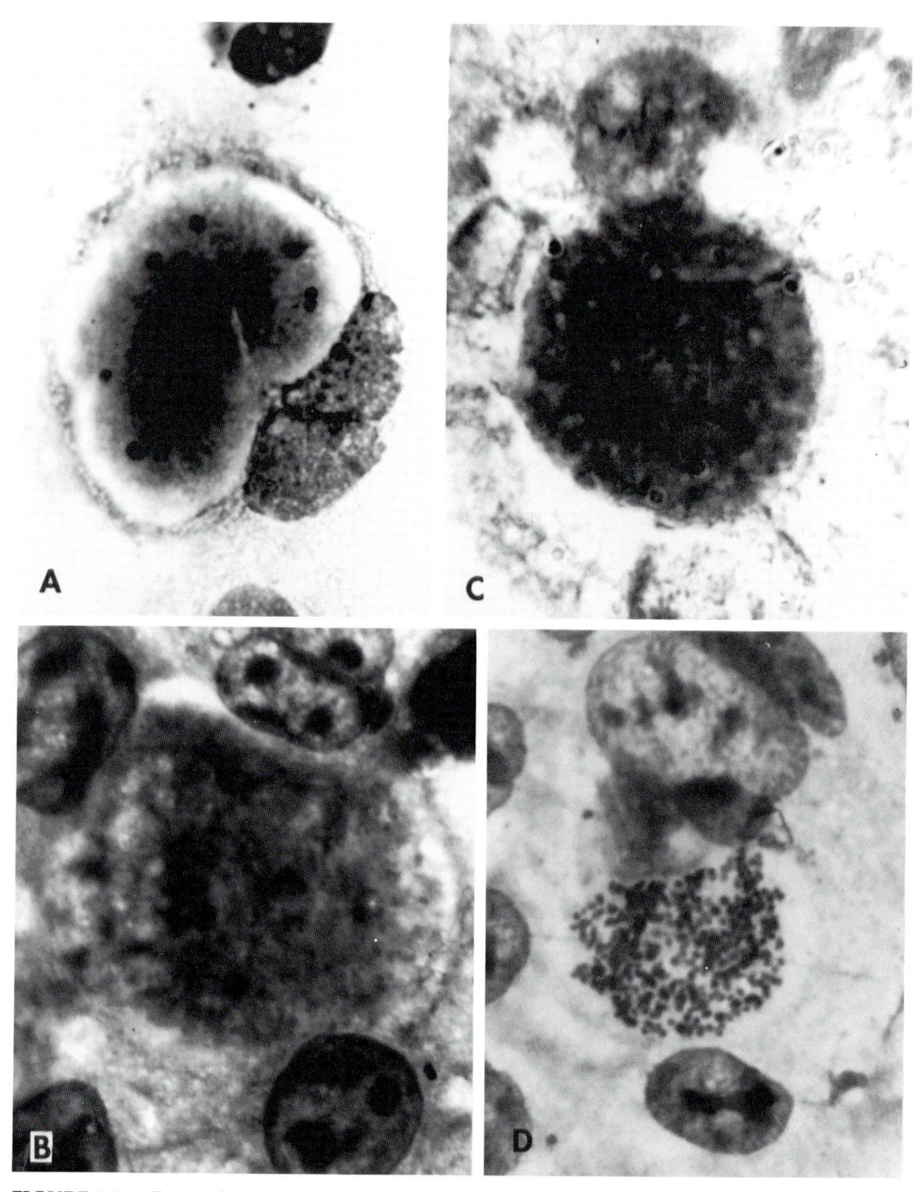

FIGURE 8.7. Exoerythrocytic schizonts of mammalian malarial parasites that have developed *in vitro* from sporozoites. (A) *Plasmodium vivax* 7 days in primary human liver cells. (B) *P. vivax* in hepatoma cell line 6-2. (C) *P. falciparum* 7 days in primary human liver cell culture. (D) *P. berghei* after 72 hr in hepatoma cell line G2-A16, showing numerous fully formed merozoites. All from Giemsa-stained material. ×1700. (A and C from Mazier *et al.*, 1984a,b, respectively; original prints courtesy of Dr. D. Mazier. D from Hollingdale *et al.*, 1983; B and D courtesy of Dr. M. Hollingdale.)

The parasitic stages present more difficulties, especially with regard to the production of fertile adults. Only a relatively few species of parasitic nematodes have been cultured from egg to adult, and with still fewer have fertile adults been produced so as to permit continuous *in vitro* propagation. I have already discussed the continuous culture of the nematode parasites of insects in the genus *Neoaplectana*. These rhabditoid nematodes may be considered facultative rather than obligate parasites. Among the parasites of vertebrates, the closest approach to continuous *in vitro* culture has been with the strongyloid parasite of the abomasum of ruminants *Ostertagia ostertagi*. Infectious free-living larvae obtained from fecal cultures were cleaned and sterilized and then exsheathed in the culture medium using a stimulatory gas mixture of 60% N_2 and 40% CO_2. These conditions already emphasize that cultivation of such organisms entails more than the provision of required nutrients. The appropriate external cues must be provided to trigger entrance into the next developmental stage. With nematodes like *Ostertagia* the protective sheath of the infective larval form must be shed. The stimulus here is provided by the high CO_2, absence of oxygen, a temperature of 37°C, and shaking, followed by brief aeration with compressed air and finally shaking in a sealed flask for 25 min. The exsheathed larvae were subsequently incubated in roller bottles in a medium designed to resemble rumen fluid with a gas phase of 5% CO_2 in air. After 2 days in these conditions the larvae were transferred for 6 days to a complex nutrient medium with pepsin at pH 4.5 and a gas phase of air, and then for the remainder of their development to this same medium and conditions but with pH at 6.0. F. W. Douvres has used a similar three-step method to grow *Ascaris suum* from artificially hatched second-stage larvae (see Diagram XIX) to the fourth stage in large numbers. Furthermore, a small proportion of the worms grew to become young and mature adults (Fig. 8.8). It is very important to note that the developmental cycle of nematodes involves both growth and molting, two independent but equally indispensable processes in their differentiation.

Among the nematodes the filarial worms are of special interest and importance both because of their involvement with several major human diseases (filariasis and onchocerciasis or "river blindness") and because all the developmental stages are parasitic, the larvae developing intracellularly in an insect vector (see Chapter 3). Not surprisingly, little progress has been made so far with cultivation of these intracellular larval stages. Recently, however, maturation of infective larvae to the fourth and even the fifth stage has been obtained *in vitro* for the two human filarial worms *Brugia malayi* and *B. pahangi*. Third-stage larvae removed from infected mosquitoes and placed in RPMI 1640 medium with human serum and an active culture of a rhesus monkey kidney cell line molted to the fourth stage in high proportion. Some also molted to the fifth and final stage. Although such cultures still do not maintain normal growth, the worms being smaller and growing more slowly than *in vivo*, they nevertheless already provide a useful source of worm secretion antigens for immunopathological studies.

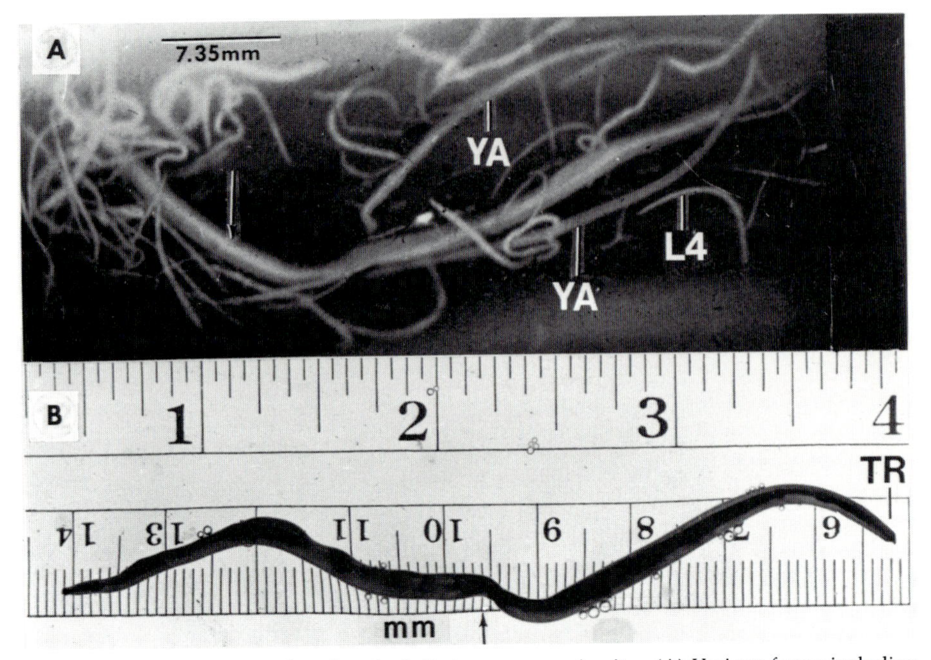

FIGURE 8.8. *Ascaris suum* larval and adult stages grown *in vitro*. (A) Various forms including fourth-stage larvae (L4), young adults (YA), and mature adult female (arrow) in culture vessel on day 61. (B) Mature adult female transferred from culture medium to balanced salt solution at day 166. Note two abnormalities in body shape: a creaselike indentation at the vulva (arrow) and the tail region (TR) blunt instead of tapered. (From Douvres and Urban, 1983. Prints courtesy of Dr. F. W. Douvres.)

The life cycles of trematodes, the flukes parasitic in intestines, liver, lungs, and blood vessels of humans and other animals, do not involve molting but are perhaps even more complex than those of most nematodes. Although attempts at *in vitro* culture of various parts of the life cycle have been done with many different species, it will suffice to discuss in detail only the work with *Schistosoma mansoni*. This blood fluke, and its close relatives *S. haematobium* and *S. japonicum*, have been extensively investigated both because they provide suitable experimental material and especially because they are causes of important human diseases.

No continuous *in vitro* development of a schistosome from egg to egg-laying adult has yet been obtained, but stepwise development of most stages beginning with a previous stage taken from an infected host has been successful. Miracidia hatched under aseptic conditions from eggs of *S. mansoni* and placed in a culture medium (Table 8.3) based in part on the composition of snail hemolymph and containing 20% human serum, transformed readily into sporocysts. After a week in culture the germinal balls of daughter sporocysts had appeared and by 10–12 days the sporocysts had grown to be 400–600 μm long, as compared to the initial miracidial length of 100 μm. There

TABLE 8.3. Basic Medium for Axenic Cultivation of *Schistosoma mansoni* Daughter Sporocysts[a]

Component	mg/liter	Component	mg/liter	Component	mg/liter
Salts		L-Alanine	1.2	D-Ca-pantothenate	1.0
NaCl	1300	L-Valine	1.1	Pyridoxal-HCl	1.0
KCl	97	L-Cystine	2.1	Thiamine-HCl	1.0
$MgCl_2 \cdot 6H_2O$	325	L-Methionine	0.9	Riboflavin	0.1
Na_2HPO_4	17	L-Isoleucine	2.6	i-Inositol	2.0
$CaCl_2 \cdot 2H_2O$	558	L-Leucine	4.4	Organic acids	
Amino acids[b]		L-Tyrosine	3.0	Malic	40
L-Serine	3.9	L-Phenylalanine	2.6	α-Ketoglutaric	30
O-Phosphorylethanolamine	1.3	L-Ornithine	4.0	Succinic	10
L-Aspartic acid	1.0	L-Lysine	4.3	Fumaric	5
L-Threonine	4.1	L-Histidine	2.5	Citric	10
DL-*O*-Phosphoserine	2.2	L-Arginine	2.7	Sugar	
L-Glutamine	7.5	Vitamins[c]		Glucose	100
L-Glutamic acid	5.5	Folic acid	1.0	Buffer	
L-Citrulline	1.9	Choline chloride	1.0	Hepes	1500
L-Glycine	2.6	Nicotinamide	1.0		

[a] Modified from Di Conza and Basch (1974).
[b] The amino acid concentrations listed here follow the analysis of hemolymph from our laboratory colony of *B. glabrata*. We further supplemented this with 1 × Eagle's Minimum Essential Medium amino acids (Grand Island Biological Co.).
[c] Eagle's Minimum Essential Medium vitamins (Grand Island Biological Co.).

was no further development *in vitro* but such sporocysts if implanted into snails (*Biomphalaria glabrata*) produced cercariae. In the same culture medium, daughter sporocysts removed from infected snails grew from 150 × 10 μm to about 350 × 20 μm in 7 days and differentiated but no cercariae were formed. Hence, the full cycle of development in the snail host has not been duplicated *in vitro*. With the next portion of the life cycle, that which occurs in the vertebrate host, excellent and important results have been obtained. Cercariae were allowed to penetrate pieces of isolated mouse skin held in a special apparatus. The schistosomula that formed (20–30% of the cercariae applied) were washed with penicillin and streptomycin and placed in a culture medium of 50% human serum and 50% Earle's balanced salt solution with 0.5% lactalbumin hydrolysate. Suspensions with about 100 schistosomula in 2 ml medium were placed in Leighton tubes which were gassed with 8% CO_2 in air, closed tightly, and incubated at 37°C. Medium was changed every 5 days and after the 5th day washed human red cells (type O) were added to give a hematocrit of 1%. *S. mansoni* developed to adults that mated but the females did not form eggs. This involves a very considerable growth and differentiation with increase in length from a fraction of a millimeter to 5–7 mm and with differentiation of adult structures, including the formation of sperm in the males. But eggs were not formed despite coupling of the males and females. In *S. haematobium* similar results were obtained except that pairing of the sexes did not occur and the females did not complete sexual maturation. Of considerable interest was the finding that individual human sera differed greatly in the amount of growth they would support of *S. haematobium* (but not of *S. mansoni*). More detailed analysis of this phenomenon might lead to significant new understanding of the factors essential for full development of these parasites. With a related schistosome of rodents, *Schistosomatium douthitti,* similar but somewhat better and more uniform growth and development were obtained. Eggs and sperm were formed, but there were no viable ova, despite the fact that *in vivo* this species will reproduce parthenogenetically.

Perhaps the best success with *in vitro* culture of helminthic parasites of vertebrates has been obtained with the cestodes. Here many species have been grown through parts of their life cycle and several through the entire life cycle. A wealth of experimental material is available for which there have so far been too few workers to exploit it fully. We will discuss three examples, the first (*Echinococcus granulosus*) because of the special developmental problem it presents, the second because it involves such a striking growth of a large tapeworm (*Spirometra mansonoides*), and the third to illustrate development of the complete life cycle *in vitro* (*Hymenolepis microstoma*).

Echinococcus granulosus (Diagram XVII), the cause of hydatid disease in sheep and people, can differentiate in nature in either of two ways. In the intermediate host, a sheep or a person, protoscolices that escape from a ruptured hydatid cyst into the body cavity differentiate to form further hydatid cysts. This indeed is one of the dangers attending surgery for removal of a hydatid cyst. In contrast, when protoscolices are ingested by the definitive host, a dog, and enter the dog's intestine, they differentiate in a strobilar direction to form adult tapeworms. In early attempts at cultivation starting with protoscolices, development that occurred *in vitro* was always in the cystic direction. J. D. Smyth and his colleagues then found that if the protoscolices were put in a diphasic rather than a completely liquid medium, they would develop in a strobilar direction. The best medium consists of a solid phase of coagulated bovine serum overlaid with a tissue culture medium with serum. The serum in the solid phase is coagulated at 75–76°C for 30–60 min to produce a relatively soft surface, which is then further perforated with a fine pipet. The nature of this surface is of prime importance for strobilar development. In the best results, worms developed almost as rapidly as in the dog intestine, segmentation being achieved in about 15 days. Adult worms were obtained with three to five proglottids and with free cells in the uterus. No fertile eggs were formed, however, and extent of development was very variable. This is a fascinating system for further study. One wonders how the nature of the surface encountered by the evaginated protoscolices can determine the direction of development.

Spirometra mansonoides is in nature a three-host parasite (Diagram XIV). The eggs hatch in water to produce swimming larvae called coracidia. These infect copepods within which they develop to a second larval form, the procercoid. This is infective to mice or other rodents where it develops in the abdominal cavity into still a third larval form, the sparganum. This is capable of growth to lengths up to 5 cm, and it is this stage that occasionally occurs in humans. If ingested by a cat, this stage (termed a plerocercoid) evaginates an adult scolex and develops rapidly in the cat intestine into a typical large tapeworm consisting of a long chain of proglottids within which eggs and sperm are formed.

Through the pioneering efforts of J. Mueller and A. K. Berntzen, all of this remarkable cycle is now available *in vitro* except for the first stage, from coracidium to procercoid. Procercoids harvested en masse from infected copepods and placed in tissue culture medium 199 with calf serum and chick embryo extract at 37°C develop rapidly into spargana. The medium, in tightly closed roller tubes, is changed daily and once a week the worms are transferred to fresh tubes and thinned out if they are too crowded. Procercoids with a length of 10 μm grow to 10-mm-long worms in 4–5 weeks. Once established, cultures of spargana continue to grow in the medium even without embryo extract and they also produce the remarkable sparganum growth factor which will be considered in detail in Chapter 12.

To get further development of the plerocercoids (either formed in culture from procercoids or obtained from infected mice), it is essential to place them in a special "evaginating solution" that replaces the environment of the cat intestine. For this cat bile can be used, or a mixture of sodium taurocholate with trypsin and glutathione in a complex culture medium. The worms are exposed to the evaginating solution for the first 72 hr; during this time the plerocercoid scolex evaginates, changes form, and a distinct neck region appears demarcated from the larval tissue of the plerocercoid, which begins to slough off. By 120 hr a definite scolex and well-developed neck free of larval tissue are present. On further maintenance with a continuous slow flow of medium, the neck elongates and segmentation begins at 8–10 days. The worms continue to grow to reach lengths of 30–40 cm in about 18 days, well within the normal range that occurs *in vivo* in the intestine of a cat (Fig. 8.9). The worms form gravid proglottids and lay fertile eggs. Eggs showed 80% viability and hatched into coracidia infective to copepods.

The medium that supports this extensive growth contains salts, sugars, amino acids, carboxylic acids, vitamins, cofactors, and a "reducing solution" with glutathione and cysteine. How much of all of these substances is essential remains to be determined and this can be done only by painstaking detailed studies designed to produce a "minimal" medium. Already it is clear that the reducing solution is very important; the redox potential must be at −160 to −190 mV at a pH of 7.2–7.4. Also essential is a gas phase with 5% CO_2 and 95% N_2. As little as 5% O_2 is deleterious. It is of great interest that plerocercoids placed directly in the optimal culture conditions but without being first exposed to the evaginating solution survive but show no differentiation or growth. The specific stimulus provided by bile and trypsin conveys the information that the worm is in a new environment where it can develop to the adult stage. What is the molecular biology at the basis of this response?

Hymenolepis microstoma is a small tapeworm having only two hosts: an

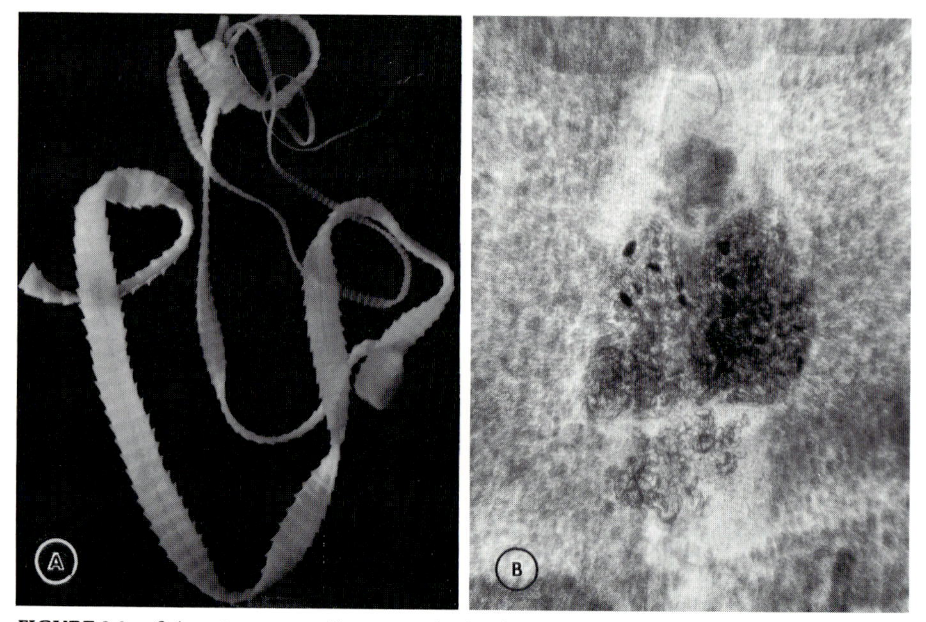

FIGURE 8.9. *Spirometra mansonoides*, grown *in vitro* from the plerocercoid to the ovigerous adult, 42 days in culture. (A) Whole worm, after relaxing in tap water and fixing in 10% formalin. Several centimeters have been removed from posterior end for detailed study. (B) Central region of a proglottid of this worm, alum cochineal stain. Note the square outer coil of the uterus, diagnostic of *S. mansonoides*, and the dark mass at the anterior notch of the uterine egg mass which is the sperm-filled vesicle. (From Berntzen and Mueller, 1972. Original prints courtesy of Dr. J. F. Mueller.)

insect, such as the flour beetle *Tribolium*, in which eggs develop into a larval form, the cysticercoid; a rodent, such as the mouse, in which the cysticercoid develops to an adult that lays eggs. This complete cycle has been propagated *in vitro* under relatively simple conditions. Eggs collected from gravid proglottids were caused to hatch by mechanical breakage of the shells followed by exposure to a solution containing trypsin with a gas phase of 5% CO_2–95% N_2. The hatched oncospheres thus obtained were cultured in an insect tissue culture medium in small sealed tubes with a gas phase of air at 28°C. Cysticercoids developed within 3 weeks (Fig. 8.10). The fully developed cysticercoid is again a resting stage that does not develop further until it is ingested by an appropriate host. As noted in Chapter 2, the enzymes and other conditions of the new host's digestive tract then play a specific role in permitting and activating a process akin to excystation. Accordingly, before further *in vitro* development of the cysticercoids could be expected, they have to be subjected to conditions producing excystment and evagination of the scolex. This is accomplished as for the plerocercoids of *Spirometra*, by exposure to a mixture containing trypsin and bile salts. This digests the cyst surrounding

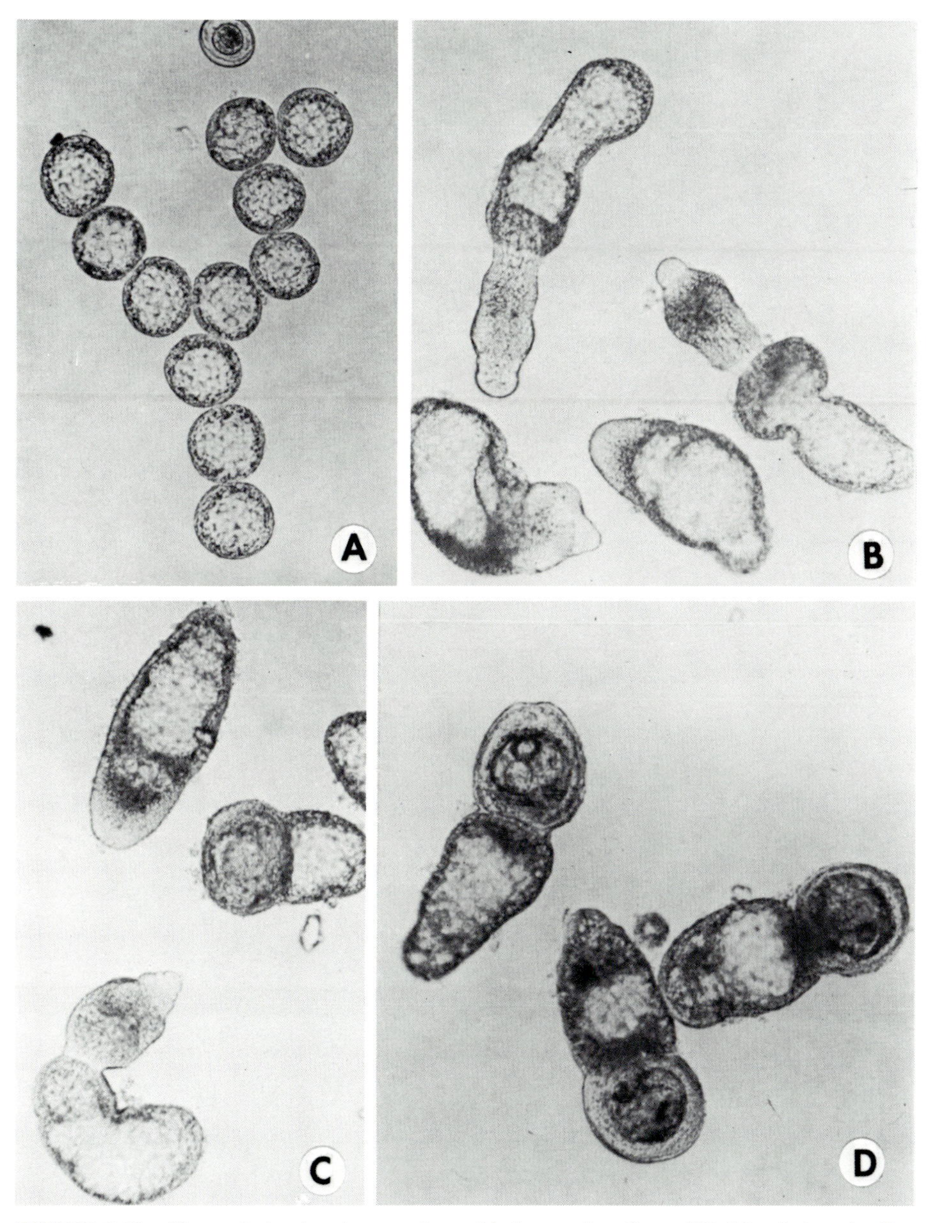

FIGURE 8.10. *Hymenolepis microstoma* cysticercoids in axenic culture. (A) After 3 days *in vitro* showing ball stages with well-demarcated cavity. Note unhatched egg at top for comparison. (B) Cysticercoids after 9 days in culture showing tripartite body with rostellum, scolex, and cavity. (C) Recently withdrawn cysticercoid and prewithdrawal stages at 13 days *in vitro*. Note rostellum development. (D) Fully developed cysticercoids after 21 days in culture. All from living material; ×160. (From Seidel, 1975.)

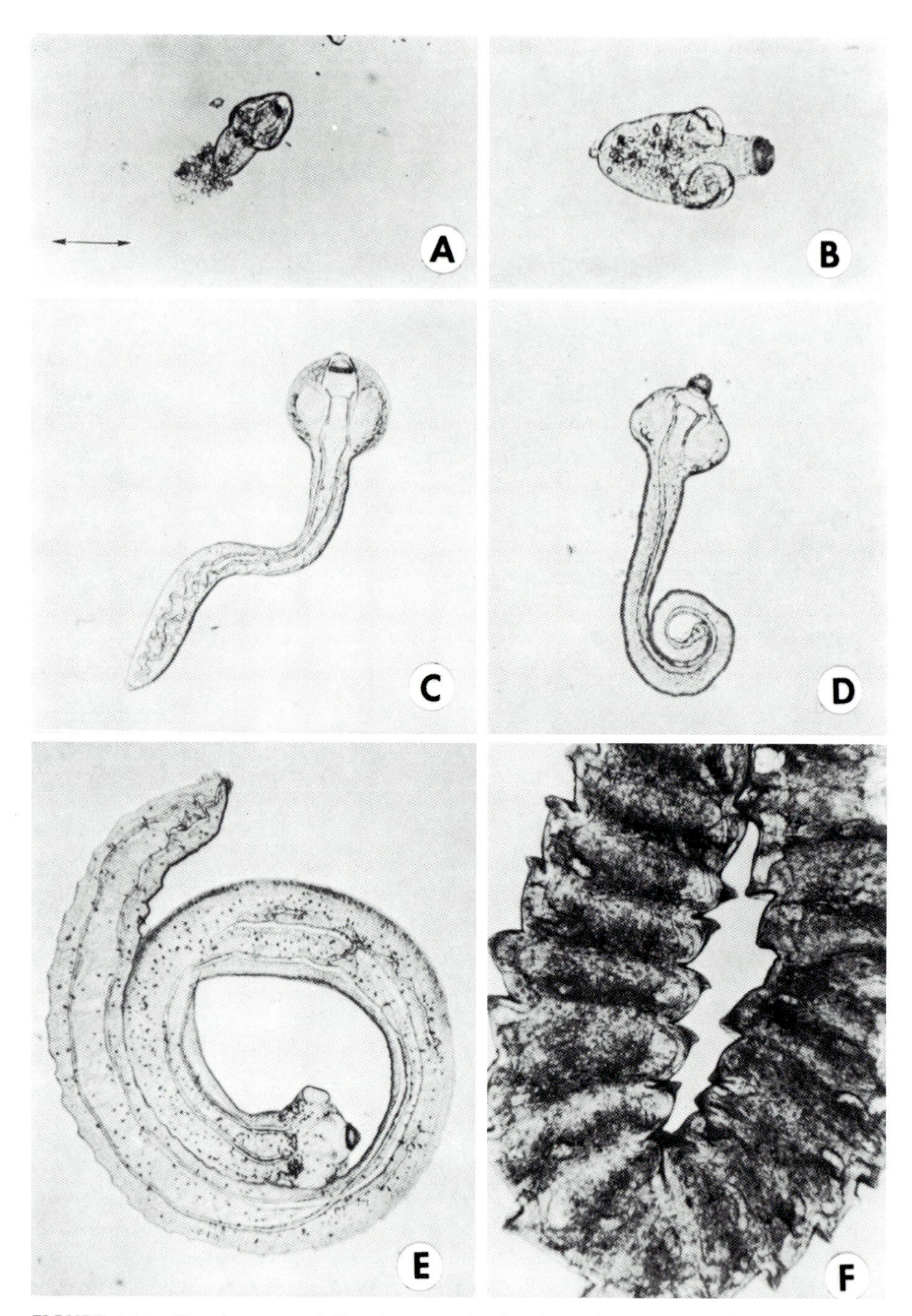

FIGURE 8.11. Development of *H. microstoma in vitro* from the cysticercoid in presence and absence of hemin. All from living organisms. Scale (on A): A, C, D = 100 μm; B, E, F = 250 μm. (A) Excysted cysticercoid. (B) Juvenile after 1 day in culture. (C) After 4 days in culture. (D)

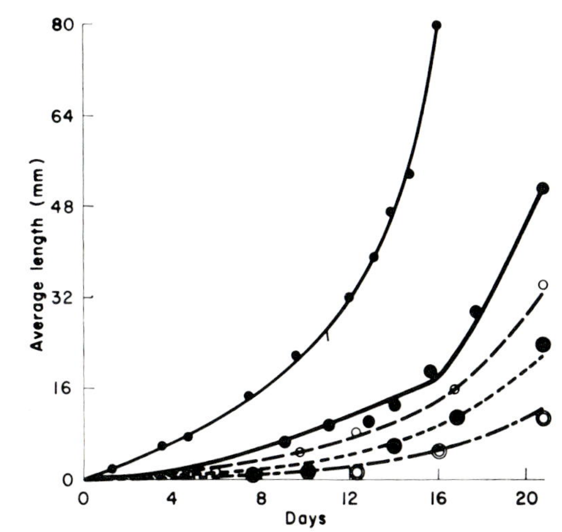

FIGURE 8.12. Growth of *H. microstoma* from the cysticercoid in different environments. (•—•) *In vivo*. (●—●) Medium with 30% horse serum. Over blood agar. (○—○) Medium with 20% horse serum. Over blood agar. (◉—◉) Medium with 30% horse serum, 0.1 mg hemin/100 ml. Over nutrient agar. (◎—◎) Medium with 20% horse serum, 0.1 mg hemin/100 ml. Over nutrient agar. (From Seidel, 1971.)

the larval worm and activates the worm so that it everts its scolex, elongates its neck, and actively pushes its way out of remaining cyst membranes. Such active young worms can then be grown to the adult stage in tubes held at 37°C with a diphasic medium consisting of a layer of nutrient agar with an overlay of triple Eagle's tissue culture medium with 30% horse serum (inactivated at 56°C for 30 min). The best growth was obtained with blood agar as the base. Even so the worms grew at a slower rate than *in vivo* (Figs. 8.11, 8.12) and fully developed eggs were not seen. Most interesting, however, were the results obtained using nutrient agar without blood and with or without a supplement of hemin. Without hemin, growth ceased after 4 days; with hemin at 0.1 mg/100 ml, strobilization began at 8 days and adult worms with gravid proglottids were present after 20 days. Protoporphyrin IX could not substitute for hemin. Hemin is the first growth factor shown to be required by tapeworms.

General Discussion on the Role of *in Vitro* Culture in Studies of Parasites

As seen in these last two chapters, cultivation *in vitro* provides a direct approach to several important aspects of parasitic life. It gives a way of determining the environmental requirements: nutrients essential for growth and reproduction, the physical environment including osmolarity, pH, reducing potential, CO_2 and oxygen tensions, solid substrate, those factors promoting

Maximum development in absence of hemin. After 2 weeks. (E) After 8 days in medium with hemin. Beginning of strobilization. (F) Fully mature adult, 20 days in culture with hemin. (From Seidel, 1971.)

differentiation of the parasite to the next stage of its life cycle essential for its transmission to new hosts, and such factors as may in part integrate the physiology of the parasite with that of the host. Among these last is the circadian rhythm exhibited by some parasites that is clearly related to the daily activity of the host and probably of adaptive value. With some species of filarial worms, the microfilariae, the larval forms destined for infection of the mosquito vector, appear in the circulating blood only at night when mosquitoes are biting. The periodicity and synchronicity of the erythrocytic stages of human malarial parasites is a striking feature of the disease. Here too F. Hawking has suggested that this has adaptive value with reference to the maturation of gametocytes infective to mosquitoes. In *in vitro* cultures *Plasmodium falciparum* soon loses the initial synchronicity it has in its host. Which of the many physiological factors that go up and down in circadian rhythm serve to synchronize the parasites in the host? Presently available culture methods provide for direct and detailed studies of this important and fascinating problem. Culture methods provide material for studies of the metabolism of the parasite and of the products of this metabolism which in turn may affect the physiology of the host. This could lead to new kinds of chemotherapeutic agents. Cultures also greatly facilitate the application of the methods of molecular biology to the study of parasitism.

Bibliography

Basch, P. F., and O'Toole, M. L., 1982, Cultivation *in vitro* of *Schistosomatium douthitti* (Trematoda: Schistosomatidae), *Int. J. Parasitol.* **12:**541–545.

Berntzen, A. K., and Müller, J. F., 1964, *In vitro* cultivation of *Spirometra mansonoides* (Cestoda) from the procercoid to the early adult, *J. Parasitol.* **50:**705–711.

Berntzen, A. K., and Müller, J. F., 1972, *In vitro* cultivation of *Spirometra* spp. (Cestoda) from the plerocercoid to the gravid adult, *J. Parasitol.* **58:**750–752.

Brun, R., and Moloo, S. K., 1982, *In vitro* cultivation of animal-infective forms of a West African *Trypanosoma vivax* stock, *Acta Trop.* **39:**135–141.

Brukot, T. R., Williams, J. L., and Schneider, I., 1984, Infectivity to mosquitoes of *Plasmodium falciparum* clones grown *in vitro* from the same isolate, *Trans. R. Soc. Trop. Med. Hyg.* **78:**339–341.

Clegg, J. A. 1965, *In vitro* cultivation of *Schistosoma mansoni, Exp. Parasitol.* **16:**133–147.

Current, W. L., and Haynes, T. B., 1984, Complete development of *Cryptosporidium* in cell culture, *Science* **224:**603–605.

Das, S. R., Rastogi, A. K., Sagar, P., and Singh, M. P., 1980, Axenic encystation of *Entamoeba invadens, Indian J. Exp. Biol.* **18:**133–136.

Davis, A. G., Huff, C. G., and Palmer, T. T., 1966, Procedures for maximum production of exoerythrocytic stages of *Plasmodium fallax* in tissue culture, *Exp. Parasitol.* **19:**1–8.

Diamond, L. S., Harlow, D. R., and Cunnick, C. C., 1978, A new medium for the axenic cultivation of *Entamoeba histolytica* and other *Entamoeba, Trans. R. Soc. Trop. Med. Hyg.* **72:**431–432.

Di Conza, J. J., and Basch, P. F., 1974, Axenic cultivation of *Schistosoma mansoni* daughter sporocysts, *J. Parasitol.* **60:**757–763.

Douvres, F. W., and Urban, J. F., Jr., 1983, Factors contributing to the *in vitro* development of *Ascaris suum* from second-stage larvae to mature adults, *J. Parasitol.* **69:**549–558.

Duszenko, M., Ferguson, M. A. J., Lamont, G. S., Rifkin, M. R., and Cross, G. A. M., 1985, Cysteine eliminates the feeder cell requirement for cultivation of *Trypanosoma brucei* bloodstream forms *in vitro, J. Exp. Med.* **162:**1256–1263.

Fawcett, D. W., Büscher, G., and Doxsey, S., 1982, Salivary gland of the tick vector of East Coast fever. III. The ultrastructure of sporogony in *Theileria parva, Tissue Cell* **14:**183–206.

Gardiner, P. R., Lamont, L. C., Jones, T. W., and Cunningham, I., 1980, The separation and structure of infective trypanosomes from cultures of *Trypanosoma brucei* grown in association with tsetse fly salivary glands, *J. Protozool.* **27:**182–185.

Glaser, R. W., and Stoll, N. R., 1938, Sterile culture of the free-living stages of the sheep stomach worm, *Haemonchus contortus, Parasitology* **30:**324–332.

Hart, D. T., Vickerman, K., and Coombs, G. H., 1981, Transformation *in vitro* of *Leishmania mexicana* amastigotes to promastigotes: Nutritional requirements and the effect of drugs, *Parasitology* **83:**529–541.

Hollingdale, M. R., Leland, P., and Schwartz, A. L., 1983, *In vitro* cultivation of the exoerythrocytic stage of *Plasmodium berghei* in a hepatoma cell line, *Am. J. Trop. Med. Hyg.* **32:**682–684.

Huff, C. G., 1964, Cultivation of the exoerythrocytic stages of malarial parasites, *Am. J. Trop. Med. Hyg.* **13:**171–177.

Lambowitz, M. B., Kumar, A. M., Grant, V. B., Kobagashi, G. S., and Medoff, G., 1981, Role of cysteine in regulating morphogenesis and mitochondrial activity in the dimorphic fungus *Histoplasma capsulatum, Proc. Natl. Acad. Sci. USA* **78:**4596–4600.

Mak, J. W., Lim, P. K. C., Sim, B. K. L., and Liew, L. M., 1983, *Brugia malayi* and *B. pahangi:* Cultivation *in vitro* of infective larvae to the fourth and fifth stages, *Exp. Parasitol.* **55:**243–248.

Mazier, D., Landau, I., Druilhe, P., Miltgen, F., Guguen-Guillouzo, C., Baccam, D., Baxter, J., Chigot, J.-P., and Gentilini, M., 1984a, Cultivation of the liver forms of *Plasmodium vivax* in human hepatocytes, *Nature* **307:**367–369.

Mazier, D., Beaudoin, R. L., Mellouk, S., Druilhe, P., Texier, B., Trosper, J., Miltgen, F., Landau, I., Paul, C., Brandicourt, O., Guguen-Guillouzo, C., and Langlois, P., 1984b, Complete development of hepatic stages of *Plasmodium falciparum in vitro, Science* **227:**440–442.

Medoff, G., Sacco, M., Maresca, B., Schlessinger, D., Painter, A., Kobayashi, G. S., and Carratu, L., 1986, Irreversible block of the mycelial-to-yeast phase transition of *Histoplasma capsulatum, Science* **231:**476–479.

Mueller, J. F., 1959, The laboratory propagation of *Spirometra mansonoides* (Mueller, 1935) as an experimental tool. III. *In vitro* cultivation of the plerocercoid larva in cell-free medium, *J. Parasitol.* **45:**561–573.

Mueller, J. F., 1974, The biology of *Spirometra, J. Parasitol.* **60:**3–14.

Pan, A. A., 1984, *Leishmania mexicana:* Serial cultivation of intracellular stages in a cell-free medium, *Exp. Parasitol.* **58:**72–80.

Pan, S. C.-T., 1978, *Trypanosoma cruzi:* Intracellular stages grown in a cell-free medium at 37°, *Exp. Parasitol.* **45:**215–224.

Pan, S. C., 1978, *Trypanosoma cruzi: In vitro* interactions between cultured amastigotes and human skin-muscle cells, *Exp. Parasitol.* **45:**274–286.

Rothman, A. H., 1959, Studies on the excystment of tapeworms, *Exp. Parasitol.* **8:**336–364.

Seidel, J. S., 1971, Hemin as a requirement in the development *in vitro* of *Hymenolepis microstoma* (Cestoda: Cyclophyllidea), *J. Parasitol.* **57:**566–570.

Seidel, J. S., 1975, The life cycle *in vitro* of *Hymenolepis microstoma* (Cestoda), *J. Parasitol.* **61:**677–681.

Smith, J. E., Meis, J. F. G. M., Ponnudurai, T., Verhave, J. P., and Moshage, H. J., 1984, *In vitro* culture of exoerythrocytic form of *Plasmodium falciparum* in adult human hepatocytes, *Lancet* **2:**757–758.

Smith, M., Clegg, J. A., and Webbe, G., 1976, Culture of *Schistosoma haematobium in vivo* and *in vitro, Ann. Trop. Med. Parasitol.* **70:**101–107.

Smyth, J. D., and Davies, Z., 1974, *In vitro* culture of the strobilar stage of *Echinococcus granulosus* (sheep strain): A review of basic problems and results, *Int. J. Parasitol.* **4:**631–644.

Tanner, M., 1980, Studies on the mechanisms supporting the continuous growth of *Trypanosoma* (*Trypanozoon*) *brucei* as bloodstream-like form *in vitro, Acta Trop.* **37**:203–220.
Trager, W., 1960, Intracellular parasitism and symbiosis, in: *The Cell,* Volume 4 (J. Brachet and A. E. Mirsky, eds.), Academic Press, New York, pp. 151–213.
Trager, W., 1978, Cultivation of *Plasmodium falciparum,* in: *Tropical Medicine: From Romance to Reality* (C. Wood, ed.), Academic Press/Grune & Stratton, New York, pp. 49–61.
Trager, W., 1983, *In vitro* growth of parasites, in: *Molecular Biology of Parasites* (J. Guardiola, L. Luzzatto, and W. Trager, eds.), Raven Press, New York, pp. 39–51.
Villalta, F., and Kierszenbaum, F., 1982, Growth of isolated amastigotes of *Trypanosoma cruzi* in cell-free medium, *J. Protozool.* **29**:570–576.
Voge, M., and Green, J., 1975, Axenic growth of oncospheres of *Hymenolepis citelli* (Cestoda) to fully developed cysticercoids, *J. Parasitol.* **61**:291–297.

CHAPTER 9

Metabolism

Energy Sources. Respiration

ATP is the key intermediate in the metabolism of parasites as it is in all other organisms. The eukaryotic parasites with which we are dealing have the enzymatic equipment for the generation of ATP by one or several pathways. Their location in the host, or in the external environment for free-living stages concerned in transmission, determines the substrates available for energy generation and in part the pathway of catabolism, whether through aerobic respiration or aerobic or anaerobic fermentation. A parasite residing in a nearly anaerobic environment may or may not require or utilize oxygen. Many such parasites are microaerophilic, requiring oxygen in small amounts. Conversely, parasites living in a relatively oxygen-rich medium, such as the vertebrate bloodstream, may nevertheless rely principally on glycolysis or other pathways rather than on the very efficient pathway involving the tricarboxylic acid cycle and the respiratory chain with a cytochrome as terminal oxidase. I will consider in detail several important examples of each type of energy-generating pathway.

Entamoeba histolytica, the pathogenic ameba of humans, lives typically in the relatively anaerobic environment of the lumen of the large intestine, but is capable of invading other tissues, notably the liver, where its environment is not anaerobic. Furthermore, when grown axenically in culture, it can be shown to require and utilize oxygen. Nonetheless, all indications are that this protozoon, which has no mitochondria and no citric acid cycle enzymes or cytochromes, meets its energy demands by glycolytic phosphorylation via the Embden–Meyerhof pathway with ethanol, acetate, and CO_2 as the major end products, and also perhaps to some extent via the Entner–Doudoroff pathway. Inorganic pyrophosphate replaces ATP in several of the glycolytic reactions. Lactate is not formed and there is no lactate dehydrogenase. How then are reducing equivalents, such as reduced nicotinamide adenine dinucleotide (NADH), reoxidized? This is not known. It may be accomplished by an iron–sulfur flavoprotein oxidase, or indirectly by a transhydrogenase. A major product of pyruvate oxidation is acetate which is formed by axenically grown amebae only under aerobic conditions. This is catalyzed by two enzymes, a pyruvate synthase for the reaction:

$$\text{pyruvate} + \text{CoA} \rightarrow CO_2 + \text{acetyl-CoA} + 2e$$

and an enzyme (not previously described from other cells) called acetyl-CoA synthase (ADP-forming) for the reaction:

$$\text{acetyl-CoA} + \text{ADP} + P_i \rightarrow \text{acetate} + \text{ATP} + \text{CoA}$$

Under aerobic conditions, oxygen serves as the final electron acceptor. The respiratory chain for *E. histolytica* may involve dehydrogenases, flavins, ferredoxins, and one or more final carriers analogous to cytochromes that may be iron–sulfur proteins or copper-containing proteins. It is of interest that *E. histolytica* under anaerobic conditions forms no acetate if it is grown axenically. If, however, it is grown with penicillin-inhibited cells of *Bacteroides symbiosus,* it then forms under anaerobic conditions not only acetate but also hydrogen. The latter is produced by a hydrogenase present in the bacteria but absent from axenically grown amebae. Hence, *Entamoeba* itself lacks the enzymatic machinery for production of hydrogen.

It differs strikingly in this respect from the trichomonad flagellates. These organisms, which also live in oxygen-poor environments, have been found to have a special organelle, now called the hydrogenosome, that carries enzymes concerned in hydrogen formation from pyruvate metabolism. This was shown first for *Tritrichomonas foetus,* a parasite causing abortion in cows, but it has also been demonstrated for *Trichomonas vaginalis,* an important venereally transmitted cause of vaginitis in women. All trichomonads have characteristic granules bounded by two membranes and usually located along the axostyle and the costa (Fig. 9.1). These structures were early found to contain α-glycerophosphate dehydrogenase and malate dehydrogenase, but they lacked catalase and NADH oxidase, thereby differing from the peroxisomes, granules of similar morphology present in many aerobic free-living protozoa and in metazoan cells. They clearly are not mitochondria, and indeed trichomonads have neither a functional tricarboxylic acid cycle nor cytochromes. D. G. Lindmark and M. Müller showed that these organelles contain a pyruvate synthase and a hydrogenase very like those characteristic of anaerobic bacteria of the genus *Clostridium.* Although some trichomonads tolerate oxygen, their metabolism is anaerobic. They obtain energy by the degradation of carbohydrates to organic acids, mainly acetate, CO_2, and, for *Tritrichomonas foetus,* succinate, for *Trichomonas vaginalis,* lactate, with the formation (under anaerobic conditions) of molecular hydrogen. Both the hydrogenases and the pyruvate synthase of the hydrogenosomes utilize as electron acceptors flavin or ferredoxin but not NAD or NADP, and both require thiol compounds for

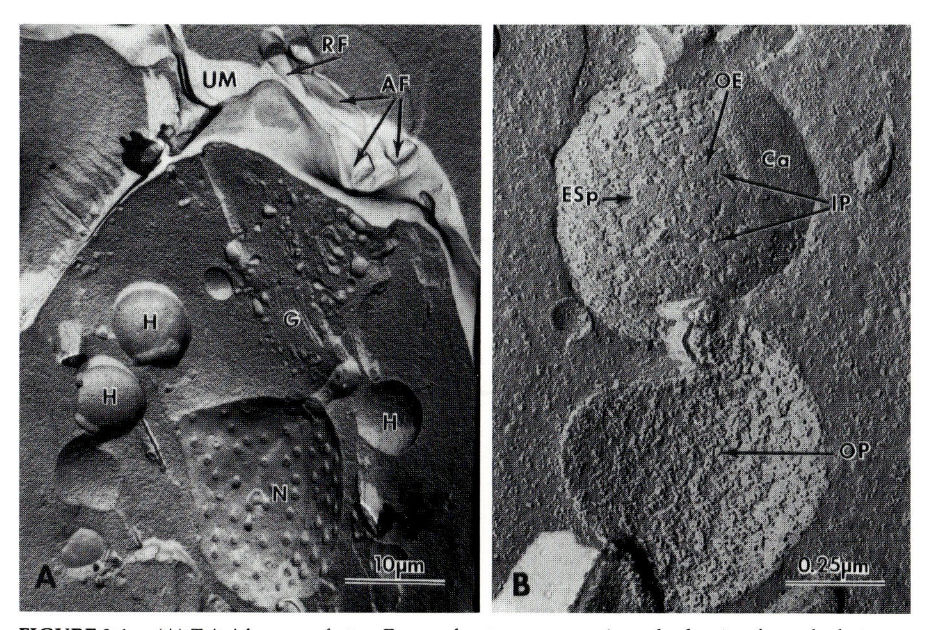

FIGURE 9.1. (A) *Tritrichomonas foetus*. Freeze-fracture preparation, shadowing from the bottom. Note the hydrogenosomes (H), Golgi (G), nucleus (N) with its pores. AF and RF, proximal portions of anterior and recurrent flagella, respectively; UM, part of undulating membrane. (B) Hydrogenosomes of *Trichomonas vaginalis*. Freeze-fracture preparation, shadowing from the right. ESp, endoplasmic space; OP, protoplasmic face of outer membrane; OE, external face of outer membrane; IP, protoplasmic face of inner membrane; Ca, caplike protrusion present on some hydrogenosomes. (From Honigberg *et al.*, 1984. Original prints courtesy of Dr. B. M. Honigberg.)

activity. In the anaerobic catabolism of pyruvate by *Tritrichomonas foetus*, as in that by clostridia, the products are acetyl-CoA, CO_2, and H_2. The pyruvate synthase, like that of *E. histolytica*, decomposes pyruvate with formation of CO_2, acetyl-CoA, and two electrons. Under anaerobic conditions, the hydrogenase produces molecular hydrogen by combining the electrons with protons.

Under aerobic conditions, the hydrogenosomes act like respiratory organelles, using oxygen as terminal electron acceptor in pyruvate metabolism. Under these conditions the respiration is greatly enhanced by CoA, which is the primary group acceptor. The addition of small amounts of CoA gives a phenomenon superficially resembling respiratory control as seen in mitochondria.

In trichomonads the usual glycolytic enzymes that change 1 mole of carbohydrate to 2 moles of phosphoenolpyruvate are present in the cytosol. Also in the cytosol are pyruvate kinase, forming pyruvate, and phosphoenolpyruvate carboxykinase and NAD-linked malate dehydrogenase, forming malate via oxaloacetate. Malate is also formed in the cytosol from succinate

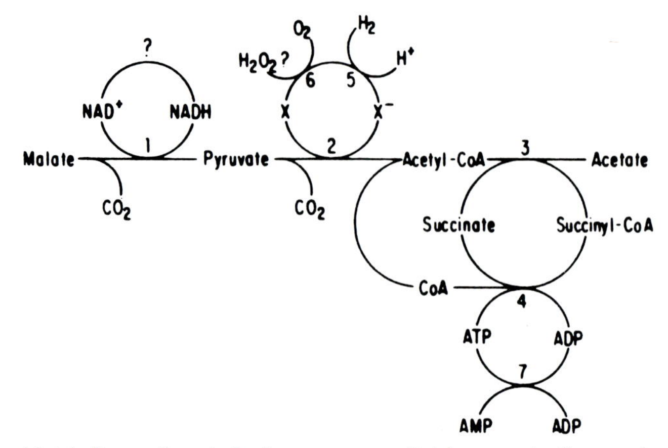

FIGURE 9.2. Metabolic reactions in hydrogenosome of trichomonads. See text for explanation. (From Müller and Lindmark, 1978.)

and fumarate. Both malate and pyruvate enter the hydrogenosomes and there participate in the reactions summarized in Fig. 9.2.

In this pathway, which accounts for about half of the carbon flow, the steps (referred to by the numbers in the map) are as follows:

1. Oxidative decarboxylation of malate to pyruvate by malate dehydrogenase (decarboxylating).
2. Oxidative decarboxylation of pyruvate (derived either from glycolysis or from Step 1) by a pyruvate ferredoxin oxidoreductase (pyruvate synthase) with CoA as acyl acceptor.
3. Transfer of the CoA from acetyl-CoA to succinate by acetate-succinate CoA transferase with release of acetate.
4. Substrate-level phosphorylation of ADP by succinate thiokinase using the thioester bond of succinyl-CoA.
5. The reducing equivalents formed in Step 2 are transferred via a ferredoxin (Fd) recently isolated in pure form from *Tritrichomonas foetus*. They are combined by hydrogenase with protons to form molecular hydrogen.
6. It is probable that under aerobic conditions, pyruvate oxidation follows the same course except that the electrons are transferred to oxygen via some terminal oxidase of unknown nature.
7. An adenylate kinase is present.

It is noteworthy that this pathway of pyruvate oxidation is very similar to that one already discussed for *E. histolytica*. In the latter, however, all the enzymes concerned are in the cytosol and there are no hydrogenosomes so that hydrogen is not formed under anaerobic conditions.

For still another type of intestinal protozoan parasite of humans, *Giardia*

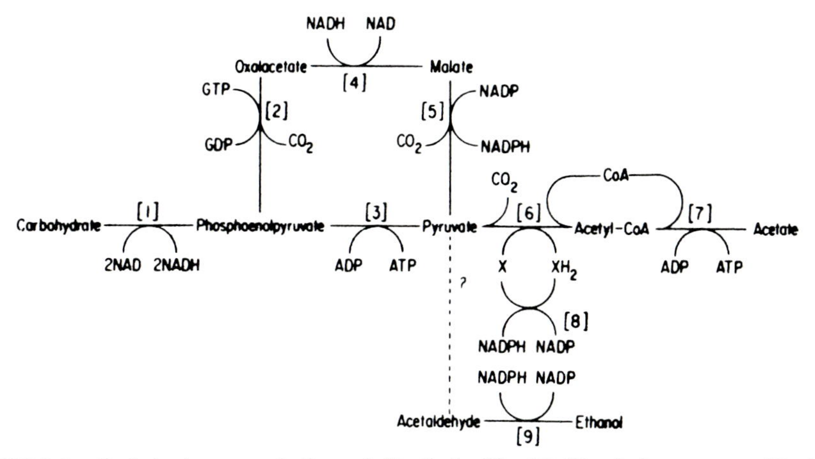

FIGURE 9.3. Carbohydrate metabolism of *Giardia lamblia*. [1] Glycolytic enzymes; [2] phosphoenolpyruvate carboxykinase (GDP); [3] pyruvate kinase (ADP); [4] malate dehydrogenase (NAD); [5] malate dehydrogenase (decarboxylating) (NADP); [6] pyruvate synthase; [7] acetyl-CoA synthase (ADP); [8] alcohol dehydrogenase (NADP); [9] NADPH oxidoreductase. (From Lindmark, 1980.)

lamblia, a very similar metabolic map has been worked out. As in *Entamoeba* and unlike the situation in trichomonads, all the enzymes are in the cystosol, and there are no hydrogenosomes. Under anaerobic conditions, ethanol, not hydrogen, is formed (Fig. 9.3).

Intestinal helminthic parasites exhibit yet another pathway of energy production, a pathway that utilizes the mitochondria but in a way entirely different from the usual pathway of the citric acid cycle and oxidative phosphorylation. This pathway was first worked out for the large intestinal nematode *Ascaris lumbricoides* by E. Bueding and H. Saz, but it was soon shown to occur in the cestode *Hymenolepis diminuta*. It is now known to be present not only in many other animal parasites that accumulate succinate (or products derived from succinate) but also in a number of free-living invertebrates and even in the anaerobic metabolism of deep-diving mammals. In this pathway (Fig. 9.4), succinate is formed by what might be considered the "backward leg" or four-carbon part of the tricarboxylic acid cycle.

Glucose is catabolized via the glycolytic pathway to phosphoenolpyruvate (PEP). Unlike the situation in the protozoa just discussed, essentially no pyruvate is formed, since in these worms the enzyme pyruvate kinase is barely detectable. Instead a PEP carboxykinase fixes CO_2 to form oxaloacetate (OAA). By means of a very active malate dehydrogenase, the OAA is then reduced to malate, the NADH formed in glycolysis being now regenerated to NAD^+. The PEP carboxykinase acts in a direction opposite to its usual role of gluconeogenesis in mammalian muscle. This reverse direction is facilitated by a sevenfold higher apparent Michaelis constant for OAA than for PEP, by high tissue concentrations of PEP since there is so little pyruvate kinase, and

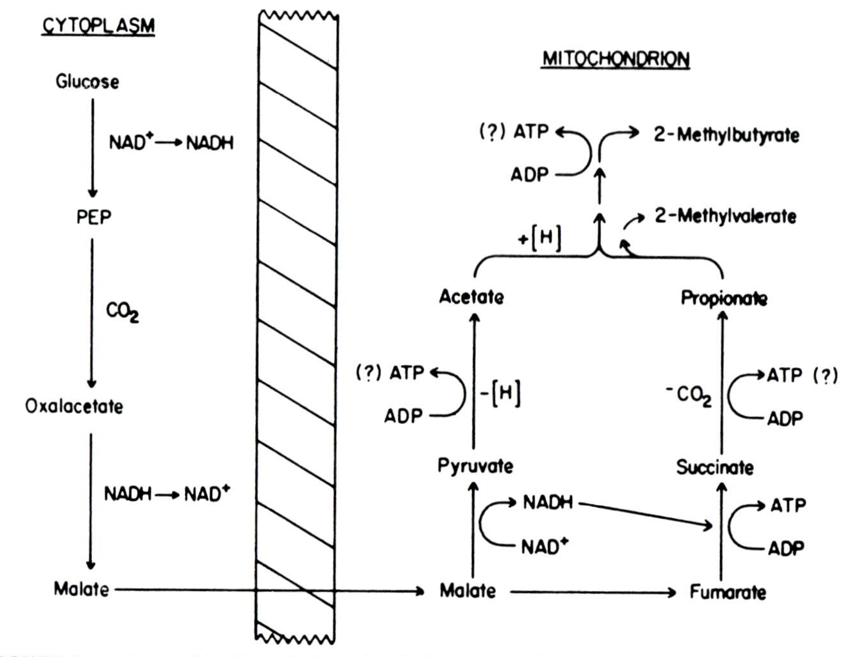

FIGURE 9.4. Anaerobic dissimilation of carbohydrate in the nematode *Ascaris lumbricoides*. See text for explanation. (From Saz, 1981.)

by the fact that the equilibrium for the malate dehydrogenase lies in the direction of malate, thus rapidly removing the OAA.

The malate then enters the mitochondrion where it serves as substrate in a dismutation reaction. One mole of malate is oxidized to pyruvate and CO_2 by an NAD^+-linked malic enzyme. This generates reducing power as NADH which serves to reduce another mole of malate to succinate via fumarate. This fumarate reductase is coupled to an electron transport-associated phosphorylation, probably at the NADH-flavoprotein (site I) level. The electron transport system couples with rhodoquinone rather than ubiquinone. This reaction is the major known source of ATP in *Ascaris* mitochondria.

There are, however, three other possible sites of ATP formation (indicated in Fig. 9.4 with question marks). The first of these is in the decarboxylation of succinate to propionate. *Ascaris* mitochondria show propionyl-CoA carboxylase, methylmalonyl-CoA mutase, and acyl-CoA transferase activities. Inorganic ^{32}P is esterified during succinate decarboxylation and both the decarboxylation and the phosphate esterification are stimulated six- to eightfold by addition of propionyl-CoA. Both stimulations are inhibited by avidin, an inhibitor of biotin-requiring enzymes such as propionyl-CoA carboxylase. Vitamin B_{12} cofactor is required in these reactions and it is significant that only those worms that contain high levels of B_{12} accumulate propionate (see also Chapter 7). In the cestode *Hymenolepis diminuta,* which lacks B_{12}, succinate but not proprionate is formed.

It is also likely that ATP is formed in a reaction sequence leading from propionate to 2-methylbutyrate, but the evidence for this is so far incomplete. Finally, energy generation might be associated with the formation of acetate from pyruvate. A pyruvate dehydrogenase complex has been isolated from *Ascaris* muscle that closely resembles the corresponding mammalian enzyme and like it is dependent on CoA, NAD^+, and pyruvate. Unlike the mammalian enzyme, however, a much lower NAD^+/NADH ratio is required to inhibit the worm enzyme. This is in keeping with the low NAD^+/NADH ratio of 0.07 to 0.7 in *Ascaris* as compared to about 10 in mammalian mitochondria.

Both *Ascaris* adults and *Hymenolepis* have been shown to be able to live in a complete absence of oxygen and *Hymenolepis* have been grown to maturity under anaerobic conditions from the cysticercoid. Although both may take up oxygen and contain some cytochromes (though very little in adult *Ascaris*), it is perfectly clear that they lack a tricarboxylic acid cycle and that they do not generate energy via an aerobic pathway of oxidative phosphorylation, even if oxygen is available to them. In some helminths, as in the large sheep tapeworm *Moniezia expansa,* a cytochrome system apparently does function under aerobic conditions. Then succinate and NADH are oxidized in the mitochondria to malate and NAD by an electron transport system linked to oxygen. The small amounts of cytochromes found in adult *Ascaris* could represent residual material from the large amounts present late in the development of an *Ascaris* larva within its egg. Here we have an example of an important metabolic switch.

The developing eggs of *Ascaris* are free-living in the external environment, where they require about 9–13 days to develop into an infective larva. This then lies dormant within the egg until activated by ingestion by a host. All stages of larval development including the dormant larva show high levels of enzymes of the tricarboxylic acid cycle whereas these are very low in adult *Ascaris*. For example, citrate synthase and aconitase show activities (as nmoles/min per mg protein at 30°C) of 1023 and 37.6, respectively, in the 10-day egg and only 9.5 and 0.8 in adult muscle. Cytochrome oxidase activity is very low in the 0-day egg but is rapidly generated early in development and remains at a high level in the infective larva, which meanwhile has usually molted within the egg to its second stage. Although the eggs hatch in the duodenum and the adult *Ascaris* lives in the intestine, development through the third and fourth larval stages involves a complex series of migrations (Diagram XIX). The second-stage larvae penetrate to the liver within 6 hr after infection, and there reach the third stage within a few days. The larvae then go to the heart and from there to the lungs where they again spend several days, becoming late third stages. These finally reach the small intestine via the trachea and pharynx by about day 9 after infection. They then molt to the fourth stage and grow to adults in the intestine. The third-stage larvae are still obligate aerobes; they require oxygen for motility, exhibit a partial Pasteur effect, and have substantial cytochrome oxidase activity. This suddently disappears as they enter the fourth stage and then remains suppressed throughout adult life. One could not ask for a better example of epigenetic control.

Such epigenetic adaptation is also exhibited by other helminths whose adults are typical homolactate fermenters. The blood flukes and filarial worms provide good examples. The adults of *Schistosoma mansoni* get their energy from the glycolytic breakdown of carbohydrates by essentially the same pathway as that in vertebrate muscle under anaerobic conditions. This is true regardless of how much oxygen may be available. In this pathway, NADH formed in the glyceraldehyde-3-phosphate dehydrogenase step is reoxidized by the reduction of pyruvate to lactate by lactate dehydrogenase. This pathway yields only 2 ATP per mole of glucose metabolized whereas in aerobic metabolism the yield is 38. In the anaerobic mitochondrial pathway of *Ascaris* discussed above, the yield is at least 3 ATP and it is probably 5 if the two other potential ATP-forming reactions occur.

When *S. mansoni* adults were maintained *in vitro* on Schiller's diphasic culture medium of nutrient blood agar overlaid with Hanks's balanced salt solution, they survived and produced viable eggs for at least 30 days and their rate of glucose utilization remained constant for at least the first 12 days. Under these conditions the amounts of glucose consumed and lactate produced were identical regardless of whether the worms were in an atmosphere of N_2 and CO_2 only or in one also containing oxygen (Table 9.1). Hence, there

TABLE 9.1. Aerobic and Anaerobic Glucose Utilization and Lactate Production by Adult *Schistosoma mansoni* over a Period of 12 Days[a,b]

Experiment no.	Days	Glucose utilization		Lactate production	
		$N_2/O_2/CO_2$	N_2/CO_2	$N_2/O_2/CO_2$	N_2/CO_2
1	1–3	3.08	3.32	5.94	6.51
	4–6	2.70	2.56	5.84	5.68
	7–9	1.90	2.02	5.13	5.17
	10–12	3.10	2.86	5.87	5.89
2	1–3	2.72	2.60	6.62	6.45
	4–6	2.32	1.99	5.76	5.22
	7–9	1.96	2.14	5.63	5.75
	10–12	2.69	3.06	5.72	5.62
3	1–3	2.92	3.02	6.97	7.10
	4–6	2.53	2.36	6.63	6.31
	7–9	2.94	2.58	5.65	5.48
	10–12	3.00	2.73	5.99	5.53

[a] Modified from Schiller *et al.* (1975).
[b] Values refer to micromoles of glucose utilized or lactate produced, respectively, per worm pair in 24 hr in a diphasic medium.

was no inhibition of glycolysis by oxygen (Pasteur effect), indicating no role for oxygen in energy metabolism. Furthermore, no differences were seen in gross morphology, motor activity, or frequency of pairing. There was, however, a striking difference in egg production. In the presence of oxygen the rate of egg-laying reached a maximum during days 4–6 when the average number of viable eggs per worm pair per day was 118, well within the range of 68 to 248 observed *in vivo.* In the absence of oxygen there was essentially no egg production. It has been found that the oxygen is required for the tanning of phenolic compounds in the formation of the eggshell. Oxygen may also be essential for motility (see Chapter 25). It seems likely that oxygen plays some similar role, entering into synthetic reactions, in other helminths whose principal energy metabolism is anaerobic, depending either on glycolysis as in schistosomes or on CO_2 fixation as in *Ascaris* and the tapeworms. A cytochrome system is present in these worms but with such low activities as to be of questionable significance in the adult stage. It may, however, be of overriding importance in larval stages, as already seen for *Ascaris.*

It would be of great interest to study the metabolism of the free-living stages of schistosomes, the miracidia and cercariae, as well as of the developmental stages in the snail vector. The cercariae are aerobic, as are most likely the miracidia.

Likewise among the parasitic protozoa living in the vertebrate bloodstream and transmitted by insect vectors, remarkable shifts in metabolic pattern occur during the life cycle. The erythrocytic stages of malarial parasites, developing as they do within an erythrocyte, might be expected to utilize oxygen for energy generation. This is not at all the case for parasites of mammalian red cells. Thus, the major human malarial parasite *Plasmodium falciparum* is in fact a homolactate fermenter, as is its host cell. This parasite, unlike the red cell, nevertheless contains appreciable amounts of cytochrome oxidase. Its function is not understood. Malarial parasites of mammals completely lack a tricarboxylic acid cycle and show no Pasteur effect. Activity of the pentose phosphate pathway is also of little significance. A detailed study of the metabolism of *P. knowlesi,* a malarial parasite of rhesus monkeys, indicated that end products of fermentation in addition to lactate were succinate, a volatile acid fraction that included formate and acetate but not CO_2, and smaller amounts of neutral volatile compounds that were not identified. Although our information is not so complete as would be desirable (and as we might expect for such an important group of parasites), it is clear that for mammalian malarial parasites the glycolytic pathway is the main source of energy. This is also true for malarial parasites of birds, despite the fact that here oxidation of pyruvate to CO_2 does occur and some enzymes of the tricarboxylic acid cycle have been demonstrated (Table 9.2).

In no case has a cytochrome-mediated electron transport chain been demonstrated. It must be realized, however, that the proper experimental conditions may not have been provided. It is difficult to measure oxygen consumption of organisms living in a cell full of hemoglobin. Malarial parasites

TABLE 9.2. Enzymes of Carbohydrate Metabolism in *Plasmodium*[a]

Enzyme	*P. gallinaceum*, *P. cathemerium*	*P. lophurae*	*P. berghei*, *P. vinckei* *P. chabaudi*	*P. knowlesi*	*P. falciparum*
Hexokinase	+[b]		+		
Glucose phosphate isomerase			+		+
Phosphofructokinase	+				
Aldolase	+				
Triose phosphate isomerase	+				
Glyceraldehyde-3-phosphate dehydrogenase	+				
Phosphoglycerate kinase		+	+		
Pyruvic kinase		+	+		
Lactic dehydrogenase	+	+	+	+	+
G6PDH		−	−	−	−
6-Phosphogluconate dehydrogenase			+	+	+
Malic dehydrogenase	+	+	±		
Succinic dehydrogenase	+	+	−		
Isocitric dehydrogenase		+	−		
Phosphoenolpyruvate carboxylase/ carboxykinase		+	+	+	
Cytochrome oxidase	+	+	+	+	+

[a] Modified from Sherman (1979).
[b] +, present; −, absent; ±, present in some strains; no symbol, no information.

removed from their host erythrocytes do not remain in good physiological condition unless kept in a complex medium with hemoglobin. In very recent work with the fluorescent compound Rhodamine 123, which labels mitochondria (and other structures with a strong ionic gradient across the membrane), H. Ginsburg and J. B. Jensen and their colleagues have observed specific staining of an organelle in erythrocytic stages of *P. falciparum* in culture, indicating the presence of an electron transport chain. A role for mitochondria in the erythrocytic stage is also indicated by the delayed growth inhibition resulting from exposure of *P. falciparum* cultures in human erythrocytes to specific inhibitors of mitochondrial protein synthesis, such as chloramphenicol. What this role of the mitochondria might be is not known.

CO_2 fixation occurs and in the rodent malaria *P. berghei* the enzymes PEP carboxylase and PEP carboxykinase have been found. It will be recalled that the latter enzyme is important in the metabolism of trichomonads and *Entamoeba*. There are also indications that malarial parasites, again like trichomonads, *Entamoeba*, and some helminths, may metabolize pyruvate to acetyl-CoA with the formation then of acetate and ATP.

Whereas host erythrocytes have an active pentose phosphate pathway which is important for the maintenance of reduced glutathione and reduced NADP and hence for prevention of accumulation of methemoglobin, this pathway in malarial parasites is nonexistent. Indeed, they lack the first enzyme in the pathway, glucose-6-phosphate dehydrogenase. They do, however, have the second enzyme, 6-phosphogluconate dehydrogenase. This paradox, and how malarial parasites obtain pentose sugars for nucleic acid synthesis, remains for future work to resolve.

Although malarial parasites depend primarily on fermentation reactions, they are not anaerobic; they must have some oxygen and they are very sensitive to cyanide. In the most detailed study done with *in vitro* cultures of *P. falciparum*, Scheibel and his associates found an optimal oxygen level of 3% O_2 with 2% CO_2. With 21% O_2, continuous growth of parasites was not obtained, about 17% O_2 being the upper limiting level. Growth still occurred down to as little as 0.5% O_2, but there was no growth in the complete absence of oxygen. The parasites are microaerophiles.

What might be the role of oxygen, and is it related to the uniform presence of cytochrome oxidase? There is now strong evidence that oxygen and cytochrome oxidase function in biosynthetic reactions especially in the *de novo* synthesis of pyrimidines (see below).

The cytochrome oxidase activity is demonstrable by electron microscope cytochemistry in the acristate mitochondria of the erythrocytic stages of mammalian malarial parasites as well as in the cristate mitochondria of erythrocytic stages of avian malaria. The activity of succinate dehydrogenase, however, could not be shown in erythrocytic stages of several species of rodent malaria but could be easily demonstrated in the oocysts and sporozoites. This enzyme can be considered indicative of tricarboxylic acid cycle activity and it is of interest that the mitochondria of oocysts are cristate. So also are the mitochondria of female gametocytes but these did not yet have succinate dehydrogenase activity. Presumably, this enzyme and perhaps the others of the tricarboxylic acid cycle appear either in the zygote (the ookinete) or in the early oocyst. Although present in the sporozoites, succinate dehydrogenase was again absent from the preerythrocytic stages in hepatic cells. There is thus strong indication that malarial parasites switch from a mainly glycolytic metabolism in the erythrocytic stages to an aerobic metabolism in the mosquito stages and back to a glycolytic one in the preerythrocytic stages. These changes are accompanied by morphological changes in the mitochondria.

Much better documented are the similar switches in metabolic pattern that occur in the life cycle of the African trypanosomes, again accompanied by corresponding changes in morphology of the mitochondria. These are especially well shown in the intensively studied parasite of cattle, *Trypanosoma brucei* (see Diagram V). The procyclic forms of this organism, that multiply extensively in the midgut of the tsetse fly, have a kinetoplast–mitochondrion which is well developed with platelike cristae. In epimastigotes in the salivary glands, the cristae are tubular and in metacyclic forms the mitochondrion is reduced to a simple tube with bulbous or tubular cristae. In the multiplicative slender bloodstream forms, the mitochondrion is further reduced and virtually without cristae, but in the stumpy forms, preadapted for infection of a tsetse fly, tubular cristae reappear. In keeping with these changes, the procyclic forms show a cytochrome-mediated aerobic respiration sensitive to cyanide whereas the bloodstream forms have a special α-glycerophosphate oxidase system insensitive to cyanide. Only enough is known about the other intermediate stages to indicate that the change from one type of metabolism to another occurs gradually over periods of several days paralleling morphological changes in the mitochondrion. Presumably, changes in temperature and in other environmental factors provide the triggers, but nothing is known as to the molecular mechanism controlling these cellular differentiations.

The metabolism of the procyclic forms of *T. brucei* has been studied through the use of the culture forms that grow axenically *in vitro* at temperatures of 27–28°C. These correspond quite closely in morphology and ultrastructure to the forms seen in the midgut of the tsetse fly and like them are noninfective. The culture trypomastigotes can oxidize glucose to CO_2 with little accumulation of organic acids and probably have a branched-chain cytochrome-mediated electron transport system. They have been considered to have a functional tricarboxylic acid cycle but there is some question as to the physiological significance of this and indeed of carbohydrate oxidation for these forms in nature. It appears that they may depend on proline as an energy source, as does their host, the tsetse fly. The tricarboxylic acid cycle might then operate mainly as a pathway of terminal oxidation for this amino acid (see Chapter 7 for nutritional role of proline for hemoflagellates).

There is no doubt, however, as to the overriding importance of glucose as the source of energy for the bloodstream forms of *T. brucei* and its subspecies

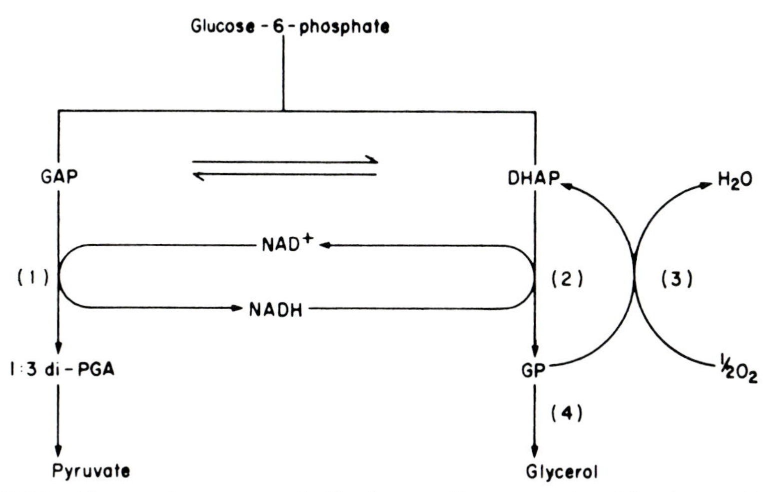

FIGURE 9.5. The reaction sequence in bloodstream *Trypanosoma brucei* for the reoxidation of NADH generated in glycolysis. GAP, glyceraldehyde-3-phosphate; DHAP, dihydroxyacetone phosphate; 1 : 3 di-PGA, 1 : 3-diphosphoglycerate; GP, L-glycerol-3-phosphate. See text for an explanation of reactions (1)–(4). (From Bowman and Flynn, 1976.)

T. b. rhodesiense and *T. b. gambiense,* the agents of African sleeping sickness. It has long been known that these organisms rapidly lose their motility and their respiratory activity when deprived of exogenous carbohydrate. They exhibit an active, incomplete oxidation of glucose with pyruvic acid as the main end product under aerobic conditions. Oxygen is consumed at the rate of 70–140 nmoles O_2/min per mg protein, about 50 times the rate of respiration of mammalian tissues. No CO_2 is formed and there is no functional tricarboxylic acid cycle although some enzymes of this pathway are present.

This rapid oxidation of glucose occurs via a pathway first discovered by Grant and Sargent and involving α-glycerophosphate oxidase (Fig. 9.5). Glyceraldehyde-3-phosphate (GAP) formed in glycolysis is oxidized to 1 : 3-diphosphoglycerate (1 : 3 di-PGA) by glyceraldehyde-3-phosphate : NAD^+ oxidoreductase with the formation of NADH (reaction 1). The NADH is reoxidized to NAD^+ by L-glycerol-3-phosphate : NAD^+ oxidoreductase with the reduction of dihydroxyacetone phosphate (DHAP) to L-glycerol-3-phosphate (GP) (reaction 2). The latter is reoxidized to DHAP by L-glycerol-3-phosphate oxidase (reaction 3). Under aerobic conditions only catalytic amounts of DHAP are needed for the reoxidation of NADH to NAD^+ required for glycolysis, and glucose is completely converted to pyruvate. ATP is formed only from the substrate-level reactions of glycolysis. Under anaerobic conditions equimolar amounts of glycerol and pyruvate are formed; DHAP serves as terminal electron acceptor and substrate amounts are reduced to GP which is acted on by a phosphatase to form glycerol (reaction 4).

The reactions of this α-glycerophosphate pathway occur in the mitochondrion of the bloodstream forms, which of course is also the site of the cytochrome-mediated electron transport system of the procyclic forms. In the bloodstream forms the latter system is turned off completely, whereas in the procyclic forms the α-glycerophosphate system is present but at a much reduced level of activity. In the transformation from bloodstream forms to procyclic forms as it occurs in culture medium, the switch from the α-glycerophosphate oxidase pathway to cytochrome-mediated respiration with a tricarboxylic acid cycle accordingly takes place in the mitochondrion. It is a gradual change and seems to take longer than the accompanying morphological changes. There may be two stages, the first characterized by appearance of new dehydrogenases that can transfer reducing equivalents to the α-glycerophosphate oxidase pathway and then to O_2. In the second stage, which appears at about 4 days in culture, the cytochrome system is formed, formation of the alternate oxidase is shut off, and respiration becomes cyanide sensitive. One stage blends gradually into the other. The alternate oxidase pathway is never completely lost but contributes less to the respiration in procyclic than in bloodstrem trypomastigotes.

At all stages the glycolytic enzymes involved in conversion of glucose and glycerol into 3-phosphoglycerate are present in membrane-bounded bodies termed glycosomes, quite unlike the situation in most cells where glycolytic enzymes are in the cell sap. The various glycolytic intermediates, however, must be distributed between glycosomes and the cell sap, from where they enter the mitochondrion.

The anaerobic dismutation of glucose to pyruvate and glycerol can be mimicked by the action of salicylhydroxamic acid, a powerful inhibitor of the glycerophosphate oxidase system. In the presence of this drug, although the α-glycerophosphate oxidase system is completely blocked, the motility of the organisms is only slightly reduced and ATP production is reduced by half. Hence, there must be another pathway of ATP formation which is at present not understood. This pathway can, however, be blocked with an excess of its product, glycerol. Indeed, the combination of glycerol with salicylhydroxamic acid has been shown to be an effective trypanocidal treatment. It must be admitted, finally, that despite a great deal of information the energy-generating metabolism of African trypanosomes is not fully understood.

In *T. cruzi* epimastigotes the tricarboxylic acid cycle and the pentose phosphate pathway are highly significant, as well as the glycolytic pathway. Much less is known about the amastigotes. In *Leishmania mexicana* the utilization of substrates differed considerably between the promastigotes and amastigotes; the latter showed a much higher uptake and metabolism of nonesterified fatty acids than the former. The rate of glucose utilization by promastigotes, however, was ten times higher than by amastigotes. This could be in part a result of the conditions under which the organisms were held for the measurements. A. J. Mukkada and colleagues have recently found that the metabolism of *L. donovani* amastigotes showed a very sharp dependence on pH; the optimal pH was 4.0–5.5 with marked decrease at higher pH. For promastigote metabolism, on the other hand, the optimal pH was around 7, the pH at which metabolic measurements are usually done. The pH 5 optimum for amastigotes is nicely in keeping with the pH of their habitat, the phagolysosome (see also Chapter 5).

It is clear that parasites in general get along well with only relatively inefficient metabolic pathways (Table 9.3). Perhaps this is because they have abundant supplies of substrates. Although many generate ATP by reactions not requiring oxygen, most can and do utilize oxygen. They respire and have terminal oxidases. In a few cases a special role for oxygen is known, as in the formation of schistosome eggs and possibly in pyrimidine biosynthesis by malarial parasites (see below). Like other cells, parasites are sensitive to the toxic effects of hydrogen peroxide, superoxide, and other active oxygen radicals formed in oxidative reactions (see also Chapter 25). Some parasites are better equipped than others with enzymes, such as superoxide dismutase and catalase, to handle these toxic metabolites. Also significant in this regard are the maintenance of sulfhydryl groups and an appropriate redox potential. Glutathione reductase, an NADP-dependent enzyme, maintains glutathione in the reduced form. In hemoflagellates, including leishmanias and trypanosomes, this enzyme turned out to require a special cofactor which has been identified as a glutathione–spermidine conjugate [trypanothione, or N^1, N^8-bis(L-γ-glutamyl-L-hemicystinyl-glycyl) spermidine].

Before leaving the subject of energy sources of parasites, it is important to consider the possibility of energy parasitism by intracellular parasites, i.e., dependence on host cell ATP. This has been shown for the prokaryotic intracellular parasites of the groups of rickettsiae and chlamydiae. Typhus rickettsiae translocate ATP somewhat after the manner of mitochondria and seem to require exogenous ATP in addition to that which they generate themselves. Chlamydiae, such as those of trachoma, are entirely dependent on host cell ATP. Studies bearing on this subject have so far been done only with the erythrocytic stages of malarial parasites. Trophozoites of the avian malaria *Plasmodium lophurae* removed from their host erythrocytes and incubated in a complex nutrient medium with duck red cell extract developed well extracellularly into multinucleate schizonts only if the medium was supplemented with ATP. Since these forms were surrounded by a parasitophorous membrane, it was suggested that external ATP might be essential to some transport function of this membrane. More recently, however, it has been found that merozoites of *P. falciparum* will undergo initial development into small ring forms extracellularly in a human red cell extract medium only if this is supplemented with di-potassium ATP and pyruvate. Since these forms have never entered an erythrocyte they are not surrounded by a parasitophorous membrane, and the ATP must function directly on the parasite. Just how remains for future study to determine. It seems likely that malarial parasites at least, and possibly other intracellular protozoa as well, make some use of the abundant ATP present in their environment.

TABLE 9.3. ATP Formation from Carbohydrate in Representative Parasites

	Stage	Habitat	Pathway	Major end products	Moles ATP/mole glucose
Tritrichomonas foetus	Trophic	Genital tract	Glycolysis + oxidative decarboxylation of pyruvate	Acetate Succinate Hydrogen	3
Entamoeba histolytica	Trophic	Intestine, liver	Same as above	Acetate Ethanol CO_2	2
Ascaris lumbricoides	Adult	Intestine	Glycolysis to PEP + dismutation of malate by malic enzyme & fumarate reductase	Acetate Propionate Methylvalerate Methylbutyrate	2–5
Hymenolepis diminuta	Adult	Intestine	Same as above	Same as above	2–5
Ascaris	Larva in egg	External environment	Glycolysis + TCA cycle	?	38(?)
Schistosoma mansoni	Adult	Portal vein	Glycolysis	Lactate	2
Plasmodium falciparum	Erythrocytic asexual	In red blood cells	Glycolysis	Lactate + others	2
Trypanosoma brucei	Blood stream	Vertebrate blood	Glycolysis + glycerophosphate oxidase	Pyruvate Glycerol	2
	Procyclic	Tsetse fly gut	Glycolysis + TCA cycle	CO_2 H_2O	38(?)

Biosynthesis, Nucleic Acids, Coenzyme A

As in other organisms, the ATP generated by metabolism in parasitic protozoa and helminths is used in a wide and diverse array of biosynthetic reactions. These have been little studied in parasites, but present indications are that in general they do not differ from corresponding biosynthetic mechanisms in other eukaryotic organisms. Protein synthesis occurs by the same mechanism on ribosomes, and the ribosomes even of the protozoa, though slightly different from mammalian ones, are nonetheless typical of eukaryotes. As we have already seen, most parasites require an exogenous source of the same 10 amino acids as free-living animals and can synthesize the other 12 or so that enter into proteins. In *in vitro* protein synthesis on ribosomes of erythrocytic malarial parasites, the reactions went better with homologous transfer RNA than with transfer RNA from rabbit reticulocytes. It will be of interest to determine the basis for this.

Lipid synthesis seems to present a rather different picture.. All parasites so far studied have a very limited capacity for synthesis of long-chain fatty acids. Mammalian cells can synthesize all their fatty acids from acetyl-CoA, so that a mammal living on a fat-free high-carbohydrate diet readily accumulates fat. This is not at all true for parasitic protozoa and helminths. At best they are capable only of limited chain elongation of already long-chain fatty acids obtained from the host. Once given the fatty acids, parasites actively synthesize triglycerides and phospholipids in amounts and ratios quite different from those of the host. Steroids, however, are not synthesized by most parasites and, as already noted (Chapter 7), must be obtained from the host (or from a culture medium). We have already noted that inability to synthesize sterol is a characteristic of many invertebrates and is not correlated with parasitism.

The synthesis of RNA and DNA in parasites seems not to be different from that in nonparasitic animals. Again like some nonparasitic invertebrates, insects for example, all parasites so far studied require an exogenous source of purine. They have only the so-called salvage pathways and adenine or hypoxanthine can serve for synthesis of all the purine nucleotides. The amastigotes and promastigotes of *L. donovani* show some interesting differences in their purine metabolism. Whereas amastigotes have an adenosine deaminase that forms inosine, promastigotes completely lack this enzyme. Promastigotes have instead an adenase that forms hypoxanthine from adenine, and this enzyme is absent in amastigotes. Moreover, the adenosine kinase activity of amastigotes is about 50 times higher (per mg protein) than that of promastigotes. With regard to biosynthesis of pyrimidines, there is much diversity. Most of the enzymes needed for *de novo* pyrimidine biosynthesis (Fig. 9.6) have been demonstrated in the hemoflagellates *Crithidia fasciculata, Trypanosoma cruzi,* and *Leishmania major,* in *Toxoplasma gondii,* in *Plasmodium,* and in *Babesia,* as well as in the helminths *Fasciola gigantica, Schistosoma man-*

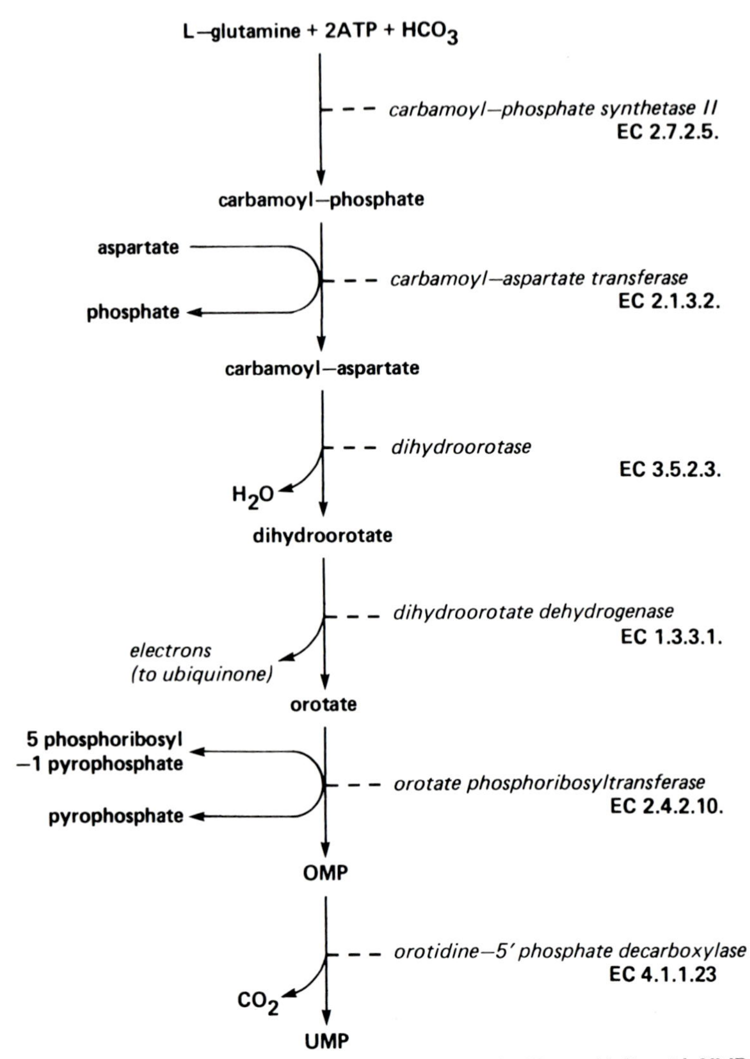

FIGURE 9.6. Pathway of *de novo* pyrimidine biosynthesis. OMP, orotidylic acid; UMP, uridylic acid. (From Hammond *et al.*, 1985.)

soni, Nippostrongylus brasiliensis, Trichiuris muris, and *Hymenolepis diminuta.* Some of these organisms also have the salvage pathway so that they incorporate exogenous thymidine. Others, such as erythrocytic malarial parasites, lack the salvage pathway; one cannot use tritiated thymidine to study DNA synthesis in these parasites. Still others have only the salvage pathway. This is the situation in *Tritrichomonas foetus* and in *Giardia lamblia.* The latter flagellate has been shown to lack completely the following four enzymes of the pyrimidine synthetic pathway: carbamoyl phosphate synthase, aspartate transcarbamoylase, dihydroorotase, and dihydroorotate dehydrogenase. Similarly, *T. foetus* does not incorporate bicarbonate, aspartate, or orotate into pyrimidine nucleotides or nucleic acids. It obtains its pyrimidine nucleotides mainly from incorporation of uracil into uridine monophosphate by the action of uracil phosphoribosyl transferase. Uracil or uridine, however, are not incorporated into DNA. The organisms lack dihydrofolate reductase and thymidylate synthase and thymidine kinase. Their only pathway for forming thymidine phosphate is by the action of thymidine phosphotransferase on thymidine and a phosphate donor such as nucleoside 5′-monophosphate. It is interesting that *Trichomonas vaginalis* was found to have only the first enzyme in the pathway, carbamoyl phosphate synthase. It may be significant that the active tachyzoite stages of the coccidian *Toxoplasma gondii* have all the enzymes whereas the oocysts of another coccidian (*Eimeria tenella*) have only the last two: orotate phosphoribosyl transferase and orotidine-5′-phosphate decarboxylase. There is room here for much further work especially with reference to changes in enzymatic equipment at different stages of the life cycle.

It is likely that at least in erythrocytic malarial parasites the oxidation of dihydroorotate to orotate is effected by molecular oxygen via cytochrome oxidase and ubiquinone according to the following scheme:

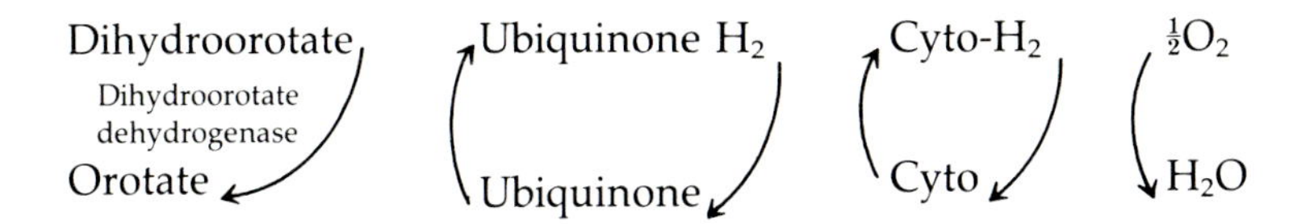

This would account for the presence of cytochrome oxidase in *Plasmodium* as well as for its oxygen requirement and great sensitivity to cyanide. All of these are related to the essential synthesis of pyrimidine.

Also essential to the synthesis of nucleic acids are the folate cofactors: these function in the transfer of one-carbon groups. Many parasitic animals, in common with many free-living ones, require an exogenous source of folate. As already seen in Chapter 7, this has been clearly shown for the few species grown axenically in a defined medium (certain flagellates and the nematode *Neoaplectana glaseri*). The hemoflagellates seem to have in addition a requirement for a pterin. On the other hand, it is a curious fact that the intracellular parasitic protozoa of the group Apicomplexa, organisms like *Toxoplasma*, other coccidia, and erythrocytic stages of malarial parasites, evidently synthesize folic acid, like many bacteria, from *p*-aminobenzoic acid, glutamic acid, and a pteridine. This was inferred early from the chemotherapeutic action of sulfanilamides against these parasites and their counteraction by *p*-aminobenzoic acid. For erythrocytic malarial parasites it is also indicated by the demonstration that *p*-aminobenzoic acid is an essential growth factor, and of the presence of a 7,8-dihydropteroate synthase. The folate synthetic pathway of malarial parasites is of great practical interest in relation to important antimalarial drugs such as sulfadoxine (interfering with the utilization of *p*-aminobenzoic acid) and pyrimethamine (an inhibitor of the parasite's dihydrofolate reductase). Whether exogenous folate can also be used, especially by drug-resistant parasites, remains an open question.

Rather different is the situation with regard to CoA and the erythrocytic stages of the avian malaria *Plasmodium lophurae*. These require preformed CoA (see also Chapter 7), a cofactor essential in both carbohydrate and lipid metabolism; they lack the enzymes for biosynthesis of CoA from pantothenate. This biosynthetic pathway, which is present in the host erythrocytes, consists of the following steps:

1. Pantothenic acid + ATP $\xrightarrow[\text{kinase}]{\text{pantothenate}}$ 4′-phosphopantothenic acid

2. 4′-Phosphopantothenic acid + cysteine $\xrightarrow[\text{phosphopantothenoyl cysteine synthase}]{\text{ATP or CTP}}$ 4′-phosphopantothenoyl cysteine

3. 4′-Phosphopantothenoyl cysteine $\xrightarrow[\text{phosphopantothenoyl cysteine decarboxylase}]{-CO_2}$ 4′-phosphopantotheine

4. 4′-Phosphopantotheine + ATP $\xrightarrow[\text{4'-phosphopantotheine kinase}]{Mg^{2+}}$ dephospho-CoA + PP_i

5. Dephospho-CoA + ATP $\xrightarrow[\text{dephospho-CoA kinase}]{Mg^{2+}}$ CoA + ADP

It is noteworthy that four of the five reactions utilize a mole of ATP so that 4 moles of ATP are needed to make 1 mole of CoA. Thus, the parasite can economize on its consumption of ATP if it can incorporate CoA from its host cell. It is not known whether all stages of malarial parasites or only the intraerythrocytic stages lack the CoA biosynthetic enzymes. It could well be that the information for these enzymes is present in the genome and is expressed in the mosquito stages but suppressed in the erythrocytic stages. It will be important to examine other kinds of intracellular parasites with regard to this and other biosynthetic pathways. They may all save energy and sponge at least to some extent on the biosynthetic machinery of their host cells.

Bibliography

Aikawa, M., Huff, C. G., and Sprinz, H., 1969, Comparative fine structure of the gametocytes of avian, reptilian and mammalian malarial parasites, *J. Ultrastruct. Res.* **26:**316–331.

Albach, R. A., and Booden, T., 1978, Amoebae, in: *Parasitic Protozoa,* Volume II (J. Kreier, ed.), Academic Press, New York, pp. 455–506.

Aomine, M., 1981, The carbohydrate transport and the utilization in protozoa, *Comp. Biochem. Physiol. A* **68:**131–147.

Asai, T., O'Sullivan, W. J., Kobayashi, M., Gero, A. M., Yokogawa, M., and Tatibana, M., 1983, Enzymes of the de novo pyrimidine biosynthetic pathway in *Toxoplasma gondii, Mol. Biochem. Parasitol.* **7:**89–100.

Bienen, E. J., Hill, G. C., and Shin, K.-O., 1983, Elaboration of mitochondrial function during *Trypanosoma brucei* differentiation, *Mol. Biochem. Parasitol.* **7:**75–86.

Blum, J. J., Yayon, A., Friedman, S., and Ginsburg, H., 1984, Effects of mitochondrial protein synthesis inhibitors on the incorporation of isoleucine into *Plasmodium falciparum in vitro, J. Protozool.* **31:**475–479.

Bowman, L. B. R., and Flynn, I. W., 1976, Oxidative metabolism of trypanosomes, in: *Biology of the Kinetoplastida* (W. H. A. Lumsden, and D. A. Evans, eds.), Academic Press, New York, pp. 435–476.

Brohn, F. H., and Clarkson, A. B., Jr., 1980, *Trypanosoma brucei brucei:* Patterns of glycolysis at 37°C *in vitro, Mol. Biochem. Parasitol.* **1:**291–305.

Brohn, F. H., and Trager, W., 1975, Coenzyme A requirement of malaria parasites: Enzymes of coenzyme A biosynthesis in normal duck erythrocytes and erythrocytes infected with *Plasmodium lophurae, Proc. Natl. Acad. Sci. USA* **72:**2456–2458.

Bryant, C., and Behm, C. A., 1976, Regulation of respiratory metabolism in *Moniezia expansa* under aerobic and anaerobic conditions, in: *Biochemistry of Parasites and Host–Parasite Relationships* (H. van den Bossche, ed.), Elsevier/North-Holland, Amsterdam, pp. 89–94.

Beuding, E., and Fisher, J., 1982, Metabolic requirements of schistosomes, *J. Parasitol.* **68:**208–212.

Cannata, J. J. B., and Cazzulo, J. J., 1984, The aerobic fermentation of glucose by *Trypanosoma cruzi, Comp. Biochem. Physiol. B* **79:**297–308.

Cazzulo, J. J., 1984, Protein and amino acid catabolism in *Trypanosoma cruzi, Comp. Biochem. Physiol. B* **79:**309–320.

Divo, A. A., Geary, T. G., Jensen, J. B., and Ginsberg, H. 1985, The mitochondrion of *Plasmodium falciparum* visualized by Rhodamine^{123} fluorescence, *J. Protozool.* **32:**442–446.

Fairbairn, D., 1970, Biochemical adaptation and loss of genetic capacity in helminth parasites, *Biol. Rev.* **45:**29–72.

Fairlamb, A. H., Blackburn, P., Ulrich, P., Chait, B. T., and Cerami, A., 1985, Trypanothione: A novel bis(glutathionyl) spermidine cofactor for glutathione reductase in trypanosomatids, *Science* **227:**1485–1487.

Ferone, R., and Roland, S., 1980, Dihydrofolate reductase: thymidylate synthase, a bifunctional polypeptide from *Crithidia fasciculata, Proc. Natl. Acad. Sci. USA* **77:**5802–5806.

Garnham, P. C. C., Bird, R. G., Baker, J. R., and Killick-Kendrick, R., 1969, Electron microscope studies on the motile stages of malaria parasites. VII. The fine structure of the merozoites of exoerythrocytic schizonts of *Plasmodium berghei yoelii, Trans. R. Soc. Trop. Med. Hyg.* **63:**328–332.

Gero, A. M., O'Sullivan, W. J., Wright, I. G., and Mahoney, D. F., 1983, The enzymes of pyrimidine biosynthesis in *Babesia bovis* and *Babesia bigemina, Aust. J. Exp. Biol. Med. Sci.* **61:**239–243.

Gero, A. M., Brown, G. V., and O'Sullivan, W. J., 1984, Pyrimidine de novo synthesis during the life cycle of the intraerythrocytic stage of *Plasmodium falciparum, J. Parasitol.* **70:**536–541.

Gutteridge, W. E., Dave, D., and Richards, W. H. G., 1979, Conversion of dihydroorotate to orotate in parasitic protozoa, *Biochim. Biophys. Acta* **582:**390–401.

Hammond, D. J., and Bowman, I. B. R., 1980, Studies on glycerol kinase, and its role in ATP synthesis in *Trypanosoma brucei, Mol. Biochem. Parasitol.* **2:**77–91.

Hammond, D. J., and Gutteridge, W. E., 1985, Purine and pyrimidine metabolism in the Trypanosomatidae, *Mol. Biochem. Parasitol.* **13:**243–251.

Hammond, D. J., Burchell, J. R., and Pudney, M., 1985, Inhibition of pyrimidine biosynthesis de novo in *Plasmodium falciparum* by 2-(4-t-butylcyclohexyl)-3-hydroxy-1,4-naphthoquinone *in vitro, Mol. Biochem. Parasitol.* **14:**97–109.

Hart, D. T., and Coombs, G. H., 1982, *Leishmania mexicana:* Energy metabolism of amastigotes and promastigotes, *Exp. Parasitol.* **54:**397–409.

Hill, B., Kilsby, J., Rogerson, G. W., McIntosh, R. T., and Ginger, C. D., 1980, The enzymes of pyrimidine biosynthesis in a range of parasitic protozoa and helminths, *Mol. Biochem. Parasitol.* **2:**123–134.

Honigberg, B. M., Volkmann, D., Entzeroth, R., and Scholtyseck, E., 1984, A freeze-fracture electron microscope study of *Trichomonas vaginalis* Donné and *Tritrichomonas foetus* (Riedmüller), *J. Protozool.* **31:**116–131.

Howells, R. E., 1970, Mitochondrial changes during the life cycle of *Plasmodium berghei, Ann. Trop. Med. Parasitol.* **64:**181–187.

Howells, R. E., 1970, Cytochrome oxidase activity in a normal and some drug-resistant strains of *Plasmodium berghei—A cytochemical study. II. Sporogonic stages of a drug-sensitive strain, Ann. Trop. Med. Parasitol.* **64:**223–225.

Kidder, G. W., Dewey, V. C., and Nolan, L. L., 1978, Transport and accumulation of purine bases by *Crithidia fasciculata, J. Cell. Physiol.* **96:**165–170.

Köhler, P., and Bachmann, R., 1980, Mechanisms of respiration and phosphorylation in *Ascaris* muscle mitochondria, *Mol. Biochem. Parasitol.* **1:**75–90.

Lindmark, D. G., 1980, Energy metabolism of the anaerobic protozoan, *Giardia lamblia, Mol. Biochem. Parasitol.* **1:**1–12.

Lindmark, D. G., and Jarroll, E. L., 1982, Pyrimidine metabolism in *Giardia lamblia* trophozoites, *Mol. Biochem. Parasitol.* **5:**291–296.

Lindmark, D. G., and Jarroll, E. L., 1984, Metabolism of trophozoites, in: *Giardia and Giardiasis* (S. L. Erlandsen and E. A. Meyer, eds.), Plenum Press, New York, pp. 65–80.

Lindmark, D. G., and Müller, M., 1973, Hydrogenosome, a cytoplasmic organelle of the anaerobic flagellate *Tritrichomonas foetus* and its role in pyruvate metabolism, *J. Biol. Chem.* **248:**7724–7728.

Looker, D. L., Berens, R. L., and Marr, J. J., 1983, Purine metabolism in *Leishmania donovani* amastigotes and promastigotes, *Mol. Biochem. Parasitol.* **9:**15–28.

Marczak, R., Gorrell, T. E., and Müller, M. 1983, Hydrogenosomal ferredoxin of the anaerobic protozoon, *Tritrichomonas foetus, J. Biol. Chem.* **258:**12427–12433.

Misset, O., and Opperdoes, F. R., 1984, Simultaneous purification of hexokinase, class-1 fructose bisphosphate aldolase, triosephosphate isomerase and phosphoglycerate kinase from *Trypanosoma brucei, Eur. J. Biochem.* **144:**475–483.

Moulder, J. W., 1983, Chlamydial adaptation to intracellullar habitats, in: *Microbiology—1983* (D. Schlessinger, ed.), American Society for Microbiology, Washington, D.C. pp. 370–374.

Mukkada, A. J., Meade, J. C., Glaser, T. A., and Bonventre, P. F., 1985, Enhanced metabolism of *Leishmania donovani* amastigotes at acid pH: An adaptation for intracellular growth, *Science* **229:**1099–1101.

Müller, M., 1975, Biochemistry of protozoan microbodies: Peroxisomes, α-glycerophosphate oxidase bodies, hydrogenosomes, *Annu. Rev. Microbiol.* **29:**467–483.

Müller, M., and Lindmark, D. G., 1978, Respiration of hydrogenosomes of *Tritrichomonas foetus.* II. Effect of CoA on pyruvate oxidation, *J. Biol. Chem.* **253:**1215–1218.

Opperdoes, F. R., Borst, P., and Fonck, K., 1976, The potential use of inhibitors of glycerol-3-phosphate oxidase for chemotherapy of African trypanosomiasis, *FEBS Lett.* **62:**169–172.

Pfefferkorn, E. R., 1978, *Toxoplasma gondii:* The enzymic defect of a mutant resistant to 5-fluorodeoxyuridine, *Exp. Parasitol.* **44:**26–35.

Reeves, R. E., Warren, L. G., Susskind, B., and Lo, H.-S., 1977,, An energy conserving pyruvate-to-acetate pathway in *Entamoeba histolytica:* Pyruvate synthase and a new acetate thiokinase, *J. Biol. Chem.* **252:**726–731.

Reyes, P., Rathod, P. K., Sanchez, D. J., Mrema, J. E. K., Rieckmann, K. H., and Heidrich, H.-G., 1982, Enzymes of purine and pyrimidine metabolism from the human malaria parasite, *Plasmodium falciparum*, *Mol. Biochem. Parasitol.* **5**:275–290.

Saz, H. J., 1981, Energy metabolisms of parasitic helminths: Adaptation to parasitism, *Annu. Rev. Physiol.* **43**:323–341.

Scheibel, L. W., and Pflaum, W. K., 1970, Carbohydrate metabolism in *Plasmodium knowlesi*, *Comp. Biochem. Physiol.* **37**:543–553.

Scheibel, L. W., Ashton, S. H., and Trager, W., 1979, *Plasmodium falciparum:* Microaerophilic requirements in human red blood cells, *Exp. Parasitol.* **47**:410–418.

Schiller, E. L., Bueding, E., Turner, V. M., and Fisher, J., 1975, Aerobic and anaerobic carbohydrate metabolism and egg production of *Schistosoma mansoni in vitro*, *J. Parasitol.* **61**:385–389.

Sherman, I. W., 1979, Biochemistry of *Plasmodium* (malarial parasites), *Microbiol. Rev.* **43**:453–495.

Smyth, J. D., 1969, *The Physiology of Cestodes*, Freeman, San Francisco.

Speck, J. F., Moulder, J. W., and Evans, E. A., Jr., 1946, The biochemistry of the malaria parasite. V. Mechanism of pyruvate oxidation in the malaria parasite, *J. Biol. Chem.* **164**:119–144.

Tanabe, K., 1983, Staining of *Plasmodium yoelii*-infected mouse erythrocytes with the fluorescent dye Rhodamine^{123}, *J. Protozool.* **30**:707–710.

Visser, N., and Opperdoes, F. R., 1980, Glycolysis in *Trypanosoma brucei*, *Eur. J. Biochem.* **103**:623–632.

Von Brand, T., 1979, *Biochemistry and Physiology of Endoparasites*, Elsevier/North-Holland, Amsterdam.

Wallach, M., and Kilejian, A., 1982, The importance of tRNA for the *in vitro* cell-free translation of messenger RNA isolated from the malaria parasite *Plasmodium lophurae*, *Mol. Biochem. Parasitol.* **5**:245–261.

Wang, C. C., and Aldritt, S., 1983, Purine salvage networks in *Giardia lamblia*, *J. Exp. Med.* **158**:1703–1712.

Wang, C. C., Verham, R., Tzeng, S. F., Aldritt, S., and Cheng, H., 1983, Pyrimidine metabolism in *Tritrichomonas foetus*, *Proc. Natl. Acad. Sci. USA* **80**:2564–2568.

Wang, C. C., Verham, R., Rice, A., and Tzeng, S., 1983, Purine salvage by *Tritrichomonas foetus*, *Mol. Biochem. Parasitol.* **8**:325–337.

Webster, H. K., and Whaun, J. M., 1981, Purine metabolism during continuous erythrocyte culture of human malaria parasites (*P. falciparum*), in: *The Red Cell* (G. J. Brewer, ed.), Liss, New York, pp. 557–570.

Weinbach, E. C., Diamond, L. S., and Claggett, E., 1976, Iron–sulfur proteins of *Entamoeba histolytica*, *J. Parasitol.* **62**:127–128.

White, A., Handler, P., Smith, E. L., Hill, R. L., and Lehman, L. R., 1978, *Principles of Biochemistry* (6th edition), McGraw–Hill, New York.

Winkler, H. H., 1976, Rickettsial permeability: An ADP–ATP transport system, *J. Biol. Chem.* **251**:389–396.

Winkler, H., 1982, Rickettsiae: Intracytoplasmic life, *ASM News* **48**:184–187.

Yamada, K. A., and Sherman, I. W., 1981, Purine metabolism by the avian malarial parasite *Plasmodium lophurae*, *Mol. Biochem. Parasitol.* **3**:253–264.

CHAPTER 10

Genetics. Developmental Biology

How different sets of genes are turned off and on in the differentiation of cells and the development of an organism is a central question of biology. Little is yet known; definitive answers are not likely until much more information becomes available regarding the organization of the genome. Parasites, with their highly specialized life cycles requiring sharp changes in both structure and function, might be expected to provide exceptional material for such studies. Yet until recently little had been done on either the genetics or the genome of parasitic animals. The rich rewards to be reaped in such work, and its bearing on both molecular biology and the biology of parasitism, have quickly become apparent in two fields now actively under investigation. Both deal with the African trypanosomes. One is concerned with the remarkable ability of these organisms, when infecting the vertebrate bloodstream, to avoid the immune responses of their host by successive and almost unlimited changes in the antigenic nature of their surface coat. The other is concerned with an equally remarkable specialized mitochondrial DNA present in trypanosomes and related flagellates. The genes and control mechanisms of the variable surface glycoproteins, and the kinetoplast DNA, each deserve a separate chapter (Chapters 18 and 11). Here I will consider sex and genetics among parasitic animals, what is known regarding their genomes, and the bearing of this on parasitism.

Sexual reproduction is common among all parasitic worms and in many of the parasitic protozoa. Often in both groups it alternates with periods of extensive asexual multiplication; this usually occurs in different hosts. In the digenean flatworms, for example, mating and the production of eggs takes place in the vertebrate host whereas one or more cycles of reproduction by budding of diploid cells proceed in the molluskan host (see Diagrams XII, XIII). There seems to be a rough inverse correlation between egg production and extent of asexual multiplication. The liver fluke *Fasciola hepatica* lays 9000 to 25,000 eggs per adult per day but produces only about 11–16 cercariae per snail host per day, whereas the blood fluke *Schistosoma mansoni* lays only about 50 to 60 eggs per day but releases 1000 to 9000 cercariae per snail per day. In schistosomes, heterosexual pairing is essential for normal growth of the female. Chemoattraction between the sexes probably depends on steroids, and the male, which encloses the female in a special gynecophoric canal, provides her with essential nutrients. In tapeworms, things are less formal—the hermaphroditic proglottids of *Hymenolepis diminuta* carry on both self- and cross-insemination. It is estimated that these worms devote 80% of their energy intake to egg production. At the same time, an asexual multiplication occurs in the neck region of the tapeworm as new proglottids are formed. Asexual multiplication of cestodes in intermediate hosts is common only in one family (Taeniidae) and is beautifully exemplified by the hydatid cysts of *Echinococcus* (Diagram XVII). These form numerous scolices, each capable of growing into a tapeworm if ingested by a predaceous definitive host (as a dog) but also able to form new cysts if introduced into another intermediate host (as a sheep). All stages of parasitic worms are diploid except for the sexual cells, the sperm and eggs. There is no evidence for haploid generations such as occur among some free-living invertebrates.

The reverse situation obtains among most parasitic protozoa for which sexual reproduction is clearly evident: the organisms are haploid except in the zygote. In all the important parasites of the group Apicomplexa, as the genera *Eimeria, Toxoplasma, Sarcocystis,* and *Plasmodium,* there is an immediate zygotic reduction division from the diploid (2 C) to the haploid (1 C). There is no 4 C stage and hence no opportunity for crossing over as occurs in classic meiosis. Such classic meiosis, however, has been shown in elegant studies by L. R. Cleveland of the cytology of the large flagellates symbiotic in the intestine of the wood-feeding roach *Cryptocercus* (see Chapter 24). The numerous species of polymastigote and hypermastigote flagellates found in this special ecological niche all show sexual phenomena related to the molting of their host. Some then multiply asexually as diploid organisms but most show typical two-division meiosis following sexual union, and then multiply in the haploid state.

Only a few species of parasitic protozoa have been used so far for conventional genetic studies involving mating of different strains and analysis of their progeny. Most of the work has been done with species of malarial parasites having convenient laboratory mosquito and vertebrate hosts. Pioneering studies carried out in the 1950s by J. Greenberg using *Plasmodium gallinaceum* in chicks and the mosquito, *Aedes aegypti,* showed that characters such as drug resistance to pyrimethamine behaved like unit genetic characters and could be transferred by crossing from one strain to another. In recent years this work has been greatly extended by D. Walliker and his associates using several species of rodent malaria in mice with *Anopheles stephensi* as the mosquito vector. The technique is straightforward but by no means simple. Two appropriate lines of parasites differing in two or more characters are used to infect mosquitoes simultaneously. This is done by allowing the mosquitoes to feed on a mixture of mouse blood containing gametocytes of the two lines in approximately equal numbers. It is assumed that random mating occurs so that equal numbers of products of self- and cross-fertilization are

TABLE 10.1. Results of Cross of Two Lines of *Plasmodium chabaudi* Having Different Isozymes of 6-Phosphogluconate Dehydrogenase (PGD) and Lactic Dehydrogenase (LDH) [Line 411AS (PGD-2, LDH-3) × Line 96AJ (PGD-3, LDH-2)][a]

Isozymes of clones	No. of clones obtained	Type	Total of each type
PGD-2, LDH-3	43	Parental	55
PGD-3, LDH-2	12	Parental	
PGD-2, LDH-2	9	Recombinant	15
PGD-3, LDH-3	6	Recombinant	

[a] Data from Walliker (1983b).

formed. The sporozoites that develop in the mosquitoes are used to infect mice. The parasites appearing in these mice are then cloned by a dilution method into other mice and the resulting infections are examined for the characters under study. A good example is provided by a cross of two lines of *Plasmodium chabaudi* differing in electrophoretic isozymes of 6-phosphogluconate dehydrogenase (PGD) and lactate dehydrogenase (LDH). One line had isozymes PGD-2 and LDH-3 whereas the other had PGD-3 and LDH-2. When these two lines were treated in the manner described above, 70 clones were obtained with the characters shown in Table 10.1. There were 15 recombinant clones to 55 of the parental types, not far from the theoretical ratio of 1 : 3 to be expected from simple Mendelian inheritance. This experiment and others like it, also show clearly that the erythrocytic stages of malaria are haploid, in keeping with other indications that meiosis occurs in the zygote early in oocyst development.

Pyrimethamine resistance in *P. yoelii*, another species of rodent malaria, is inherited in a manner similar to that for isozymes. To show this, two parental lines were used, one highly resistant to pyrimethamine and with a glucose phosphate isomerase (GPI) electrophoretic variant designated GPI-1, whereas the other line was sensitive to pyrimethamine and had an electrophoretically different GPI, designated GPI-2. The clones obtained from the cross are shown in Table 10.2. The proportion of recombinants obtained indicates that the resistance was a result of a single gene mutation. A more complex situation was observed with a strain of *P. chabaudi* highly resistant to chloroquine. When this was crossed with a sensitive line, four types of clones were obtained: some highly resistant and others fully sensitive like the two parental lines, and in addition two of different intermediate sensitivity. This suggests that the high resistance depended on several mutant genes that segregated out in the recombinants to give intermediate levels of resistance. Virulence likewise has been shown to have a genetic basis.

Studies on drug sensitivity in the coccidia *Eimeria* and *Toxoplasma* have

TABLE 10.2. Results of Cross of Two Lines of *Plasmodium yoelii* Differing in Sensitivity to Pyrimethamine and in Isozymes of Glucose Phosphate Isomerase (GPI) [Line A (Resistant, GPI-1) × Line C (Sensitive, GPI-2)][a]

Characteristics of clones	No. of clones obtained	Type	Total of each type
Resistant, GPI-1	21	Parental	51
Sensitive, GPI-2	30	Parental	
Resistant, GPI-2	13	Recombinant	20
Sensitive, GPI-1	7	Recombinant	

[a] Data from Walliker (1983b).

again shown that recombination occurs and that meiosis takes place soon after zygote formation so that all other stages are haploid. *Toxoplasma gondii* is particularly suitable for experimental genetic studies, since much of the work can be done with the tachyzoite stage (see Diagram VII) of the parasites growing intracellularly in fibroblasts maintained in tissue culture. Experiments with crosses, however, have still to be done in the cat host, since, as noted earlier, sexual stages of most coccidia do not develop in tissue culture. Elmer R. Pfefferkorn is exploiting this system to good effect. When two drug-resistant mutants of *Toxoplasma* were crossed by feeding them to a cat, the resulting oocysts were found to contain sporozoites resistant to both drugs as well as the two parental types each resistant to one. These three types were easily demonstrated by a plaque assay in which fibroblast cultures were infected and exposed to each drug separately or to both together. The fourth type to be expected, sensitive to both drugs, was also demonstrated by "blind" cloning in drug-free medium followed by separate testing with the two drugs. Twelve percent of the clones obtained were resistant to both drugs, exactly the theoretical value to be expected for two unlinked markers if the organisms are haploid except in the zygote stage and if approximately equal numbers of gametes are produced by each parental type. The influence of this last parameter, the proportion of gametes contributed by each parental type, was evaluated experimentally by analysis of the sporozoites released from sporulated oocysts and plaque-assayed with each drug separately (Fig. 10.1). It is clear that if each parent contributes about half of the gametes, the expected proportion of about 12% doubly resistant progeny is obtained.

Unlike the Apicomplexa just discussed, trypanosomes and perhaps other flagellates of the large group of the Kinetoplastida are diploid. There are two lines of evidence to support this. First, an analysis of the distribution and frequency of isozymes of eight enzymes among natural populations of *Trypanosoma brucei* showed that these could best be accounted for on the basis of a randomly mating population of diploid organisms in Hardy–Weinberg equilibrium. Cytophotometric methods then showed an average nuclear DNA content for nonreplicating cells of *T. brucei* of 0.097 pg. This is not much over twice the value of 0.041 pg/nucleus found in DNA–DNA renaturation exper-

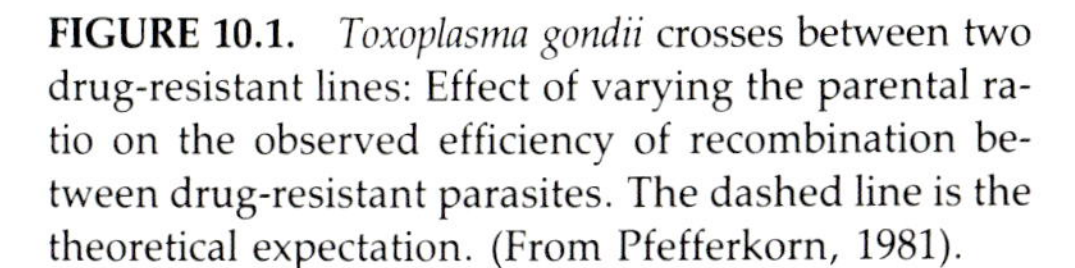

FIGURE 10.1. *Toxoplasma gondii* crosses between two drug-resistant lines: Effect of varying the parental ratio on the observed efficiency of recombination between drug-resistant parasites. The dashed line is the theoretical expectation. (From Pfefferkorn, 1981).

iments, a technique that gives the haploid amount. In the cytophotometric work no haploid cells were found, either in bloodstream forms from rats or in procyclic forms from culture. In cells having two kinetoplasts but one nucleus, and hence about to divide, the nuclear DNA was 0.181 pg/nucleus, close to twice the value of 0.097 in nondividing cells. The metacyclic forms, however, have been clearly shown to be haploid, strongly supporting the inference that sexual recombination occurs in these parasites. Furthermore, Leo Jenni and his colleagues have now directly demonstrated such sexual recombination. When tsetse flies were simultaneously infected with two clones of *T. brucei* differing in isoenzyme patterns, it was possible to recover from the flies, after cyclical development, clones with hybrid isoenzyme patterns. This should facilitate genetic analysis of the African trypanosomes and of their variant surface glycoprotein genes (see Chapter 18).

For *Trypanosoma cruzi,* the agent of American trypanosomiasis or Chagas disease, a study of 15 enzymes in 121 stocks from a wide geographic range has shown that the organisms are indeed diploid but their reproduction is basically clonal, with little if any sexual recombination occurring. In this very large sample population there were only 43 genotype patterns and of these 16 differed by only one allele. Furthermore, the same clone, at least as judged by these isozyme patterns, was often present in distant sites and in different hosts. Whether the biochemical heterogeneity with regard to isozymes is in any way related to heterogeneity in pathogenicity and other biological properties remains an open question of considerable medical importance.

In bacterial genetics the phenomenon of transformation has provided a valuable tool, especially when sexual recombination did not occur. This depends on incorporation into the genome of an exogenously supplied DNA. There is as yet no clear-cut example of this phenomenon among the protozoa. The closest approach is a series of observations and experiments by B. M.

Honigberg with the flagellate *Trichomonas gallinae*. Culture strains of this highly pathogenic parasite of pigeons lose first their pathogenicity and later their infectivity. The infectivity for pigeons and some measure of pathogenicity as measured by a mouse assay could be restored by exposing the noninfective cultures to *both* DNA and RNA from a virulent strain. It will be both interesting and important to find out what is going on here.

The genome size and complexity have now been determined for a sufficient number and variety of parasitic animals to show that their genome is neither smaller nor less complex than that of free-living animals. It is reasonable to suppose that any loss of biosynthetic function is compensated for by all the special adaptations a parasite must have to get to its host, to enter it, to survive and multiply within it, and to get its progeny disseminated to other hosts. Table 10.3 gives some comparative data. It will be seen that most of the parasites have a genome size of about 1×10^8 base pairs (bp) to 2×10^8 bp, very like that of *Drosophila* (but also not very different from that of salmon!). Similarly, the percentage of guanine + cytosine is like that of corresponding groups of free-living organisms. In a number of parasitic protozoa this is very low (only 20% in several species of *Plasmodium*), but it is also low in some free-living protozoa. Not enough is known to ascribe any special significance to this.

Of special interest in Table 10.3 are the figures for *Ascaris* and *Parascaris*. Both of these large parasitic nematodes show much less DNA in their somatic nuclei than in their germline nuclei. This striking diminution of chromatin occurring in the first division of the fertilized egg, a phenomenon by no means restricted to parasites, was first observed in *Ascaris* many years ago. Recent work with *Ascaris* has shown that the percentage of repetitive DNA (22%) is the same as the percentage of germline-limited DNA. Only in the germline DNA is heterochromatin present. The germline DNA shows a satellite with an AT content of 65% and a 123-bp repeat unit, of which about 0.1% is retained in the somatic genome. This indicates that the cleavage leading to DNA diminution is within the satellite DNA itself. Further study of this material cannot fail to be of significance, for we have here a sequence of DNA that is a major component of the germline genome but largely absent from the somatic genome.

The relative amounts of repetitive DNA in eukaryote genomes may give some indication of relative complexity. Here, too, the bare figures as given in Table 10.3 cannot be taken too seriously, since much depends on exactly how they were determined. Special interest attaches to the results reported for *Plasmodium berghei* and based on renaturation kinetics as measured spectrophotometrically. The wide range in percent repetitive DNA was determined for two lines, one of which with 3% repetitive DNA no longer produced gametocytes whereas the other, with 18% repetitive DNA, did form gametocytes. A gametocyte-forming line was then passaged by syringe every 4 days for 120 passages, and its infectivity for mosquitoes and the proportion of repetitive DNA were determined at five times during this period. Both

TABLE 10.3. Genome Size and Complexity of Representative Parasitic and Nonparasitic Organisms (*E. coli* Genome = 4.2×10^6 bp)[a]

Group	Species	DNA/ nucleus (pg)	Genome size (bp)	% repetitive DNA
Parasitic protozoa	*Leishmania donovani*	0.10	0.9×10^8	—
	Trypanosoma brucei	0.097	—	—
	T. cruzi	—	2.5×10^8	—
	Plasmodium lophurae	0.13	0.9×10^8	—
	P. falciparum	—	0.2 to 3.8 $\times 10^8$	10
	P. berghei	—	0.2×10^8	3–18
	Eimeria tenella (oocyst)	5.8	—	—
	E. tenella (haploid amount)	0.075	—	—
	Toxoplasma gondii (haploid amount)	0.096	—	—
	Sarcocystis cruzi (haploid amount)	0.216	—	—
	Trichomonas vaginalis	0.53	—	—
	Entamoeba histolytica	0.45	—	—
Free-living protozoa	*Acanthamoeba castellanii*	10.6	—	—
	Dictyostelium discoideum	—	0.45×10^8	30
	Tetrahymena pyriformis	—	1.4×10^8	10–20
Fungi	*Neurospora crassa*	—	0.24×10^8	8
	Saccharomyces carlsbergensis	—	0.14×10^8	11
Parasitic helminths	*Ascaris lumbricoides*			
	Germline	0.32	2.3×10^8	22
	Soma	0.25	—	—
	Parascaris equorum			
	Germline	1.2–2.1	—	—
	Soma	0.25	—	—
	Trichinella spiralis	—	2.5×10^8	42
	Hymenolepis diminuta	—	1.4×10^8	16
	Schistosoma mansoni	0.26	2.7×10^8	—
Free-living helminths	*Caenorhabditis briggsae*	—	1.2×10^8	9
	Dugesia tigrina	—	0.7×10^8	50
Other metazoa	Human	6.4	32×10^8	25
	Rattus norvegicus	3.2	22×10^8	—
	Salmon	—	2.8×10^8	58
	Drosophila melanogaster	0.12	1.6×10^8	10
	Limulus polyphemus	2.8	17.9×10^8	75

[a] Data from various sources. See Bibliography.

values decreased, with a high coefficient of correlation. Furthermore, there was a correlation between the infectivity for mosquitoes and the proportion of repetitive DNA in two clones derived (by dilution) from the passage line. It is suggested that the repetitive DNA may play a role both in gametocytogenesis and in differentiation of gametocytes into viable fertile gametes. However, a study of repetitive DNA in *P. falciparum* failed to reveal any significant differences in amount between a gametocyte-forming clone and a non-gametocyte-forming clone, both derived from the same isolate.

Clones of *P. falciparum* from the same isolate do show small differences in the organization of repetitive DNA sequences; much larger differences are seen among isolates from different geographical regions (Fig. 10.2). These differences may correspond to differences in antigenic structure and so become of interest in relation to immunological studies. Several genes of *P. falciparum* and of other malarial parasites have been cloned by recombinant DNA methods and expressed in bacteria and yeast (see Chapter 19 for detailed discussion). The same has been done for several other kinds of parasites, notably for the African trypanosomes in relation to the phenomenon of antigenic variation (see Chapter 18).

Antigenic variation is an extreme example of the switching on and off of particular genes. Throughout their life cycles, however, parasites must be able to switch off one set of genes and switch on another, sometimes at short notice. Only in this way can they respond appropriately to the sometimes sudden changes in their environment. For example, in those nematodes (e.g., hookworms, Diagram XX) that have a free-living infective third-stage larva, the expression of genes specifically concerned with the next or first parasitic stage must be suppressed. This suppression is probably at the level of DNA transcription and it is likely that juvenile hormone is the agent responsible. In these organisms the stimulus releasing suppression is mainly CO_2 (see Chapter 3); this may act on both gene activity and the endocrine system with carbonic anhydrase serving as the receptor. There are obviously analogies here with the function of juvenile hormone in the development and metamorphosis of insects.

When a bloodstream form of an African trypanosome is ingested into the midgut of a tsetse fly, the genes responsible for formation of the enzymes of the cytochrome-mediated respiratory pathway must be turned on (see Chapter 9). Some of these may actually reside in the kinetoplast DNA (see Chapter 11). For *Trypanosoma cruzi* some evidence has been obtained for specific substances, probably lectins, that interact with surface glycoproteins to effect transformation from epimastigote to trypomastigote in the vector bug (Diagram IV). The leishmanial amastigote-to-promastigote transformation (see Chapter 8) provides a particularly good example amenable to experimental approach. In the change of *Leishmania mexicana* from amastigote to promastigote, new surface antigens appear and there is a dramatic increase in both α- and β-tubulin, associated with the formation of the flagellum. Unlike the usual situation in eukaryotic cells, the α-tubulin seems to be under posttran-

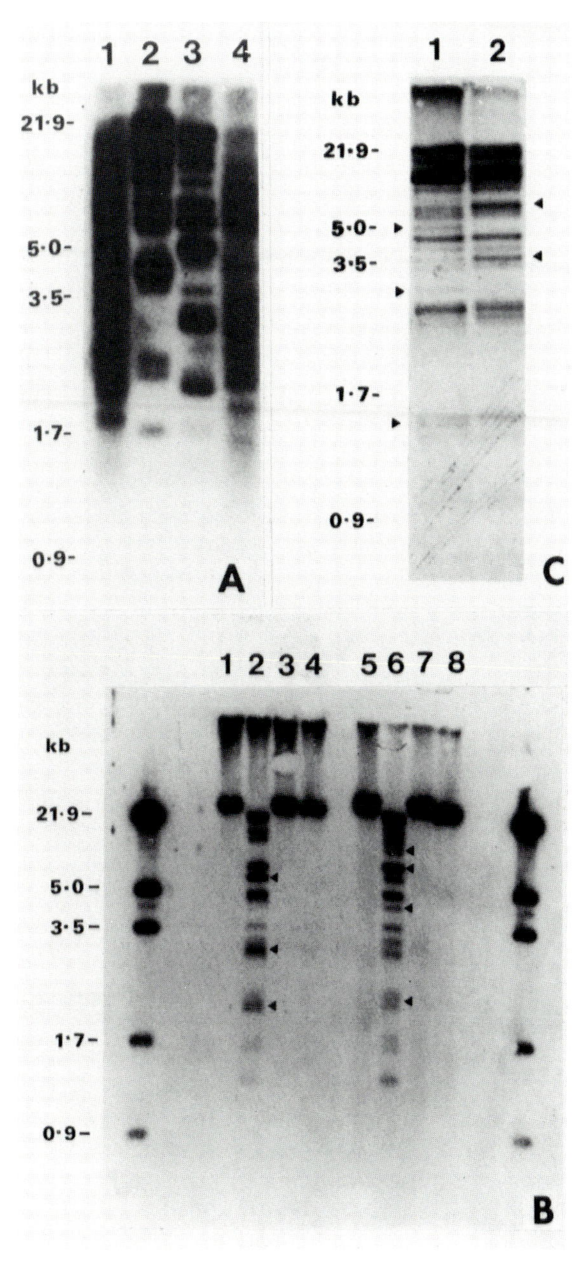

FIGURE 10.2. (A) Comparison of repetitive DNA in two isolates of *Plasmodium falciparum*. Digested genomic DNA (10 μg/lane) was fractionated on 0.8% agarose gel and analyzed by transfer hybridization using as probe a repetitive falciparum DNA clone (2a) prepared in phage lambda. Restriction enzymes were *Hin*dIII (lanes 2 and 3) and *Sau*3A (lanes 1 and 4). Lanes 1 and 2: isolate FCR-3/Gambia; lanes 3 and 4: isolate Honduras I/CDC. Note the marked differences between the *Hin*dIII-digested materials (lanes 2 and 3). Marker sizes shown on left. (B) Differences in repetitive DNA of two clones derived from the Honduras I/CDC isolate. Lanes 1–4: clone HB-3, gametocyte-producing; lanes 5–8: clone HB-2, non-gametocyte-producing. Genomic DNA (5 μg/lane) was digested, electrophoresed through 0.8% agarose, and a nitrocellulose replica probed with the DNA clone 2a. Enzymes were: *Pst*I, lanes 1 and 5; *Hin*dIII, lanes 2 and 6; *Bam*HI, lanes 3 and 7; *Eco*RI, lanes 4 and 8. Marker sizes shown on left. Prominent differences are marked by arrowheads. (C) Comparison of *Hin*dIII-digested genomic DNA from two clones of isolate FCR-3/Gambia. Lane 1: clone D-4 (knobless); lane 2: clone A-2 (knobby). Experimental details as for B. (From Bhasin *et al.*, 1985. Original prints courtesy of Dr. C. Clayton.)

scriptional control. With the β-tubulin, however, a single mRNA of 3600 nucleotides present in the amastigotes is replaced in the promastigotes with three species of mRNA of 2800, 3500, and 4400 nucleotides, the smallest being the most abundant.

In the formation of gametocytes of malarial parasites, either in an infected host or in culture *in vitro*, a certain small proportion of the merozoites develop into gametocytes. In these, something must activate whole regions of the genome that otherwise remain quiescent and unexpressed, resulting in a cell entirely different from the asexually reproducing form. The trigger here must be a particularly subtle one, since the gametocytes develop in the same host or in the same culture along with asexual forms. I have already noted (Chapter 8) that *in vitro* conditions slightly unfavorable to the asexual forms support the formation of gametocytes of *P. falciparum*. With *P. falciparum* in people there is a definite time in the course of the infection when gametocytogenesis begins, suggesting that the signal for this might be related to immune reactions of the host. In keeping with this is the intriguing observation that a certain proportion of lymphocytes from infected children added to cultures of *P. falciparum* together with serum from infected children greatly stimulated gametocyte formation. Neither the infected serum alone, nor the lymphocytes with normal serum, produced this effect. It has been found that the rodent malaria *P. chabaudi* will show up to 50% gametocytes when inoculated from an infected mouse into a splenectomized rat, a partially unfavorable host (this species shows no development in intact rats). One must assume some effect of the rat host on the erythrocyte, for the gametocytes developed as much in the introduced mouse erythrocytes as they did in rat erythrocytes. Subtle changes in physiology of particular erythrocytes may determine the developmental path of merozoites that have entered them. In keeping with this is the observation for *P. falciparum* that doubly infected cells very rarely have one gametocyte and one asexual parasite; almost always there are two gametocytes or two asexual forms.

Once gametocytes are triggered to gamete formation by being taken into the gut of a mosquito, additional new genes are activated to initiate the sporogonic cycle. Similarly, when the sporozoites enter a hepatic cell, still other sets of genes become active, and still others with the return to the erythrocyte.

The molecular biology that may help to explain these phenomena is only just begun. For malarial parasites a number of genes have been cloned. Most of these are of interest in relation to the antigens of the parasite and the possible development of a vaccine (see Chapter 19). Genes have also been cloned for trypanosomes and leishmanias and for schistosomes. Whereas helminthic parasites show typical chromosomes to which it will be possible to assign various characters, chromosomes of the small parasitic protozoa have not been seen by either light or electron microscopy. A powerful new method, however, has recently been developed for yeast which also isolates chromosomes of malaria, trypanosomes, and leishmanias in an electric field. The pulsed-field gradient gel electrophoresis separates chromosome-sized DNA molecules up to at least 2000 kilobases (kb) in length. It has been shown that the DNA is not degraded and that these molecules represent full-length chromosomes. In this way it has been found that *P. falciparum* has at least seven chromosomes, and work has already begun to localize on the chromosomes

TABLE 10.4. Main Features of the Molecular Karyograms of Nine Kinetoplastid Species[a]

Species	Chromosome size			Gene location								
	Range of migrating chromosomes (kb)	No. of chromosomes >700 kb[b]	No. of chromosomes <700 kb	Miniexon[c]			Tubulin[d]			rRNA[e]		
				Slot	≥2000 kb	≤2000 kb	Slot	≥2000 kb	≤2000 kb	Slot	≥2000 kb	≤2000 kb
Trypanosoma brucei	50–2000	—[f]	~100	+[i]			+			+	+	
T. equiperdum	50–2000	—[f]	~10	+				+		+	+	
T. vivax	2000	—[f]	0	+			+			+	+	
T. cyclops	700–4000[g]	>20	0		+				+	+	+	+
T. rangeli	700–4000	>20	0	+			+			+		
Leishmania t. minor	700–4000	>20	0	ND[h]				+	+		+	
Herpetomonas m. muscarum	700–4000	>20	0		+			+		+	+	
Leptomonas ctenocephali	700–4000	>20	0	ND			+		+	+	+	
Crithidia fasciculata	700–2000	>10	0	+			ND			ND		

[a] Modified from Van der Ploeg *et al.* (1984).
[b] Tentative estimates of the total number of chromosomes were obtained by summing the number of chromosomal bands and the copy number of each band estimated by their relative intensities after ethidium staining.
[c] Hybridization was with a synthetic oligonucleotide complementary to part of the miniexon of variant surface glycoprotein mRNA.
[d] Hybridization with an α- and β-tubulin probe.
[e] Hybridization with a probe containing a complete repeat of the rRNA transcription unit of *T. brucei*.
[f] The exact number of chromosomes cannot be determined as a large fraction of the DNA remains close to the slot.
[g] The extrapolation to 4000 kb is tentative.
[h] ND, not done.
[i] +, hybridization in that area.

particular genes that have been cloned. *Leishmania major* has at least 17 chromosomes. The α- and β-tubulin genes have been mapped and found to be in multiple unlinked loci. Their arrangement is different from that in other organisms; this may be related to their different expression during the two main stages of the life cycle.

Molecular karyograms of this type have been reported for a number of other kinetoplastid flagellates (Table 10.4). It will be noted that except for *Crithidia fasciculata,* all those for which a determination was possible had at least 20 chromosomes. Also noteworthy is the fact that the presence of minichromosomes (< 700 kb) was not correlated with antigenic variation. Thus, although *Trypanosoma brucei, T. equiperdum,* and *T. vivax* all show antigenic variation, the number of minichromosomes ranged from about 100, to 10, to 0. This is a matter to which I will return in Chapter 18. It will also be seen that ribosomal and tubulin genes were distributed over different portions of the genomes.

Further work along these and related lines may be expected to help to define the nature and role of particular parasite proteins and the control of developmental changes in the life cycle. The techniques of molecular biology are also finding important practical applications in new methods for diagnosis of parasitic infections and in attempts to develop vaccines (see Chapter 19).

Bibliography

Bahr, S. F., 1966, Quantitative cytochemical study of erythrocytic stages of *Plasmodium lophurae* and *Plasmodium berghei, Mil. Med.* **131**(Suppl.):1064–1070.

Bhasin, V. K., Clayton, C., Trager, W., and Cross, G. A. M., 1985, Variations in the organization of repetitive DNA sequences in the genomes of *Plasmodium falciparum* clones, *Mol. Biochem. Parasitol.* **15:**149–158.

Birago, C., Bucci, A., Dore, E., Frontali, C., and Zenobi, P., 1982, Mosquito infectivity is directly related to the proportion of repetitive DNA in *Plasmodium berghei, Mol. Biochem. Parasitol.* **6:**1–12.

Bone, N., Gibson, T., Goman, M., Hyde, J. E., Langsley, G. W., Scaife, J. G., Walliker, D., Yankofsky, N. K., and Zolg, J. W., 1983, Investigations of the DNA of the human malaria parasite *Plasmodium falciparum* by *in vitro* cloning into phage λ, in: *Molecular Biology of Parasites* (J. Guardiola, L. Luzzatto, and W. Trager, eds.), Raven Press, New York, pp. 125–134.

Borst, P., and Fairlamb, A. H., 1976, DNA of parasites with special reference to kinetoplast DNA, in: *Biochemistry of Parasites and Host–Parasite Relationships* (H. van den Bossche, ed.), Elsevier/North-Holland, Amsterdam, pp. 169–191.

Borst, P., and Hoeijmakers, J. H. J., 1979, Review—Kinetoplast DNA, *Plasmid* **2:**20–40.

Borst, P., van der Ploeg, M., van Hoek, J. F. M., Tas, J., and James, J., 1982, On the DNA content and ploidy of trypanosomes, *Mol. Biochem. Parasitol.* **6:**13–23.

Castro, C., Craig, S. P., and Castaneda, M., 1981, Genome organization and ploidy number in *Trypanosoma cruzi, Mol. Biochem. Parasitol.* **4:**273–282.

Cleveland, L. R., 1956, Brief accounts of the sexual cycles of the flagellates of *Cryptocercus, J. Protozool.* **3:**161–180.

Cornelissen, A. W. C. A., and Walliker, D., 1985, Gametocyte development of *Plasmodium chabaudi* in mice and rats: Evidence for host induction of gametocytogenesis, *Z. Parasitenkd.* **71:**297–303.

Cornelissen, A. W. C. A., Overdulve, J. P., and van der Ploeg, M., 1984, Determination of nuclear DNA of five eucoccidian parasites, *Isospora (Toxoplasma) gondii, Sarcocystis cruzi,*

Eimeria tenella, E. acervulina and *Plasmodium berghei,* with special reference to gametogenesis and meiosis in *I.* (T.) *gondii, Parasitology* **88:** 531–553.

Davidson, E. H., Galau, G. A., Angerer, R. C., and Britten, R. J., 1975, Comparative aspects of DNA organization in metazoa, *Chromosoma* **51:**253–259.

Dore, E., Birago, C., Frontali, C., and Battaglia, P. A., 1980, Kinetic complexity and repetitivity of *Plasmodium berghei* DNA, *Mol. Biochem. Parasitol.* **1:**199–208.

Fairbairn, D., 1970, Biochemical adaptation and loss of genetic capacity in helminth parasites, *Biol. Rev.* **45:**29–72.

Fong, D., and Chang, K.-P., 1982, Surface antigenic change during differentiation of a parasitic protozoan, *Leishmania mexicana:* Identification by monoclonal antibodies, *Proc. Natl. Acad. Sci. USA* **79:**7366–7370.

Fong, D., Wallach, M., Keithly, J., Melera, P. W., and Chang, K.-P., 1984, Differential expression of mRNAs for α- and β-tubulin during differentiation of the parasitic protozoan *Leishmania mexicana, Proc. Natl. Acad. Sci. USA* **81:**5872–5886.

Hope, I. A., Mackay, M., Hyde, J. E., Goman, M., and Scaife, J., 1985, The gene for an exported antigen of the malaria parasite *Plasmodium falciparum* cloned and expressed in *Escherichia coli, Nucleic Acids Res.* **13:**369–379.

Hough-Evans, B. R., and Howard, J., 1982, Genome size and DNA complexity of *Plasmodium falciparum, Biochim. Biophys. Acta* **698:**56–61.

Jenni, L., Marti, S., Schweizer, J., Betschart, B., Le Page, R. W. F., Wells, J. M., Tait, A., Paindavoine, P., Pays, E., and Steinert, M., Hybrid formation between African trypanosomes during cyclical transmission, 1986, *Nature,* in press.

Kemp, D. J., Corcoran, L. M., Coppel, R. L., Stahl, H. D., Bianco, A. E., Brown, G. V., and Anders, R. F., 1985, Size variation in chromosomes from independent cultured isolates of *Plasmodium falciparum, Nature* **315:**347–350.

Kilejian, A., and MacInnis, A. J., 1976, Density distribution of DNA from parasitic helminths with special reference to *Ascaris lumbricoides, Rice Univ. Stud.* **62:**161–174.

Lanar, D. E., Levy, L. S., and Manning, J. E., 1981, Complexity and content of the DNA and RNA in *Trypanosoma cruzi, Mol. Biochem. Parasitol.* **3:**327–341.

Leon, W., Fouts, D. L., and Manning, J., 1978, Sequence arrangement of the 16S and 26S rRNA genes in the pathogenic haemoflagellate *Leishmania donovani, Nucleic Acids Res.* **5:**491–504.

Long, E. O., and Dawid, I. B., 1980, Repeated genes in eukaryotes, *Annu. Rev. Biochem.* **49:**727–764.

McCutchan, T. F., Dame, J. B., Miller, L. H., and Barnwell, J., 1984, Evolutionary relatedness of *Plasmodium* species as determined by the structure of DNA, *Science* **225:**808–811.

McCutchan, T. F., Simpson, A. J. G., Mullins, J. A., Sher, A., Nash, T. E. Lewis F., and Richards, C., 1984, Differentiation of schistosomes by species, strain, and sex by using cloned DNA markers, *Proc. Natl. Acad. Sci. USA* **81:**889–893.

Michel, F. (ed.), 1984, *Modern Genetic Concepts and Techniques in the Study of Parasites, Tropical Disease Research Series* 4, Schwabe, Basel.

Morel, C. M. (ed.), 1984, *Genes and Antigens of Parasites: A Laboratory Manual* (2nd edition), UNDP/World Bank/WHO Special Programme for Research and Training in Tropical Diseases, Fundaçáo Oswaldo Cruz, Rio de Janeiro, Brazil.

Moritz, K. B., and Roth, G. E., 1976, Complexity of germline and somatic DNA in *Ascaris, Nature* **259:**55–57.

Nollen, P. M., 1983, Patterns of sexual reproduction among parasitic helminths, *Parasitology* **86:**99–120.

Padua, R. A., 1981, *Plasmodium chabaudi:* Genetics of resistance to chloroquine, *Exp. Parasitol.* **52:**419–426.

Pereira, M. E. A., Andrade, A. F. B., and Ribeiro, J. M. C., 1981, Lectins of distinct specificity in *Rhodnius prolixus* interact selectively with *Trypanosoma cruzi, Science* **211:**597–599.

Pfefferkorn, E. R., 1981, Molecular genetics of toxoplasma: *Toxoplasma gondii* as a model parasite for genetic studies, in: *Modern Genetic Concepts and Techniques in the Study of Parasites* (F. Michel, ed.), Schwabe, Basel, pp. 195–210.

Pollack, Y., Katzen, A. L., Spira, D. T., and Golenser, J., 1982, The genome of *Plasmodium falciparum.* I. DNA base composition, *Nucleic Acids Res.* **10:**539–546.

Price, P. W., 1980, *Evolutionary Biology of Parasites,* Princeton University Press, Princeton, N.J.

Roger, W. P., and Petronijevic, T., 1982, The infective stage and the development of nematodes, in: *Biology and Control of Endoparasites* (L. E. A. Symons, A. D. Donald, and J. K. Dineen, eds.), Academic Press, New York, pp. 3–28.

Saithill, T. W., and Samaras, N., 1985, The molecular karyotype of *Leishmania major* and mapping of α and β tubulin gene families to multiple unlinked chromosome loci, *Nucleic Acids Res.* **13:**4155–4169.

Searcy, D. G., and MacInnis, A. J., 1970, Measurements by DNA renaturation of the genetic basis of parasitic reduction, *Evolution* **24:**796–806.

Sher, A., and Snary, D., 1982, Specific inhibition of the morphogenesis of *Trypanosoma cruzi* by a monoclonal antibody, *Nature* **300:**639–640.

Short, R. B., Menzel, M. Y., and Pathak, S., 1979, Somatic chromosomes of *Schistosoma mansoni, J. Parasitol.* **65:**471–473.

Simpson, A. J. G., Sher, A., and McCutchan, T. F., 1982, The genome of *Schistosoma mansoni:* Isolation of DNA, its size, bases and repetitive sequences, *Mol. Biochem. Parasitol.* **6:**125–137.

Sinden, R. E., 1983, Sexual development of malarial parasites, *Adv. Parasitol.* **22:**153–216.

Sinden, R. E., 1983, The cell biology of sexual development in plasmodium, *Parasitology* **86:**7–28.

Smalley, M. E., and Brown, J., 1981, *Plasmodium falciparum* gametocytogeneisis stimulated by lymphocytes and serum from infected Gambian children, *Trans. R. Soc. Trop. Med. Hyg.* **75:**316–317.

Sober, H. A. (ed.), 1970, *Handbook of Biochemistry: Selected Data for Molecular Biology* (2nd edition), The Chemical Rubber Co., Cleveland, Ohio, pp. A1–L108.

Soldo, A. T., Brickson, S. A., and Larin, F., 1981, The kinetic and analytical complexities of the DNA genomes of certain marine and freshwater ciliates, *J. Protozool.* **28:**377–383.

Streeck, R. E., Mority, K. B., and Beer, K., 1982, Chromatin diminution in *Ascaris suum:* Nucleotide sequence of the eliminated satellite DNA, *Nucleic Acids Res.* **10:**3495–3502.

Tait, A., 1980, Evidence for diploidy and mating in trypanosomes, *Nature* **287:**536–537.

Tait, A., 1983, Sexual processes in the Kinetoplastida, *Parasitology* **86:**29–57.

Tibayrenc, M., Cariou, M.-L., and Solignac, M., 1981, Parasitologie animale—Interprétation génétique des zymogrammes de flagellés des genres *Trypanosoma* et *Leishmania, C. R. Acad. Sci.* **292:**623–625.

Tibayrenc, M., Ward, P., Moya, A., and Ayala, F. J., 1986, Natural populations of *Trypanosoma cruzi,* the agent of Chagas disease, have a complex multiclonal structure, *Proc. Natl. Acad. Sci. USA* **83:**115–119.

Van der Ploeg, L. H. T., Cornelissen, A. W. C. A., Barry, J. D., and Borst, P., 1984, Chromosomes of Kinetoplastida, *EMBO J.* **3**:3109–3115

Van der Ploeg, L. H. T., Smits, M., Ponnudarai, T., Vermeulen, A., Meuwissen, J. H. E. T., and Langsley, G., 1985, Chromosome-sized DNA molecules of *Plasmodium falciparum, Science* **229**:658–661.

Villalba, E., and Ramirez, J. L., 1982, Ribosomal DNA of *Leishmania brasiliensis:* Number of ribosomal copies and gene isolation, *J. Protozool.* **29:**438–441.

Walliker, D., 1983a, The genetic basis of diversity in malarial parasites, *Adv. Parasitol.* **22:**217–259.

Walliker, D., 1983b, Genetics of parasites, in: *Molecular Biology of Parasites* (J. Guardiola, L. Luzzatto, and W. Trager, eds.), Raven Press, New York, pp. 53–61.

Wesley, R. D., and Simpson, L., 1973, Studies on kinetoplast DNA. III. Kinetic complexity of kinetoplast and nuclear DNA from *Leishmania tarentolae, Biochim. Biophys. Acta* **319:**267–280.

Whitfield, P. J. (ed), 1983, *The Reproductive Biology of Parasites, Parasitology* **86.**

Whitfield, P. J., and Evans, N. A., 1983, Parthenogenesis and asexual multiplication among parasitic platyhelminths, *Parasitology* **86:**121–160.

Zamenhof, S., and Eichhorn, H. H., 1967, Study of microbial evolution through loss of biosynthetic functions: Establishment of "defective" mutants, *Nature* **216:**456–458.

Zampetti-Bosseler, F., Schweizer, J., Pays, E., Jenni, L., & Steinert, M., 1986, Evidence for haploidy in metacyclic forms of *Trypanosoma brucei*, *Proc. Natl. Acad. Sci. USA*, in press.

CHAPTER 11

The Kinetoplast and Kinetoplast DNA

In previous chapters I have referred to the remarkable changes in morphology and especially in physiology exhibited in the life cycles of the African trypanosomes (see Diagram V), causative agents of sleeping sickness in humans and of various diseases of cattle and other domestic animals. Especially striking is the switch in metabolism from the glycerophosphate oxidase system of the bloodstream form to the cytochrome-mediated respiration of the insect form (procyclic), accompanied by loss of infectivity. These changes seem to be mediated by changes in the single mitochondrion. This mitochondrion is unique in having a special region, always immediately adjacent to the basal body of the flagellum, that contains "the most unusual DNA in nature" (as P. Borst has aptly termed it). There is so much DNA that it can be seen by light microscopy after staining with Feulgen or Giemsa stain. Indeed, this was the first extranuclear DNA ever noted, long before DNA had been demonstrated in mitochondria and long before anyone suspected that the kinetoplast is actually a part of the mitochondrion of trypanosomes.

We know now that a kinetoplast and kinetoplast DNA are found in all the many species of flagellates of a large order, the Kinetoplastida. A few relatively minor groups of this order are free-living; the bacteria-feeding freshwater organism *Bodo caudatus* is typical of these. The great majority of species are parasitic, many in the alimentary tract or body cavities of insects and other invertebrates; these have a relatively simple one-host life cycle. Many others, including the African trypanosomes, *Trypanosoma cruzi* (the agent of Chagas' disease in South America), and the leishmanial parasites, alternate between a vertebrate and an invertebrate host, with a complex life cycle involving shifts in morphology and metabolism. In relation to parasitism, the questions of interest are: has the possession of the kinetoplast been particularly useful in enabling these organisms so successfully to exploit the parasitic mode of life; does the kinetoplast DNA play a main role in the transformations in the life cycle? And of course we would like to know how this strange DNA is replicated and what it does and how.

Kinetoplast DNA (kDNA) accounts for anywhere from 5 to 20% of total

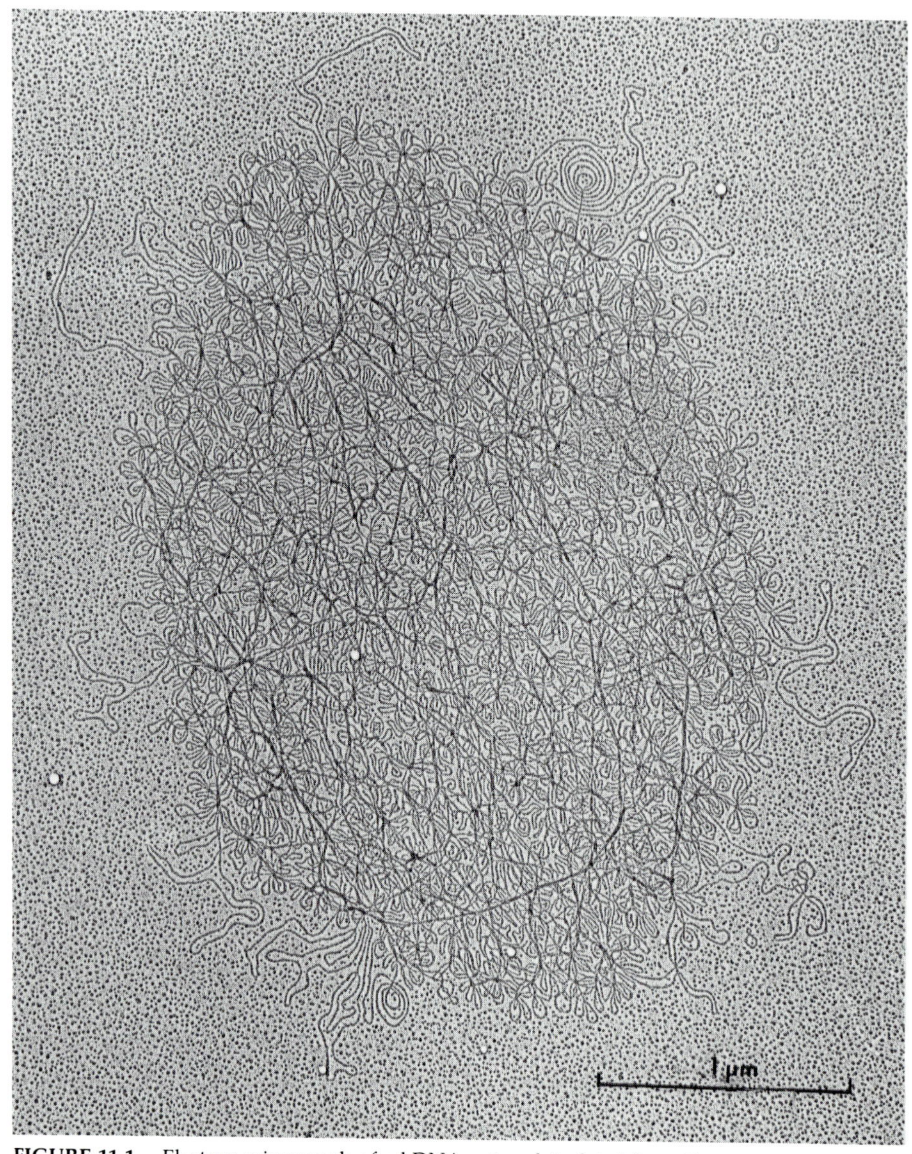

FIGURE 11.1. Electron micrograph of a kDNA network isolated from *Trypanosoma brucei*. This was spread in a protein monolayer and shadowed with platinum. (Original print courtesy of Dr. P. Weijers. See Borst and Hoeijmakers, 1979.)

cellular DNA. Because of its relatively high AT content, it generally appears as a distinct lighter band in cesium chloride density gradient centrifugation. The kDNA, with a molecular weight of up to 4×10^{10}, is easily isolated. When appropriately spread and prepared for electron microscopy, it shows the beautiful structure illustrated in Fig. 11.1. This is clearly a large network of small circles of DNA, highly catenated. That they are typically arranged

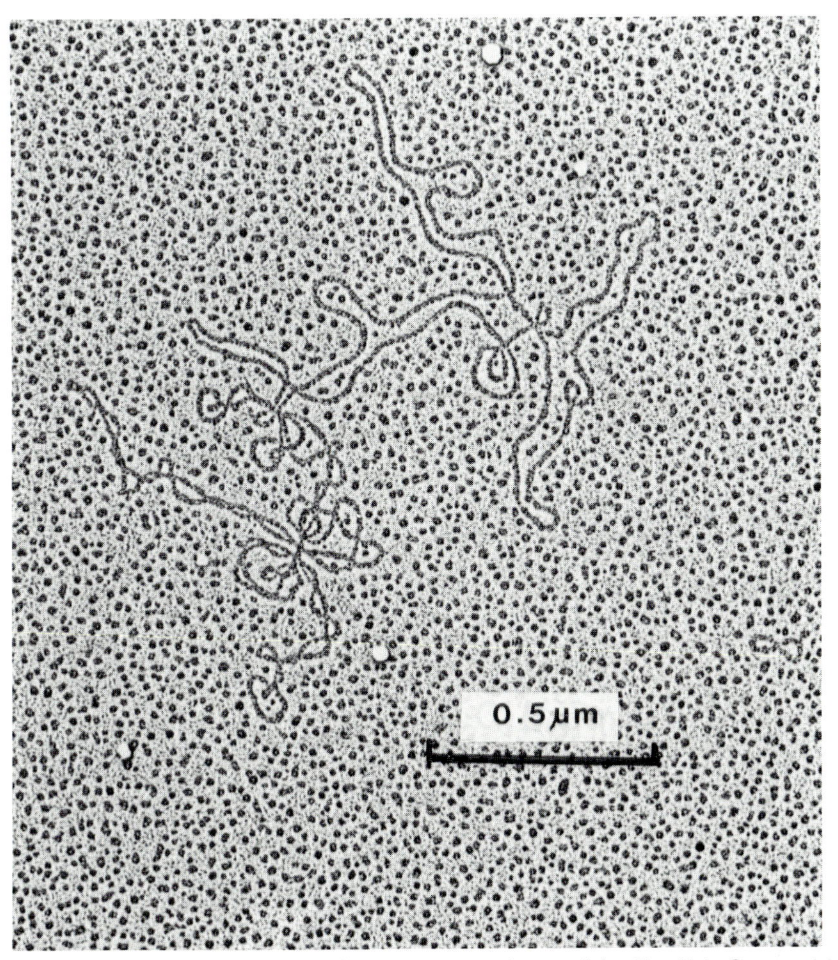

FIGURE 11.2. Two maxicircles from the same preparation used for Fig. 11.1. One maxicircle is supercoiled, the other one is open. (Original print courtesy of Dr. P. Weijers.)

in a single disk-shaped layer of vertically oriented circles is shown by electron micrographs of thin sections through an intact kinetoplast (see Fig. 11.3). Hence, the circumference of the minicircles determines the thickness of the disk. A single network contains about 10,000 of these minicircles. They are all the same size in any one species, but range in size among species from about 0.3 μm in length in *Leishmania tarentolae* and *Trypanosoma brucei* to about 0.8 μm in *Crithidia fasciculata,* a one-host parasite of mosquitoes. Catenated into the thousands of minicircles is, in most species, a much smaller number (20 to 50) of much larger circles, the maxicircles (Fig. 11.2).

Maxicircles are about 10 to 12 μm long and hence very much like the typical circular mitochondrial DNA of many eukaryotic cells. They are present in 20 to 50 identical copies per network. They are most readily seen as "edge loops" in electron micrographs of spreads (Fig. 11.1) but their actual distri-

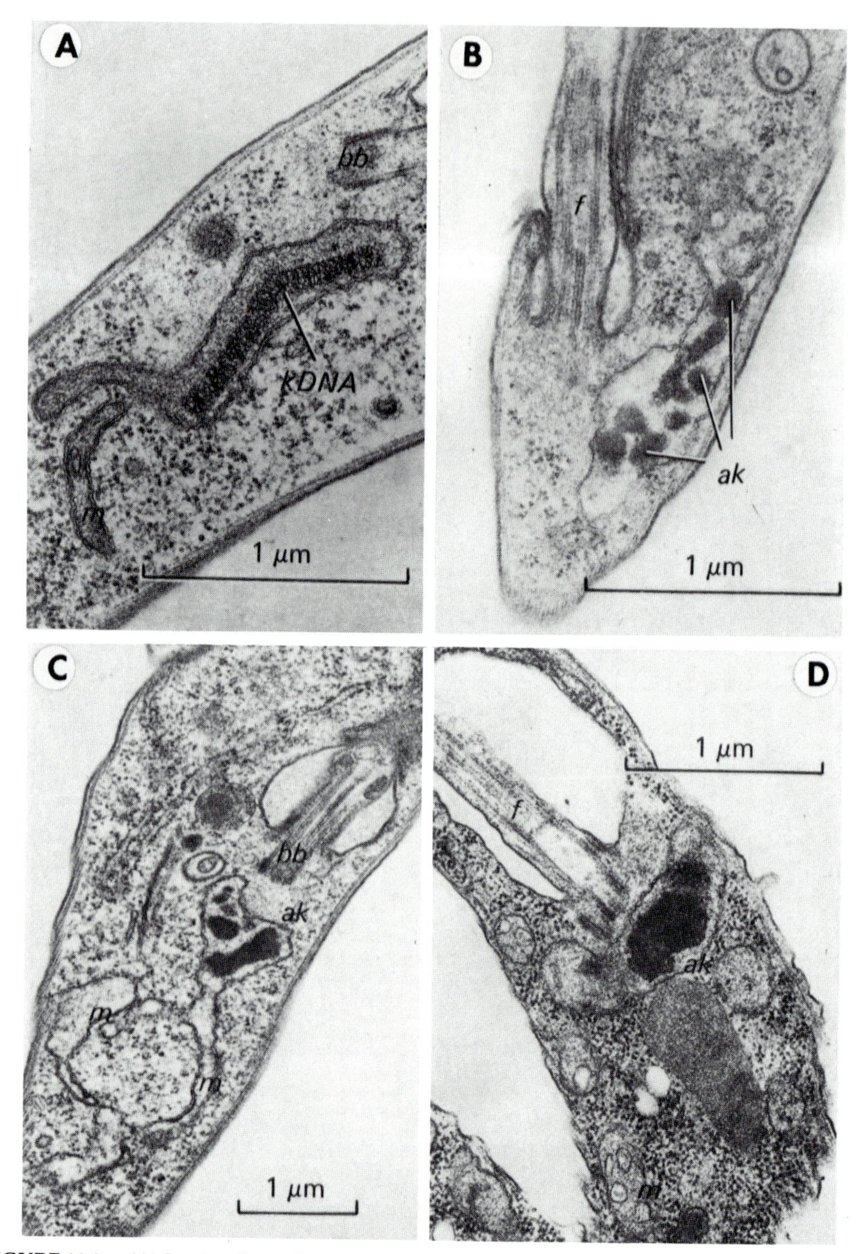

FIGURE 11.3. (A) Section through the kinetoplast region of the mitochondrion (m) of the normal strain of *Trypanosoma equiperdum*. The kDNA appears as an electron-dense fibrous band bounded by the double mitochondrial membrane. Note that the kinetoplast is adjacent to the basal body (bb) in the normal strain. ×45,800. (B) Section through the kinetoplast region of the mitochondrion at the base of the flagellum (f) in the spontaneously dyskinetoplastic strain of *T. equiperdum*. Numerous electron-dense clumps of altered kDNA (ak) are seen. ×45,800. (C) Section through the kinetoplast of a cell from the acriflavine-induced dyskinetoplastic strain of *T. equiperdum*. Clumps of altered kDNA (ak) in the portion of the mitochondrion (m) adjacent to the basal body

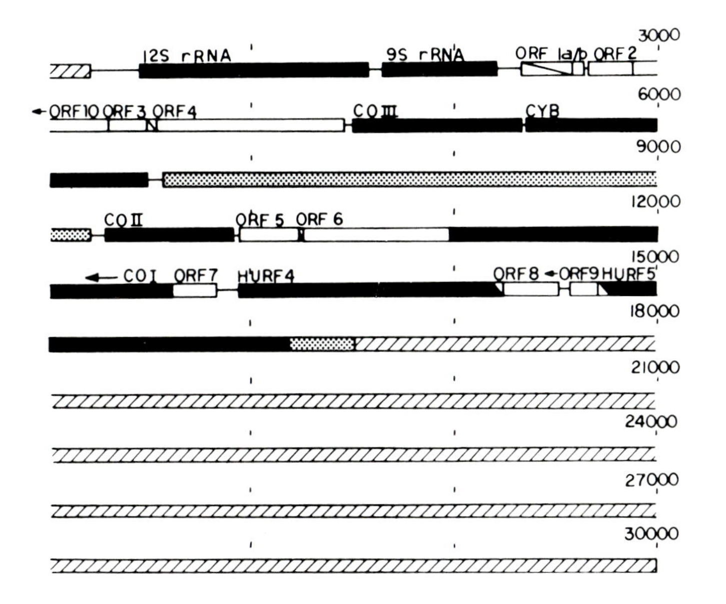

FIGURE 11.4. Genomic organization of the maxicircle DNA of *L. tarentolae*. The nontranscribed divergent region is indicated by cross-hatching, and the identified genes are indicated by dark shading. Unidentified open reading frames (ORFs) are blank. All identified genes are transcribed left to right except for COI (arrow). The portions of the transcribed region that have not been sequenced are indicated by stippling [i.e., between cytochrome *b* (CYB) and COII and at the 3′ end of HURF5]. Numbers at right indicate base pairs from the *Eco*RI site. COI, COII, COIII, cytochrome oxidase subunits I, II, III, respectively; HURF4, HURF5, unidentified reading frames homologous to human mitochondrial unidentified reading frames 4 and 5, respectively. (From de la Cruz *et al.*, 1984. Print courtesy of Dr. L. Simpson.)

bution in the network is not known. They can be removed with restriction enzymes without causing any change in network structure, except for the lack of "edge loops."

It is now clear that these molecules are transcribed. In *Leishmania tarentolae* for example, most of the maxicircle is transcriptionally active. The principal products are 9 and 12S rRNA, but a considerable variety of other transcripts also are formed, especially from the so-called "120 region" of the maxicircle. Moreover, sequences homologous to the yeast mitochondrial structural genes for cytochrome oxidase subunits I and II, ATPase 6, and cytochrome *b* have now been demonstrated on *L. tarentolae* maxicircles by low-stringency hybridization using probes prepared from mitochondrial DNA of yeast petite mutants. Furthermore, by sequencing of about 80% of the transcribed region of maxicircle DNA of *L. tarentolae* and comparison of the translated amino acid sequences with those of known mitochondrial genes from other organ-

(bb) are seen. ×21,000. (D) Cell from an acriflavine-treated culture of *Crithidia fasciculata* sectioned longitudinally through the flagellum (f) and the kinetoplast region of the mitochondrion (m) showing a large clump of altered kDNA (ak). ×34,500. (All from Hajduk, 1979.)

isms, it has been possible to localize the genes for cytochrome oxidase subunits I, II, and III, for cytochrome *b*, and for two human unidentified reading frames (Fig. 11.4). It is possible that the unidentified open reading frame ORF-4 is a highly divergent ATPase 6. It is noteworthy that *L. tarentolae* genes have diverged more from the homologous yeast and human genes than the yeast and human genes have diverged from each other. With these and similar data for some other species, especially *Trypanosoma brucei,* the evidence is strong that the maxicircles correspond to the mitochondrial DNA of other eukaryotic cells; their function is to code for mitochondrial ribosomal RNA and for certain structural mitochondrial proteins.

The function of the minicircles remains elusive. The thousands of small circles that make up a network in any one species are all the same size but show from some to very considerable sequence heterogeneity (Table 11.1). Present evidence indicates that these circles all have a constant region, with extensive sequence homology (more than 90% of its length), and a variable region. In the variable region the circles differ over more than 80% of their length. Whereas *T. brucei* has about 300 different classes of variable regions, *Crithidia fasciculata* has 10 to 20 and *L. tarentolae* has 5 to 10. In *T. equiperdum* there is no heterogeneity of the minicircles, a situation I will discuss later. The number and relative proportion of the variable regions may change slowly with time. Thus, in *L. tarentolae* comparison of the restriction enzyme pattern of a strain kept frozen with that of the same strain in continuous culture for 4 years showed two altered bands out of about 19 restriction fragments.

Although it has been thought that minicircles are not transcribed, and indeed that they *cannot* be transcribed, recent work casts some doubt on this notion. An RNA homologous to minicircles was found in *Crithidia acanthocephali,* a parasite of the alimentary tract of flies. This RNA was about 240 nucleotides in length, as compared with the minicircle length of 2500 bp. In *T. lewisi* minicircles there are sequences ordered in the same way as prokaryotic promoters; hence, these could serve as initiation site recognized by RNA

TABLE 11.1. Properties of kDNA in Some Representative Parasitic Kinetoplastids[a]

Species	Network molecular weight × 10^{-10}	Maxicircle size μm	Maxicircle size kb	Minicircle size μm	Minicircle size kb	Minicircle sequence classes
Crithidia fasciculata	0.72	12	38	0.79	2.5	10–20
Leishmania tarentolae	0.6	11	31	0.29	0.87	5–10
Trypanosoma cruzi	2.1	—	33	0.49	1.44	<20
T. brucei	0.4	8.3	20–22	0.32	1.0	~300
T. equiperdum	0.36	— (or absent)	23	0.31	1.0	1

[a] Data from Borst and Hoeijmakers (1979) and Englund (1980).

polymerase. Very recently, Battaglia and his associates have reported an experiment that seems to show protein synthesis coded for by minicircles from *T. lewisi*. The pBR322 plasmid containing a minicircle cut by *Bam*HI restriction enzyme was used to transform an *E. coli* strain that forms the so-called minicells. These are small anucleate bodies continuously produced during growth of a mutant strain of *E. coli* K12. They are useful for examining RNA and protein synthesis in the absence of chromosomal synthesis. Minicells were isolated from the transformed line and from a line containing only the unaltered pBR322 plasmid (without the minicircle insert). They were incubated with [^{35}S]methionine and then subjected to gel electrophoresis and autoradiography. A protein band of 19,000 daltons was present in the material from the transformed line but absent from the control.

Perhaps most convincing is the recent work of Shlomai and Zadok with *C. fasciculata*. The kDNA minicircles of this organism were obtained by decatenation of networks with *Crithidia* type II topoisomerase and purified by preparative electrophoresis. Minicircle fragments obtained with restriction endonuclease were inserted into the *Sma*I site of pORF expression vectors. These contained the 5′ end of the *E. coli ompF* gene, providing a strong promoter, a translation initiating site, and a signal sequence for export from the cytoplasm. Coupled to it was the *E. coli lacZ* gene lacking its own 5′ end. kDNA fragments between *ompF* and *lacZ* supported realigning of these genes in frame so that β-galactosidase was expressed. This permitted identification of the clones with an expressed open reading frame from the kDNA. Gel electrophoresis of the bacteria with plasmids carrying the minicircle fragments revealed a new peptide. Its molecular weight of 135,000 agrees with the predicted one for the tribrid protein fusion—ompF–kDNA–β-galactosidase. This peptide was used to raise antibodies in rabbits. The sera were absorbed with polymerized extracts of bacteria having pORF vectors lacking the kDNA insert but expressing β-galactosidase. These sera showed strong immunofluorescence with living *C. fasciculata*, suggesting that the kDNA coded for a surface protein (Fig. 11.5). Clearly, much still remains to be done. Even if minicircles do code for some RNAs and proteins, this in itself would not account for their presence in such large numbers and their organization in so complex a network. No satisfactory hypothesis has yet been proposed.

We think we know, however, how they replicate, and this is a fascinating process only recently worked out. The following picture has emerged. In *C. fasciculata* the DNA replicates during an S phase that occurs at 3.8 to 4.8 hr of the 6.2-hr cell cycle. Before replication, the kDNA network consists of about 5000 minicircles. The minicircles at the periphery become nicked and replicate first, then those in the next adjacent region, and so on into the center. Since each minicircle replicates once to produce one daughter, the size of the network increases until 10,000 minicircles are present, all of which are nicked. Covalent closure of the minicircles and cleavage of the network into two then occur at about the same time, and the double membrane of the surrounding mitochondrial structure also divides. Just how all this is brought about is not known. The replication of each minicircle occurs in the following way. A

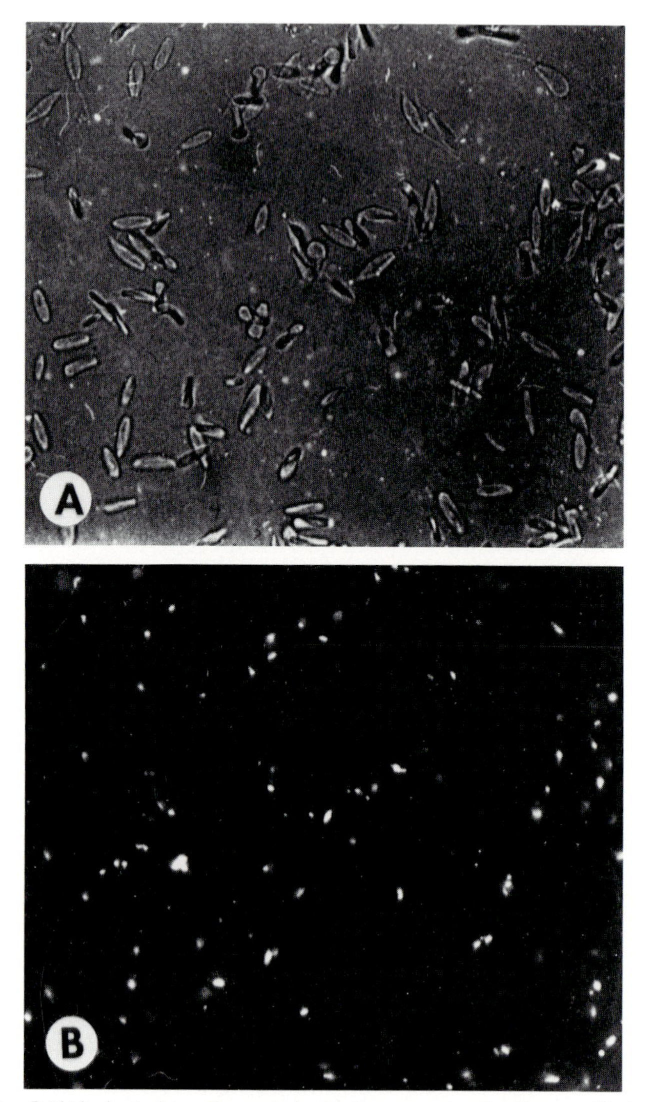

FIGURE 11.5. *Crithidia fasciculata* cells reacted with fluorescent antibodies specific for *E. coli*–kDNA tribrid fused protein. (A) As seen by light microscopy. (B) The same field by fluorescence microscopy showing the many positively labeled cells. Both ×400. (From Shlomai and Zadok, 1984).

topoisomerase, whose activity is strictly dependent on ATP, decatenates the rings (with hydrolysis of the ATP), a reaction that depends on a narrow optimal concentration of KCl at 140 mM. The *free* minicircle then replicates and shows a Cairns-type replication intermediate. A novel type of intermediate has been found consisting of a highly gapped free minicircle with the newly synthesized strand composed of fragments only 20–110 nucleotides in length and all heavy strand. This represents only one daughter circle; the other with a nascent light strand has a different structure. The two daughter minicircles are then reattached to the network by a reattachment enzyme. The same topoisomerase isolated from *Crithidia* that effects decatenation is also responsible for catenane formation. The latter again requires ATP, but is favored by a lower concentration of KCl (10–40 mM) and in addition requires the presence of a protein that aggregates DNA. *In vitro* this protein could be replaced by spermidine., It is interesting that the topoisomerase of *Crithidia* was barely detectable in crude whole cell extracts but was active in cell supernatants freed of DNA, cell membranes, and large insoluble particles. The presence of a replicon on the minicircles of *L. tarentolae* has been demonstrated by insertion into the YIp^5 plasmid of yeast, a plasmid that lacks a replicon and that cannot replicate in its natural host. Maxicircles seem to replicate by the rolling circle model and appear as free linear intermediates prior to recatenation into the network. Since the maxicircles code for mitochondrial ribosomal RNA and for mitochondrial structural proteins, and since each hemoflagellate cell has a single chondriome, there must be a mechanism to ensure transmission to each daughter chondriome of an appropriate number of maxicircles suitably attached in the network of minicircles. This likewise remains to be worked out. We do know that the network is doubled in size after the circles have replicated and reattached. It then constricts in the middle and splits in two halves, each going to one daughter cell at cell division. The kinetoplast region of the mitochondrion is somehow attached to the basal body of the flagellum even though there is no visible connection, and this attachment may help to bring about the appropriate separation of the two halves of the mitochondrion containing the daughter kDNA network.

The hypothesis that the kDNA of African trypanosomes in particular, but also of other related flagellates having two hosts and a complex life cycle, is essential for the morphological and physiological changes occurring in the life cycle is an attractive one but far from proven. It rests on the following evidence. As we have already seen (Chapter 9) the blood stream forms of *T. brucei* obtain energy from the fermentation of glucose to pyruvate with a special glycerophosphate oxidase system serving to replenish the NAD. In this stage the chondriome is a poorly developed simple tubular structure. When such bloodstream forms are introduced into the midgut of a fly, or into appropriate culture medium, there develop the procyclic forms with an extensive cristate mitochondrion, an active cytochrome-mediated respiration, a complete tricarboxylic acid cycle, and a repressed glycerophosphate oxidase system. Since the kDNA resides within the mitochondrion and since the maxicircles have been shown to be involved in synthesis of some of the usual mitochondrial proteins, it is reasonable to suppose that this extensive mitochondrial elaboration is controlled by the kDNA. In keeping with this is the recent finding that four transcripts from the maxicircle genes of *T. brucei* are differentially expressed in bloodstream and procyclic forms. Two of these are more abundant in bloodstream forms, one is more abundant in procyclic forms, and one is present only in procyclic forms. These latter two correspond to the genes for cytochrome oxidase (CO) subunits I and II, respectively.

Evidently, COI is transcribed more in procyclic than in bloodstream forms and COII is transcribed only in the procyclic. What determines this transcriptional control is not known.

The idea that mitochondrial elaboration is controlled by the kDNA is further supported by information from mutant forms, both naturally occurring and artificially produced, in which a kinetoplast is not visible by light microscopy. Such forms were called akinetoplastic, but with the application of electron microscopy to these organisms, it was soon apparent that the double-membraned chondriome was present in all; only the DNA was either absent or modified so that it could not be seen by light microscopy. These forms, now termed "dyskinetoplastic" to indicate an abnormal kinetoplast, occur naturally with high frequency especially in two species of trypanosomes that otherwise very closely resemble *Trypanosoma brucei.* These are *T. equiperdum,* causing a disease of horses and donkeys, and *T. evansi,* causing disease in camels and a variety of other ungulates. Both of these species naturally show high proportions of organisms lacking a kinetoplast (as seen by light microscopy) and in both strains readily arise, under laboratory passage conditions in rodents, that are 100% so. In South America there is a disease of horses caused by trypanosomes all of which lack a visible kinetoplast. This trypanosome, called *T. equinum,* is probably a strain of *T. evansi.* The striking feature of all three of these trypanosomes is that they do not have cyclical development in tsetse, or any other vector host. *T. equiperdum* is transmitted directly from horse to horse venereally; it is present in the mucous membrane of the genital tract of both sexes. *T. evansi* and *T. equinum* are transmitted by biting flies (e.g., Tabanidae), which serve as "flying pins." The organisms are transmitted when a fly with a freshly contaminated proboscis from feeding on an infected animal moves to and bites an uninfected animal. There is no infection of the fly or development in it. All three of these species cannot be grown in culture axenically at 26–28°C under the conditions suitable for growth of *T. brucei.* Together with their high propensity for loss of the kinetoplast, they have evidently lost the ability to develop the mitochondrial structures that characterize the procyclic forms and that are associated with growth in the tsetse fly, or in axenic culture at 26°C. It is also of great interest that in strains of *T. evansi* and *T. equiperdum* with a kinetoplast, the minicircles are homogeneous (see Table 11.1), suggesting that perhaps the sequence heterogeneity of the minicircle of the kDNA network is essential to mitochondrial development. It should be noted that the one-host trypanosomatid flagellates (e.g., *Crithidia* sp.) always have a well-developed chondriome, and a large kDNA network with the minicircles showing a fair degree of heterogeneity, though not so extensive as in *T. brucei* (Table 11.1).

Artificial mutants lacking kDNA or with altered kDNA have been produced through the use of certain dyes, notably acriflavine and ethidium bromide. These are the same dyes which induce in yeast the cytoplasmic petite mutants that are unable to form mitochondria and that either lack mitochondrial DNA completely or have an altered mitochondrial DNA. When acriflavine is applied to flagellates such as *Crithidia* with one host, or to culture forms like the promastigotes of *L. tarentolae* or epimastigotes of *T. cruzi,* the dyskinetoplastic forms that appear are capable only of severely limited reproduction; no culture conditions have yet been found that will support continued growth of these forms. On the other hand, if the dye is used to treat an animal infected with the bloodstream forms of *T. brucei,* or a closely related form, a dyskinetoplastic strain can be produced that can be propagated indefinitely by animal passage, just like the naturally occurring dyskinetoplastic strains of *T. equiperdum, T. evansi,* and *T. equinum.* Again, just like these, the dye-induced strains will not infect tsetse flies or grow axenically in culture media at 26°C.

Just to make the situation still more complicated, there are some mutants of the *T. brucei* group that appear to have a "normal" kDNA but cannot infect tsetse flies or grow in axenic culture. These mutants, however, are characterized by a mitochondrial ATPase that is insensitive to oligomycin. This is also the situation in dyskinetoplastic trypanosomes and, again, this is the situation in the cytoplasmic petite mutants of yeast. In all those strains of trypanosomes that not only have kDNA but also can infect the tsetse fly, the mitochondrial ATPase, like that of typical mitochondria, is sensitive to oligomycin. These properties are summarized for several subspecies and strains in Table 11.2. It will be seen that three strains of *T. brucei* that are not infective to tsetse appear to have normal kDNA but lack oligomycin-sensitive ATPase. Presumably, these have maxicircles that are stable even though nonfunctional. One strain of *T. equiperdum* (ATCC 30019) has an altered maxicircle in which one segment of 7.5 kb has been deleted, and the minicircles lack sequence heterogeneity. The first two strains of *T. evansi* listed lack maxicircles. They have minicircles that do not show sequence heterogeneity. The last four strains, two of *T. evansi* and one each of *brucei* and *equiperdum,* lack the kDNA network; they have neither maxicircles nor minicircles. Some strains of this type may nevertheless contain some DNA with the same buoyant density in cesium chloride as kDNA. The material seems to be scattered in amorphous patches through the chondriome as revealed by fluorescence with 4′,6-diamidino-2-phenylindole and by the presence of unstructured dense material in electron micrographs (Figs. 11.3, 11.6). This is the situation in the naturally dyskinetoplastic *T. equinum.* The kDNA of this organism consists of two populations of circles, one very small (length 0.16–0.22 μm) and hence even smaller than *T. brucei* minicircles. The other includes circles varying in length from 0.5 to 10 μm. They seem not to be catenated. On the other hand, many strains of trypanosomes lacking a kDNA network have no kDNA whatever. They may still show some amorphous material in electron micrographs of the chondriome, as is the case in *L. tarentolae* or *C. fasciculata* made dyskinetoplastic with acriflavine (Fig. 11.3). There is again a parallel situation with the cytoplasmic petite mutants of yeast: some have no mitochondrial DNA whereas others have the normal amount but evidently nonfunctional.

All of this fascinating information still leaves us without definitive an-

TABLE 11.2. The State of kDNA in Strains of *Trypanosoma brucei* and Related Species[a]

Organism	Strain[b]	Infective to tsetse	Oligomycin sensitive ATPase	DAPI[c] stain kDNA	kDNA network present	Maxicircles (size μm)	Minicircles present	Minicircles sequence heterogeneity
T. brucei	427	+	+	+	+	6	+	+
	31	−	−	+	+	6	+	+
	LUMP 127	−	−	+	+	6	+	+
	EATRO 1244	−	−	+	+	6	+	+
T. equiperdum	ATCC 30019	−	?	+	+	4	+	−
T. evansi	AMB 3	−	−	+	+	−	+	−
	ILRAD B32	−	?	?	+	−	+	−
	SAK	−	?	±	−	−	−	
T. equiperdum	ATCC 30023	−	?	−	−	−	−	
T. brucei	LUMP 1027	−	−	?	−	−	−	
T. evansi	AMB 2	−	−	−	−	−	−	

[a] Data from Borst *et al.* (1981).
[b] Strain 427, the only one listed that is infective to tsetse flies, served as control.
[c] DAPI (4′6-diamidino-2-phenylindole), a DNA-specific fluorescent dye reaction visible in the chondriome.

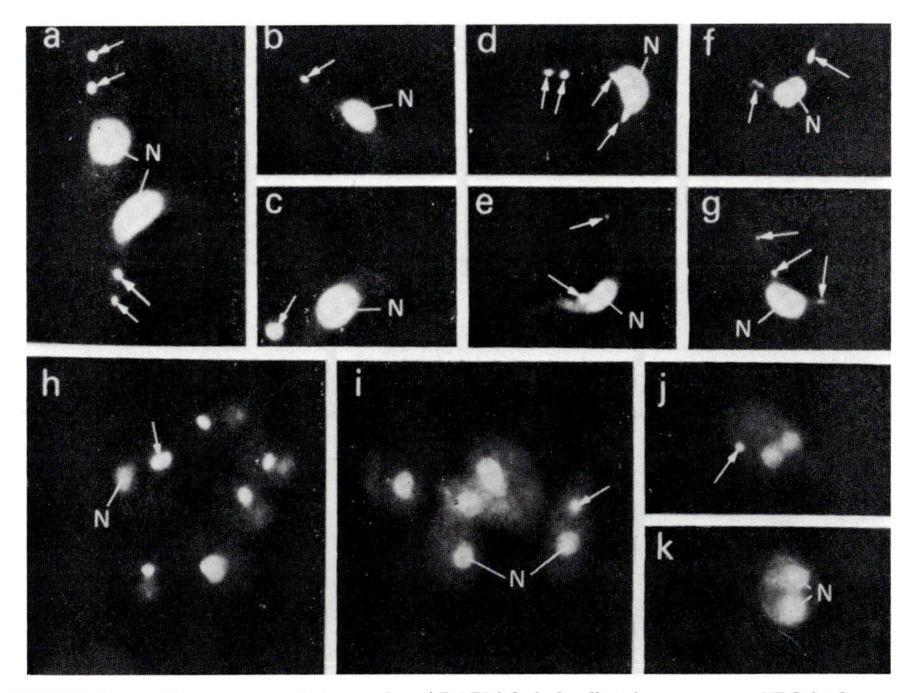

FIGURE 11.6. Fluorescence photographs of DAPI-labeled cells taken using an XBO high-pressure xenon lamp, Zeiss UG_1 BG_{32} exciter filters, and a Zeiss GG_{400} barrier filter. ×5100. (a–c) Kinetoplastic strain of *T. equiperdum* showing a kinetoplast (arrow) and a nucleus (N). (d, e) Spontaneously dyskinetoplastic strain of *T. equiperdum* with fluorescent cytoplasmic granules (arrows) and nucleus (N). (f, g) Acriflavine-induced dyskinetoplastic strain of *T. equiperdum* also showing cytoplasmic (arrows) and nuclear (N) fluorescence. (h) Untreated *C. fasciculata* showing nucleus (N) and kinetoplast (arrow). (i) Acriflavine-treated *C. fasciculata* showing lack of kinetoplast fluorescence in all but one cell (arrow). (j) Acriflavine-treated *C. fasciculata* with two nuclei and a single kinetoplast (arrow). (k) Acriflavine-treated *C. fasciculata* with two nuclei (N) and no kinetoplast. (From Hajduk, 1979.)

swers. It would seem of special importance to determine the functions of the minicircle network. What is known of the sequence of minicircles would indicate that they could code for peptides having about 50 to 70 amino acids, or a molecular weight around 10,000. As already noted, at least one protein of *C. fasciculata* evidently is coded for by minicircle DNA. If the minicircles also have a noncoding function, no one has yet made a good guess as to what this might be.

Meanwhile, it is worth noting that kDNA might be useful in the diagnosis of certain parasitic disease. Two methods have been suggested, both depending on the sequence heterogeneity of minicircles and both readily applicable to field conditions. In one, strains and clones of *T. cruzi* were characterized by the patterns of restriction endonuclease products of their kDNA minicircles. In the other, *Leishmania* species were identified by hybridization of kDNA to organisms in impression smears from cutaneous lesions.

Bibliography

Battaglia, P. A., del Bua, M., Ottaviano, M., and Pouzi, M., 1983, A puzzle genome: Kinetoplast DNA, in: *Molecular Biology of Parasites* (J. Guardiola, L. Luzzatto, and W. Trager, eds.), Raven Press, New York, pp. 107–124.

Benne, R., de Vries, B. F., van den Burg, J., and Klaver, B., 1983, The nucleotide sequence of a segment of *Trypanosoma brucei* mitochondrial maxi-circle DNA that contains the gene for apocytochrome *b* and some unusual unassigned reading frames, *Nucleic Acids Res.* **11:**6925–6941.

Borst, P., and Hoeijmakers, J. H. J., 1979, Review—Kinetoplast DNA, *Plasmid* **2:**20–40.

Borst, P., Hoeijmakers, J. H. J., and Hajduk, S. L., 1981, Structure, function and evolution of kinetoplast DNA, *Parasitology* **82:**81–93.

Cuthbertson, R. S., 1981, Kinetoplast DNA in *Trypanosoma equinum, J. Protozool.* **28:**182–188.

de Bruijn, M. H. L., 1983, *Drosophila melanogaster* mitochondrial DNA, a novel organization and genetic code, *Nature* **304:**234–241.

de la Cruz, V. F., Neckelmann, N., and Simpson, L., 1984, Sequences of six genes and several open reading frames in the kinetoplast maxicircle DNA of *Leishmania tarentolae, J. Biol. Chem.* **259:**15136–15147.

de la Cruz, V. F., Lake, J. A., Simpson, A. M., and Simpson, L., 1985, A minimal ribosomal RNA: Sequence and secondary structure of the 9S kinetoplast ribosomal RNA from *Leishmania tarentolae, Proc. Natl. Acad. Sci. USA* **82:**1401–1405.

Englund, P. T., 1978, The replication of kinetoplast DNA networks in *Crithidia fasciculata, Cell* **14:**157–168.

Englund, P. T., 1980, Kinetoplast DNA, in: *Biochemistry and Physiology of Protozoa* (2nd edition), Volume 4 (M. Lewandovsky and S. H. Hutner, eds.), Academic Press, New York, pp. 333–383.

Englund, P. T., Hajduk, S. L., and Marini, J. C., 1982, Molecular biology of trypanosomes, *Annu. Rev. Biochem.* **51:**695–726.

Englund, P. T., Hajduk, S. L., Marini, J. C., and Plunkett, M. L., 1982, The replication of kinetoplast DNA, in: *Mitochondrial Genes* (P. Slonimski, P. Borst, and G. Attardi, eds.), Cold Spring Harbor Laboratory, Cold Spring Harbor, N.Y., pp. 423–433.

Fairlamb, A. H., Weislogel, P. O., Hoeijmakers, J. H. J., and Borst, P., 1978, Isolation and characterization of kinetoplast DNA from bloodstream form of *Trypanosoma brucei, J. Cell Biol.* **76:**293–309.

Fouts, D. L., and Wolstenholme, D. R., 1979, Evidence for a partial RNA transcript of the small circular component of kinetoplast DNA of *Crithidia acanthocephali, Nucleic Acids Res.* **6:**3785–3804.

Hajduk, S. L., 1979, Dyskinetoplasty in two species of trypanosomatids, *J. Cell Sci.* **35:**185–202.

Hajduk, S., and Cosgrove, W. B., 1979, Kinetoplast DNA from normal and dyskinetoplastic strains of *Trypanosoma equiperdum, Biochim. Biophys. Acta* **561:**1–9.

Kallinokova, V. D., 1977, *Cell Organelle—The Kinetoplast* [in Russian], Nauka, Moscow.

Kidane, G. Z., Hughes, D., and Simpson, L., 1984, Sequence heterogeneity and anomalous electrophoretic mobility of kinetoplast minicircle DNA from *Leishmania tarentolae, Gene* **27:**265–277.

Kitchin, P. A., Klein, V. A., Fein, B. I., and Englund, P. T., 1984, Gapped minicircles: A novel replication intermediate of kinetoplast DNA, *J. Biol. Chem.* **259:**15532–15539.

Marini, J. C., Levene, S. D., Crothers, D. M., and Englund, P. T., 1982, Bent helical structure in kinetoplast DNA, *Proc. Natl. Acad. Sci. USA* **79:**7664–7668.

Morel, C., Chiari, E., Camargo, E. P., Mattei, D. M., Romanha, A. J., and Simpson, L., 1980, Strains and clones of *Trypanosoma cruzi* can be characterized by pattern of restriction endonuclease products of kinetoplast DNA minicircles, *Proc. Natl. Acad. Sci. USA* **77:**6810–6814.

Muhich, M. L., Simpson, L., and Simpson, A. M., 1983, Comparison of maxicircle DNAs of *Leishmania tarentolae* and *Trypanosoma brucei, Proc. Natl. Acad. Sci. USA* **80:**4060–4064.

Opperdoes, F. R., Borst, P., and de Rijke, D., 1976, Oligomycin sensitivity of the mitochondrial ATPase as a marker for fly transmissability and the presence of functional kinetoplast DNA in African trypanosomes, *Comp. Biochem. Physiol. B* **55:**25–30.

Rezepkina, L. A., Maslov, D. A., and Kolesnikov, A. A., 1984, DNA of the crithidia-oncopelti kinetoplast: Principles of the structural organization of minicircular DNA molecules, *Biochemistry (USSR)* **49**:371–379.

Shlomai, J., and Zadok, A., 1983, Reversible decatenation of kinetoplast DNA by a DNA topoisomerase from trypanosomatids, *Nucleic Acids Res.* **11**:4019–4034.

Shlomai, J., and Zadok, A., 1984, Kinetoplast DNA minicircles of trypanosomatids encode for a protein product, *Nucleic Acids Res.* **12**:8017–8028.

Simpson, A. M., Simpson, L., and Livingston, L., 1982, Transcription of the maxicircle kinetoplast DNA of *Leishmania tarentolae*, *Mol. Biochem. Parasitol.* **6**:237–252.

Simpson, L., 1979, Isolation of maxicircle component of kinetoplast DNA from hemoflagellate protozoa, *Proc. Natl. Acad. Sci. USA* **76**:1585–1588.

Simpson, L., 1986, Kinetoplast DNA in trypanosomid flagellates, *Int. Rev. Cytol.* **99**:119–179.

Simpson, L., Spithill, T. W., and Simpson, A. M., 1982, Identification of maxicircle DNA sequences in *Leishmania tarentolae* that are homologous to sequences of specific yeast mitochondrial structural genes, *Mol. Biochem. Parasitol.* **6**:253–264.

Stuart, K., 1980, Cultivation of dyskinetoplastic *Trypanosoma brucei*, *J. Parasitol.* **66**:1060–1061.

Stuart, K., 1983, Mitochondrial DNA of an African trypanosome, *J. Cell. Biochem.* **23**:13–26.

Stuart, K., 1983, Kinetoplast DNA, mitochondrial DNA with a difference, *Mol. Biochem. Parasitol.* **9**:93–104.

Wirth, D. F., and Pratt, D. M., 1982, Rapid identification of *Leishmania* species by specific hybridization of kinetoplast DNA in cutaneous lesions, *Proc. Natl. Acad. Sci. USA* **79**:6999–7003.

Zaitseva, G. N., Mett, I. L., Maslov, D. A., Lunina, L. D., and Kolesnikov, A. A., 1979, The presence of genes of ribosomal and transfer RNAs in the kinetoplast DNA of two *Crithidia* species, *Biokhimiya* **44**:2073–2082.

CHAPTER 12

Parasite-Induced Modifications of the Host

Growth Factors. Effects on Behavior. Parasitic Castration

Parasite-induced modifications of the host include pathogenesis, symbiosis, and immunity. First, however, I will discuss effects on the host that do not fit in any of these three categories. A good example of such an effect is provided by the sparganum growth factor (see also Chapter 8). J. F. Mueller noted in 1963 that mice used to propagate the larval stage of the tapeworm *Spirometra mansonoides* (Diagram XIV) became exceptionally large (Fig. 12.1). It soon became apparent that this was not a matter of simple obesity, but rather involved true skeletal and muscular growth.

In the life cycle of *Spirometra* a swimming coracidium hatched from an egg infects a copepod in which it develops to the next larval stage, a procercoid, When this is ingested by any species of vertebrate, except a fish, it develops into the sparganum, characterized by a small head or scolex and a narrow ribbonlike body that elongates at a rate of 1–5 mm/day, eventually reaching a length of several feet. In nature the water snake *Natrix* is the principal host for this stage, but mice and rats serve well as laboratory hosts. The head or scolex can be removed and injected into a new host, where it will again grow into a sparganum. Animals can be infected by feeding them with infected copepods by stomach tube or by injecting scolices under the skin.

A rodent implanted with several scolices soon begins to show accelerated growth the extent of which is dose dependent. Twelve scolices in a mouse, or 36 in a hamster, produce a maximal effect, which persists through the life of the host. Normally there is little or no tissue reaction to the worms. In occasional animals which did not show the growth enhancement, the worms were found encapsulated. By implanting worms into rats made hypothyroid by treatment with propylthiouracil, it was found that not only did the worms grow as well as in normal rats but, surprisingly, the host rats gained weight rapidly. Their growth curve approximated that of the normal male rat. This

FIGURE 12.1. Growth-promoting effect of larval (sparganum) form of the tapeworm *Spirometra mansonoides*. All four rats, from the same initial lot, were hypophysectomized when they weighed 90 g. The two rats which are here much larger received 20 larval scolices of *S. mansonoides* about 1 month later. Photo taken 6 months later. The infected rats outweigh the controls by a factor of 3 or 4×. (From Mueller, 1974. Original print courtesy of Dr. J. Mueller.)

led to the development of a suitable assay method with hypophysectomized rats. Such animals showed a rapid growth response to the injections of spargana, giving a result within 48 hr.

With this assay it was found that sparganum growth factor (SGF) has a molecular weight of about 70,000. SGF seems to be secreted from the worm.

It is present in the serum of an infected animal as well as in medium in which worms have been incubated. It resembles pituitary growth hormone and competes with growth hormone for binding sites on rabbit liver cell membrane. Its effect on tibial cartilage is like that of growth hormone and, like growth hormone, it induces production of sulfation factor (somatomedin). It differs from growth hormone, however, in that it does not mobilize fatty acids. After injection it will persist in the blood up to 10 days, whereas growth hormone disappears after a few hours. It cannot be identical with growth hormone since there is no cross-reactivity in radioimmunoassay. Like growth hormone, however, it does induce production of liver ornithine decarboxylase in the hypophysectomized rat.

Hypophysectomized sparganum-infected rats often reach normal adult size, but they do not become aggressive or mature sexually, since they lack gonadotropic hormone. The spargana also exert a peculiar effect on diabetic rats. Whereas alloxan-diabetic rats ordinarily sicken progressively and die, those implanted with a few spargana or injected with SGF had a healthy appearance and grew normally. Their blood sugar levels, however, showed that they were still diabetic.

The binding of the factor to rabbit liver cell microsomal membranes has permitted development of an assay based on competition with radiolabeled human growth hormone. With this assay C. K. Phares has recently found that the factor can be solubilized with Triton X-100 from homogenates of the plerocercoids. This has provided much higher yields of active material than could formerly be obtained and should lead to purification of the factor.

SGF clearly must take a place alongside other growth factors—its chemical structure and especially its mode of action are of fundamental biological importance. But its role in parasitism is less clear. It seems to be produced only by the one species *S. mansonoides;* various related worms do not show such a factor. It does not seem to have an effect on the normal intermediate host in nature, the water snake. If rodents were the natural intermediate host, one could suppose that an enlarged overweight rodent might be an easier prey for a cat, the definitive host of the tapeworm, thereby conferring a selective advantage to the worm.

This is not such a far-fetched idea. For any parasite that depends on predation of one host by another for its life cycle, any slight weakening of the prey resulting from the parasitic infection would certainly enhance the likelihood of this individual falling victim to the predator. It is easy to imagine that a mouse with numerous cysts of *Toxoplasma* in its brain would be more easily caught by a cat than an uninfected one. Similarly, a sheep infected with hydatid cysts of the cestode *Echinococcus* might well lag behind healthier members of the herd and fall victim to a dog or another canid. Mice infected with *Trichinella spiralis* are less active than uninfected ones, and the infection leads to a loss of behavioral dominance among the males.

Such effects, though having a selective advantage, are relatively nonspecific. There are now known, however, more interesting positive effects on behavior that clearly facilitate the ingestion of a parasitized intermediate host by the definitive host. Good examples are provided by trematodes of the genus *Dicrocoelium* and related forms. These worms, which resemble the sheep liver fluke *Fasciola hepatica*, have a somewhat more complex life cycle involving three hosts rather than two. The definitive host is a vertebrate in which eggs are produced having a resistant shell. When these are ingested by a terrestrial snail, a cycle of asexual multiplication ensues within the snail that terminates with the emergence of many cercariae. These become entangled in slime from the snail to form small balls that stick to vegetation and in which the cercariae can survive for a period of time. The slime balls with their contained cercariae are eaten by ants. The cercariae develop within the ant into metacercariae. These nearly mature flukes then reach sexual maturity after they have been ingested, along with the ant, by the vertebrate host. This last link in the chain is facilitated by the remarkably altered behavior of ants infected with these metacercariae. Two slightly different situations have been described. In carpenter ants (*Campanotus*) infected with a species of *Brachylecithum*, the infected workers become somewhat lethargic and lose their normal photophobic response. They tend to wander around in the sunshine on exposed rocks. They also become somewhat obese, and in general make very easy targets for the birds, e.g., robins, that prey on them and that are definitive hosts for the worm. Even more striking is the effect of species of *Dicrocoelium* in any of several species of formicine ants. Ants infected with these metacercariae show essentially normal behavior during the day but toward evening, as the temperature falls, they crawl up on blades of grass, seize the grass firmly with their mandibles, and remain in a torpid condition until morning. This behavior clearly makes them especially likely to be ingested during evening or early morning grazing by a herbivore such as a sheep, the definitive host of the worm. The metacercariae of *Dicrocoelium* often lodge in or near the brain or the subesophageal ganglion of the ant, but it is difficult to imagine that a mechanical injury to the nervous system could have such a specific effect on behavior. More likely there is some neurohormonal influence.

Just such an influence has been postulated to explain another behavioral effect of parasitism, also remarkably purposeful from the parasites' point of view. Amphipods such as *Gammarus lacustris* or *Hyalella aztecus* generally hide in darkened regions in their freshwater habitat. Under laboratory conditions in an aquarium, uninfected *Hyalella* spent all their time in a darkened portion and *Gammarus* nearly all. When the same species of amphipods were infected with the cystacanths of a species of acanthocephalan worm (*Polymorphus paradoxus* in *Gammarus* or *Corynosoma constrictum* in *Hyalella*), they spent all of their time in the light zone. The amphipods bearing cystacanths were highly photophilic whereas uninfected ones were strongly photophobic. Even when disturbed, many of the infected amphipods tended to skim along the surface of the water whereas uninfected ones dove down to hide in the dark. The infected ones would be more likely to be ingested by their next host, such as a duck. Acanthocephala are a separate group of parasitic worms of uncertain

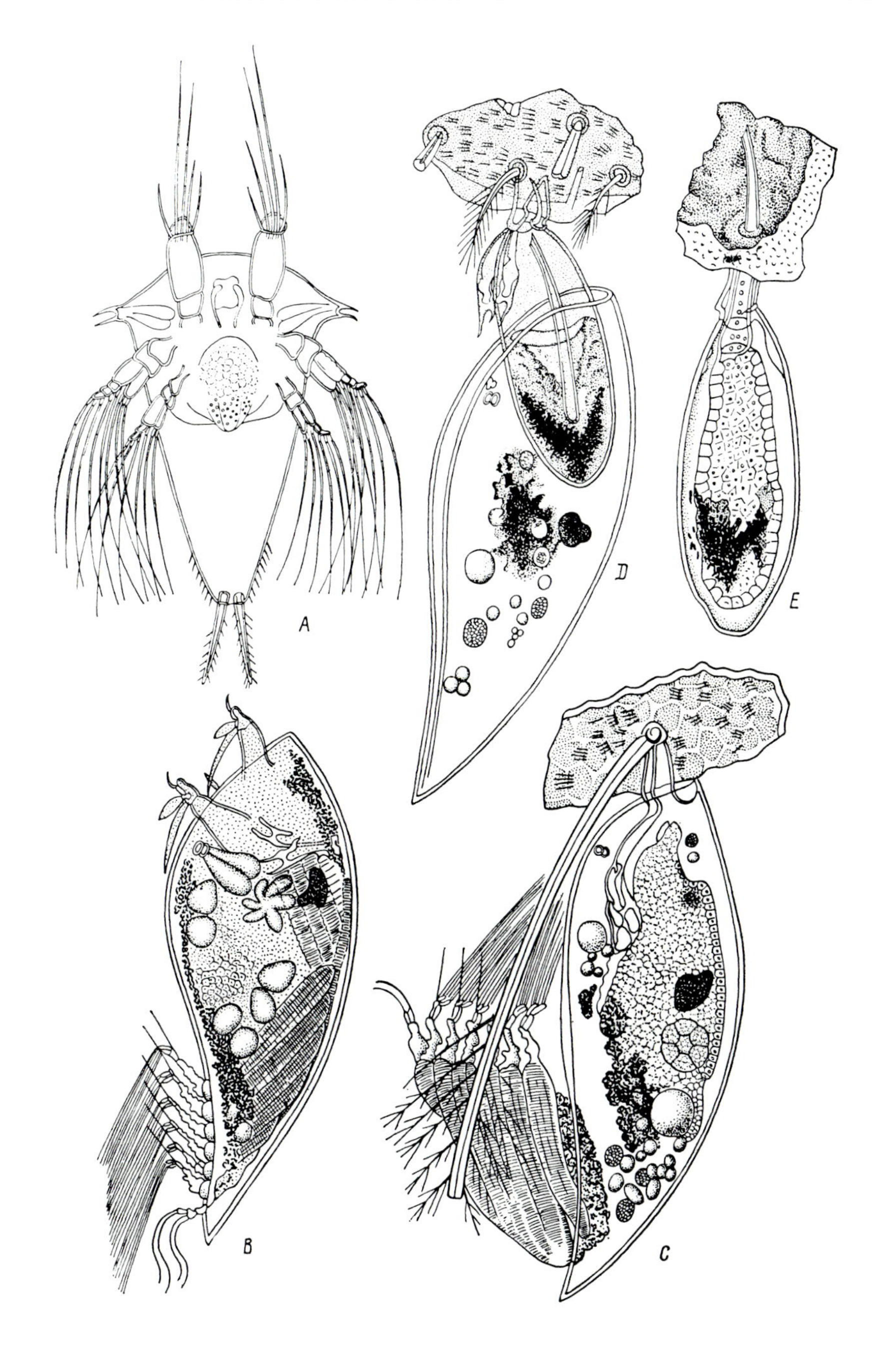

FIGURE 12.2. Life cycle of *Sacculina carcini.* (A) Nauplius. (B) Cypris. (C) Cypris attached by its antenna to shell of host crab and discarding its appendages. (D) Cypris molting and changing into an unsegmented oval sac, the kentrogon. (E) Penetration of kentrogon into the crab host. (From Dogiel, 1966.)

phylogenetic position with affinities to both the nematodes and the platyhelminths. The definitive host is typically a vertebrate. Insects and crustacea then serve as intermediate hosts for terrestrial and aquatic species, respectively. For the latter, eggs excreted from the vertebrate host (such as a duck) infect an amphipod, such as *Gammarus.* It is of special interest that the response of the amphipods to light was not changed when they were infected with acanthellae, the earlier developmental stage of the worm, which is not infective to the vertebrate. Only when these had matured to the infective cystacanth stage, a process that requires about 2 months, was the behavior of the amphipods modified. There can be little question of the adaptive value for the parasite of this parasite-induced modification of host behavior. How it is brought about remains for future work to discover. Future work likewise will reveal many other instances of subtle interaction resulting in behavioral modifications of parasitized hosts.

Less subtle and far more striking is the phenomenon of parasitic castration. A consideration of this will also serve to introduce briefly examples of two vast groups of parasites of great biological interest but of no medical importance and hence rarely mentioned in texts on animal parasitism (however, see Dogiel, 1966). The classic instance of parasitic castration is that of crabs by another crustacean, *Sacculina.* Parasitism is common among crustacea, ranging from facultative ectoparasites to the extreme exemplified by species of the genus *Sacculina.* Thus, *S. carcini* hatches from its eggs to produce a free-swimming nauplius, a typical larval crustacean. This develops and molts to another free-swimming larval form, the cypris. The cypris then begins parasitic life by attaching to a crab of the genus *Carcinus.* It now sheds all its appendages and then molts to become an unsegmented oval sac (Fig. 12.2). This form penetrates into the crab and develops an extensive system of branches extending even into the appendages of the host (Fig. 12.3). These presumably serve for uptake of nutrient. Finally, there appears on the undersurface of the crab's abdomen a saclike outgrowth of the parasite. Here eggs and sperm

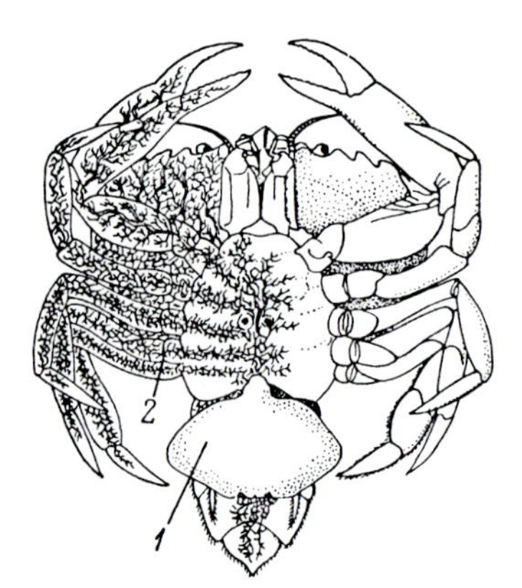

FIGURE 12.3. A crab carrying a mature *Sacculina* (1) on the ventral side of its abdomen. The highly branched absorptive stalk (2) is shown inside the crab. (From Dogiel, 1966.)

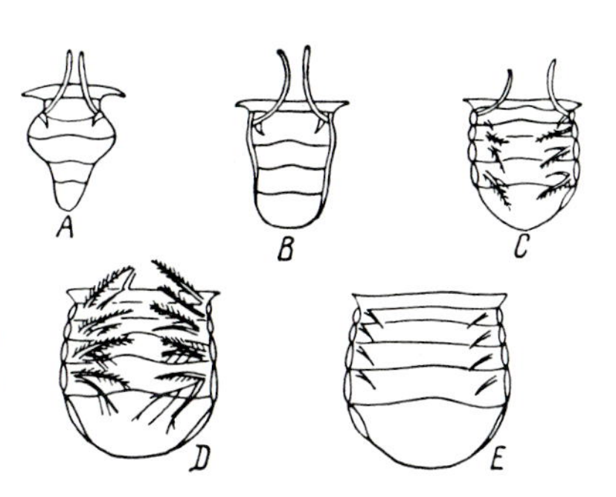

FIGURE 12.4. The effects of parasitization by *Sacculina* on the morphology of the abdomen of the crustacean *Inachus mauritanicus*. (A, D) Normal male and female. (B, C) Males infected with *Sacculina*. (E) A female infected with *Sacculina*. Note especially the femalelike morphology in C. (From Dogiel, 1966.)

are formed, the organisms being hermaphroditic. Within the shelter of the mantle of the sac, the fertilized eggs develop into the nauplii, which escape to start another generation.

The metabolism of the crab is completely altered, especially in the males. These develop yellow blood with a high fat content, like that of a mature female, and they even acquire some of the external morphological characteristics of the female (Fig. 12.4). Host crabs of neither sex can reproduce. The injection of extracts prepared from the internal root system of a sacculinid into pubertal male crabs resulted in a reduction in the testes, their invasion by hemocytes, and total inhibition of spermatogenesis.

Parasitic insects of the order Strepsiptera provide another example. Bees of the genus *Andrena* often carry a female strepsipteran of the genus *Stylops*. This is visible as a small saclike structure protruding from the bee's abdomen. In it are produced active larval *Stylops* that escape and attach to larval bees in the nest, ordinarily one or two parasites per larva. The *Stylops* lives first as an internal parasite. With the metamorphosis of the bee it also metamorphoses to protrude from the adult bee's abdomen. The males then emerge as active winged forms. The females remain attached in their host's abdomen. After insemination the eggs develop and the female degenerates into a sac from which the larval *Stylops* emerge to repeat the cycle. Unlike the more typical situation with the many parasites of other insects (mostly in the orders

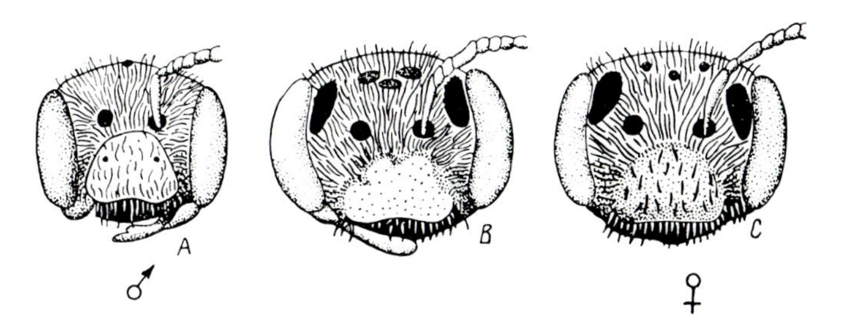

FIGURE 12.5. Effects of parasitization by a strepsipteran on the bee *Andrena solidaginis*. (A) Head of a normal male. (B) Head of a parasitized stylopized male. (C) Head of a normal female. (From Dogiel, 1966.)

Diptera and Hymenoptera), where the host insect is consumed from within and killed, the stylopized bees live but are castrated. Males acquire secondary sex characters of the females and vice versa; the parasitized bees mostly appear as intersexes (Fig. 12.5).

It seems likely that with both Strepsiptera/bees and sacculinids/crabs, the effects on the host are exerted via a hormonelike substance. Before this can be understood in detail, we need to know more about hormonal control of secondary sexual characters in arthropods. What, if any, advantage the parasite derives from castration of its host remains conjectural.

Bibliography

Anokhin, I. A., 1966, The 24-hour rhythm in ants invaded by metacercariae of *Dicrocoelium lanceatum*, *Dokl. Akad. Nauk SSSR* **166:**757–759.

Bethel, W. M., and Holmes, J. C., 1973, Altered evasive behavior and responses to light in amphipods harboring acanthocephalan cystacanths, *J. Parasitol.* **59:**945–956.

Bethel, W. M., and Holmes, J. C., 1974, Correlation of development of altered evasive behavior in *Gammarus lacustris* (Amphipoda) harboring cystacanths of *Polymorphus paradoxus* (Acanthocephala) with the infectivity to the definitive host, *J. Parasitol.* **60:**272–274.

Carney, W. P., 1969, Behavioral and morphological changes in carpenter ants harboring dicrocoeliid metacercariae, *Am. Midl. Nat.* **82:**605–611.

Chang, T. W., Raben, M. S., Mueller, J. F., and Weinstein, L., 1973, Cultivation of the sparganum of *Spirometra mansonoides in vitro* with prolonged production of sparganum growth factor, *Proc. Soc. Exp. Biol. Med.* **143:**457–459.

Day, J. H., 1935, The life history of Sacculina, *Q. J. Microsc. Sci.* **77:**549–583.

Dogiel, V. A., 1966, *General Parasitology* (revised and enlarged by Y. I. Polyansky and E. M. Kheisin; translated by Z. Kabata), Academic Press, New York.

Herberts, C., Andrieux, N., and de Frescheville, J., 1980, Influence du parasite *Sacculina carcini* Thompson au début de son développement sur les fractions hémolymphatiques et épidermiques du crabe *Carcinus maenas* Linné, *Can. J. Zool.* **58:**572–579.

Holmes, J. C., and Bethel, W. M., 1972, Modification of intermediate host behavior by parasites, *Zool. J. Linn. Soc.* **51**(Suppl. 1)**:**123–149.

Moore, J., 1984, Altered behavioral responses in intermediate hosts—an acanthocephalan parasite strategy, *Am. Nat.* **123:**572–577.

Mueller, J. F., 1974, The biology of Spirometra, *J. Parasitol.* **60:**3–14.

Phares, C. K., 1982, The lipogenic effect of the growth factor produced by plerocercoids of the tapeworm, *Spirometra mansonoides,* is not the result of hypothyroidism, *J. Parasitol.* **68:**999–1003.

Phares, C. K., 1984, A method for solubilization of a human growth hormone analogue from plerocercoids of *Spirometra mansonoides, J. Parasitol.* **70:**840–842.

Rau, M. E., 1984, Loss of behavioural dominance in male mice infected with *Trichinella spiralis, Parasitology* **88:**371–373.

Reinhard, E. G., 1956, Parasitological review: Parasitic castration of crustacea, *Exp. Parasitol.* **5:**79–107.

Rubiliana, C., 1985, Response by two species of crabs to a rhizocephalan extract, *J. Invert. Pathol.* **45:**304–310.

Shizume, K., and Takano, K., (eds.), 1980, *Growth and Growth Factors,* University of Tokyo Press, Tokyo, pp. 193–201.

Thompson, S. N., 1983, Biochemical and physiological effects of metazoan endoparasites on their host species, *Comp. Biochem. Physiol. B* **74:**183–211.

CHAPTER 13

Modification of Host Cells Produced by Intracellular Protozoa

Obligate intracellular parasites depend on the integrity of their host cell. They do not damage the host cell so severly as to cause its death before they have completed their development within it. Many kinds of intracellular parasites actually stimulate growth or multiplication or both of their host cells. They may cause extensive hypertrophy and hyperplasia. Some have been shown to have a distinct mitogenic effect.

Most striking are the tumorlike outgrowths that occur on fish infected with certain species of microsporidia and that R. Weissenberg has termed xenomas. These are seen in the three-spined stickleback infected with *Glugea anomala* or *G. hertwigi* or in flounder with *G. stephani*. A mesenchymal cell invaded by the microsporidian sporoplasm is stimulated to extensive growth accompanied by repeated amitotic division of its nucleus. A cell originally 10 μm in diameter grows to a structure 3–4 mm in diameter with hundreds of nuclei mostly in the cortical cytoplasm (Figs. 13.1, 13.2). Lying in the more central parts of this structure, but without any sharp demarcation, is the parasite schizont, a polynucleate elongate cylindrical body. The hypertrophied parasite-containing cell is encapsulated by cuticular cells, surrounded by vascularized connective tissue layers. When growth of the parasite is completed, a vacuole is delimited around the plasmodium, which then divides and forms the resistant spores (see Chapter 3). With other species, the host cell nucleus does not divide but undergoes extensive hypertrophy along with the cytoplasm. In the anglerfish *Lophius* infected with the microsporidian *Nosema lophii*, the parasites specifically invade the nervous system where hypertrophy of the ganglion cells results in the formation of numerous nodules.

Although concerned with parasitism among plants rather than among animals, it is relevant here to consider a remarkable situation recently discovered in the marine red alga *Polysiphonia* parasitized by the red alga *Choreocolax*. It had long been known that certain cells of the host alga enlarge up

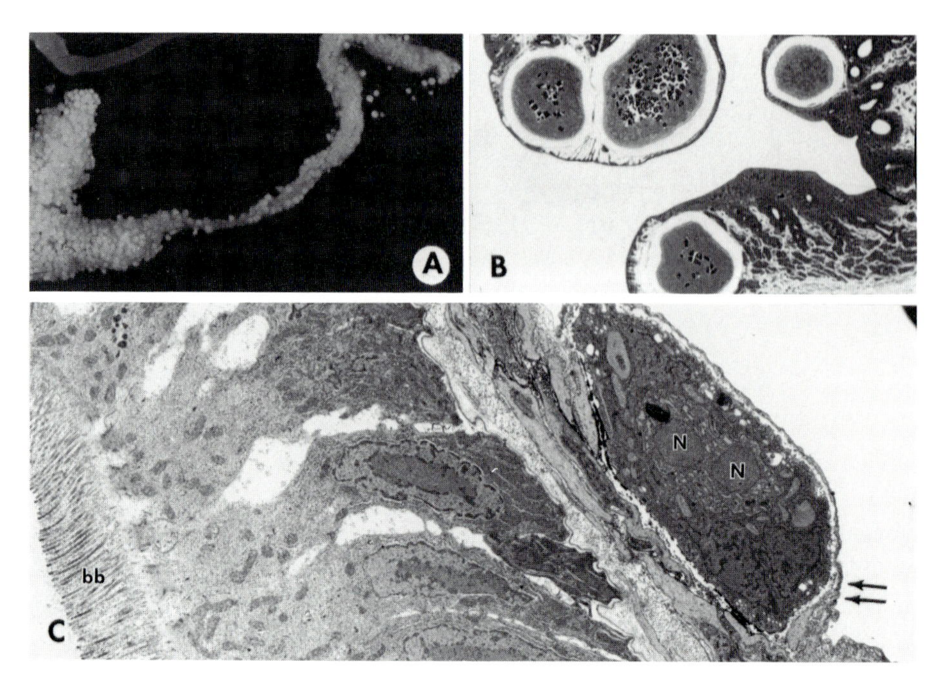

FIGURE 13.1. Xenomas caused by the microsporidian *Glugea stephani* in the winter flounder *Pseudopleuronectes americanus*. (A) Intestine of 1-year-old winter flounder with numerous xenomas that have matured to cysts about 1 mm in diameter. A portion of uninfected intestine is seen in the upper left corner. ×3. (B) Section of intestine of flounder showing four middle-stage xenomas 20 days after infection. Note the clear space around each; this contains much mucinlike material. Spore formation has begun in the central region. ×100. (C) An early stage xenoma in the intestinal submucosa of a 10-mm juvenile flounder, 10 days after infection. Note the parasite with two nuclei (N), lying in a host cell covered with a thin layer of clear mucinlike material bounded by an endothelial cell (arrows). The mucosal brush border is at the left (bb). EM. ×10,000. (From Weidner, 1985. Original prints courtesy of Dr. E. Weidner.)

to 20-fold, and show a decrease in size of the central vacuole with concomitant increase in nuclei and in cytoplasmic constituents including mitochondria, plastids, and photosynthetic storage products. It has now been shown that these cells actually contain small planetic nuclei derived from the parasite and transferred to the host cell by means of special conjunctor cells (Fig. 13.3). These parasite nuclei can survive in the host cell for the full 5 to 8 weeks of the infective cycle and it may be that they are directly responsible for the hypertrophy and other changes in the host cell.

Extensive hypertrophy of the host cell and nucleus is also seen in the tissues of ducks infected with *Leucocytozoon simondi*, representative of a group of parasites closely related to the malarial plasmodia. Like the latter, *Leucocytozoon* undergoes sexual union and subsequent sporogony in an insect vector, a species of *Simulium* (blackfly). Sporozoites introduced by the bite of the fly develop into schizonts in hepatic cells of the liver, which may show a

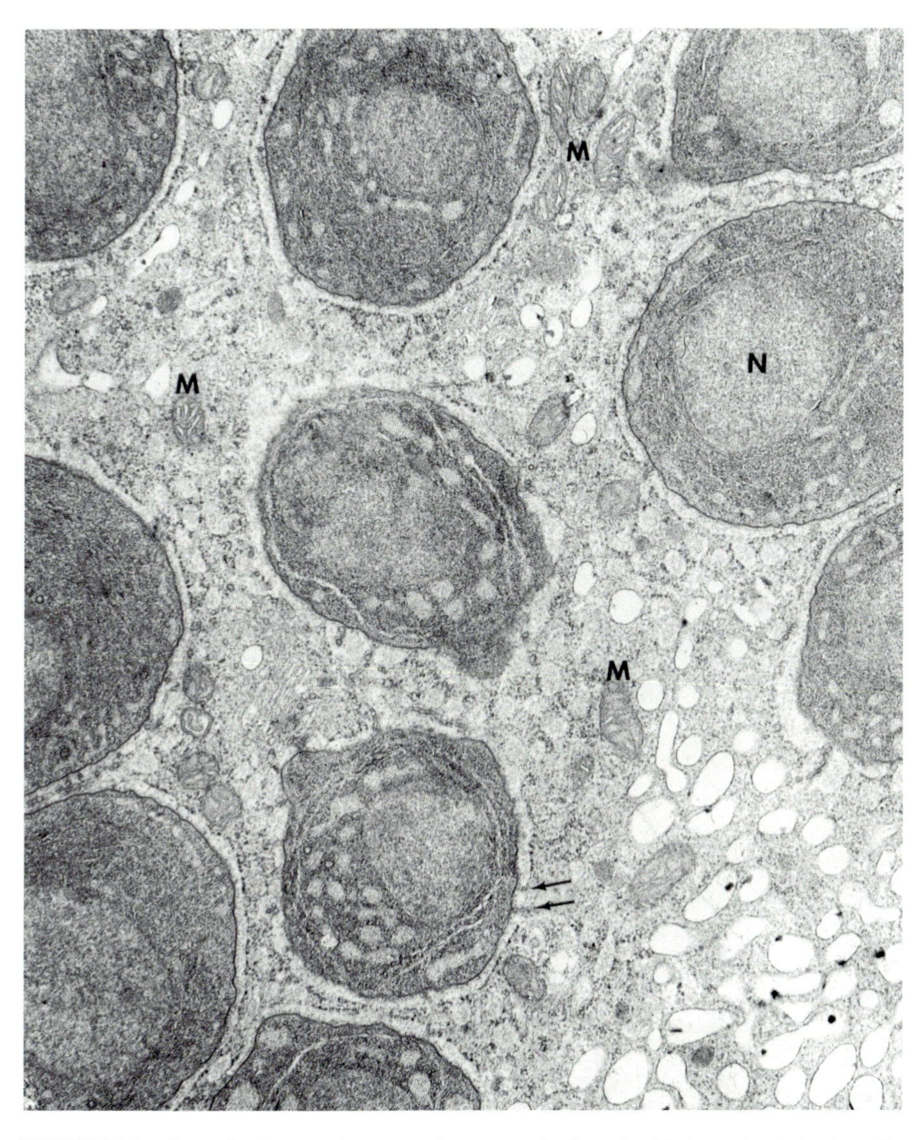

FIGURE 13.2. *G. stephani* vegetative stages in xenoma in flounder 20 days after infection. The numerous parasites, each with a single nucleus (N), lie at this stage directly in the host cytoplasm. Note filopodia (arrows) connecting parasite and host cytoplasm. These become conspicuous shortly before spore formation. Also note the numerous host cell mitochondria (M). TEM. ×30,000. (Original print courtesy of Dr. E. Weidner.)

moderately enlarged nucleus. Merozoites from these schizonts enter erythrocytes to become male and female gametocytes. Some of the products of the hepatic schizonts, however, enter cells throughout the body, especially vascular endothelium, and form secondary schizonts. These produce enormous hypertrophy of both host cell and nucleus (Fig. 13.4) and grow into mega-

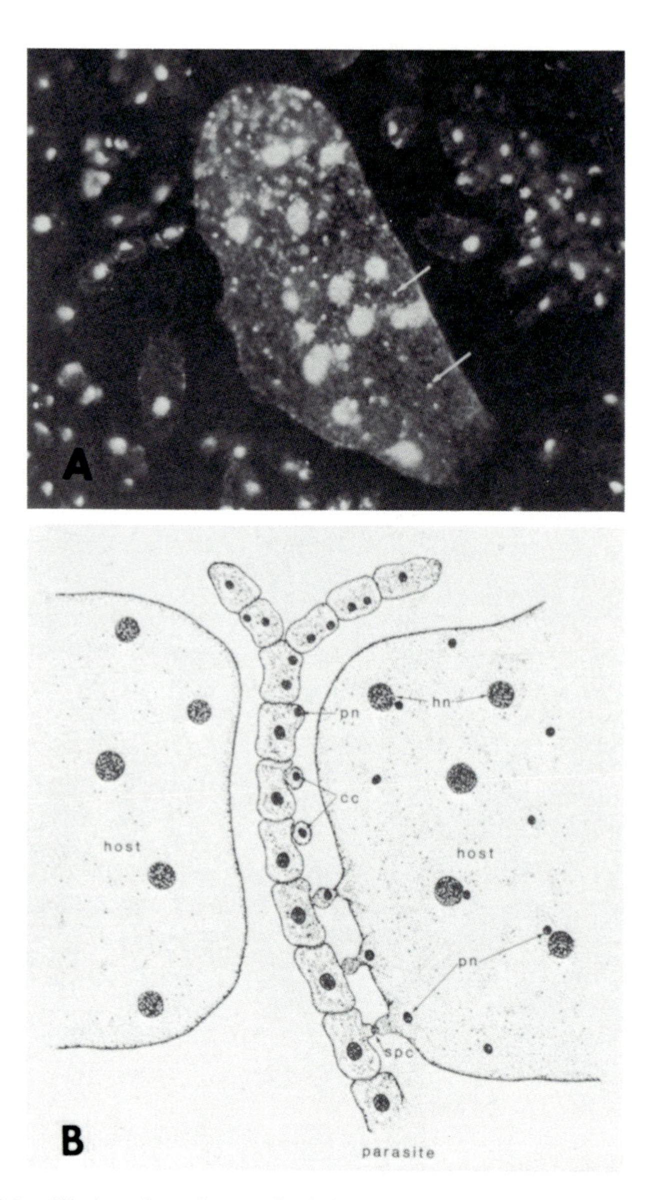

FIGURE 13.3. (A) A multinucleate cell of the red alga *Polysiphonia* containing many small planetic nuclei of the parasitic alga *Choreocolax*. This host cell was contained within the parasite reproductive tissue. Squashed and stained with 4′,6-diamidino-2-phenylindole. Epifluorescence. ×64. (B) Diagram to show conjunctor cell formation and planetic nuclear transfer from the parasitic alga *Choreocolax* to the host *Polysiphonia*. An apically dividing parasite filament is growing between two host cells. By means of conjunctor cells (cc), it is forming secondary pit connections (spc). In one host cell several parasite nuclei (pn) have been injected, and the host nuclei (hn) have increased in size and DNA content. (From Goff and Coleman, 1984).

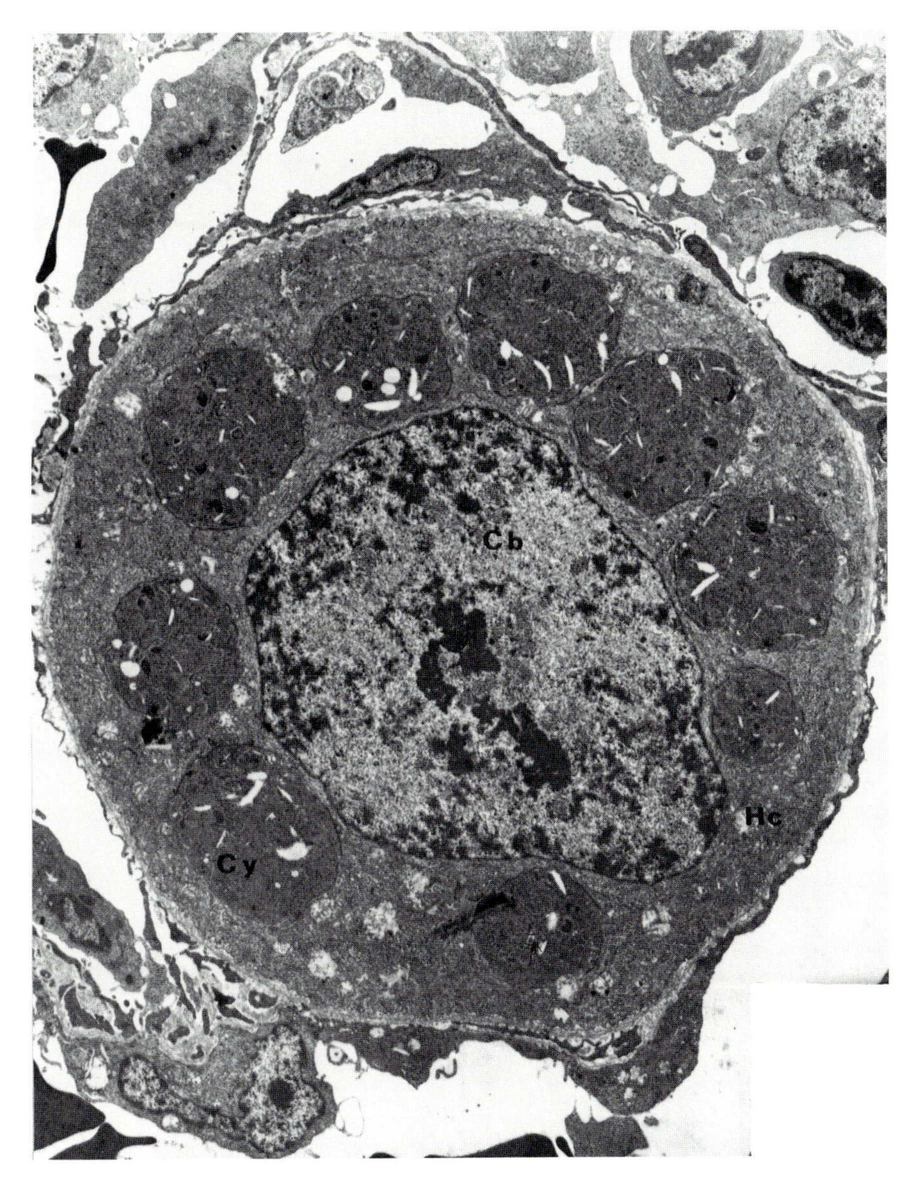

FIGURE 13.4. *Leucocytozoon simondi* in reticuloendothelial cell of duck. A young megaloschizont with the characteristic enlarged host cell nucleus (Cb). Several round cytomeres (Cy) lie in hypertrophied cytoplasm of host cell (Hc). TEM. ×7000. (Original print courtesy of Dr. S. Desser. See Fallis *et al.*, 1974).

loschizonts 100–200 μm in diameter, each producing a million or so merozoites 1 μm in diameter. These merozoites invade erythrocytes and erythroblasts to form male and female gametocytes. The host nucleus may increase in volume some 200-fold. How polyploid it becomes is not known.

Similarly, when *Theileria parva* (Diagram IX) develops in certain cells of particular acini of the salivary glands of its tick vector, it causes massive hypertrophy. At the same time the host cell accumulates early in infection large stores of glycogen, which is not present at all in uninfected cells. As the parasite syncytium develops, branching throughout the cell, the host cell endoplasmic reticulum and secretory granules are broken down by autophagy, but the mitochondria and the nucleus remain in good condition to the very end when the infective sporozoites of the parasite are fully formed (Figs. 13.5, 13.6).

Neither *Leucocytozoon* nor the tick stages of *Theileria* nor any of the microsporidia producing marked hypertrophy have as yet been placed under controlled *in vitro* conditions that might permit closer study of these remarkable effects. Such study is, however, now possible with stages of *Theileria* in cattle. These exert a strong mitogenic effect on bovine lymphoblasts. If a suspension of such cells from an uninfected cow is exposed to infection *in vitro* by sporozoites of *Theileria parva* (cause of East Coast fever) from the salivary glands of an infected tick, only those cells that become infected are capable of continuous growth *in vitro*. Similarly, in cell cultures prepared from infected cows, the already infected lymphoblasts show continuous growth *in vitro*. Each infected cell normally contains a macroschizont of the parasite (Fig. 13.7) with 10–17 nuclei. When the host cell divides, the parasite is also divided in two so that each daughter host cell is infected with a parasite that again grows to have 10–17 nuclei. If the parasites are killed, as with Aureomycin, the host cells soon stop growing. On the other hand, if the division of the lymphoblasts was inhibited without inhibition of the parasites, as with colchicine or with a minimal concentration of actinomycin D, a great increase in number of parasite nuclei per cell occurred. With colchicine there were up to 100 theilerias per lymphoblast, and the host cells showed a cytopathogenic effect. The association between *Theileria* and the bovine lymphoblast as it exists in tissue culture can be regarded as a symbiotic or mutualistic one (see Chapter 24) even though *T. parva* is highly pathogenic to the intact bovine host. It would be of great interest to know the biochemical nature of the mitogenic effect. Could it be that the parasite produces a protein like the platelet-derived growth factor, which in turn is nearly identical with the transforming protein of simian sarcoma virus? Are protein kinases involved in these effects? The *Theileria*–bovine lymphocyte system would lend itself well to such studies. So also would another more recently established *in vitro* system, that of *Babesia equi* in horse lymphocytes. In the lymphoblastoid cell lines transformed by *T. parva* infection, the cells have been shown to express antigens that are recognized by effector cells from immune animals. Little progress has yet been made toward the identification of these antigens; they may not be theilerial in origin.

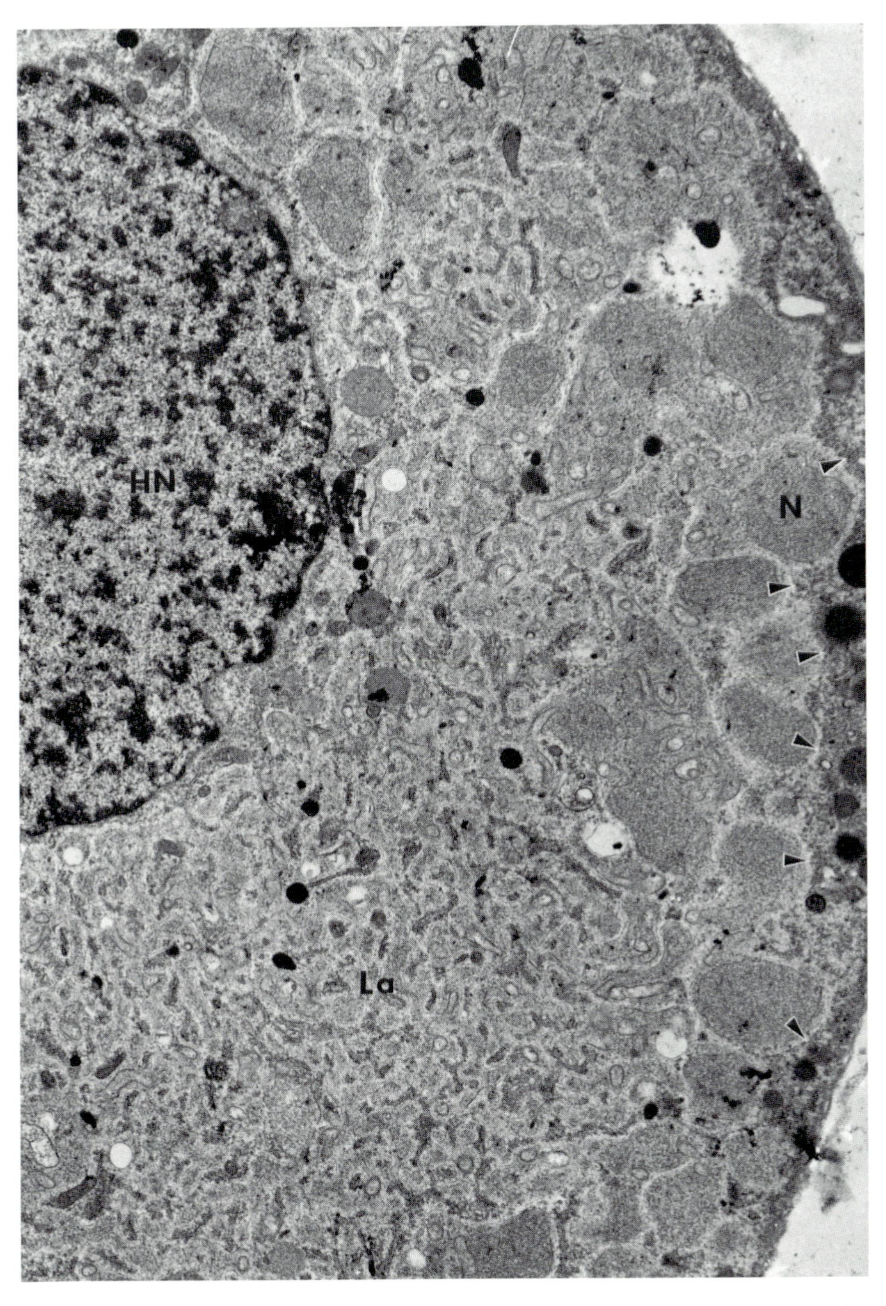

FIGURE 13.5. A portion of a salivary gland cell of the tick *Rhipicephalus appendiculatus* infected with the sporogonic stage of *Theileria parva*. This low-power transmission electron micrograph shows part of the greatly hypertrophied host cell and its nucleus (HN) as well as the parasite that penetrates throughout the cytoplasm as a "labyrinth" (La), an intricate meshwork of interconnected processes. The proliferating nuclei (N) of the parasite lie in lobules at its periphery, which is indicated by arrowheads at the right. ×11,200. (From Fawcett *et al.*, 1982. Original print courtesy of Dr. D. W. Fawcett.)

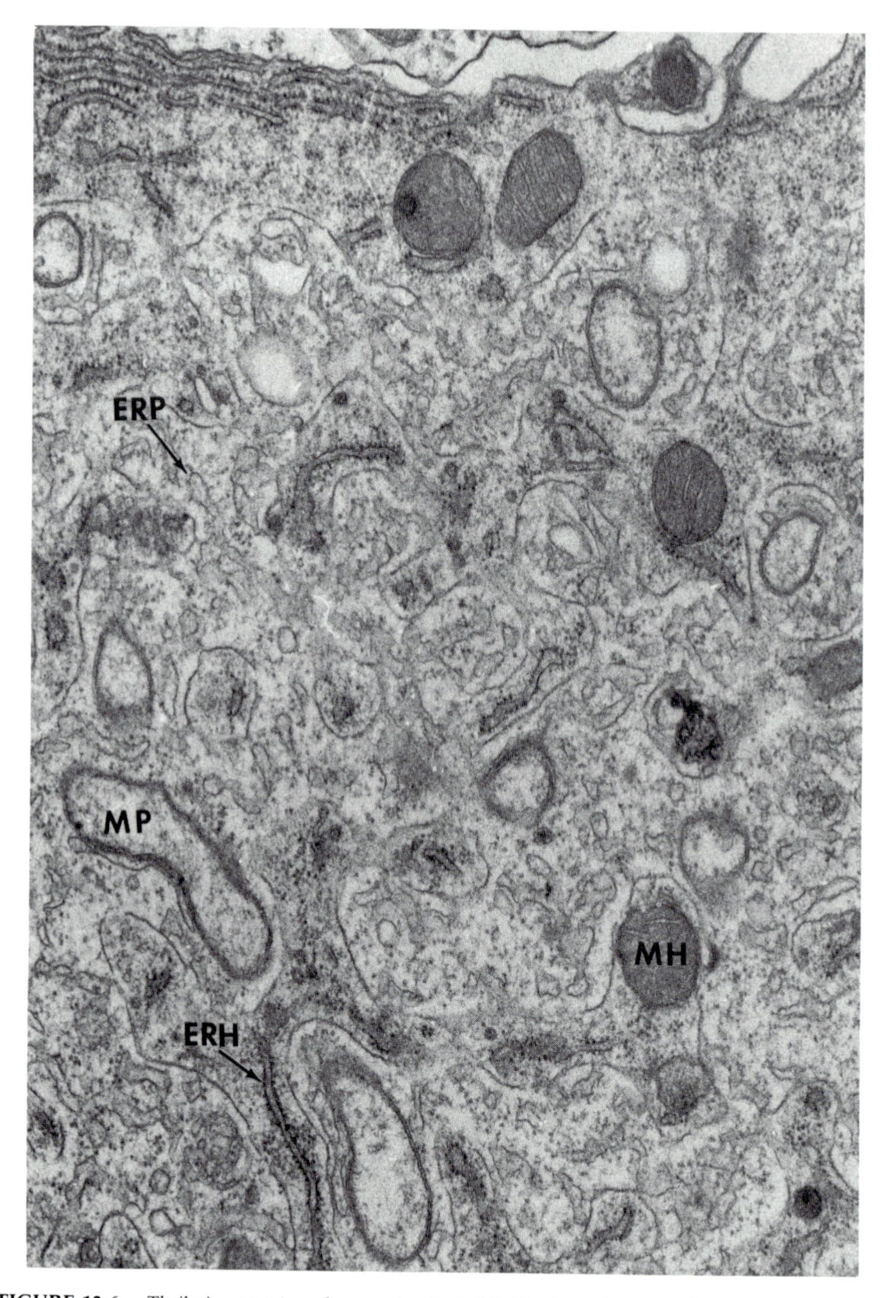

FIGURE 13.6. *Theileria parva* in salivary gland of tick *Rhipicephalus appendiculatus*. A portion of the labyrinth (see Fig. 13.5) at higher magnification to show intimate intermingling of parasite and host cytoplasm, which are nonetheless separated by the continuous parasite plasma membrane. Parasite mitochondria (MP) have electronlucent matrix and no cristae. Host mitochondria (MH) have multiple cristae projecting into a dense matrix. Endoplasmic reticulum of parasite (ERP) is a network of branching tubules limited by thin membranes. Host endoplasmic reticulum (ERH) appears as elongated profiles covered with heavily stained ribosomes. TEM. ×40,000. (From Fawcett *et al.*, 1982. Original print courtesy of Dr. D. W. Fawcett).

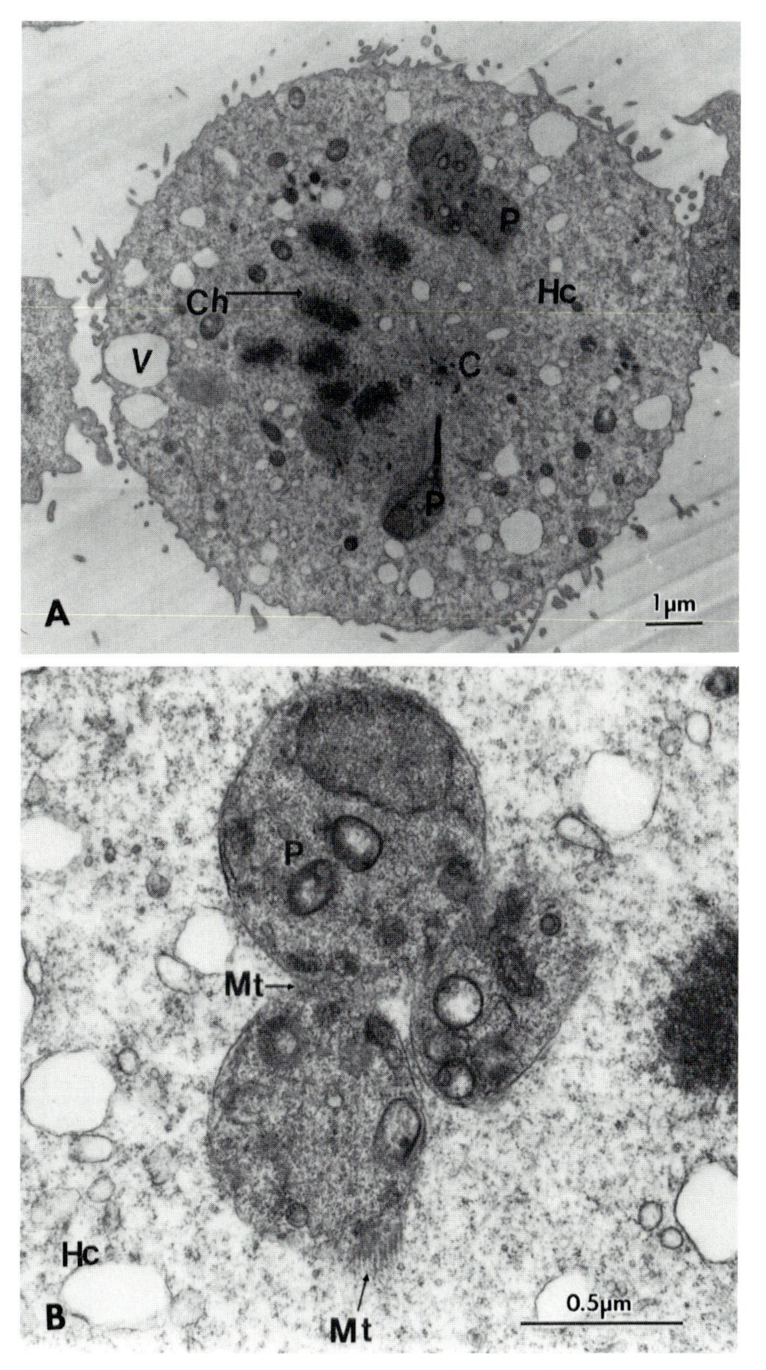

FIGURE 13.7. (A) *Theileria annulata* in a bovine lymphoblastoid cell in mitosis. The chromosomes (Ch) and parasites (P) are drawn together on the spindle of microtubules radiating from a centriole (C). (B) The group of three parasites seen in A at higher magnification to show columns of microtubules (Mt) attached to their pellicle. TEM. ×15,000. (From Musisi *et al.*, 1981. Original prints courtesy of Dr. F. L. Musisi.)

A much more complete story is unfolding with regard to malarial antigens appearing on the surface of erythrocytes infected with plasmodia. Effects on the host cell begin just as soon as the invading malarial merozoite enters an erythrocyte. I have already reviewed the process of entry (Chapter 3) and noted that the invaginating red cell membrane loses its intramembranous particles and that the polarity of membrane enzymes is reversed. The parasitophorous membrane lies closely apposed to the plasma membrane of the developing parasite and grows with the growth of the parasite. How completely the membrane is changed is indicated by the fact that it no longer reacts with antibody to the erythrocytes. It is perhaps not so surprising that the parasitophorous membrane, though derived from the erythrocyte membrane, should be so completely altered. But what about the rest of the erythrocyte? Here too there are far-reaching changes. There are structural changes visible by light microscopy as well as by electron microscopy. Best known at the light microscope level are Schüffner's dots appearing in erythrocytes infected with *Plasmodium vivax* or certain related primate malarias. In the electron microscope these correspond to caveolae (Fig. 13.8) in the red cell surface of unknown nature and function. With all species of malaria the infected erythrocytes show in their cytoplasm membrane-lined clefts (Fig. 13.9B) visible by transmission electron microscopy. These may originate from the parasitophorous membrane and, again, their role is not known.

More is known about a conspicuous surface alteration found on eryth-

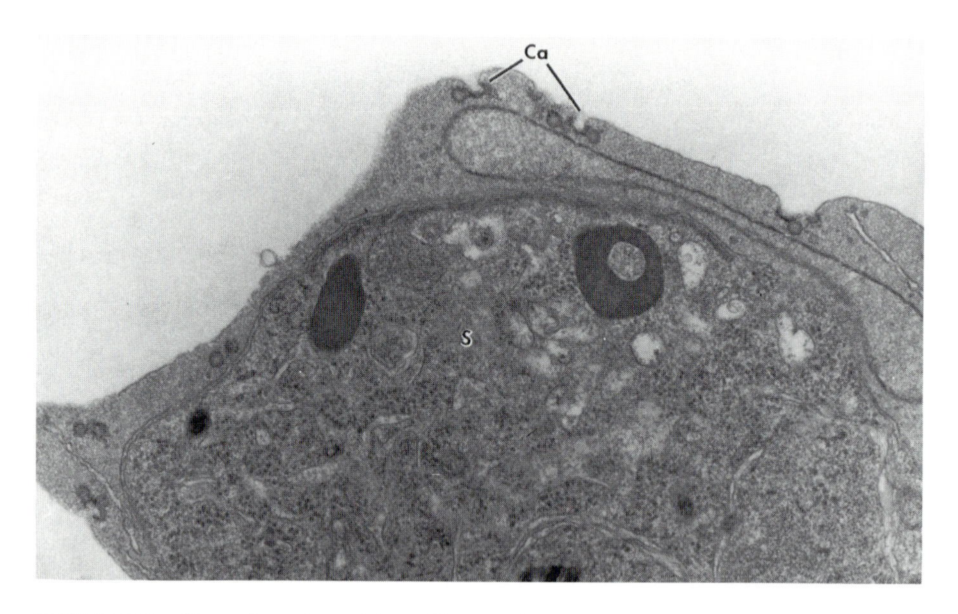

FIGURE 13.8. *Plasmodium vivax* schizont (S) in erythrocyte of the owl monkey *Aotus trivirgatus*. Note caveolae (Ca) at surface of host cell. These correspond to Schüffner's dots seen by light microscopy. TEM. ×47,000. (Original print courtesy of Dr. H. N. Lanners.)

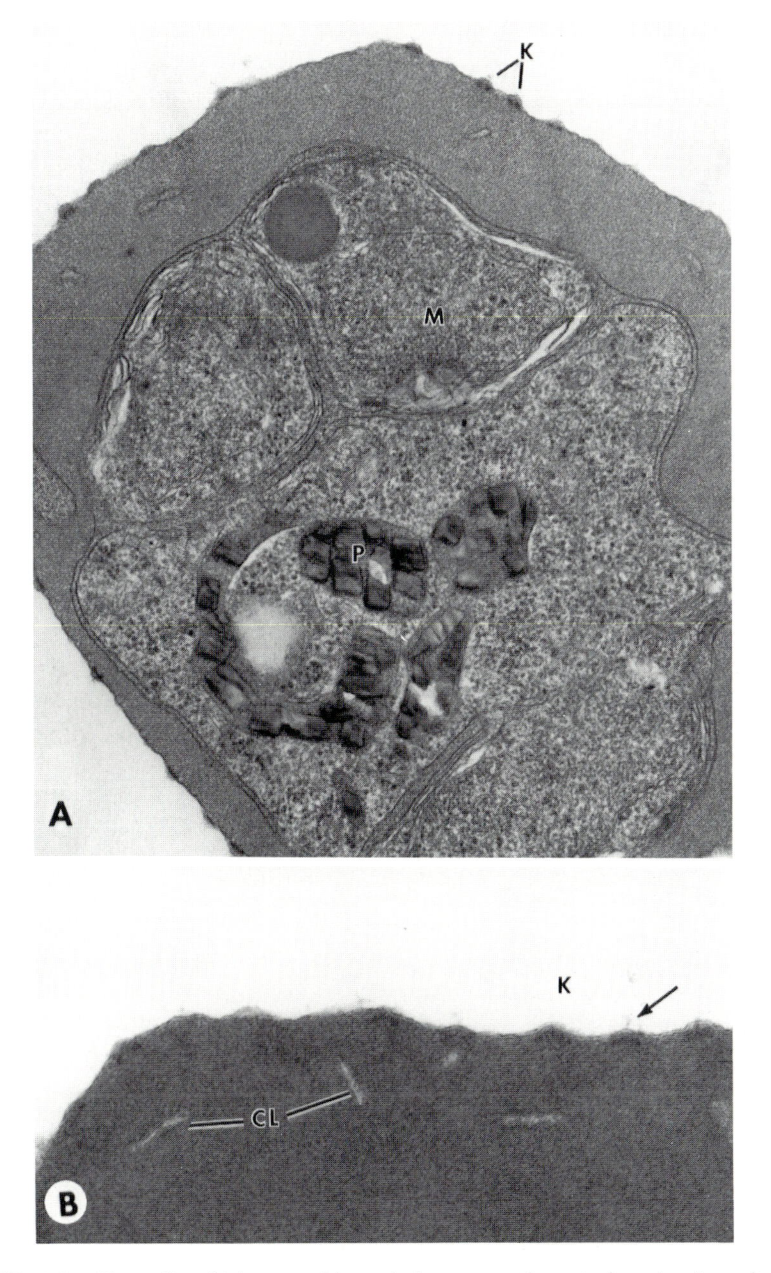

FIGURE 13.9. *Plasmodium falciparum* schizont in human erythrocyte from *in vitro* culture. (A) Note the knobs (K) on the surface of the erythrocyte. The parasite shows forming merozoites (M) and pigment clumps (P). ×37,400. (B) Enlarged view of a portion of another infected erythrocyte to show the dense material lying beneath the red cell membrane at each knob (K). Note the faint plume of material (at arrow) extending from some knobs. This may represent cytoadherent material. Also shown are the membrane-lined clefts (CL) in the cytoplasm of the erythrocyte. ×79,200. (Original prints courtesy of Dr. H. N. Lanners.)

rocytes infected with trophozoites or schizonts of *P. falciparum* (Fig. 13.9) and certain other species of primate malaria. In all of these infections, only ring stages (young uninucleate forms) are found in the peripheral blood. The later stages of the 48-hr asexual cycle are sequestered in the capillaries of the heart, brain, and other organs. They adhere to the capillary endothelium. This is a great advantage to the parasite. Only while it is a small ring does it circulate through the spleen; the later stages, much more likely to be recognized and destroyed by splenic cells, do not circulate at all. They are in this way protected from the spleen, the major organ of defense against erythrocytic parasites (see Chapter 15).

It is clear that adherence to endothelium is a function of the knobs. Furthermore, an elegant demonstration of the role of this adherence in the blocking of capillary vessels and hence in the pathology of malignant tertian malaria has recently been provided. It depended first on the observation that in continuous cultures of *P. falciparum* variants may arise that do not form knobs, and then on the isolation from such cultures of so-called knobby and knobless clones that breed true in culture. When the Baez preparation of rat mesoappendix was perfused with the cells from a knobby culture, they could be seen to adhere to the walls of venules in such a way as eventually to block the venules, whereas no such adherence or blockage occurred if the perfusion was with uninfected red cells or cells from a knobless culture or cells from a knobby clone but containing only young rings and hence lacking the knobs (Fig. 13.10). Knobby (K^+) clones inoculated into an intact *Aotus* monkey generally show only ring stages in the peripheral blood; K^- clones give no patent parasitemia. In a splenectomized *Aotus*, however, K^- clones appeared to multiply better than K^+ ones and showed all stages of the parasites. Injection of immune serum into an intact *Aotus* infected with a K^+ line caused the release of late-stage parasites into the peripheral blood. Similarly, in an *in vitro* test in which K^+ but not K^- cells adhere to a layer of endothelial cells, or of a melanoma line, immune serum prevents this adherence. This convenient *in vitro* test has also been used to show that some culture lines having knobs nevertheless fail to adhere. Evidently, the knob must be present, but in itself it is not sufficient, suggesting two factors: the structural knob and a cytoadherence factor.

The nature of the knob material is clearly of great interest and significance. Following on the original finding of A. Kilejian that K^+ lines have a histidine-rich protein that is absent from K^- lines, it has now been shown that a protein of 92,000 daltons is formed by the parasite and somehow integrated into the structure of the erythrocyte membrane to form the knob. This protein is synthesized between 9 and 21 hr after invasion of the red cell by the merozoite (Fig. 13.11). The protein is not formed by K^- clones of parasites and these organisms fail to show the corresponding mRNA.

The knob protein is by no means an isolated instance of insertion of a parasite protein into the surface of erythrocytes infected with malarial parasites. In two species of rodent malarias a species-specific antigen of 250,000

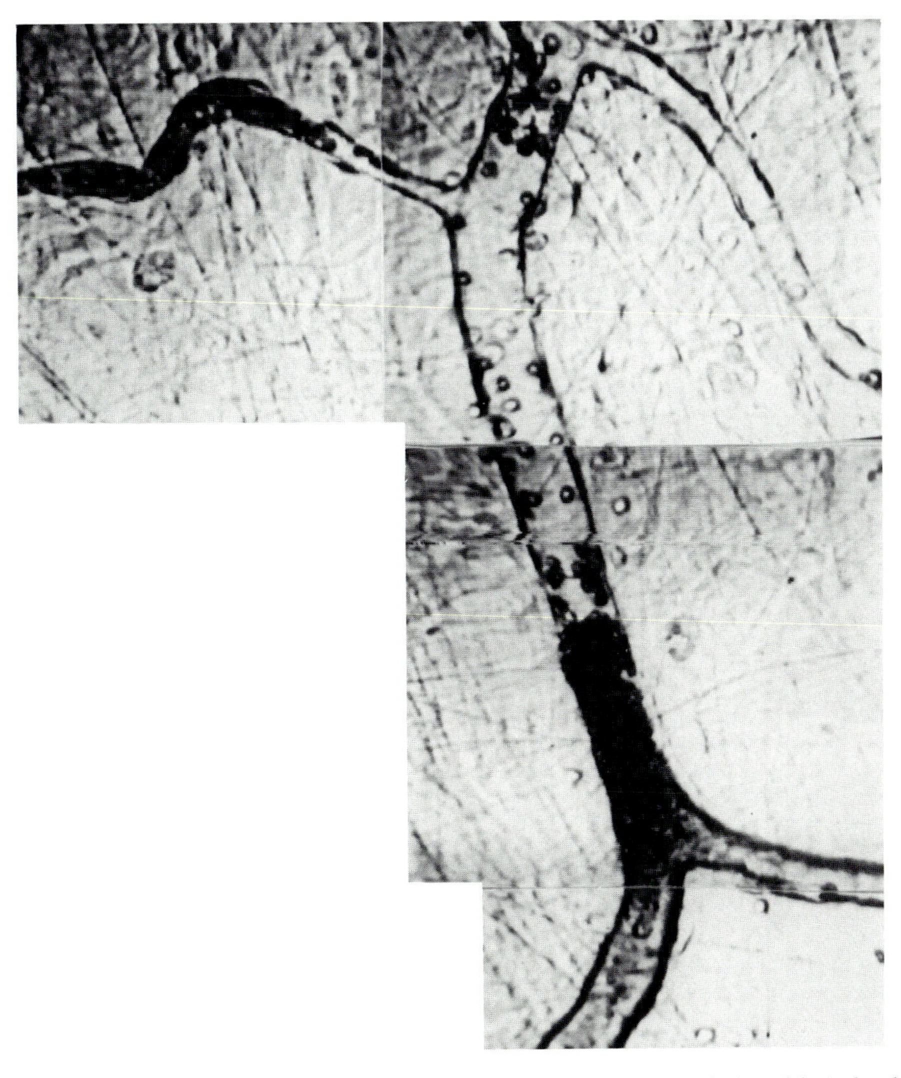

FIGURE 13.10. Microcirculatory obstruction by *P. falciparum*, as seen in perfusion of the isolated rat mesoappendix (Baez preparation) with knobby erythrocytes infected with the clone A-2 of a Gambian isolate (FCR-3). Note the two regions of complete obstruction found in the venules after perfusion with Ringer's solution. This results from adherence of the knobby erythrocytes to the wall of the venule. Such adherence did not occur and no obstructions developed when perfusion was done with knobless erythrocytes infected with the mutant clone D-4 of the same isolate of *P. falciparum*. (From Raventos-Suarez *et al.*, 1985. Prints courtesy of Dr. R. L. Nagel.)

daltons has been demonstrated on the surface of schizont-infected erythrocytes. In *P. falciparum* growing *in vitro* there is evidence for eight glycoproteins of parasite origin on the surface of the infected human erythrocytes. The variant specific antigens of the monkey malaria *P. knowlesi* have been shown to be formed by the parasites and located on the surface of erythrocytes

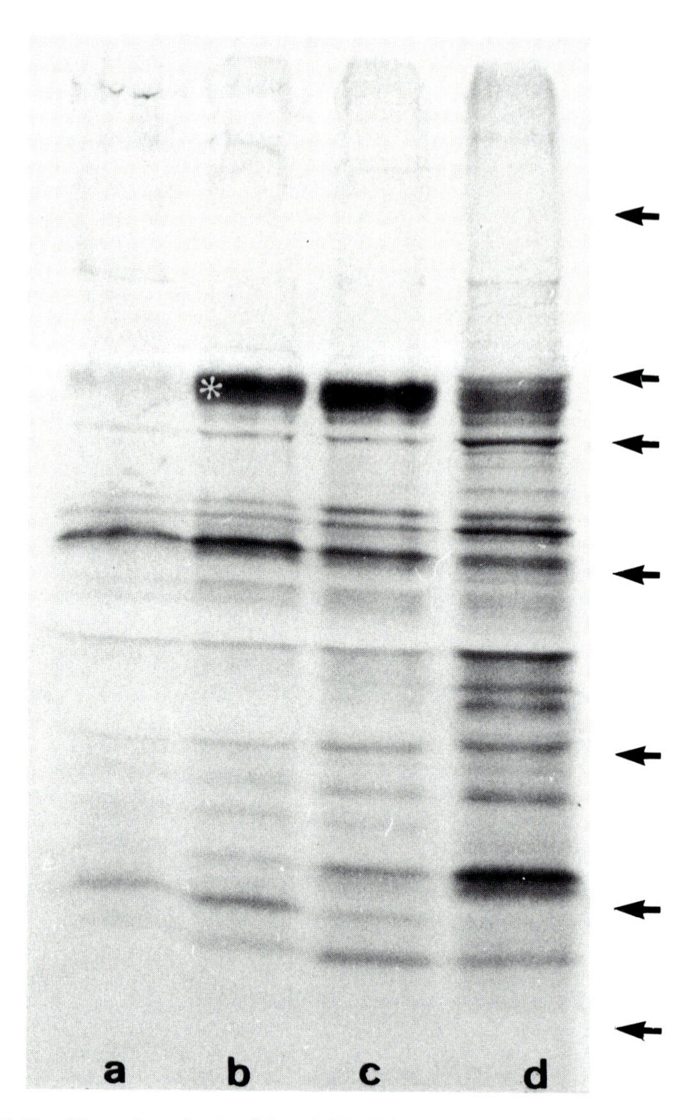

FIGURE 13.11. Time of synthesis of the 92,000-dalton protein present in knobs on the surface of erythrocytes infected with *P. falciparum* (clone A-2, knobby, of isolate FCR-3/Gambia). The synchronized cultures were exposed to [^{14}C]proline for 6-hr periods as follows. Lane a: 3 to 9 hr after merozoite invasion; lane b: 9 to 15 hr after; lane c: 15 to 21 hr after; lane d: 21 to 27 hr after. At 27 hr, when all the parasites were in the trophozoite stage, the membranes from the cells were prepared and subjected to sodium dodecyl sulfate polyacrylamide gel electrophoresis. The resulting fluorogram is shown. The 92,000-dalton protein is marked with an asterisk. Its synthesis was most active between 9 and 21 hr after invasion. The arrows indicate positions of molecular weight markers as follows, from top to bottom: 200,000, 94,000, 68,000, 43,000, 30,000, 18,000, 12,300. (From Vernot-Hernandez and Heidrich, 1984. Print courtesy of Dr. H. G. Heidrich.)

infected with schizonts. In membrane from *P. berghei*-infected mouse red cells a phosphorylated protein appears that is not present in membrane from uninfected cells. What are the functions of these various proteins? The knob protein of *P. falciparum* is essential for sequestration of late stages and their consequent protection from the spleen. The variant antigens of *P. knowlesi* play a role in evasion of the host's immune response. The other parasite proteins associated with the erythrocyte surface could well be concerned with specific changes in the permeability of the infected erythrocytes. Such changes in permeability are well documented. In general, there is an increased permeability, of adaptive value to the parasite in permitting more extensive entry of nutrients and escape of waste products, especially of the large amounts of lactic acid formed by the parasites. Permeability changes could result from changes in the lipid bilayer as well as in membrane proteins. Such changes have been demonstrated in rhesus monkey erythrocytes infected with *P. knowlesi*. In normal rhesus red cells, 70% of the phosphatidylcholine and 15–20% of the phosphatidylethanolamine were in the external layer, whereas a reverse situation obtained in the infected cells, with only 20–30% of the phosphatidylcholine and 50% of the phosphatidylethanolamine in the external layer.

How such changes are brought about by the parasite and how parasite proteins are transported to and become associated with the red cell membrane are questions for future work. Certainly, the parasite synthesizes a great deal of membrane. Some of this appears in the infected cell cytoplasm as the membrane-bounded clefts. Perhaps these help to move parasite proteins to the periphery. As a result of their intrinsic structure, the proteins may have special sequences that cause them to be targeted for particular cell compartments. This is the case with proteins of mitochondria and chloroplasts that are made in the cytosol and then inserted into these organelles. The sequences of a number of malarial proteins are being determined (see also Chapter 19) and they may well furnish exciting revelations for cell biologists as well as for parasitologists.

Already there is beginning to be evidence for a similar situation with intracellular parasites other than malaria. Studies of cell membranes have shown that they are in a state of continuous dynamic interchange. Labeled constituents of lysosomes move to the plasma membrane in 5–10 min, and there is a similar centripetal movement from the plasma membrane to the phagolysosome. There is every reason to expect similar continuous interchange of membrane at the membrane of a phagolysosome enclosing a *Leishmania* or at the parasitophorous membrane surrounding an *Eimeria*. This rapid movement of membrane polypeptides must be mediated by vesicles but these vesicles have not been visualized. We must begin to think of the infected cell as a new and somewhat integrated entity even though, as in malaria, it may be eventually destroyed when the parasite completes its development.

Bibliography

Aley, S. B., Barnwell, J. W., Daniel, W., and Howard, R. J., 1984, Identification of parasite proteins in a membrane preparation enriched for the surface membrane of erythrocytes infected with *Plasmodium knowlesi, Mol. Biochem. Parasitol.* **12:**69–84.

Aley, S. B., Sherwood, J. A., and Howard, R. J., 1984, Knob-positive and knob-negative *Plasmodium falciparum* differ in expression of a strain-specific malarial antigen on the surface of infected erythrocytes, *J. Exp. Med.* **160:**1585–1590.

Barnwell, J. W., Howard, R. J., and Miller, L. H., 1982, Altered expression of *Plasmodium knowlesi* variant antigen on the erythrocyte membrane in splenectomized rhesus monkeys, *J. Immunol.* **128:**224–226.

Bennett, V., 1982, The molecular basis for membrane–cytoskeleton association in human erythrocytes, *J. Cell. Biochem.* **18:**49–65.

Berman, J. D., and Dwyer, D. M., 1981, Expression of Leishmania antigen on the surface membrane of infected human macrophages *in vitro, Clin. Exp. Immunol.* **44:**342–348.

Bulla, L. A., Jr., and Cheng, T. C. (eds.), 1976, *Comparative Pathobiology,* Volume 1, Plenum Press, New York.

Chaimanee, P., and Yuthavong, Y., 1979, Phosphorylation of membrane proteins from *Plasmodium berghei*-infected red cells, *Biochem. Biophys. Res. Commun.* **87:**953–959.

Chua, V.-H., and Schmidt, G. W., 1978, Post-translational transport into intact chloroplasts of a precursor to the small unit of ribulose-1,5-bisphosphate carboxylase, *Proc. Natl. Acad. Sci. USA* **75:**6110–6114.

Deuel, T. F., Huang, J. S., Huang, S. S,. Stroobant, P., and Waterfield, M. D., 1983, Expression of a platelet-derived growth factor-like protein in simian sarcoma virus transformed cells, *Science* **221:**1348–1350.

Duffus, W. P. H., Wagner, G. C., and Preston, J. M., 1978, Initial studies on the properties of a bovine lymphoid cell culture line infected with *Theileria parva, Clin. Exp. Immunol.* **34:**347–353.

Fallis, A. M., Desser, S. S., and Khan, R. A., 1974, On species of leucocytozoon, *Adv. Parasitol.* **12:**1–67.

Fawcett, D. W., Büscher, G., and Doxsey, S., 1982, Salivary gland of the tick vector of East Coast fever. III. The ultrastructure of sporogony in *Theileria parva.* IV. Cell type selectivity and host cell responses to *Theileria parva, Tissue Cell* **14:**183–206, 397–414.

Fremount, H. N., and Rossan, R. N., 1974, The sites of sequestration of the Uganda–Palo Alto strain of *Plasmodium falciparum*-infected red blood cells in the squirrel monkey, *Saimiri sciureus, J. Parasitol.* **60:**534–536.

Friedman, M. J., 1978, Erythrocytic mechanism of sickle cell resistance to malaria, *Proc. Natl. Acad. Sci. USA* **75:**1994–1997.

Gasser, S. M., Damm, G., and Schatz, G., 1982, Import of proteins into mitochondria: Energy-dependent uptake of precursors by isolated mitochondria, *J. Biol. Chem.* **257:**13034–13041.

Ginsburg, H., Krugliak, M., Eidelman, O., and Cabantchik, Z. I., 1983, New permeability pathways induced in membranes of *Plasmodium falciparum* infected erythrocytes, *Mol. Biochem. Parasitol.* **8:**177–190.

Goff, L. J., and Coleman, A. W., 1984, Transfer of nuclei from a parasite to its host, *Proc. Natl. Acad. Sci. USA* **81:**5420–5424.

Hadley, T. J., Leech, J. H., Green, T. J., Daniel, W. A., Wahlgren, M., Miller, L. H., and Howard, R. J., 1983, A comparison of knobby (K+) and knobless (K−) parasites from two strains of *Plasmodium falciparum, Mol. Biochem. Parasitol.* **9:**271–278.

Howard, R. J., 1982, Alterations in the surface membrane of red blood cells during malaria, *Immunol. Rev.* **61:**67–107.

Howard, R. J., Barnwell, J. W., Kao, V., Daniel, W. A., and Aley, S. B., 1982, Radioiodination of new protein antigens on the surface of *Plasmodium knowlesi* schizont-infected erythrocytes, *Mol. Biochem. Parasitol.* **6:**343–367.

Hullinger, L., Wilde, J. K. H., Brown, C. G. D., and Turner, L., 1964, Mode of multiplication of *Theileria* in cultures of bovine lymphocytic cells, *Nature* **203:**728–730.

Kilejian, A., 1979, Characterization of a protein correlated with the production of knob-like protrusions on membranes of erythrocytes infected with *Plasmodium falciparum*, *Proc. Natl. Acad. Sci. USA* **76**:4650–4653.

Kilejian, A., 1983, Immunological cross-reactivity of the histidine-rich protein of *Plasmodium lophurae* and the knob protein of *Plasmodium falciparum*, *J. Parasitol.* **69**:257–261.

Kilejian, A., 1984, the biosynthesis of the knob protein and a 65 000 dalton histidine-rich polypeptide of *Plasmodium falciparum*, *Mol. Biochem. Parasitol.* **12**:185–194.

Kutner, S., Baruch, D., Ginsburg, H., and Cabantchik, Z. I., 1982, Alterations in membrane permeability of malaria-infected human erythrocytes are related to the growth stage of the parasite, *Biochim. Biophys. Acta* **687**:113–117.

Langreth, S. G., and Peterson, E., 1985, Pathogenicity, stability, and immunogenicity of a knobless clone of *Plasmodium falciparum* in Colombian owl monkeys, *Infect. Immun.* **47**:760–766.

Langreth, S. G., Reese, R. T., Motyl, M. R., and Trager, W., 1979, *Plasmodium falciparum:* Loss of knobs on the infected erythrocyte surface after long-term cultivation, *Exp. Parasitol.* **48**:213–219.

Lanners, H. N., and Trager, W., 1984, Comparative infectivity of knobless and knobby clones of *Plasmodium falciparum* in splenectomized and intact *Aotus monkeys*, Z. Parasitenkd. **70**:739–745.

Leech, J. H., Barnwell, J. W., Aikawa, M., Miller, L. H., and Howard, R. J., 1984, *Plasmodium falciparum* malaria: Association of knobs on the surface of infected erythrocytes with histidine-rich protein and the erythrocyte skeleton, *J. Cell Biol.* **98**:1256–1264.

Moulton, J. E., Krauss, H. H., and Malmquist, W. A., 1971, Growth characteristics of *Theileria parva*-infected bovine lymphoblast cultures, *Am. J. Vet. Res.* **32**:1365–1370.

Musisi, F. L., Bird, R. G., Brown, C. G. D., and Smith, M., 1981, The fine structural relationship between *Theileria* schizonts and infected bovine lymphoblasts from cultures, *Z. Parasitenkd.* **65**:31–41.

Pearson, T. W., Lundin, L. B., Dolan, T. T., and Stagg, D. A., 1979, Cell-mediated immunity to *Theileria*-transformed cell lines, *Nature* **281**:678–680.

Perkins, M., 1982, Surface proteins of schizont-infected erythrocytes and merozoites of *Plasmodium falciparum*, *Mol. Biochem. Parasitol.* **5**:55–64.

Raventos-Suarez, C., Kaul, D. K., Macaluso, F., and Nagel, R. L., 1985, Membrane knobs are required for the microcirculatory obstruction induced by *Plasmodium falciparum*-infected erythrocytes, *Proc. Natl. Acad. Sci. USA* **82**:3829–3833.

Rehbein, G., Zweygarth, E., Voigt, W. P., and Schein, E., 1982, Establishment of *Babesia equi*-infected lymphoblastoid cell lines, *Z. Parasitenkd.* **67**:125–127.

Rothman, J. E., and Lenard, J., 1977, Membrane asymmetry: The nature of membrane asymmetry provides clues to the puzzle of how membranes are assembled, *Science* **195**:743–753.

Stagg, D. A., Chasey, D., Young, A. S., Morzaria, S. P., and Dolan, T. T., 1980, Synchronization of the division of *Theileria* macroschizonts and their mammalian host cells, *Ann. Trop. Med. Parasitol.* **74**:263–265.

Stagg, D. A., Dolan, T. T., Leitch, B. L., and Young, A. S., 1981, The initial stages of infection of cattle cells with *Theileria parva* sporozoites *in vitro*, *Parasitology* **83**:191–197.

Takahashi, Y., and Sherman, I. W., 1978, *Plasmodium lophurae:* Cationized ferritin staining, an electron microscope cytochemical method for differentiating malarial parasite and host cell membranes, *Exp. Parasitol.* **44**:145–154.

Trager, W., 1960, Intracellular parasitism and symbiosis, in: *The Cell*, Volume 4 (J. Brachet and A. E. Mirsky, eds.), Academic Press, New York, pp. 151–213.

Trager, W., 1982, Effects of malarial parasites on their host erythrocytes, *Microbiology* **1983**:384–386.

UNDP/World Bank/WHO Special Programme for Research and Training in Tropical Diseases, 1979, *The Membrane Pathobiology of Tropical Diseases*, Proceedings of the Meeting held in Titisee, GFR, 4–8 October 1978, Schwabe, Basel.

van den Broeck, G., Timko, M. P., Kausch, A. P., Cashmore, A. R., van Montagu, M., and Herrera-Estrella, L., 1985, Targeting of a foreign protein to chloroplasts by fusion to the transit peptide from the small subunit of ribulose 1,5-bisphosphate carboxylase, *Nature* **313**:358–363.

Vernot-Hernandez, J. P., and Heidrich, H. G., 1984, Time-course of synthesis, transport and incorporation of a protein identified in purified membranes of host erythrocytes infected with a knob-forming strain of *Plasmodium falciparum, Mol. Biochem. Parasitol.* **12**:337–350.

Wallach, D. F. H., and Conley, M., 1977, Altered membrane proteins on monkey erythrocytes infected with simian malaria, *J. Mol. Med.* **2**:119–136.

Wallach, M., Cully, D. F., Haas, L. O. C., Trager, W., and Cross, G. A. M., 1984, Histidine-rich protein genes and their transcripts in *Plasmodium falciparum* and *P. lophurae, Mol. Biochem. Parasitol.* **12**:85–94.

Weidner, E., 1985, Early morphogenesis of *Glugea*-induced xenomas in laboratory-reared flounder, *J. Protozool.* **32**:269–275.

CHAPTER 14

Innate Resistance

It is common knowledge that each kind of organism has a relatively limited number of kinds of parasites that may be associated with it. Reciprocally, each kind of parasite has a more or less limited range of hosts, some parasites showing an exquisite specificity for one or very few host species. These properties clearly must be results of the different genetic characters of the organisms concerned. In plant pathology, where a great many different lines of both host plant and, for example, a fungal parasite are readily bred, all the data fit the so-called gene-for-gene hypothesis. This holds that a positive gene in the host plant and a corresponding positive gene in the fungal parasite produce incompatibility. Lack of either gene would permit the parasitic relationship. Furthermore, if one parasite/host gene pair specifies incompatibility, that gene pair will be epistatic to all parasite/host gene pairs that would permit a compatible relationship. If the genes *Px* of the parasite and *Rx* of the host determine an incompatible relationship, any mutation of *Px* to *px* resulting in loss of recognition of *Rx* should lead to a compatible relationship. Such mutations, i.e., to increased virulence, constitute loss of a specificity; they would be expected to occur readily and indeed they do. Mutation in the plant, however, from the susceptible state *rx* to *Rx* would be a gain of specificity. These would be expected to be rare, and they are. This hypothesis has never been tested with animal parasites, though a few systems are now available that could be developed for this purpose. Purebred lines of mice vary greatly in their innate susceptibility to different kinds of parasites. This has been particularly well shown for protozoa of the genera *Plasmodium* and *Leishmania*. For acute susceptibility of mice to *Leishmania donovani*, the gene determining this (*Lsh*) has been mapped on chromosome 1. It is in the same vicinity as the aldehyde oxidase locus but the two are not identical. It is distinct from *H-2*, *Ig-1*, and all other mapped histocompatibility and alloantigen-determining loci. The mechanism by which *Lsh* determines susceptibility to leishmaniasis is not known.

Far more interesting than the demonstration of a genetic background to susceptibility and resistance is the elucidation of the physiological mechanisms that determine compatibility or incompatibility. These can be relatively trivial or highly subtle. If an oocyst of a coccidian parasite of chickens requires some mechanical action such as it receives in the chicken gizzard before it will be stimulated to excyst, it clearly will not infect a mammal even though the sporozoites might be quite capable of infecting mammalian cells. If a vector insect infected with a protozoan or worm parasite never feeds on certain species of hosts, these hosts will not be infected even though they might be susceptible. Such an ecological barrier can be of great epidemiological significance and will be considered in a later chapter. Here I will concentrate on hormonal and cellular mechanisms of innate resistance, as distinct from the mechanisms that come into play with the development of an acquired immunity, recognizing at the same time that there is not a sharp boundary.

Nowhere is there a better illustration of the interplay between a parasite and the genetic composition of its host than in the relationship between *Plasmodium falciparum* malaria of humans and human red cell variants. Falciparum malaria, as a parasite that is responsible for a high infant mortality, has had a profound selective effect on human evolution. Wherever malaria is or has been highly prevalent, there exist in large proportion different genetic mutants involving the red cell—either its hemoglobin or other aspects of its physiology. It is becoming clear now that each of these in some way renders the red cell less suitable for the development within it of the malarial parasite. Some of these mutations at the same time have deleterious effects on the host. So well adjusted are the parasites to a normal erythrocyte that not many changes are possible that would be deleterious to the parasite without at the same time affecting at least to some extent red cell function. Best known and most studied of these red cell mutants is sickle hemoglobin—HbS. HbS differs from the normal HbA in only one amino acid in the two β chains, the α chains remaining unchanged. A valine is present in place of a glutamic acid. In the homozygous *SS* condition, where all the hemoglobin is HbS, a serious disease results. HbS when it becomes deoxygenated in the tissues tends to aggregate in long fibrous paracrystalline arrays. These distort the red cell. This distortion, together with an increased stickiness of the cells, cause local blockage of circulation and tissue death. Homozygotes rarely survive to reproductive age. And yet the *S* gene is present in up to 15 or 20% of some populations in West Africa (Fig. 14.1). For a gene lethal in the homozygous state to survive to this extent, it must confer some special advantage to the heterozygotes. In these *SA* individuals, HbS constitutes about 40% of the hemoglobin in each erythrocyte. Their cells do not become distorted except under extreme conditions of anoxia and they do not suffer any disease related to their HbS. What advantage do they have over *AA* people with only the normal hemoglobin? In view of the striking geographical correlation between frequency of the *S* gene and intensity of malarial infection, J. B. S. Haldane long ago suggested that *SA* people must have relatively higher resistance to malaria than *AA* paople. This is now well established. *SA* children generally have less severe falciparum infection than *AA* children; this suffices to give them a survival advantage and to maintain the *S* gene despite its lethality in

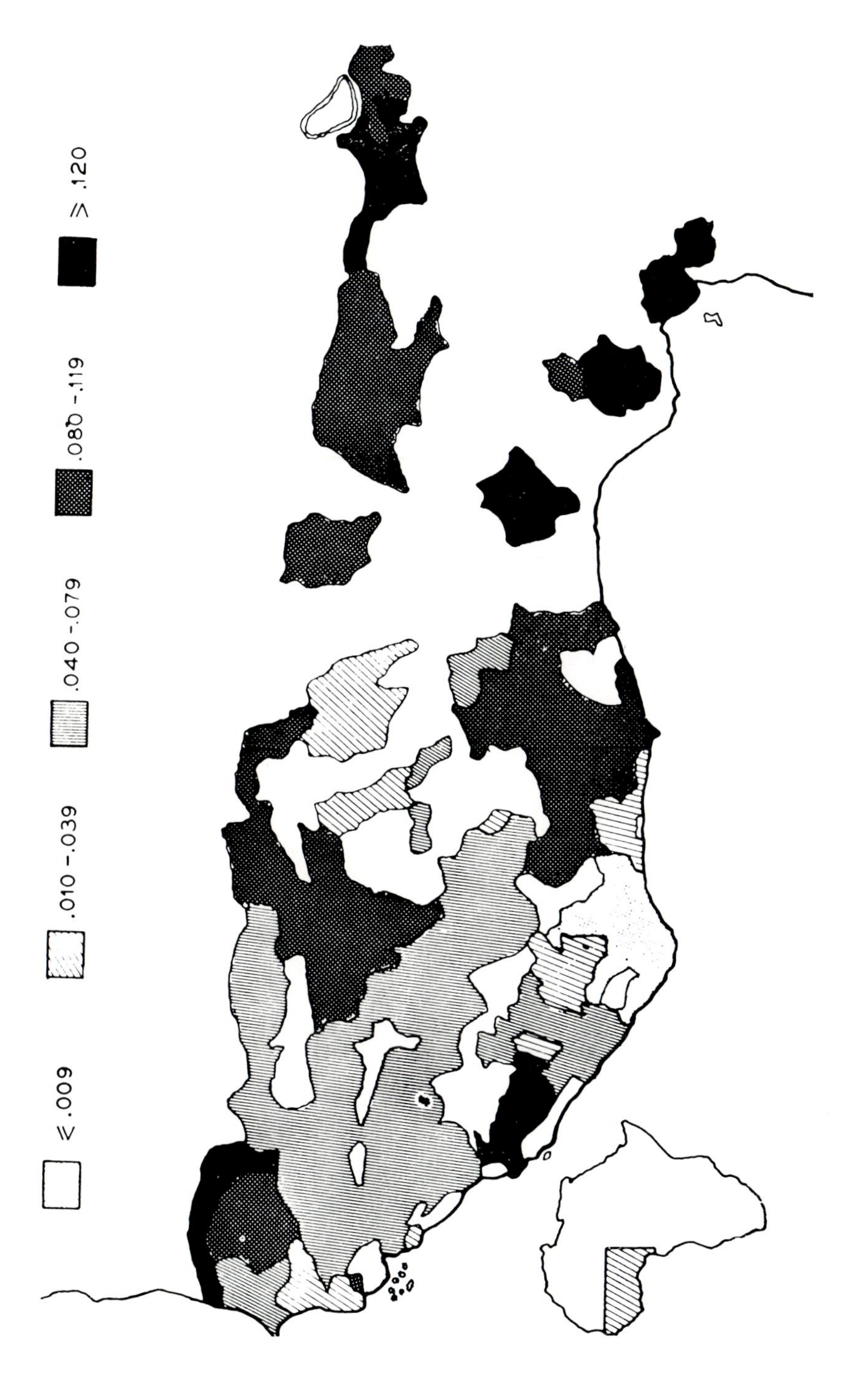

FIGURE 14.1. The distribution and frequency of the $Hb\beta^S$ gene in West Africa. (From Neel, 1961.)

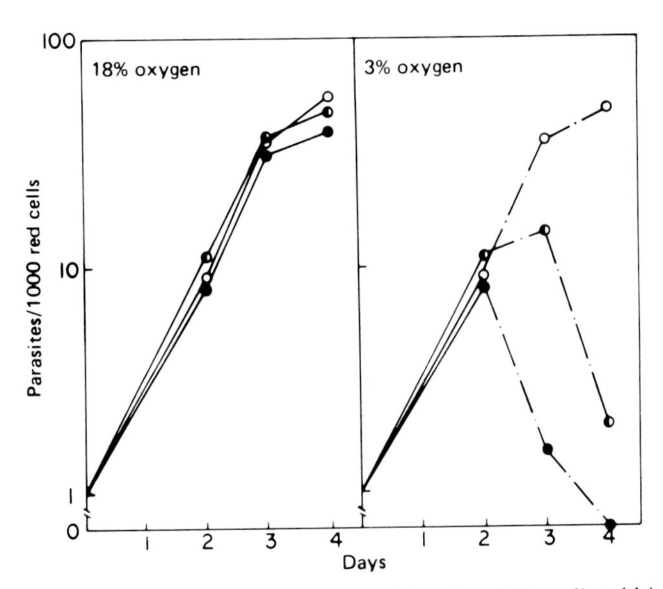

FIGURE 14.2. Effect of sickle hemoglobin on growth *in vitro* of *Plasmodium falciparum* cultures in normal erythrocytes (*AA*, ○), or erythrocytes from individuals homozygous (*SS*, ●) or heterozygous (*SA*, ◐) for sickle hemoglobin. The cultures were grown for 2 days in an atmosphere of 18% O_2/3% CO_2/balance N_2 and then either maintained in this gas mixture (left) or shifted to 3% O_2/3% CO_2/balance N_2 (right). (From Friedman, 1978.)

the homozygous *SS* state. Furthermore, the physiological basis for this enhanced resistance is now understood. When *P. falciparum* was cultured *in vitro* with an atmosphere of 3 or 5% O_2, which occurs normally in the deep tissues and which is favorable to the parasites growing in normal *AA* red cells, in *SA* cells the parasites failed to grow and soon died (Fig. 14.2). This growth failure is probably a result of the known loss of potassium from *SA* cells when they are kept at reduced oxygen tensions. In the *in vivo* situation, red cells infected with trophozoites or later stages are sequestered in capillaries of various organs because they adhere to endothelial cells by their surface knobs (see Chapter 13). They are thus kept in regions of relatively lower O_2 tension. This combined with the lower pH of an infected cell would tend to produce sickling of infected *SA* cells and loss of potassium, with consequent destruction of many of the parasites and, as a result, a reduced parasitemia.

With other red cell variants the picture is less complete but the overall effect is similar. Deficiency of the enzyme glucose-6-phosphate dehydrogenase (G6PD) is geographically correlated with high incidence of malaria (Fig. 14.3). *In vitro* experiments again show poorer growth in the mutant cells especially under prooxidant conditions. G6PD of the red cell plays a role in the maintenance of reduced glutathione and regeneration of NADPH (the reduced coenzyme nicotinamide adenine dinucleotide phosphate) essential for protection against oxidative damage. At least some species of malarial

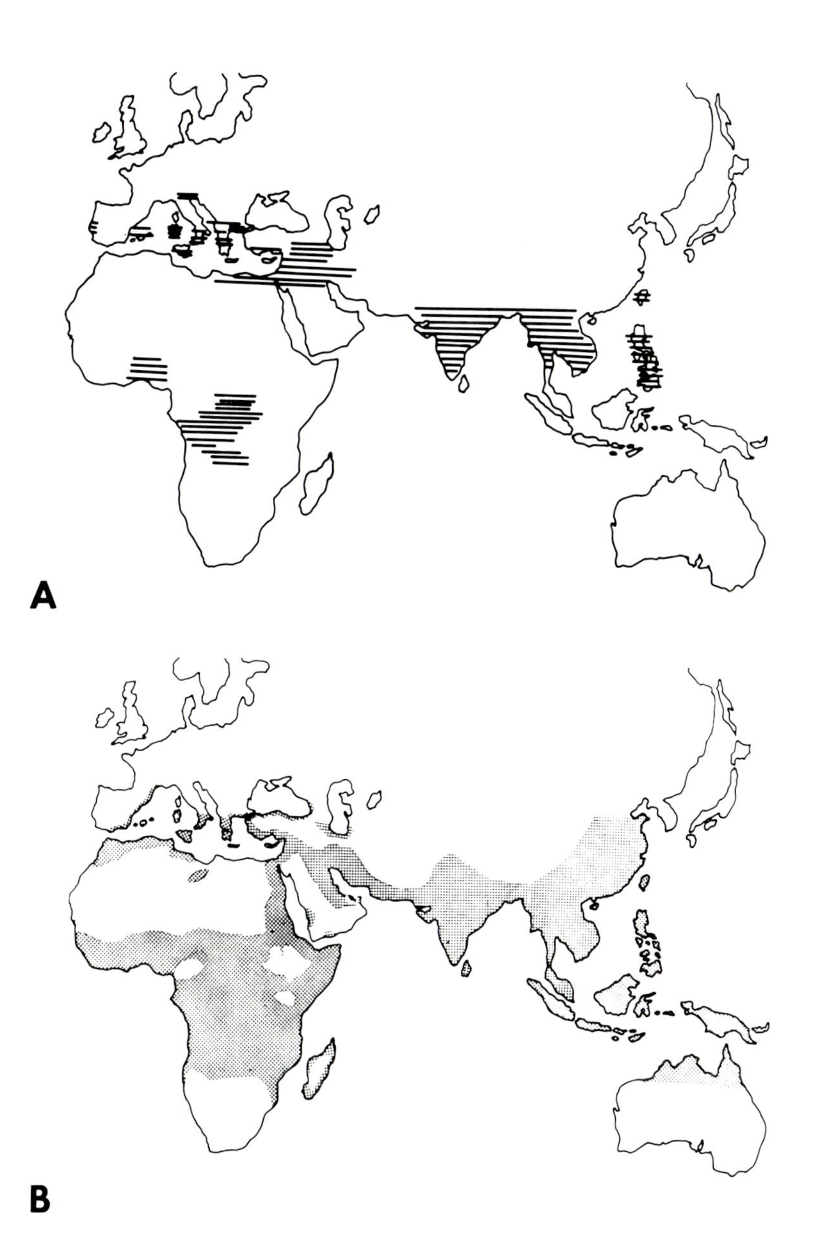

FIGURE 14.3. (A) Distribution of glucose-6-phosphate dehydrogenase deficiency in the eastern hemisphere. (B) Distribution of falciparum malaria in the eastern hemisphere in 1930. At present there is no malaria in Europe, North Africa, or Australia, and very little falciparum malaria in China. (From Motulsky and Campbell-Kraut, 1961.)

parasites have no G6PD of their own (see Chapter 9). In any case, in the absence of the red cell enzyme they become more liable to damage by oxidants such as H_2O_2 and the superoxide radical. The price the host pays for this protection is that individuals with G6PD deficiency are sensitive to certain drugs (e.g., primaquine) and to toxic compounds in certain foods, as in fava beans. The protection conferred by G6PD deficiency may be especially significant in regions where fava beans are a staple of the diet. *In vitro* a component of fava beans, isouramil, has been found to render G6PD-deficient red cells altogether unsuitable for development of *P. falciparum* whereas it has no effect on growth of the parasites in normal red cells. The thalassemias, which result from defects in globin chain synthesis, are again geographically correlated with endemic malaria. Here the membrane lipids are unusually susceptible to irreversible oxidant damage, and this may be responsible for the relative resistance of heterozygotes to falciparum malaria.

Still another type of red cell mutant, common in New Guinea, a highly malarious region, is called ovalocytosis. *In vitro* experiments show that ovalocytic erythrocytes are poorly invaded by merozoites of *P. falciparum* (Table 14.1). It is assumed that some membrane change accounts for this, but the nature of the change is not known.

The absence of a blood group membrane antigen, the Duffy factor, is correlated with resistance of human red cells to invasion *in vitro* by merozoites

TABLE 14.1. Invasion of Normocytic and Ovalocytic Erythrocytes of *Plasmodium falciparum* in Culture[a,b]

Source of recipient red cells	% ovalocytes	Invasion index (mean ± S.E.M.)
Controls (3 Caucasians and 5 Melanesians	3.4 ± 0.9	100 ± 4.72
Ovalocytes		
Ov 1	79	3.31 ± 0.83
Ov 2	80	3.87 ± 1.32
Ov 3	78	3.23 ± 0.31
Ov 4	80	1.09 ± 0.42
Ov 5	86	2.25 ± 0.89
Ov 6	80	5.33 ± 1.76
Ov 7	51	3.07 ± 0.92
Ov 8	64	30.78 ± 9.16
Ov 9	96	2.87 ± 2.96
Ov 10	82	2.25 ± 0.90
Mean		2.65 ± 0.34

[a] Modified from Kidson *et al.* (1981).

[b] The recipient erythrocytes were labeled with fluorescein isothiocyanate and exposed to a suspension of schizonts under culture conditions. The following day the number of rings in labeled cells was determined. The invasion index for ovalocytes is defined as: (fraction parasitized ovalocytes/mean fraction parasitized normocytes) × 100, in which both ovalocytic and normocytic samples are infected in parallel from the same batch of schizonts.

of the monkey malaria *P. knowlesi*. Duffy-positive cells (as occur in about 70% of the white population in the United States) are readily invaded; Duffy-negative cells (as occur in over 95% of the black population in West Africa) are not invaded. This observation led to another correlation: *P. vivax* is not known to occur in West Africa where the native population is almost entirely Duffy-negative. Furthermore, in a small group of volunteers inoculated with vivax malaria, Duffy-negative blacks did not become infected. These correlations suggest but do not prove that lack of the Duffy antigen confers resistance to vivax malaria.

Many similar situations must exist with any widespread infectious agent. For the most part we know little about them. We know something about malaria because it is such a ubiquitous infection and because it affects the red cells, where mutants are relatively easily detected. In addition to cellular variations in susceptibility to a parasite, there must exist factors in the body fluids that influence this. Certainly the complement system of vertebrate plasma is such a factor. An invading organism that activates the complement cascade cannot survive. This is nicely illustrated by the differential infectivity of various forms of *Trypanosoma cruzi*, the agent of Chagas' disease. As already noted, the epimastigote forms that develop in culture, or in the midgut of the vector triatomid bug, are noninfective. This is so because they activate the complement pathway and are lysed. The metacyclic tryposmastigote forms that develop in the rectum of the bug or the trypomastigotes that develop intracellularly in a vertebrate host or in tissue culture, do not activate complement and are not affected by this early defensive system. Complement plays a special role in interaction with antibodies in acquired immunity. This will be considered in a later chapter.

The African trypanosomes provide a particularly interesting example of host–parasite compatibility that depends on a factor in the plasma. *T. brucei gambiense* and *T. b. rhodesiense*, the agents of two types of African sleeping sickness in humans, and *T. b. brucei*, the cause of nagana of domestic animals, are morphologically identical. They are considered as subspecies of *T. brucei*. It was found very early, however, that most isolates of *T. brucei* obtained from cattle or from wild ungulates in Africa, were quickly killed and lysed in human serum *in vitro* whereas trypanosomes from human infections were not. Furthermore, those few isolates from animals that survived in human serum were shown to be infective to humans and therefore were *T. b. rhodesiense*. Recent work demonstrates that it is the high-density lipoprotein (HDL) of human serum that is specifically responsible for the killing effect on *T. b. brucei*. Human HDL, and also that of baboons which are likewise resistant to *T. brucei*, has this effect, but not the HDL of rats, rabbits, or cattle. However, neither human nor baboon HDL has an effect on *T. b. rhodesiense*. There are here two questions of great biological interest and of possible practical importance. First, what is the key difference between human (or baboon) HDL and rat or rabbit HDL responsible for the killing of *T. brucei* by the former? Second, what is the difference in the plasma membrane between *T. b. brucei* and *T. b. rhodesiense* making the former susceptible to the action of human HDL whereas the latter is not affected? These are difficult questions. There are indications that the phospholipid composition of the HDL and the phosphatidylethanolamine metabolism of the trypanosomes are involved.

In addition to the innate cellular and serological factors affecting compatibility of a host with a parasite, there occur very soon after any inflammatory stimulus increases in concentration of certain components of vertebrate serum. These constitute the acute-phase response. Best known of these is the C-reactive protein. This protein was discovered over 40 years ago in the serum of acutely ill pneumonia patients because it reacted with the pneumococcal cell wall polysaccharide. In normal serum it is present at the low level of under 0.2 mg/dl. It increases quickly in response to bacterial infections or some kinds of trauma and may reach levels well over 10 mg/dl. There is probably a humoral mediator formed by monocytes, macrophages, or the reticuloendothelial system that induces biosynthesis in the liver of the C-reactive protein. This protein is increased in acute falciparum malaria. It can activate complement and initiate phagocytosis by macrophages. It binds specifically to injured or necrotic tissue. Presumably it helps to limit infections, but its exact role remains to be determined.

Another acute-phase protein, α_1-acid glycoprotein, increases three- to fourfold in malarial infection, as well as in bacterial infections. This protein resembles a cell membrane sialoglycoprotein, leading to the interesting hypothesis that it could serve as a nonspecific competitor for cell surface and thereby block the binding and subsequent invasion of infective agents. In *in vitro* experiments with *Plasmodium falciparum*, conflicting results have been obtained, perhaps because it is difficult to make uniform and pure preparations of this material.

Still another serum protein increased in the acute-phase response is the so-called tumor necrosis factor, thought to be formed by macrophages. There is evidence that this factor is cytotoxic for *P. falciparum* in culture as well as for *Babesia microti* in mice. This again would be a humoral factor tending to hold down the rate of increase of various infective agents until effective specific immunity is acquired (see also Chapter 19).

The nutritional status of a host may markedly affect both its innate resistance and its ability to develop acquired immunity. One tends to associate partial starvation or specific nutritional deficiencies with decreased resistance. And indeed deficiencies in biotin or in zinc interfere with development of certain immune responses, a subject to which I will return in a later chapter. On the other hand, the surprising observation has been made that under famine conditions in a malarious region, partially starved people brought to a hospital and provided with adequate food have shown exacerbations of their malarial infection. Malarial parasites require exogenous sources of iron and riboflavin and there is evidence that these are the main factors responsible. In the partially starved individuals the levels of these essential factors are too low. When higher levels are supplied with an adequate diet, the growth of the parasites is stimulated before the immune system can recover. In experimental malarial infections of animals similar effects can be shown. If chickens are made pantothenate deficient, or mice deficient in *p*-aminobenzoic acid, otherwise lethal malarial infections are rendered innocuous; the parasites cannot develop in the absence of sufficient levels of required growth factors (see also Chapter 7). Human malarial parasites (*P. falciparum*) when grown in erythrocytes *in vitro* are very sensitive to iron deprivation. Their growth was inhibited completely by as little as 15 μM desferrioxamine. Evidently the large amounts of iron in the heme of the red cell are not available to them. Desferrioxamine has also been observed to inhibit intracellular growth of *Trypanosoma cruzi* in macrophages (Fig. 14.4), and their inhibition could be reversed with referrated desferrioxamine. These findings suggest that host responses of the nature that occur in some bacterial infections, where iron is bound in such a way as to be less available to the microorganism, might also operate in malaria and other parasitic infections.

No discussion of innate resistance can be complete without reference to the spleen. Its protective role is especially clear in infections of the red blood cell and is evident before as well as during and after the development of an acquired immunity. Hosts that are not susceptible to various species of *Babesia* or *Plasmodium* can be made fully susceptible by removal of the spleen. Hosts already susceptible generally become more so if first splenectomized. This role of the spleen was dramatically demonstrated with the first report of fatal cases of human babesiosis in splenectomized individuals. In much work with experimental malaria, splenectomized animals have been used to obtain heavy infections with species of parasites ordinarily giving very light or even no infections in the experimental host. Thus, chimpanzees and squirrel monkeys

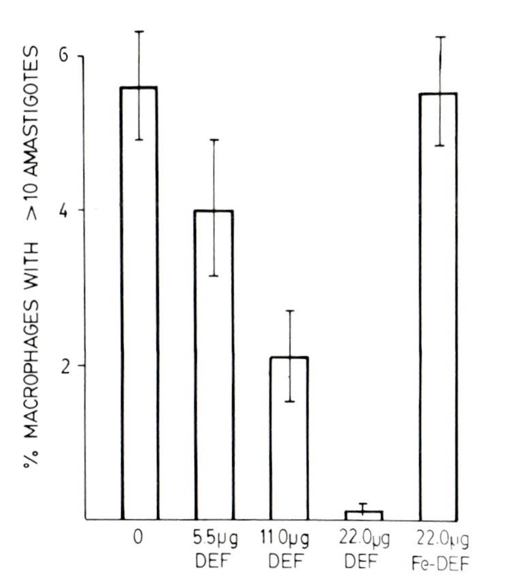

FIGURE 14.4. Role of iron in intracellular growth of *Trypanosoma cruzi*. Effects of desferrioxamine (DEF) on the proportion of cultured peritoneal exudate cells containing a large burden of *T. cruzi* amastigotes. By Student's *t* test, the following *p* values were obtained: >0.2 for 5.5 μg DEF/ml; ≤ 0.02 for 11 μg; ≤ 0.001 for 22 μg. Fe-DEF, DEF referrated with equimolar concentration of ferric ammonium sulfate. (From Loo and Lalonde, 1984. Print courtesy of Dr. R. Lalonde.)

(*Saimiri sciureus*) may be rendered susceptible to human malarial parasites if splenectomized before inoculation. Splenectomized rhesus monkeys show tenfold higher infections with the monkey parasite *P. inui* than do intact animals. The avian malaria *P. gallinaceum* will multiply in splenectomized Norway rats. Since all the blood percolates repeatedly through the complex channels of the spleen, we must assume that in the intact resistant host the incompatible species of parasite is somehow recognized and removed just as would be a defective erythrocyte. Perhaps the question is not how this recognition occurs but rather why it does not occur with those species of parasites to which the host is not resistant. Since the spleen is also of major importance in the development of acquired immunity by vertebrate hosts, and since it has recently been found to exert certain peculiar effects on parasite characteristics, I will proceed to a detailed discussion of this remarkable organ in the following chapter.

Bibliography

Allison, A. C., 1954, Protection afforded by sickle-cell trait against subtertian malarial infection, *Br. Med. J.* **1**:290–294.

Bradley, D. J., Taylor, B. A., Blackwell, J., Evans, E. P., and Freeman, J., 1979, Regulation of *Leishmania* populations whithin the host. III. Mapping of the locus controlling susceptibility to visceral leishmaniasis in the mouse, *Clin. Exp. Immunol.* **37**:7–14.

Castelino, D., Saul, A., Myler, P., Kidson, C., Thomas, H., and Cooke, R., 1981, Ovalocytosis in Papua New Guinea—Dominantly inherited resistance to malaria, *Southeast Asian J. Trop. Med. Public Health* **12**:1–7.

Clark, I. A., Wills, E. J., Richmond, J. E., and Allison, A. C., 1977, Suppression of babesiosis in BCG-infected mice and its correlation with tumor inhibition, *Infect. Immun.* **17**:430–438.

Diamond, L. S., Harolow, D. R., Phillips, B. P., and Keister, D. B., 1979, *Entamoeba histolytica:* Iron and nutritional immunity, *Arch. Invest. Med.* **9**(Suppl. 1):329–338.

Edirisinghe, J. S., Fern, E. B., and Targett, G. A. T., 1981, The influence of dietary protein on the development of malaria, *Ann. Trop. Paediatr.* **1**:87–91.

Ellingboe, A. H., 1979, Inheritance of specificity: The gene-for-gene hypothesis, in: *Recognition and Specificity in Plant Host–Parasite Interactions* (J. M. Daly and I. Uritani, eds.), Japan Scientific Societies Press, Tokyo, and University Park Press, Baltimore, pp. 3–15.

Friedman, M. J., 1978, Erythrocytic mechanism of sickle cell resistance to malaria, *Proc. Natl. Acad. Sci. USA* **75**:1994–1997.

Friedman, M. J., 1981, The biology of inherited resistance to malaria, in: *Biochemistry and Physiology of Protozoa,* Volume 4 (M. Lewandowsky and S. H. Hutner, eds.), Academic Press, New York, pp. 463–493.

Friedman, M. J., 1983, Control of malaria virulence by α-acid glycoprotein (oroso-mucoid), an acute-phase (inflammatory) reactant, *Proc. Natl. Acad. Sci. USA* **80**:5421–5424.

Friedman, M. J., Roth, E. F., Nagel, R. L., and Trager, W., 1979, *Plasmodium falciparum:* Physiological interactions with the human sickle cell, *Exp. Parasitol.* **47**:73–80.

Gloria-Bottini, F., Falsi, A. M., Mortera, J., and Bottini, E., 1980, The relations between G-6-PD deficiency, thalassemia and malaria: Further analysis of data from Sardinia and the Po Valley, *Experientia* **36**:541–543.

Golenser, J., Miller, J., Spira, D. T., Navok, T., and Chevion, M., 1983, Inhibitory effect of a fava bean component on the *in vitro* development of *Plasmodium falciparum* in normal and glucose-6-phosphate dehydrogenase deficient erythrocytes, *Blood* **61**:507–510.

Greenberg, J., and Kendrick, L. P., 1958, Parasitemia and survival in mice infected with *Plasmodium berghei:* Hybrids between Swiss (high parasitemia) and Str (low parasitemia) mice, *J. Parasitol.* **44:**492–498.

Haidaris, C. G., Haynes, J. D., Meltzer, M. S., and Allison, A. C., 1983, Serum containing tumor necrosis factor is cytotoxic for the human malaria parasite *Plasmodium falciparum, Infect. Immun.* **42:**385–393.

Haldane, J. B. S., 1949, Disease and evolution, *Rec. Sci.* **19**(Suppl.):68–76.

Hempelmann, E., and Wilson, R. J. M., 1981, Detection of glucose-6-phosphate dehydrogenase in malarial parasites, *Mol. Biochem. Parasitol.* **2:**197–204.

Hendrickse, R. G., Hason, A. H., Olumide, L. O., and Akinkunmi, A., 1971, Malaria in early childhood, *Ann. Trop. Med. Parasitol.* **65:**1–20.

Kidson, C., Lamont, G., Saul, A., and Mursa, G. T., 1981, Ovalocytic erythrocytes from Melanesians are resistant to invasion by malaria parasites in culture, *Proc. Natl. Acad. Sci. USA* **78:**5829–5832.

Klainer, A. S., Clyde, D. F., Bartelloni, P. J., and Beisel, W. R., 1968, Serum glycoproteins in experimentally induced malaria in man, *J. Lab. Clin. Med.* **72:**794–802.

Livingstone, F. B., 1971, Malaria and human polymorphisms, *Annu. Rev. Genet.* **5:**33–64.

Loo, V. G., and Lalonde, R. G., 1984, Role of iron in intracellular growth of *Trypanosoma cruzi, Infect. Immun.* **45:**726–730.

Luzzato, L., 1979, Genetics of red cells and susceptibility to malaria, *Blood* **54:**961–976.

Maggioni, G., Antognoni, G., Agostino, R., Businco, L., Corbo, R. M., Scacchi, R., Gallo, M. P., Gloria, F., Lucarelli, P., Palmarino, R., Spano, B., and Bottini, E., 1975, Genetic studies on the Sardinian population. I. Erythrocyte acid phosphatase polymorphism: Relationship with past malarial morbidity and other genetic and environmental factors, *Studi Sassar. Sez. 2* **53:**229–240.

Martin, S. K., Miller, L. H., Alling, D., Okoye, V. C., Esan, G. J. F., Osunkoya, B. O., and Deane, M., 1979, Severe malaria and glucose-6-phosphate-dehydrogenase deficiency: A reappraisal of the malaria/G-6-P.D. hypothesis, *Lancet* **1:**524–526.

Motulsky, A. G., and Campbell-Kraut, J. M., 1961, Population genetics of glucose-6-phosphate dehydrogenase deficiency of the red cell, in: *Proceedings of the Conference on Genetic Polymorphisms and Geographic Variations in Disease* (B. S. Blumberg, ed.), Grune & Stratton, New York, pp. 159–180.

Neel, J. V., 1961, The geography of hemoglobinopathies, in: *Proceedings of the Conference on Genetic Polymorphisms and Geographic Variations in Disease* (B. S. Blumberg, ed.), Grune & Stratton, New York, pp. 102–122.

Pasvol, G., Weatherall, D. J., and Wilson, R. J. M., 1978, Cellular mechanism for the protective effect of haemoglobin S against *P. falciparum* malaria, *Nature* **274:**701–703.

Pollack, S., 1983, Annotation—Malaria and iron, *Br. J. Haematol.* **53:**181–183.

Raik, P., and Voller, A., 1984, Serum C-reactive protein levels and falciparum malaria, *Trans. R. Soc. Trop. Med. Hyg.* **78:**812–813.

Ree, G. H., 1971, C-reactive protein in Gambian Africans with special reference to *P. falciparum* malaria, *Trans. R. Soc. Trop. Med. Hyg.* **65:**574–580.

Rifkin, M. R., 1983, Interaction of high-density lipoprotein with *Trypanosoma brucei:* Effect of membrane stabilizers, *J. Cell. Biochem.* **23:**57–70.

Rogers, H. J., 1983, Role of iron chelators, antibodies and iron-binding proteins in infection, *Microbiology* **1983:**334–337.

Roth, E. F., Jr., Raventos-Suarez, C., Perkins, M., and Nagel, R. L., 1982, Glutathione stability and oxidative stress in *P. falciparum* infection *in vitro:* Responses of normal and G6PD deficient cells, *Biochem. Biophys. Res. Commun.* **109:**355–362.

Thurnham, D. I., Oppenheimer, S. J., and Bull, R., 1983, Riboflavin status and malaria in infants in Papua New Guinea, *Trans. R. Soc. Trop. Med. Hyg.* **77:**423–424.

Wood, W. G., Clegg, J. B., and Weatherall, D. J., 1977, Developmental biology of human hemoglobins, *Prog. Hematol.* **10:**43–90.

CHAPTER 15

The Spleen

The spleen of vertebrates is an organ with a commanding regulatory function over both hematopoietic and immune systems. It provides both a special structure and a source of special cells. The structure (Fig. 15.1) consists of a stroma of connective tissue cells supporting a vasculature around which are organized the lobules that are the functional units of the spleen. A branching system of arteries is surrounded by the periarterial lymphatic sheath and lymphatic nodules that together constitute the white pulp. The lymphocytes of the periarterial lymphatic sheath are derived from the thymus (hence are varied types of T cells), whereas those of the lymphatic nodules are rich in B cells. Many of the arterial vessels terminate in the marginal zone, a dense reticular meshwork, and have dense nodules of macrophages near their ends. Other arteries terminate in the cords of the red pulp, which consists of a reticular meshwork rich in macrophages. Embedded in this are the thin-walled venous sinuses or venules that drain blood from the spleen. It is assumed that blood monocytes entering the cords from arterial terminations are selectively held in the reticular meshwork and transformed into macrophages. An erythrocyte traverses the splenic pulp through the reticular meshwork of the marginal zone or the red pulp and then must pass between the endothelial cells lining the venous sinuses to enter these and get back in circulation (Fig. 15.2). Thus, the erythrocyte is subjected to a barrage of macrophages, releasing hydrolytic enzymes and probably other materials, and then to a mechanical testing, for the slits between endothelial cells are only about 0.5 μm wide; only normally pliant flexible erythrocytes can squeeze through. Erythrocytes with Heinz bodies or intraerythrocytic parasites may be delayed here or may even break in two, with the flexible part entering the sinus as a spherocyte and the parasite being retained (Fig. 15.3) to be phagocytosed by macrophages. It is clear that the structure of the spleen together with the phagocytic cells within it provide an efficient mechanism for removal of foreign particles and abnormal cells, and particularly for screening out erythrocytes that have lost their normal flexibility as a result of containing a parasite or for other reasons. As already discussed, it is easy to understand the role

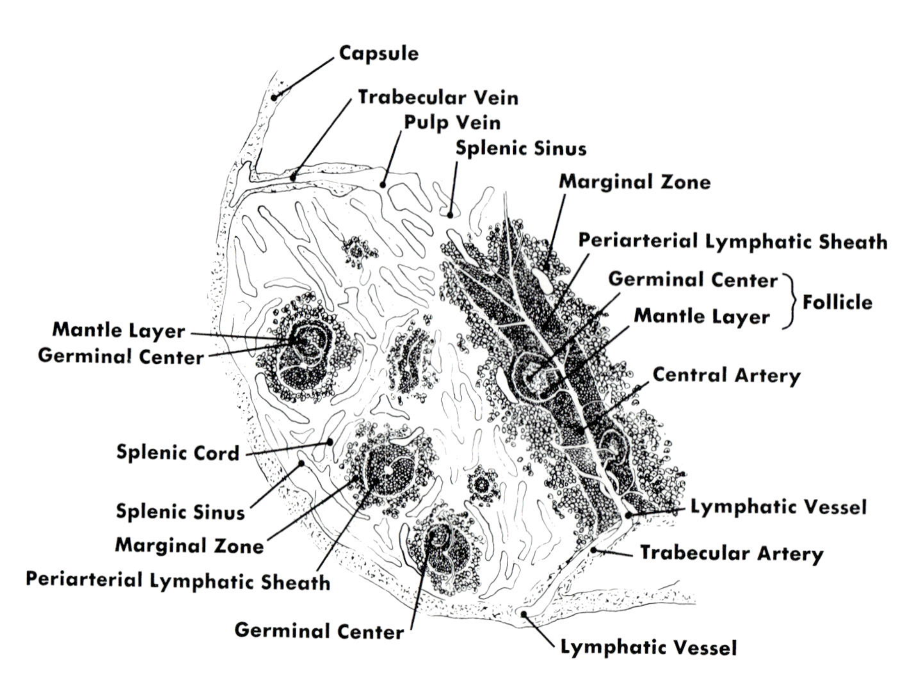

FIGURE 15.1. Diagrammatic representation of the organization of the human spleen. (From Weiss, 1979. Original print courtesy of Dr. L. Weiss.)

of the spleen in removing infected red cells and hence in innate resistance to the many species of blood parasites to which a given species of host is not susceptible. The real question is why this does not occur with those species to which the host is susceptible (Fig. 15.4). Of course, it does occur to some extent, but not enough to prevent infection. As an acquired immunity develops, more parasites are removed in the spleen and other lymphoid organs.

The special importance of the spleen in malaria is emphasized by the fact that in humans who have become naturally immunized to malaria, splenectomy has generally resulted in exacerbation of parasitemia with potentially fatal consequences. It is also indicated by the rapid enlargement of the spleen, which is often palpable within a few days of onset of the illness. With successful therapy the spleen regresses. After an acute fatal attack of malaria, the spleen, which in an adult man normally weighs about 150 to 200 g, may weigh 500 to 700 g, and weights up to 5000 g have been recorded after prolonged infections with vivax malaria. The spleen index based on the extent of splenic enlargement in children 2–14 years old, provides a useful measure of endemicity of malaria. There are many other parasitic and infectious diseases that produce some splenic enlargement but most of these are temporary. In nonmalarious communities there will be no more than 5% of enlarged spleens among the children aged 2–14 years old. It is furthermore of great interest that the tropical splenomegaly syndrome has been found to respond

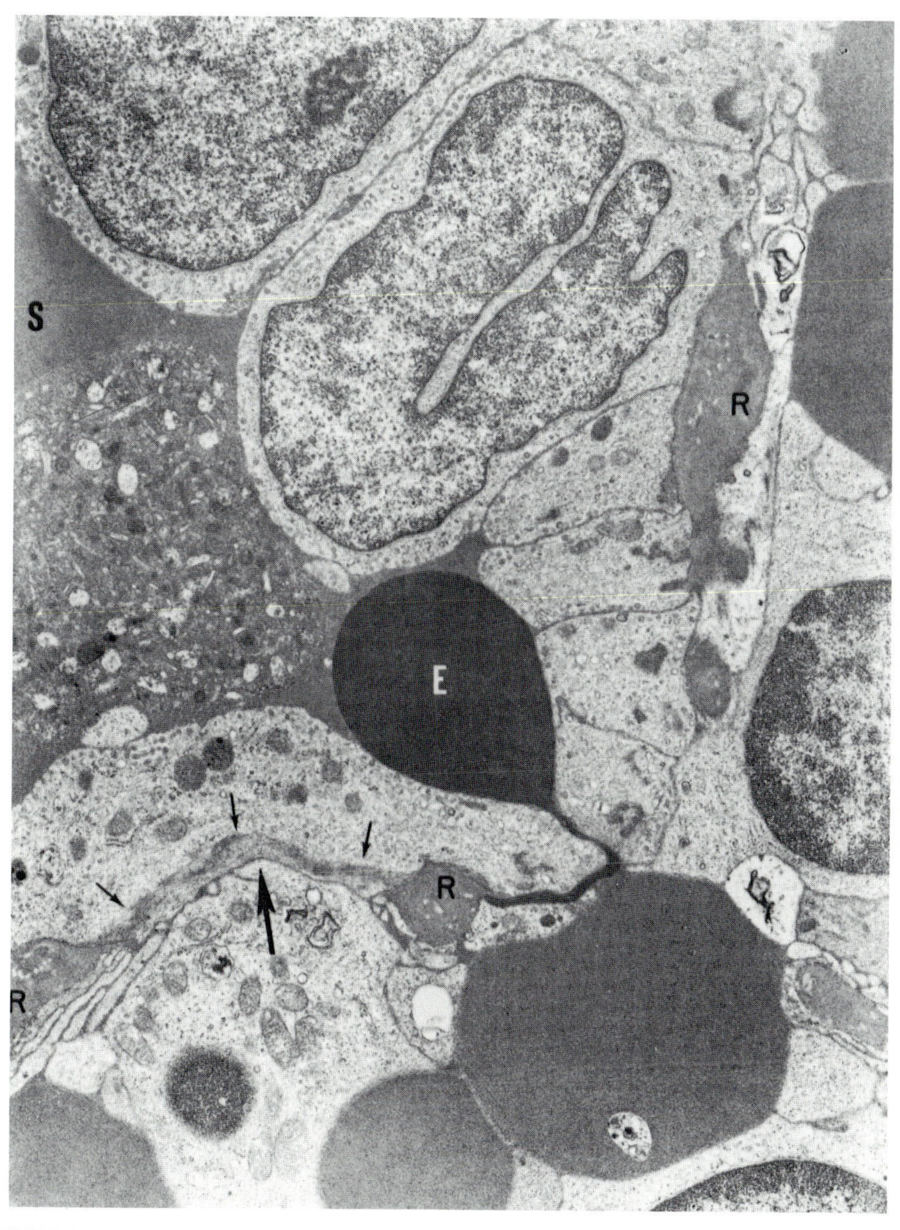

FIGURE 15.2. A part of the human splenic sinus (S). An erythrocyte (E) in passage across the sinus wall appears squeezed in between the endothelial cells and a portion of its taillike cytoplasm remains outside the endothelium. On the abluminal side of the endothelium, the "dense cytoplasmic web" is bound to the reticular fibers (R). It is commonly observed that this web (small arrows) is stretched between the reticular fibers. When the cells (large arrow) in the splenic cord push against the endothelial cells, the web (small arrows) appears to hold the endothelial cell to the reticular fibers. EM. ×10,000. (Original print courtesy of Dr. Leon Weiss.)

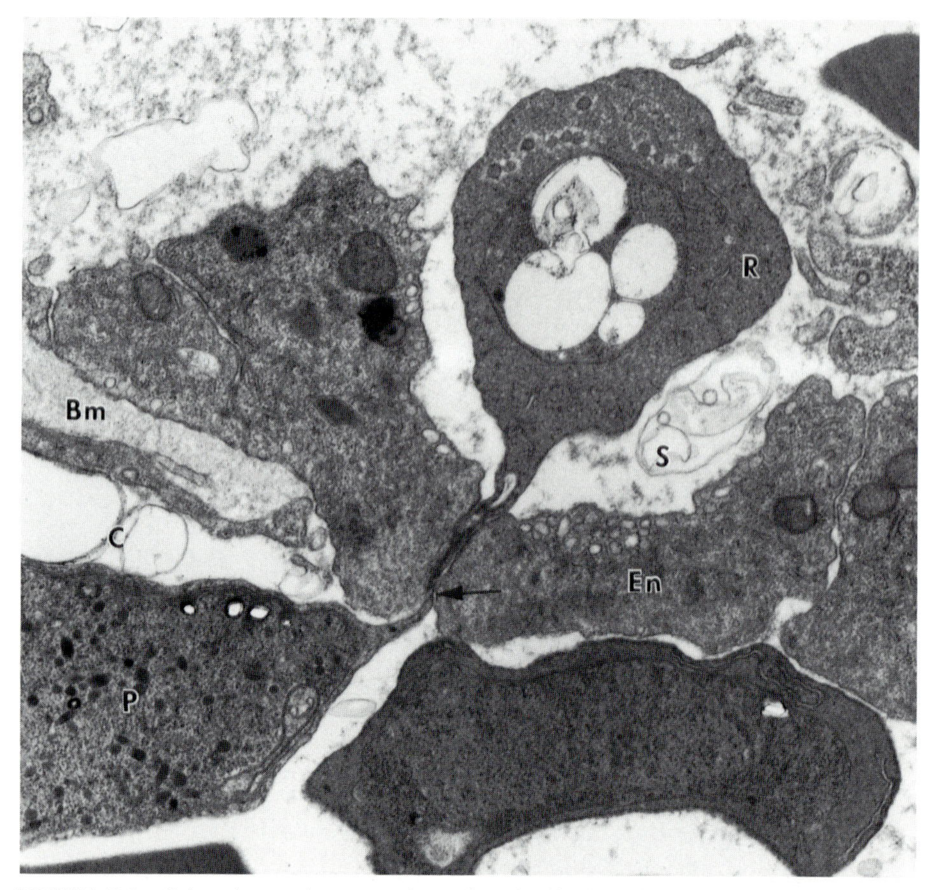

FIGURE 15.3. Spleen from a rhesus monkey infected with *Plasmodium knowlesi*, removed 6 days after inoculation when there were 360,000 parasites per mm^3 of blood. A doubly parasitized red cell is seen partly within sinus (S) and partly in cord (C). Portion of red cell (R) containing smaller parasite has squeezed through fenestration in basement membrane (Bm) and between endothelial cells (En) lining sinus, while part of cell with large parasite (P) is trapped in cord. Small break (arrow) is seen in the stalk that connected the two parts of the red cell. EM. ×18,300. (From Schnitzer *et al.*, 1973. Original print courtesy of Dr. B. Schnitzer.)

(though slowly) to treatment with antimalarials. This finding has led to the conclusion that this syndrome, which is found only in certain hyperendemic malarious regions (such as the coast of Papua New Guinea), arises through an abnormal immune response to recurrent malarial infection. In these individuals, serum IgM levels become disproportionately elevated even in young children, and continue to rise markedly. The IgM consists mainly of immunofluorescent antimalarial antibody and is a main constituent of cold-precipitable immune complexes apparently responsible for the major symptoms of the tropical splenomegaly syndrome. A more appropriate descriptive name

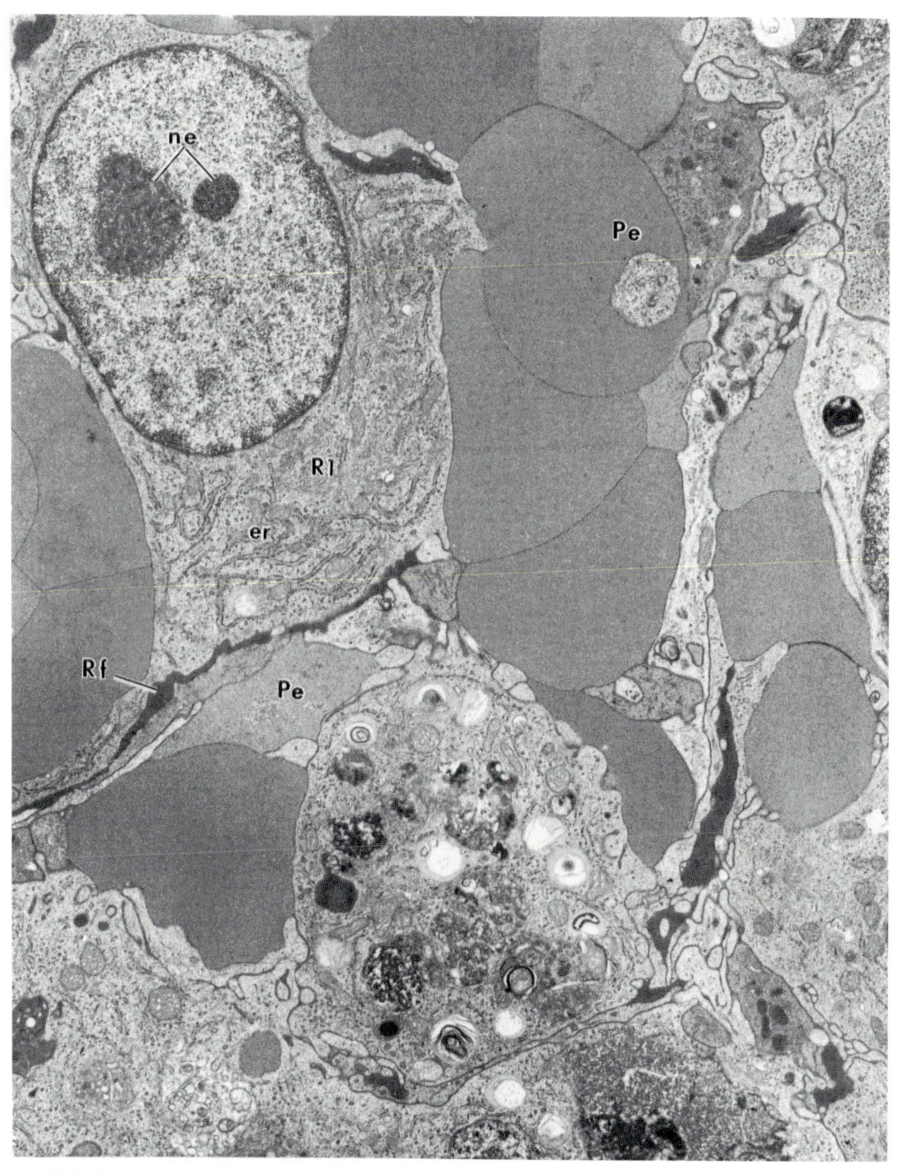

FIGURE 15.4. Red pulp 3 days after onset of nonlethal *Plasmodium berghei* malaria in mouse. The reticular cells branch and surround reticular fibers (Rf) which are quite dark in this preparation. Note that one reticular cell (R1) shows early activation: nucleoli (ne) are present and its rough endoplasmic reticulum (er) is increased in concentration. The reticular cells form a filtration bed which contains erythrocytes, two of which are parasitized (Pe), and macrophages. EM. ×9000. (Original print courtesy of Dr. Leon Weiss.)

for this syndrome is hyperreactive malarial splenomegaly. What determines this abnormal immune response is unknown.

Splenectomy is not necessarily deleterious, and indeed is an effective therapy for some hematological disorders. If the spleen has been removed, some of its functions are assumed by peripheral lymph nodes. This is particularly true of those functions that depend primarily on the various lymphoid cells of the spleen rather than on the spleen's special structure. With particulate antigens (e.g., parasites) present in the blood, however, the inductive phase of an immune response occurs mainly in the spleen, since such particles are unlikely to find their way to peripheral lymph nodes. Furthermore, only the spleen is effective in the removal of mildly damaged or moderately abnormal erythrocytes. This is the reason why splenectomy is so effective a treatment for certain blood disorders, notably hereditary spherocytosis. In this disease the red cell membrane is affected and the normally functioning spleen removes many of the resulting spherocytes to produce a severe anemia. If the spleen is removed, the spherocytes persist in the circulation and function well enough to permit normal life. Even in a normal person, splenectomy, as in a case of trauma, is followed by changes in morphology of the red cells. Cells with Howell–Jolly bodies (small darkly stained inclusions) appear in the circulating blood as well as siderocytes, cells with a diffuse basophilia, and nucleated erythrocytes. Presumably all of these are ordinarily eliminated by the spleen.

The spleen also provides a microenvironment in which occur cell sorting, cell interactions, and cell differentiation. As lymphocytes cross the white pulp, the T and B cells separate into the T and B cell zones, respectively. T cells are the primary cells mediating splenic effects on other immunological domains. The T cells in turn are apparently primed by the dendritic cells of the spleen. All responses altered by asplenia, whether congenital or produced by splenectomy, involve functional T cells, although some of them may also depend in part on B cells. Among these responses are B and T cell cooperation, responses of bone marrow and lymph node cells to phytohemagglutinins, homing of thymocytes to the spleen, and suppressor cell maturation. This last response, suppressor cell maturation and the elaboration of suppressor factor, is of special interest and helps to account for apparently contradictory effects of splenectomy. For example, in tumor-bearing hosts, splenectomy has given both beneficial and detrimental effects. In neonates the spleen has a high suppressive activity that may be important in the establishment of self-tolerance. The suppressive activity of the spleen is well shown with respect to skin grafts. Female C3H mice do not ordinarily reject skin grafts from males. If, however, the male skin grafts were applied 2 weeks after splenectomy, 40% of the splenectomized females rejected the grafts, and all grafts showed at least transient signs of crisis.

Similar suppressive activity may account for the interesting effects of splenectomy of rhesus monkeys on the course of infection with the quartan malarial parasite *Plasmodium inui,* a species closely resembling the human quartan parasite *P. malariae*. Intact monkeys infected with *P. inui* develop benign infections characterized by a low parasitemia and marked chronicity, with infections persisting up to 13 years, much like the prolonged chronicity (of up to 50 years) shown by *P. malariae* in humans. If monkeys were splenectomized before inoculation with *P. inui*, they developed rapidly fatal infections with a high parasitemia, the usual decrease in resistance to blood parasites. In monkeys that were infected while intact and then splenectomized 2–3 months later, the parasitemia rose to the same level as in those splenectomized before infection but there were no deaths. The presence of the spleen during the few months of the benign infection had in some way provided clinical tolerance to the effects of a high parasitemia. Most remarkable, however, was the finding that these animals spontaneously cured themselves of their infection within the first year; they did not show the prolonged chronic infections characteristic of the intact monkeys. This emphasizes the dual role of the spleen, protective on the one hand against acute infection but later promoting some form of immunological tolerance to permit the long-term maintenance of a low-grade infection.

Even more remarkable effects of the spleen on malarial parasites have recently been described, effects for which there is as yet no adequate explanation. In infections of rhesus monkeys (*Macaca mulatta*) with *Plasmodium knowlesi*, a variant surface antigen is expressed on the surface of erythrocytes containing late-stage parasites. This antigen can be readily detected by the schizont-infected cell agglutination produced by serum from a recovered monkey (see Chapter 19). When a clone having a defined surface antigen was inoculated into a splenectomized monkey, the agglutination titer with antiserum to that antigen decreased by 4- to 16-fold. On a second passage in a splenectomized monkey, the parasites were not agglutinated by any sera tested, including those from the animals infected with these nonagglutinating parasites. The spleen seems to play some role in the expression of this antigen on the surface of the infected erythrocyte. Similarly, with *P. falciparum* in squirrel monkeys (*Saimiri sciureus*), passage through splenectomized hosts resulted in loss of the property of cytoadherence normally exhibited by cells containing trophozoites or later stages and mediated by the knobs on the cell surface (see Chapter 13). Knobs, however, were present; evidently they lacked a second component essential to their function of adherence. When such a nonadhering but still knobby clone was then passed back into an intact squirrel monkey, parasites that did bind to cells reappeared after 20 days. Again the spleen seems somehow to modify the nature of the knob, which contains at least one parasite-derived protein. Furthermore, completely knobless clones of *P. falciparum* which produced only a barely detectable infection in intact *Aotus trivirgatus* gave higher infections in splenectomized *Aotus* than did corresponding knobby ones.

Spleen cells, as distinct from the intact spleen, have been found particularly effective in the transfer of resistance in mice to the blood parasite *Babesia microti*. The resistance could be induced either specifically by a prior infection or nonspecifically by injection with BCG or with *Corynebacterium parvum*. It probably depends on the activity of natural killer (NK) cells. BCG or *C. parvum* injections are known to augment the activity of NK cells in the spleen. These cells produce a soluble mediator that somehow interferes with the growth of the parasites, possibly by affecting transport systems at the surface of the infected erythrocyte.

It is clear that this kind of activity of the spleen cannot be discussed apart from a discussion of all the complex processes of acquired immunity, the subject to which we now turn our attention.

Bibliography

Barnwell, J. W., Howard, R. J., and Miller, L. H., 1982, Altered expression of *Plasmodium knowlesi* variant antigen on the erythrocyte membrane in splenectomized rhesus monkeys, *J. Immunol.* **128:**224–226.

Bryceson, A., Fakunle, Y. M., Fleming, A. F., Crane, G., Hutt, M. S. R., de Cock, K. M., Greenwood, B. M., Marsden, P., and Rees, P., 1983, Malaria and splenomegaly, *Trans. R. Soc. Trop. Med. Hyg.* **77:**879.

David, P. H., Hommel, M., Miller, L. H., Udeinya, I. J., and Oligino, L. D., 1983, Parasite sequestration in *Plasmodium falciparum* malaria: Spleen and antibody modulation of cyto-adherence of infected erythrocytes, *Proc. Natl. Acad. Sci. USA* **80:**5075–5079.

Inaba, K., and Steinman, R. M., 1985, Protein-specific helper T-lymphocyte formation initiated by dendritic cells, *Science* **229:**475–479.

Schnitzer, B., Sodeman, T. M., Mead, M. L., and Contacos, P. G., 1973, An ultrastructural study of the red pulp of the spleen in malaria, *Blood* **41:**207–218.

Taliaferro, W. H., and Mulligan, H. W., 1937, The histopathology of malaria with special reference to function and origin of macrophages in defense, *Indian Med. Res. Mem.* **29:**1–138.

UNDP/World Bank/WHO Special Programme for Research and Training in Tropical Diseases, 1979. *The Role of the Spleen in the Immunology of Parasitic Diseases*, Schwabe, Basel.

Weiss, L. P., 1979, The spleen, in: *The Role of the Spleen in the Immunology of Parasitic Diseases*, UNDP/World Bank/WHO Special Programme for Research and Training in Troical Diseases, Schwabe, Basel, pp. 7–20.

Wintrobe, M. M., Lec, G. R., Boggs, D. R., Bithell, T. C., Athens, J. W., and Foerster, J., 1974, *Clinical Hematology* (7th edition), Lea & Febiger, Philadelphia.

Wyler, D. J., and Gallin, J. I., 1977, Spleen-derived mononuclear cell chemotactic factor in malaria infections: A possible mechanism for splenic macrophage accumulation, *J. Immunol.* **118:**478–484.

Wyler, D. J., Miller, L. M., and Schmidt, L. H., 1977, Spleen function in quartan malaria (due to *Plasmodium inui*): Evidence for both protective and suppressive roles in host defense, *J. Infect. Dis.* **135:**86–93.

CHAPTER 16

Immunity in Invertebrates

The science of immunology, which has recently undergone an explosive development, has dealt almost exclusively with the mechanisms of acquired immunity in vertebrate animals, especially in humans and other mammals. It must be realized that the great bulk of animal life consists of invertebrates. These must have effective mechanisms for dealing with invading parasites. In the light of evolutionary theory one would expect these mechanisms to bear some relation to those of vertebrates. Moreover, invertebrates serve as intermediate hosts for parasites of medical and economic importance; their relative susceptibility and resistance to these parasites are highly relevant to practical problems. The degree of susceptibility of mosquitoes to malarial parasites is genetically controlled. For certain mosquito–malarial parasite combinations it has been shown that a single gene is largely responsible for innate resistance, but the effector mechanisms for this are unknown. The same is true for innate susceptibility of mosquitoes to filarial larvae, and also for the susceptibility of snails to the developmental stages of schistosomes. If more were known regarding the detailed genetic basis of such susceptibility of vectors, it might be possible to develop practical applications for control of the parasites, a matter to which I will return in a later chapter.

In both insects and snails, the two groups most studied, and probably in all other metazoan invertebrates, phagocytic cells of the hemolymph play a major role in defense against foreign invaders. In mollusks, as in the oyster, hemocytes ingest and digest bacteria and most other microorganisms. They serve a dual role, both in defense and in nutrition. Larger parasites in resistant hosts become encapsulated by hemocytes and destroyed by their cytotoxic action. This has been well shown with the snail *Biomphalaria glabrata,* the main intermediate host in the western hemisphere for *Schistosoma mansoni.* In a resistant strain of snail, the miracidia penetrate successfully but the sporocysts become encapsulated and killed by hemocytes. The reaction can be demonstrated *in vitro.* Sporocysts placed with hemocyte-containing hemolymph from snails of the resistant line quickly became surrounded by the cells and within 12 hr showed beginning degeneration. By 24 hr the tegument had been lifted

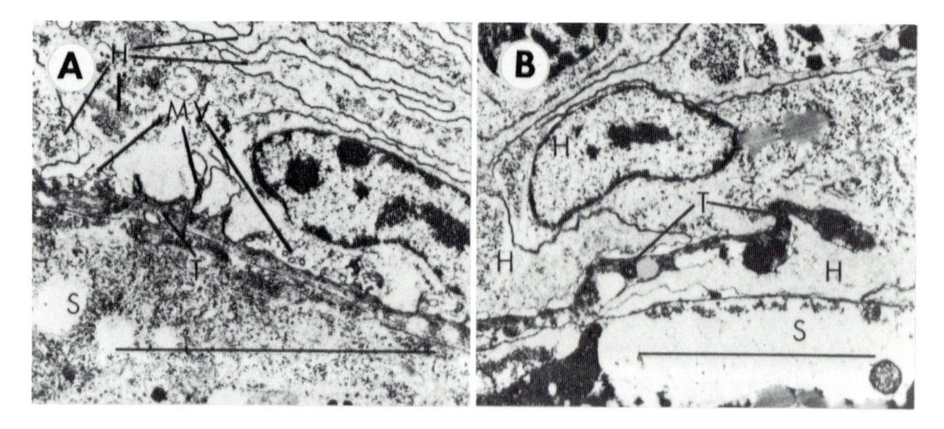

FIGURE 16.1. Sporocysts of *Schistosoma mansoni* were cultivated synxenically for 24 hr with hemolymph from a susceptible strain (PRalb; A) or a resistant strain (10RZ; B) of the snail *Biomphalaria glabrata*. Shown are electron micrographs of the sporocyst surface. S, sporocyst; H, hemocyte; MV, microvilli on sporocyst. Note that in A, although hemocytes adhere to the sporocyst, it is undamaged, even the microvilli being unaffected. In B, however, the closely applied hemocyte is phagocytosing the tegument (T) which is lifting off the sporocyst. Bars = 5 μm. (From Bayne *et al.*, 1980.)

off and was being phagocytosed and the sporocysts were degenerating (Fig. 16.1). Sporocysts placed with the hemocyte-containing hemolymph of susceptible snails did not become surrounded by cells and remained in good condition, as did sporocysts placed in the cell-free plasma from either susceptible or resistant snails. The reaction is dependent on the cells and much resembles that of eosinophils in a vertebrate host with acquired immunity to the schistosomula of *S. mansoni* (see Chapter 21). How the hemocytes of a resistant snail recognize the sporocysts as foreign, whereas those of a susceptible snail do not, is the question of interest. This might well depend on the relationship of particular surface antigens or lectins of the parasite to similar molecules on the surface of the hemocytes. Indeed, it has been shown that sporocysts of *S. mansoni* form antigens that closely resemble antigens on hemocytes of ordinary susceptible snails of the species *B. glabrata*. Sporocysts formed *in vitro* and hence never in contact with snail tissue, were used to produce an antiserum in rabbits. The IgG from this serum was then shown to give a positive indirect fluorescent reaction with normal hemocytes from susceptible *B. glabrata*. This is a true case of molecular mimicry, where the parasite has genes for proteins of the same epitope as proteins in the host. It would be of interest to determine whether the same reaction would occur with hemocytes from a resistant line of snail.

Encapsulation by hemocytes is a common reaction to any large parasite or other foreign object that has gained access to the body cavity. Again the question of interest is why certain parasites in certain hosts escape encapsulation. For example, larvae of the rat tapeworm *Hymenolepis diminuta* develop in the hemocoels of their normal hosts, certain beetles, and also in a useful experimental host, the locust *Schistocerca gregaria*. If, however, the larvae are introduced into another orthopteran insect, the roach *Periplaneta americana*, they quickly become encapsulated by hemocytes and destroyed. In this insect they are recognized as foreign objects, whereas in the locust or in the natural beetle hosts they are not. The hemolymph of the roach will agglutinate the hatched oncospheres of the tapeworm, whereas the hemolymph of the locust will not. Perhaps the agglutinin mediates attachment of the hemocytes.

The insect parasites of other insects must have ways to prevent this encapsulation reaction. Several have been demonstrated. The cynipid wasp *Pseudocoila bochei* when laying its eggs in *Drosophila* larvae apparently injects at the same time something that suppresses the hemocytic reaction to any foreign object. With the ichneumon wasp *Nemeritis canescens* parasitizing caterpillars of the moth *Ephestia*, the general hemocytic reaction to a foreign object is not suppressed. Instead the eggs of *Nemeritis* have a coating on their surface consisting of particles entangled in long projections from the chorion. These particles prevent adherence of hemocytes to the surface of the egg. An even more selective method has been developed by the parasitoid wasp *Leptopilina heterotoma* which parasitizes larvae of the fly *Drosophila melanogaster*. Foreign objects in the hemocoel of *Drosophila* larvae are ordinarily encapsulated by one type of hemocyte, the lamellocyte. This does not happen to the eggs of *Leptopilina*. It has been found that as the female wasp oviposits in a fly larva, it injects a secretion from an accessory gland that specifically affects the lamellocytes. They become elongated and lose their normal adhesive properties; accordingly they cannot form a capsule around the wasp egg. As the wasp larva develops, the lamellocytes are destroyed. But no other kinds of blood cells are affected and the host larva develops in a normal way up to puparium formation. The nature of this active substance, which has been named lamellolysin, will be of great interest, in view of its remarkable specificity.

Microbicidal activity has been demonstrated in fluids and tissue extracts from a variety of invertebrates, and in a few cases the active principle has been purified. Hemagglutinins are common and are widely distributed but their significance for the organisms in which they occur is not known. These are all substances naturally present, as distinct from antibodies formed in vertebrates as a specific immunological response to a foreign protein. Although nothing quite like the antibody response of vertebrates has been demonstrated in an invertebrate, there is now an ever increasing number of reports of functionally analogous responses. I will consider a few of these.

In several crustacea, such as the crayfish (*Parachaeraps*), the lobster (*Homarus*), and the spiny lobster (*Panalarus*), injection of killed bacteria or endotoxin induces a relatively nonspecific increase in bactericidal activity of the hemolymph. The marine worm *Sipunculus nudus* when injected with bacteria formed a lytic substance active against infective ciliates. Extracts of snails already infected with *Schistosoma mansoni* immobilized the miracidia of this fluke, whereas extracts of uninfected snails did not, suggesting the presence of a mechanism to prevent superinfection.

Perhaps the most interesting and most significant results have been obtained in recent work with insects, especially with pupae of the large moth *Hyalophora cecropia*. In cecropia, and also in other insects such as larvae of the wax moth *Galleria mellonella*, a primary injection either with living nonpathogenic bacteria or with heat-killed pathogenic bacteria is followed, after a lag period, by a relatively nonspecific protection against a secondary infection. At the same time it can be shown that the hemolymph has acquired a potent antibacterial activity (Fig. 16.2). Cecropia moth pupae in their winter diapause can be immunized in this way. In response to the immunizing injection, about a dozen proteins are synthesized, but the antibacterial activity seems to reside mainly in three. These three major cecropins, designated A, B, and D, have been purified and their sequence has been determined. One of them, cecropin A, a 37-residue peptide amide, has been synthesized by the solid-phase method of Merrifield. All of the cecropins are small basic peptides with molecular weights around 4000 and with considerable sequence homology. Cecropin A has 7 lysine residues, 5 residues each of alanine and isoleucine, 4 of glutamic acid, glycine, and valine, 2 of aspartic acid, and 1 of threonine, proline, leucine, phenylalanine, arginine, and tryptophane. Cecropin A was lethal to gram-negative as well as gram-positive bacteria at concentrations ranging from 0.3 μM for *Escherichia coli* to 3.5 μM for *Pseudomonas aeruginosa*. Which cells synthesize these powerful antibacterial peptides has not been determined.

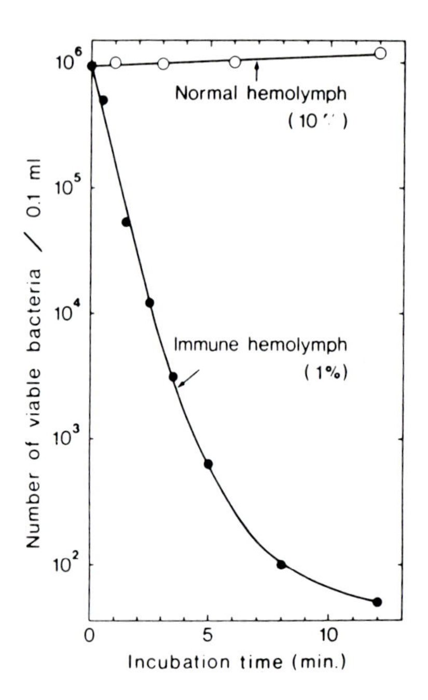

FIGURE 16.2. Antibacterial activity of hemolymph from cecropia pupae. Normal hemolymph was from an untreated pupa, immune hemolymph was from a pupa vaccinated with about 10^6 viable cells of *Enterobacter cloacae* and hemolymph was withdrawn 7 days later. The test organism, *E. coli* D31, was incubated at room temperature in phosphate buffer with hemolymph in the concentrations indicated. At different times aliquots were removed for assay of viable counts. Most, but not all, of the antibacterial activity in immune hemolymph is due to the P9 proteins (the cecropins). (From Boman and Hultmark, 1981.)

Nor is it known how they act on bacteria. For a somewhat similar small protein, formed in hemolymph of flesh fly larvae (*Sarcophaga*) in response to pricking of the body wall with a needle and also highly bactericidal, it has been shown that the bacterial cell membrane is quickly injured.

The deposition of melanin is usually associated with the defense mechanisms of arthropods. It is of interest in this connection that the "black spores" seen by Ronald Ross in his early attempts to infect mosquitoes with malarial parasites actually represented abortive oocysts that failed to develop and became melanized. Melanin deposition depends on the enzyme phenol oxidase. It has recently been found that activation of the prophenol oxidase system in the wax moth *Galleria* enhances by sixfold the phagocytosis of bacteria by plasmatocytes. In these experiments, laminarin, a glucan prepared from fungal cell walls, stimulated one type of hemocyte, the granulocytes which contain phenol oxidase, to discharge their contents. This discharge presumably coated the bacteria to enhance their recognition by the other type of hemocyte, the plasmatocyte, resulting in increased phagocytosis.

There is every reason to suppose that similar systems exist in other kinds of insects and perhaps in all invertebrates. Indeed, on general principles one would expect all organisms to have systems to inhibit microbial development and in addition to be able to respond to injury by enhancement of such systems. Both of these situations find their highest development in the immune system of vertebrates.

Bibliography

Andreu, D., Merrifield, R. B., Steiner, H., and Boman, H. G., 1983, Solid-phase synthesis of cecropin A and related peptides, *Proc. Natl. Acad. Sci. USA* **80:**6475–6479.

Bayne, C. J., and Stephens, J. A., 1983, *Schistosoma mansoni* and *Biomphalaria glabrata* share epitopes: Antibodies to sporocysts bind host snail hemocytes, *J. Invert. Pathol.* **42:**221–223.

Bayne, C. J., Buckley, P. M., and de Wan, P. C., 1980, Macrophage-like hemocytes of resistant *Biomphalaria glabrata* are cytotoxic for sporocysts of *Schistosoma mansoni in vitro, J. Parasitol.* **66:**413–419.

Boman, H. G., and Hultmark, D., 1981, Cell-free immunity in insects. *Trends Biochem. Sci.* **Nov.:**306–309.

Cheng, T. C., 1977, Biochemical and ultrastructural evidence for the double role of phagocytosis in molluscs: Defense and nutrition, in: *Comparative Pathobiology,* Volume 3 (L. A. Bulla, Jr., and T. C. Cheng, eds.), Plenum Press, New York, pp. 21–30.

Damian, R. T., 1979, Molecular mimicry in biological adaptation, in: *Host–Parasite Interfaces* (B. B. Nickol, ed.), Academic Press, New York, pp. 103–125.

Hultmark, D., Engström, A., Bennich, H., Kapur, R., and Boman, H. G., 1982, Insect immunity: isolation and structure of cecropin D and four minor antibacterial components from cecropia pupae, *Eur. J. Biochem.* **127:**207–217.

Lackie, A. M., 1981, Humoral mechanisms in the immune response of insects to larvae of *Hymenolepis diminuta* (Cestoda), *Parasite Immunol.* **3:**201–208.

Maramorosch, K., and Shope, R. E. (eds.), 1975, *Invertebrate Immunity,* Academic Press, New York.

Okada, M., and Natori, S., 1984, Mode of action of a bactericidal protein induced in the haemolymph of *Sarcophaga peregrina* (flesh-fly) larvae, *Biochem. J.* **222:**119–124.

Ratcliffe, N. A., Leonard, C., and Rowley, A. F., 1984, Prophenoloxidase activation: Nonself recognition and cell cooperation in insect immunity, *Science* **226:**557–559.

Rizki, R. M., and Rizki, T. M., 1984, Selective destruction of a host blood cell type by a parasitoid wasp, *Proc. Natl. Acad. Sci. USA* **81:**6154–6158.

Salt, G., 1980, A note on the resistance of two parasitoids to the defence reactions of their insect hosts, *Proc. R. Soc. London Ser. B* **207:**351–353.

Yoshino, T. P., Cheng, T. C., and Renwrantz, L. R., 1977, Lectin and human blood group determinants of *Schistosoma mansoni:* Alteration following *in vitro* transformation of miracidium to mother sporocyst, *J. Parasitol.* **63:**818–824.

CHAPTER 17

The Immune System of Vertebrates in Relation to Parasitic Infections

A complex system of cells and humoral substances protects the vertebrate organism against invading parasites, foreign materials of all kinds, and perhaps also against the abnormal cells that give rise to tumors and cancer. Some of these cells and substances have been known for a long time, such as macrophages, complement, and antibodies. Others, such as NK cells and the interleukins and interferons, have only recently been demonstrated. It is likely that still others remain to be discovered. Enough is known to show clearly that the system functions in two ways. First, it provides some measure of immediate protection, and this is effective against the great majority of invading organisms. This is a part of the mechanism of innate resistance. Second, it provides for a remarkable, greatly enhanced specific resistance against those parasites that evade the mechanisms of innate resistance and succeed in establishing themselves. This is acquired immunity. Depending on the nature and effectiveness of the acquired immunity and on the nature of the parasites' further responses, a whole spectrum of associations becomes possible. At one extreme the acquired immunity is insufficient to control the parasite. If the parasite is a microorganism, it continues to multiply and destroys the host. If it is a metazoan parasite, such as a helminth, the extent of damage will depend in part on the number of worms to which the host is exposed. At the other extreme the acquired immunity quickly wipes out the parasite. In between are the most interesting situations. For example, some protozoan parasites persist at low levels for many years—they coexist with their host. Similarly, some helminths continue to live and produce eggs for many years in a host which nevertheless is immune to further infection by the invasive stages of worms of the same species.

The effector systems providing protection against parasites are both cellular and humoral. Among the effector cells are monocytes, macrophages, polymorphonuclear leukocytes (neutrophils, eosinophils, and basophils), NK cells, and several kinds of lymphocytes derived from the thymus (T cells) (see Table 17.1). Among the humoral systems are complement and antibodies. An-

TABLE 17.1. Characteristics of Effector Cells[a]

	T cells	Monocytes or macrophages	Polymorphonuclear leukocytes	Natural killer cells
Size	Small (9–12 μm)	Large (16–20 μm)	Large (12–18 μm)	Medium (12–15 μm)
Ratio cytoplasm to nucleus	Low	High	High	High
Nucleus	Round	Markedly indented	Multilobed	Slightly indented
Adherence to surface	–	+	+	–
Phagocytosis	–	+	+	–
Receptors for sheep erythrocytes (human cells)	High-affinity receptors	–	–	+ on about 50%; have low-affinity receptors
Receptors for IgG	Less than 10% have receptors	+	+	+
Antigens				
Human	Most react with 9.6, OKT3 Subsets react with OKT4, OKT8	Most react with OKM1, anti-asialo GM1 Subsets react with anti-Ia	Most react with OKM1, anti-asialo GM1	Most react with OKM1, anti-asialo GM1, OKT10 Subsets react with 9.6, anti-Ia
Mouse	All express Thy 1, Lyt 1	Most express Mac 1, asialo GM1, Mph 1		Most express GM1, NK1, NK2, Lyt 1, Ly 5, Qa5, ? Mph 1

[a] Modified from Herberman and Ortaldo (1981).

tibodies are formed by still another type of lymphocyte, the B cells (derived from bone marrow). In reality, none of these acts alone; there is a complex interplay among all of them. The polymorphonuclear leukocytes, particularly the neutrophils, and monocytes and macrophages all act in the absence of antibody, and so do the NK cells. The activity of each is much enhanced by antibody and other responses in acquired immunity. T cells, on the other hand, have no spontaneous cytotoxic or other activity but must be activated, usually by exposure to specific antigens on accessory cells such as macrophages. Thus, the polymorphonuclear cells, monocytes and macrophages, and NK cells act in innate as well as in acquired immunity. The same is true of complement. The so-called alternative pathway is activated by many kinds of microorganisms and culminates in damage to their membranes. One interesting example is provided to *Trypanosoma cruzi*. Epimastigote forms such as occur in the midgut of the vector bug activate the alternative pathway and cannot initiate infection. Metacyclic forms, the infective stage in the hindgut of the bug, or bloodstream trypomastigote forms do not activate the alternative pathway. In the presence of specific antibody, however, they will activate the classical pathway (see below) and be lysed by complement. Similarly, the activity of all the cells active in innate resistance is much enhanced in the presence of antibodies. This is in part a direct effect of the antibodies acting as opsonins. But to an even greater extent this results from the activation of T cells by the specific antigen. Such activated T cells produce a wide array of soluble mediators called lymphokines. The nature of most of these is unknown. Included among them are the interferons and interleukins 1, 2, and 3.

Interleukin-1 is released by macrophages undergoing an immune response. Its biological activities include not only activation of bone marrow, of neutrophils, and of T cells but also stimulation of hepatocytes to secrete acute-phase proteins (see Chapter 14) and an effect on the hypothalamus to raise the thermostat set point and produce fever. *In vitro* it stimulates mouse thymocytes to incorporate [^{3}H]thymidine, and this provides a convenient assay method. This effect may occur via stimulation of the secretion of interleukin-2, the T-cell growth factor. Interleukin-2 provides for the amplification of T cells. *In vitro* it is an essential growth factor for continuous growth of T cells. Interleukin-3 similarly supports *in vitro* growth of pre-B-cell clones that can be induced to mature into antibody-secreting cells. T cells are activated by exposure to antigens on accessory cells such as macrophages, which also present the major histocompatibility complex. T cells activated in this way will then cause lysis of target cells bearing *both* the specific antigen (such as a virus) and the same histocompatibility complex. The mechanism for lysis is not known but may involve serine proteases and osmotic alterations in the target cells. Interferons and various other lymphokines activate macrophages and NK cells. The latter especially are activated very quickly, within a few hours, and show augmented cytotoxic activity. They also produce interferons and interleukins, providing mechanisms for positive self-regulation. At the same time other kinds of T cells produce factors suppressing reactivity. All of these different reactions go on simultaneously, interacting with each other to create a complex situation difficult to sort out in detail. Nevertheless, there is now good evidence that one of the requisites for tumor induction by carcinogenic agents may be interference with host defenses, perhaps especially those mediated by NK cells. NK cells also seem to play a role in natural and acquired resistance to microbial infections including the protozoan parasite *Babesia microti* and possibly also in malaria, a subject to which I will return (Chapter 19).

Oxidative mechanisms are of prime importance in the killing of protozoan parasites both by phagocytic cells, such as neutrophils and macrophages, and through the action of Nk cells. In the phagocytic cells, superoxide and H_2O_2

are formed in the oxidation of NADPH by NADPH oxidase, and peroxidase is delivered from lysosomes into the phagocytic vacuole. Whereas neutrophils from a nonimmune host will ingest and rapidly kill by oxidative means parasites such as the amastigotes of *Leishmania donovani,* macrophages do not kill these organisms unless they have been activated by lymphokines secreted by T cells in response to specific antigens. Macrophages activated by such lymphoctye mediators show enhanced microbicidal capacity, mainly as a result of enhanced oxidative mechanisms. They also show many other changes including increased membrane activity, increased pinocytosis, increased adenylate cyclase, increased cytoplasmic granules, increased calcium influx, increased glucose transport, increased collagenase, increased cyclic guanosine monophosphate, and others.

The killing activity of NK cells may also be mediated in part by oxidative mechanisms even though they do not phagocytose the parasites. For example, malarial parasites are killed within their host erythrocytes by factors of unknown nature produced by NK cells and possibly bearing some relation to tumor necrosis factor. That this killing effect could be a result at least in part of the interaction of polyamines and polyamine oxidase to form toxic aldehydes and H_2O_2 is indicated by effects observed with *Plasmodium falciparum* in *in vitro* cultures. In the presence of both polyamine oxidase and a polyamine, such as spermine, growth of the parasites in culture was sharply inhibited, most of them appearing as the abnormal so-called crisis forms (see also Chapter 19).

Although these oxidative mechanisms readily become apparent and lend themselves to experimental work, there is also much evidence indicating killing effects of cells on other cells exerted via effects on their membranes. The details of such effects by killer cells are not yet understood. But that major humoral killing mechanism, complement, has been almost completely elucidated.

Complement is a group of about 20 serum proteins whose interactions lead to an attack mechanism on cell membranes. Activation of the system can be initiated by either of two pathways: (1) the classical pathway, which can be started only by antibody after it has been aggregated by reaction with antigen; (2) the alternative pathway, which can be started by other substances, such as polysaccharides, as well as by aggregated antibody. In the classical pathway, aggregated antigen–antibody complex, or antibody bound to the surface of a cell, binds component 1 (C1) activating it through its three subcomponents. One of these C1's when activated is a proteolytic enzyme that hydrolyzes components C2 and C4 to give products that combine to yield another protease, $C\overline{42}$. This converts component C3 into an active form, $C\overline{423}$, which then activates C5. In the alternative pathway the first events are not yet clear. However, C3 binds to factor B and is hydrolyzed by factor D to give $C\overline{3B}$, which is a C3 convertase. The addition of one or more activated $C\overline{3}$ molecules to the $C\overline{3B}$ complex changes the specificity to a C5 convertase, and this in turn activates C5. Factor D is present in blood as the active enzyme.

Also present in blood are a number of controlling proteins which inhibit or inactivate different components of the system.

The activation by cleavage of C5 is the last enzymatic step in both pathways. A complex is then formed, probably by self-assembly, that includes C5b6789. These supermolecular complexes represent C5b–8 monomers with varying numbers of C9 molecules. Their formation is accompanied by a unique transition of the molecules from a hydrophilic to an amphiphilic state. The convex surfaces of the complexes are hydrophobic and carry lipid-binding sites that effect insertion into cell membranes. The concave surfaces are hydrophilic, permitting passage of water and ions. In this way channels are produced in cell membranes, disrupting their osmotic control. The C5b formed by C5 convertase spontaneously associates first with C6 and C7, and the C5b67 complex inserts into lipid bilayers and binds C8. C5b678 then serves as binding substrate for C9. Defined channellike structures visible by electron microscopy have a ratio of C5b–8 to C9 molecules of 1 to 6–9 and a functional pore size of about 5–7 nm (Fig. 17.1). All recent data support this channel model for the action of complement on membranes originally proposed by Manfred Mayer.

It is a fascinating fact that at least three components of complement are coded for by structural genes that are part of the major histocompatibility complex, located on chromosome 17 in the mouse and chromosome 6 in humans. In this complex are the genes coding for highly polymorphic histocompatibility antigens that are present on the surface of all nucleated cells and are responsible for tissue graft rejection. Also in this complex are the *Ir* genes controlling immune responses, and the genes coding for Ia antigens, found in B lymphocytes. Furthermore, receptors for activated complement components are present on lymphocytes and macrophages. The lymphocyte C3 receptors may play an important role in the interactions of different classes of lymphocytes. Thus, complement may be important in integration of the immune response as well as as an effector mechanism.

Interacting with complement and with all the cellular effector systems are the antibodies, certainly one of the most important mechanisms of acquired immunity. These are immunoglobulin molecules specifically targeted against surface antigens and secretions and excretions of the parasites. An animal can make over a million different kinds of antibody molecules all with the same basic structure. This consists of two light and two heavy chains held together by disulfide bridges. Each of the four chains has a constant region and a variable region. The latter provides for the almost infinite heterogeneity permitting an antibody to many different kinds of molecules. In the three-dimensional structure of the variable region, the areas of greatest diversity are clustered together to form the combining site for antigen.

The immunoglobulins are produced and secreted by B lymphocytes, but these cells cannot respond directly to an antigen. Most antigens must be presented by macrophages to T lymphocytes and in general there must be histocompatibility between these cells. The antigen plus a T-cell factor will

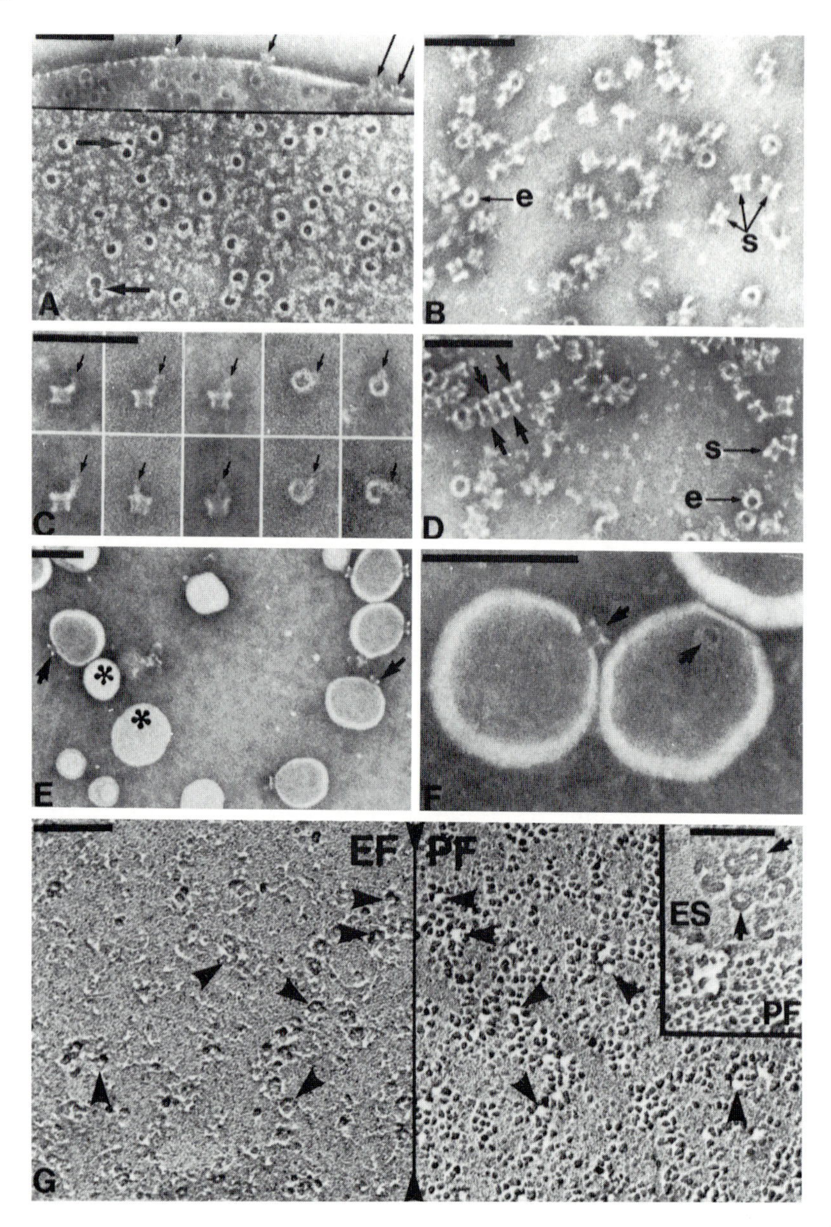

FIGURE 17.1. (A) Negatively stained, complement-lysed erythrocytes. C5b–9(m) complexes are seen as numerous circular "classical" lesions over the membrane together with some "twinned" forms (bold arrows). The C5b–9(m) complexes are seen as 10-nm-high cylindrical projections along the bent edge of the ghost membrane at the top (arrows). The light rim representing the sharply bent membrane in tangential view is attenuated or interrupted at the site of attachment of the complexes. (B) Negatively stained preparation of isolated C5b–9(m) complexes in detergent solution. The complex has the basic structure of a 15-nm-high, thin-walled cylinder, rimmed by an annulus at one end. The cylinder is seen in various levels of tilt between side views (s) and axial projection (e). (C) Selection of C5b–9(m) complexes exhibiting a small appendage (arrows),

then stimulate B cells having an immunoglobulin receptor for that antigen to proliferate and to form plasma cells that secrete the specific antibody. The antigen must first be internalized and processed by the specific B cells. It is then presented to T cells in a manner restricted by the major histocompatibility complex. The T cells so recognized become the helper T cells essential for immunoglobulin release by the plasma cells.

Five main classes of immunoglobulins are known. These differ in their heavy chains and in their properties and functions. Generally first to be formed in an initial immunological response are the IgM antibodies with a molecular weight of 850,000. Usually but not always these antibodies reach a peak within about a week and then rapidly decline. The other four classes of immunoglobulins have molecular weights of 155,000 to 200,000. The IgG antibodies are most prevalent especially in secondary responses to antigen. In some parasitic infections, such as malaria and leishmaniasis, they reach very high levels bearing little relation to the degree of protective immunity. Four subclasses of IgG, designated IgG1, IgG2, IgG3, and IgG4, exist in normal serum. These have different antigenic determinants on their heavy chains and differ in biological properties. For example, IgG1 and IgG3 seem to be most efficient both in activation of complement and in binding to Fc receptors on macrophages and granulocytes. The IgA antibodies are produced in two separate systems, one providing them in the circulation and various internal secretions such as the pleural and peritoneal fluids, the other providing them to external secretions including those of the respiratory tract and the gastrointestinal tract. The last are important in relation to intestinal parasites. IgE or reaginic antibodies are present in relatively minute amounts but can mediate life-threatening acute allergic reactions. They are often associated with helminthic infections. Also present in low concentrations in the serum

---

often seen on the annulus, particularly by low-electron-dose image recording. This stalk carries antigenic determinants of C5 and C6. (D) "Poly-C9" formed by prolonged incubation of purified human C9 in detergent-free buffer solution at 37°C. The C9 oligomers exhibit a cylindrical structure closely resembling the C5b–9(m) complex except for the absence of appendages on the annulus. Occasionally, small ordered arrays of cylinders are seen (arrows), associated at the putative apolar terminus opposite the annulus. (E, F) C5b–9(m) complexes (arrows) generated on erythrocytes, purified, and reincorporated into phosphatidylcholine liposomes. Vesicles that escaped incorporation of a complex (asterisks) are characteristically impermeable to the stain. Typically, the complexes project 10 nm exterior to the plane of the membrane. (G) Complementary freeze-etch replicas of antibody-sensitized sheep erythrocyte, lysed with human complement. Fracture EF-face (left frame) exhibits numerous ring-shaped structures, interpreted to represent the intramembranous portion of C5b–9(m) cylindrical complexes. The rings are complementary to circular defects in the lipid plateau of the inner membrane leaflet (PF-face). A number of complementary lesions are labeled by arrowheads. Inset at upper right shows C5b–9(m) annuli (arrows) on the etched true outer surface (ES) of a proteolytically stripped ghost membrane. 25° rotary shadowing with Pt. Bars = 100 nm. Sodium silicotungstate was used for negative staining in A–F. (From Bhakdi and Tranum-Jensen, 1984. Original prints courtesy of Drs. S. Bhakdi and J. Tranum-Jensen.)

are the IgD antibodies. They constitute a major surface globulin of peripheral blood lymphocytes but little is yet known as to their functional signficance.

As a result of complex chromosomal events controlling the sequences especially in the variable regions of the immunoglobulins, a particular sequence becomes expressed attached to a constant region. The differentiated B lymphocyte then expresses a surface receptor with a single antibody-combining site defined by the particular combination of variable regions from light and heavy chains. This antibody surface receptor renders the lymphocyte subject to stimulation by an antigen that will combine with it. This stimulation requires interaction with helper T cells and results in proliferation of the cell as well as in secretion of antibodies to this antigen. In this way the immune response is expanded. This is the basis of the clonal selection theory of Burnet

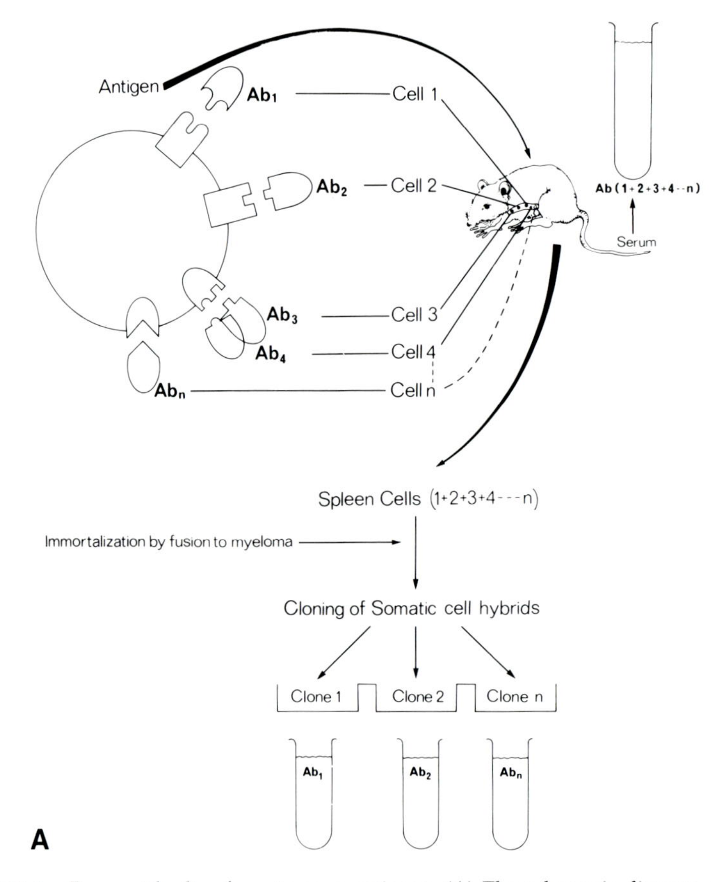

FIGURE 17.2. Pure antibodies from impure antigens. (A) The schematic diagram shows that different antigens on a cell surface are recognized by different antibodies. Single antigens can be recognized by different antibodies (Ab_3 and Ab_4) and the overlap in the antigenic determinant could be such that different antibody molecules recognize exactly the same determinant. Each antibody is made by a cell but the products are all mixed in the serum, so that the antiserum of

proposed many years ago and so beautifully confirmed by the demonstration of monoclonal antibodies by Milstein and Kohler. Whereas biochemistry could not purify antibodies because of their complexity and similarities, a biological method now provides pure antibodies from impure antigens (Fig. 17.2). As we will see, monoclonal antibodies provide a powerful tool in the antigenic analysis of such complex organisms as the animal parasites.

The immune system, like any other aspect of the physiology of an or-

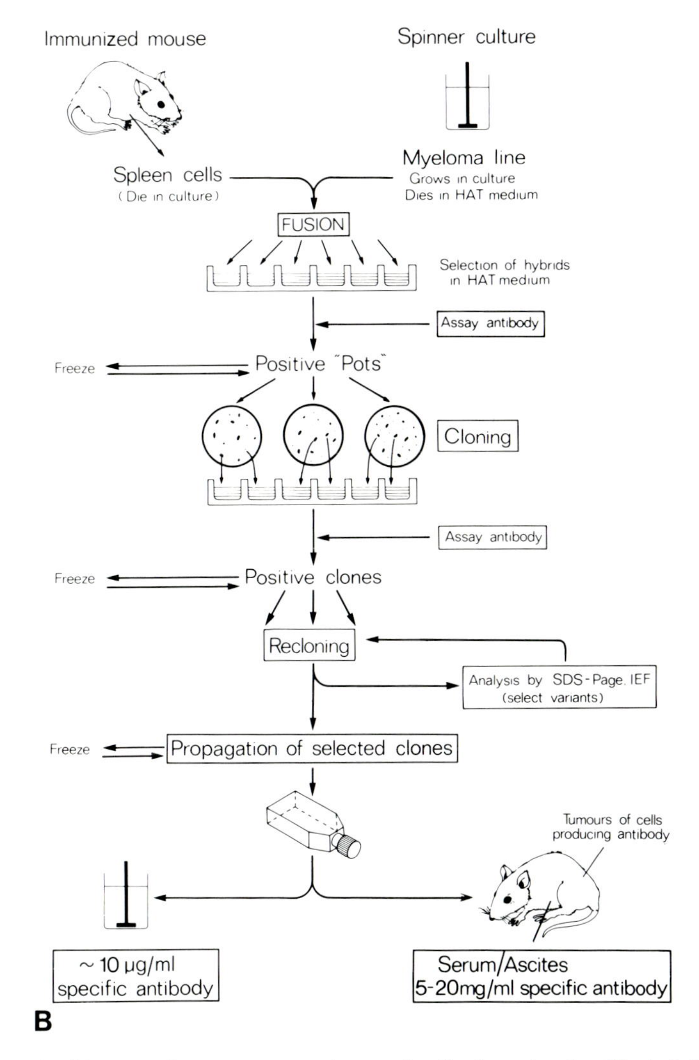

an immune animal is a very heterogeneous mixture of antibody molecules. The hybrid myeloma method permits the separation of each antibody molecule by immortalization of antibody-producing cells and fractionation of clones as shown in B. (B) Immortalization of antibody-producing cells, cloning, and preparation of monoclonal antibodies. (From Milstein, 1981. Print courtesy of Dr. C. Milstein.)

ganism, is affected by environmental conditions, notably by nutrition. Malnutrition has been observed to enhance certain parasitic infections, such as those with *Entamoeba histolytica*, *Giardia lamblia*, *Leishmania*, and *Toxoplasma*, and some of the intestinal helminths. The effects are often complex, even if specific nutritional deficiencies are considered. I have already noted how iron and riboflavin deficiency may interfere directly with development of malarial parasites. On the other hand, iron deficiency and protein deficiency separately or together delay greatly the immunological rejection by rats of the nematode *Nippostrongylus brasiliensis*. This fits well with studies in which malnourished children have been found more susceptible to infections with gut parasites. A tenfold greater incidence of the nematode *Strongyloides stercoralis* has been found in severely malnourished children. These results can all be interpreted as depending on interference with effective immunological reactions to the parasites.

Biotin deficiency has been found to decrease resistance of chickens to an avian malarial parasite. This may affect both innate and acquired resistance, since biotin, and the fatty acids in the synthesis of which it is a cofactor, are essential in the cytotoxic T-cell response and possibly also in a variety of other immunological cellular reactions. Biotin deficiency is a rare occurrence in humans. On the other hand, zinc deficiency has been observed as a result of a genetic defect in absorption of this metal and has been correlated with a severe dyscrasia of the immune system. It is of great interest that zinc deficiency in mice has recently been found to enhance their susceptibility to their normally benign flagellate parasite, *Trypanosoma musculi*. This is apparently a result of a delay in formation of the reproduction-inhibiting antibody ablastin (see Chapter 18) as well as in the formation of the terminal lytic antibody. These antibodies will be considered in more detail in a subsequent chapter.

In summary then, a parasite that has succeeded in penetrating the vertebrate host's mechanisms of innate resistance soon finds arrayed against it activated NK cells and macrophages, cytotoxic T cells specifically armed to recognize the parasite's antigens and to attack it, a variety of specific antibodies that react directly with antigens in the parasite, and also activate complement to both lyse the parasite's membrane and enhance its susceptibility to phagocytosis. The details of these responses to the parasite and how the parasite in turn responds and attempts to overcome or evade them vary with the parasite and host. At the same time these responses and counterresponses largely determine the pathogenic effects of the infection. I will therefore next consider a series of particular parasite–host associations that illustrate a variety of immunological responses.

Bibliography

Allison, A. C., and Clark, I. A., 1978, Macrophage activation and its relevance to the immunology of parasitic diseases, in: *Immunity in Parasitic Diseases* (A. Capron, P. H. Lambert, B. Ogilvie, and P. Péry, eds.), INSERM, Paris, pp. 147–160.

Bhakdi, S., and Tranum-Jensen, J., 1984, Mechanism of complement cytolysis and the concept of channel-forming proteins, *Philos. Trans. R. Soc. London Ser. B,* **306:**311–324.

Burnet, M., 1959, *The Clonal Selection Theory of Acquired Immunity,* Vanderbilt University Press, Nashville, Tenn.

Burnet, M., 1962, Role of the thymus and related organs in immunity, *Br. Med. J.* **2:**807–811.

Chandra, R. K., 1984, Parasitic infection, nutrition and immune response, *Fed. Proc.* **43:**251–255.

Cohen, S., and Warren, K. S. (eds.), 1982, *Immunology of Parasitic Infections* (2nd edition), Blackwell, Oxford.

David, J. R., 1982, Immune effector mechanisms against parasites, in: *Immunology of Parasitic Infections* (2nd edition), (S. Cohen and K. S. Warren, eds.), Blackwell, Oxford, pp. 74–98.

Dinarello, C. A., 1984, Interleukin-1 and the pathogenesis of the acute phase response, *N. Engl. J. Med.* **311:**1413–1418.

Duncombe, V. M., Bolin, T. D., Davis, A., and Kelly, J. D., 1979, The effect of iron and protein deficiency on the development of acquired resistance to reinfection with *Nippostrongylus brasiliensis* in rats, *Am. J. Clin. Nutr.* **32:**553–558.

Frank, W. (ed.), *Immune Reactions to Parasites,* Fischer, Stuttgart.

Golub, E. S., 1977, *The Cellular Basis of the Immune Response,* Sinauer, Sunderland, Mass.

Good, R. A., 1981, Nutrition and immunity, *J. Clin. Immunol.* **1:**3–11.

Greenwood, B. M., and Whittle, H. C., 1981, *Immunology of Medicine in the Tropics,* Arnold, London.

Halliwell, B., 1984, Oxygen radicals—A commonsense look at their nature and medical importance, *Med. Biol.* **62:**71–77.

Halliwell, B., and Gutteridge, J. M. C., 1984, Oxygen toxicity, oxygen radicals, transition metals and disease, *Biochem. J.* **219:**1–14.

Herberman, R. B., and Ortaldo, J. R., 1981, Natural killer cells: Their role in defenses against disease, *Science* **214:**24–30.

Kohler, G., and Milstein, C., 1975, Continuous cultures of fused cells secreting antibody of predefined specificity, *Nature* **256:**495–497.

Kung, J. T., Mackenzie, C. G., and Talmage, D. W., 1979, The requirement for biotin and fatty acids in the cytotoxic T-cell response, *Cell. Immunol.* **48:**100–110.

Lackman, L. B., 1983, Human interleukin 1: Purification and properties, *Fed. Proc.* **42:**2639–2645.

Lanzavecchia, A., 1985, Antigen-specific interaction between T and B cells, *Nature* **314:**537–539.

Lee, C. M., Humphrey, P. A., and Aboko-Cole, G. F., 1983, Interaction of nutrition and infection: Effect of zinc deficiency on resistance to *Trypanosoma musculi, Int. J. Biochem.* **15:**841–847.

Mayer, M. M., 1978, Complement, past and present, *Harvey Lect.* **72:**139–175.

Metcalf, D., 1985, The granulocyte–macrophage colony-stimulating factors, *Science* **229:**16–22.

Milstein, C., 1981, 12th Sir Hans Krebs Lecture—From antibody diversity to monoclonal antibodies, *Eur. J. Biochem.* **118:**429–436.

Mitchell, G. F., 1979, Effector cells, molecules and mechanisms in host-protective immunity to parasites, *Immunology* **38:**209–223.

Mitchell, G. F., Anders, R. F., Brown, G. V., Handman, E., Roberts-Thomson, I. C., Chapman, C. B., Forsyth, K. P., Kahl, L. P., and Cruise, K. M., 1982, Analysis of infection characteristics and anti-parasite immune responses in resistant compared with susceptible hosts, *Immunol. Rev.* **61:**137–188.

Müller-Eberhard, H. J., 1980, Complement reaction pathways, in: *Progress in Immunology IV,* Volume 3 (M. Fongereau and J. Dausset, eds.), Academic Press, New York, pp. 1001–1024.

Palacios, R., Henson, G., Steinmetz, M., and McKearn, J. P., 1984, Interleukin 3 supports growth of mouse pre-B-cell clones *in vitro, Nature* **309:**126–131.

Playfair, J. H. L., 1978, Effective and ineffective immune responses to parasites: Evidence from experimental models, *Curr. Top. Microbiol. Immunol.* **80:**37–64.

Porter, R. R., 1977, The eleventh Hopkins memorial lecture: The biochemistry of complement, *Biochem. Soc. Trans.* **5:**1659–1674.

Porter, R. R., 1984, Introduction to the complement system, *Philos. Trans. R. Soc. London Ser. B* **306:**279–281.

Santoro, F., Bernal, J., and Capron, A., 1979, Complement activation by parasites: A review, *Acta Trop.* **36**:5–14.

Scala, G., Allavena, P., Djeu, J. Y., Kasahara, T., Ortaldo, J. R., Herberman, R. B., and Oppenheim, J. J., 1984, Human large granular lymphocytes are potent producers of interleukin-1, *Nature* **309**:56–59.

Smith, K. A., 1984, Interleukin 2, *Annu. Rev. Immunol.* **2**:319–333.

Soulsby, E. J. L., 1980, Immunological reactions to parasites, in: *Scientific Foundations of Veterinary Medicine* (A. T. Phillipson, L. W. Hall, and W. R. J. Pritchard, eds.), Heinemann, London, pp. 381–390.

Thorne, K. J. I., and Blackwell, J. M., 1983, Cell-mediated killing of protozoa, *Adv. Parasitol.* **22**:43–151.

Tschopp, J., Podack, E. R., and Müller-Eberhard, H. J., 1982, Formation of transmembrane tubules by spontaneous polymerization of the hydrophilic complement protein C9, *Nature* **298**:534–537.

Whitlow, M. B., Ramm, L. E., and Mayer, M. M., 1985, Penetration of C8 and C9 in the C5b–9 complex across the erythrocyte membrane into the cytoplasmic space, *J. Biol. Chem.* **260**:998–1005.

CHAPTER 18

Immune Reactions to Trypanosomes and How They Are Evaded. Ablastin. Antigenic Variation

The immunology of trypanosome infections provides examples of two phenomena of great general biological interest and importance. One is the formation of a reproduction-inhibiting antibody, ablastin, that controls trypanosome populations without killing them. The other is antigenic variation, enabling certain species of trypanosomes to evade the immune responses of their host.

There exist a large number of species of trypanosomes more or less specifically associated with particular vertebrate hosts, ranging from fish to mammals, and essentially nonpathogenic. Good examples of this kind of association are provided by *Trypanosoma lewisi* of the rat and *T. musculi* of the mouse. Both have a limited host range; *T. lewisi* will not normally infect mice, nor will *T. musculi* infect rats. Transmission is via a developmental cycle in ectoparasites again specifically associated with the particular host. Many years ago W. H. Taliaferro observed that in rats infected with *T. lewisi* there was a short period (up to 7–8 days) of active multiplication of the parasites which ceased before an overwhelmingly large population of trypanosomes had developed. A first trypanocidal crisis then occurred on about day 10. Nondividing so-called adult trypanosomes persisted, however, for another 2 or 3 weeks, when there was a second trypanolytic reaction with removal of the parasites from the blood. This course of events can be viewed as a useful mutual adaptation for survival of both host and parasite. A limited multiplication of the parasite prevents significant pathogenic effects to the host. At the same time, the prolonged survival of nondividing trypanosomes provides for an extended period in which the vector rat fleas can become infected, thereby both perpetuating and disseminating the parasite population. Clearly, the reproduction-inhibiting mechanism is of major biological significance. That it is an antibody there can no longer be any doubt. Largely as the result of the work of P. A. D'Alesandro, it has been shown to be an IgG formed very early in the infection and rapidly coating the surface of the trypanosomes. This latter fact is the reason why the antibody could not be adsorbed with trypanosomes grown in ordinary immunocompetent rats—their surface was already coated with it. If, however, the *T. lewisi* were grown in rats that had been made immunodeficient (e.g., with cortisone or irradiation) so that they did not form ablastin, the trypanosomes readily adsorbed out the specific IgG from ablastic immune serum. The cycle has been completed by using as an immunoadsorbent column a preparation of Sepharose beads to which surface coat IgG eluted from trypanosomes has been covalently linked. This specifically removed proteins from an extract of trypanosomes, and these partially purified antigens stimulated an ablastic immunity in rats. Ablastin probably exerts its trypanostatic activity by interfering with active uptake of nutrients at the cell membrane. It has been shown to inactivate a sodium–potassium ATPase, after the manner of ouabain, which will indeed mimic the effect of ablastin.

The trypanocidal antibody (an IgG) responsible for the first crisis is specific for dividing forms and very recently formed adults so that a population of nondividing trypanosomes survives. This is finally eliminated by the second trypanocidal antibody, an IgM. This IgM, possibly a rheumatoid factor-like protein, is actually directed against the ablastic IgG with which these adult trypanosomes are now coated. Neither dividing nor adult trypanosomes are resistant to ablastin. A dividing population placed under appropriate *in vitro*

TABLE 18.1. Reproductive Activity in Cultures Inoculated with Reproducing and Inhibited (Adult) Trypanosomes after Incubation Overnight at 37°C with Normal and Ablastic Sera Collected from Rats at Various Times after Infection with *Trypanosoma lewisi*[a]

	Percentage dividing and small forms obtained from an initial inoculum of:	
Type of rat serum in culture	Reproducing trypanosomes[b]	Inhibited (adult) trypanosomes[c]
Normal	45.0	77.1
Ablastic 19th day[d]	5.0	0
Ablastic 1st day	38.5	71.1
Ablastic 2nd day	35.8	73.1
Ablastic 3rd day	28.4	69.7
Ablastic 4th day	24.8	52.2
Ablastic 5th day	20.0	17.3

[a] Modified from D'Alesandro (1975).
[b] Obtained on day 4 of infection and showing 16.2% dividing and small forms initially.
[c] Obtained on day 12 of infection—nonreproducing.
[d] Collected between the first and terminal trypanocidal crises when the ablastic effect is maximal; the other ablastic sera were collected before the first crisis when the effects of inhibition first became evident.

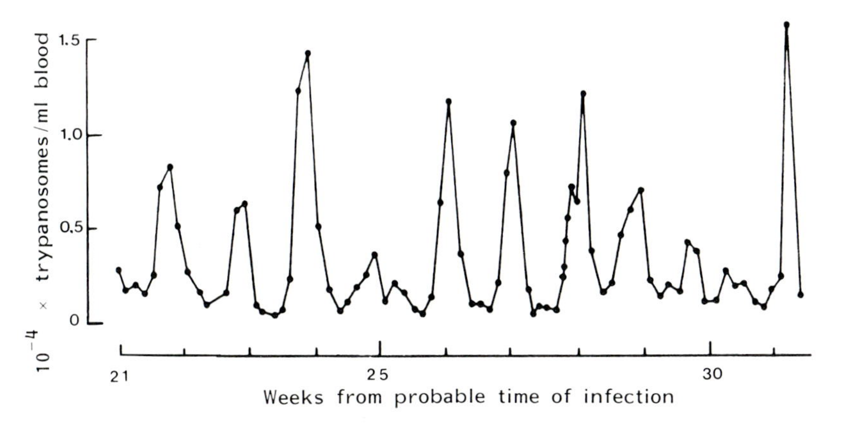

FIGURE 18.1. Fluctuation of parasitemia in a case of human trypanosomiasis (*Trypanosoma gambiense*). (From Cross, 1978.)

conditions will cease to divide if ablastin is added; an adult population that would begin multiplying in the absence of ablastin will not do so if ablastin is present (Table 18.1).

In lactating rats and their young nursing pups (up to about 10–14 days of age), *T. lewisi* infections are aborted with the first crisis. This has been shown to be a result of the natural presence in serum and milk of the uninfected mother rat of a rheumatoid factor-like IgM. This evidently acts in the same way as the IgM ordinarily formed late in the infection.

Whereas an ablastic type of initial immunological response may well be the rule in the vast majority of trypanosome infections, there is no evidence for such a response to those species of trypanosomes that are of medical and economic importance in Africa. These are represented by only five main species: *Trypanosoma brucei brucei, T. vivax,* and *T. congolense* causing disease in cattle and *T. b. gambiense* and *T. b. rhodesiense* causing two forms of human sleeping sickness (see Diagram V). All except *T. b. gambiense* are found in wild animals, but very little is known regarding the course of infection or the immunological reactions in these natural hosts. There are some data to indicate a relatively benign course of infection with *T. congolense* in previously uninfected eland and waterbuck. In domestic animals and in humans, and also in experimental laboratory hosts such as rabbits or rats, the infections are characterized by a more or less prolonged relapsing chronicity. A population of trypanosomes develops; there is a trypanocidal crisis in which most are eliminated but some survive and then multiply to produce another population (Fig. 18.1). This in turn is mostly killed off but is followed by a third population increase, and so on. This continues until the host dies or until no further increase in trypanosomes occurs. Each successive population of trypanosomes differs antigenically from the preceding one, requiring the host to mount a series of immune responses. Infections initiated with a single trypanosome produce multiple variants. Over 100 have been shown for *T.*

equiperdum, a venereally transmitted parasite of horses closely related to the *T. brucei* group. This mechanism provides for a prolonged residence of parasites in a single host, thereby increasing the number of tsetse flies that can become infected. Survival of the host can result if the repertoire of variant antigens is exhausted before pathological effects of the relapsing parasitemias have become too severe. Or it may be that at least some hosts in addition to developing antibodies to the variant antigens are able to produce effective antibodies to some common but minor surface constituent, thus providing cross-variant protection. There are indications that such minor components exist but are ignored by the usual experimental hosts and perhaps also by the hosts, such as humans and cattle, to which these trypanosomes are highly pathogenic.

In these hosts the variant antigens are of overwhelming importance. They have been found to exist as a dense surface coat about 12–15 nm thick covering the plasma membrane of the bloodstream forms of the trypanosomes. The surface coat, and the variable antigen type, are lost during development in the fly but reappear on the metacyclic forms that are accordingly equipped for life in the vertebrate host. The surface coat can be readily removed by treatment with trypsin; organisms so treated are lysed by normal serum as a result of activation of the alternative complement pathway. The coat, constituting about 5–10% of the total protein of the organism, apparently consists of a single glycoprotein. The protein portion is capable of extensive modulation. What seems to happen is that a population with a particular variant surface glycoprotein (call it VSG-1) develops in a newly infected animal. There appear in the population, for reasons not known at present, very small numbers of trypanosomes having a different VSG, VSG-2. When antibody has developed to the dominant VSG-1, the trypanosomes with this VSG are destroyed, but not those with VSG-2. These now multiply until antibody to VSG-2 in turn destroys them. But meanwhile some organisms with VSG-3 have arisen. And so it goes. The immunofluorescent reaction with monospecific antiserum raised against a particular VSG is especially useful (Fig. 18.2) for identification of the antigenic type.

Careful biochemical work has shown that the carbohydrate content of the VSGs, though somewhat variable, is not responsible for their individuality. Rather this depends on an immense variation in their amino acid sequences. Despite this variation, however, the VSGs show considerable sequence homology, particularly in their C-terminal portion (Table 18.2).

Like the immunoglobulins, they comprise a family of related proteins. Also like the immunoglobulins, they have a flexible "hinge" region easily cleaved by proteolysis. They show, furthermore, a striking conservation of cysteine residues, especially of two groups of four in the carboxy portion of the molecule. Of the approximately 500 amino acids in a nascent VSG, about 20–30 are in a signal or leader peptide, about 350 in the relatively more variable N-terminal part, about 120 in the C-terminal part, and about 20 in the tail. Whereas the N-terminal portion of the molecule extends into the external

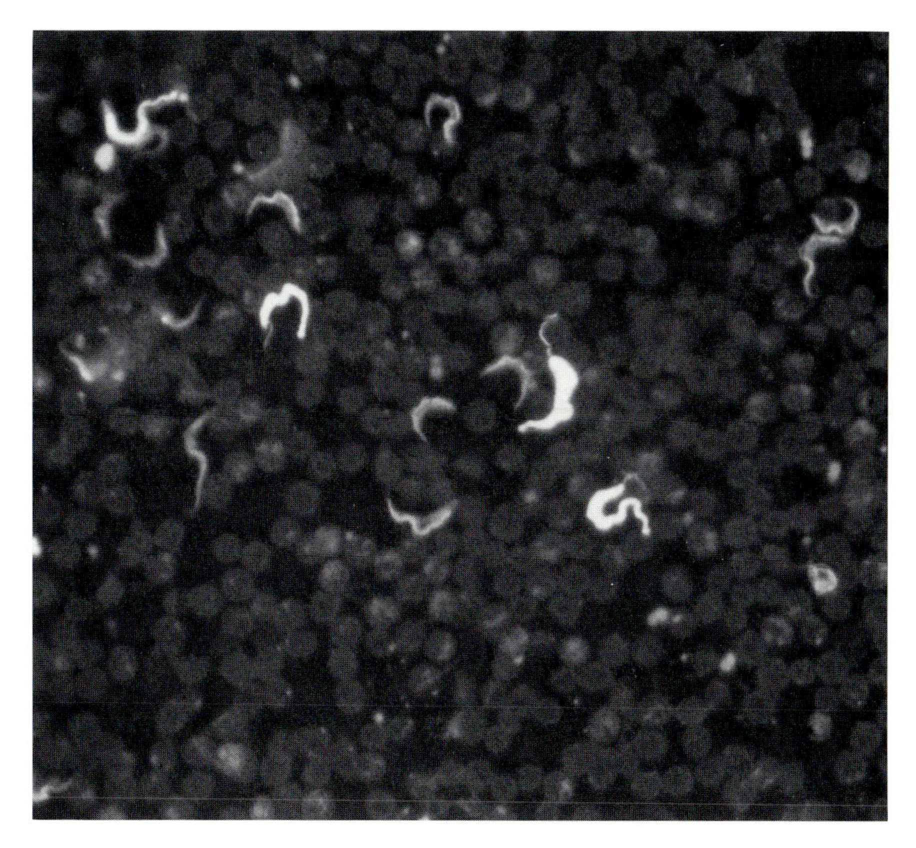

FIGURE 18.2. *T. brucei* in the blood of an infected mouse. The trypanosomes have been labeled by immunofluorescence using a monospecific antiserum against a single antigenic type (AnTat 1.7). Note the threadlike structures, plasmanemes, extending from several of the organisms. ×1200. (From Hajduk, 1984. Original print courtesy of Dr. S. L. Hajduk.)

medium, the C-terminal is bound to the plasma membrane through covalently linked myristic acid. This is the portion of the molecule that is cleaved, by a phospholipase C, to liberate the soluble VSG. One carbohydrate moiety, containing *N*-acetylglucosamine and mannose, is attached near the C-terminus; the other, containing in addition galactose and lipid, is at the extreme C-terminus. Present information would support the diagrammatic representation shown in Fig. 18.3.

We now know that there is a separate gene for each variant surface glycoprotein. This is a phenotypic phenomenon, a matter entirely of what controls the expression of one of this family of genes at one time. This is the big question for which there is yet no answer—what triggers expression of a particular VSG gene? But a great deal has been learned. First, it can be said that antibody is not the essential factor. With the development by H. Hirumi of a method for growing the bloodstream forms of *T. brucei* in culture, in

TABLE 18.2. Carboxy Termini of Mature VSGs[a,b]

MIT	1.1000BC	G K T G D K H N C A F R K G K D G K E E P E K E K C C D	
	118A	T P G K S A D C G F R K G K D G E T D E P D K E K C R N	***
	121	W E G E T C K D	***
	117A	D K C K G K L E D T C K K E S N C K W E N N A C K D	***
ANT	1.1	P A E K C T G K K K D D C K D G – – C K W E A E T C K D	
	1.8	T T D K C K G K L E D T C K K E S N C K W E G E T C K D	
ILT	1.3	K C K D K K K D D C K S P – D C K W E G E T C K D	
MIT	1.1006BC	R D R E K M E T T N T T A S	
	221	N T N T T G S S	***
	055	A E T N T T G S	***
	060	A N T T G S	***
ILT	1.1	A K R V A E Q A A T N Q E T E G K D G K T T N T T G S	
IOT	1.2	D K E E A K K L E E K T E Q N D S K T V T T N T T G S	
TXT	1	E E A A E N Q E G K K E K T S N T T A S	
TXT	5.28	T E K K D E K S	

[a] Modified from Cross (1984).
[b] Asterisked sequences were determined by amino acid sequencing. Other alignments were deduced by homology from DNA sequences.

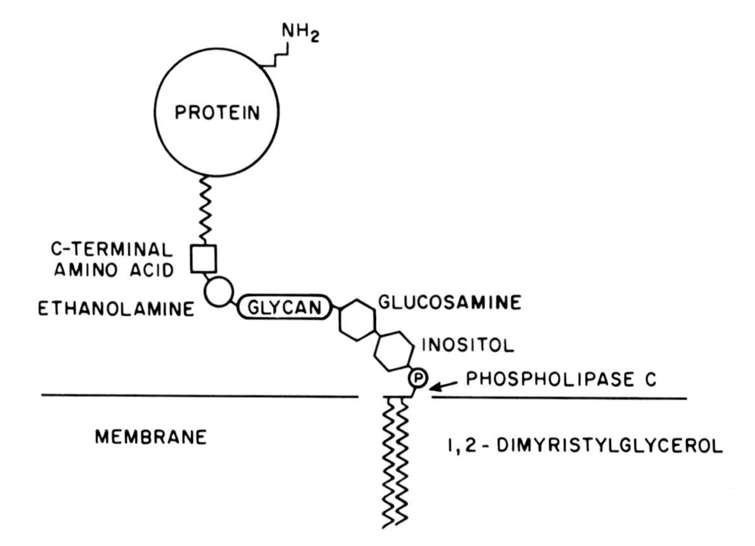

FIGURE 18.3. Diagrammatic representation of the *T. brucei* surface coat glycoprotein and its insertion into the membrane. When the trypanosomes are lysed, an endogenous phospholipase C cleaves the molecule at the inositol phospholipid bond. (Original print courtesy of Dr. G. A. M. Cross. See Low *et al.*, 1986.)

association with a feeder layer of fibroblasts, it became possible to show that antigenic variation occurred under these conditions in the complete absence of antibody. *In vitro,* however, the new variants appeared later than in an infected animal. This could result from a slower growth rate *in vitro* and from the fact that elimination of the initial variant by the host's immune response would make it easier to detect newly arisen variants *in vivo.* Eighteen clones of a single variable antigen type (VAT) 052 were maintained in culture and tested at weekly intervals by the immunofluorescent reaction of living trypanosomes with antisera against the purified glycoprotein of VAT 052 and of a VAT 221 that always appeared in the first relapse population of an animal infected with VAT 052. In 9 of the 18 clones, VAT 221 appeared within 18 to 46 days, and there were also present trypanosomes that did not react with either antiserum, indicating the presence of additional VATs.

As illustrated by these results, there is some order in the sequence of occurrence of the different VATs. This in itself argues strongly against the possibility that antigenic variation is a random mutational event, and in support of the idea that it depends on a phenotypic control of expression of a series of related genes. Application of the methods of recombinant DNA has proved beyond doubt that this is so. Indeed, in view of current interest in gene expression in eukaryotes and the lack of understanding of its control, antigenic variation in trypanosomes is now attracting much attention as a model system that may provide information of general significance with regard to genomic rearrangements in eukaryotes. Studies with recombinant probes prepared from the cDNA from mRNA of different VSGs have shown clearly that each VSG is the product of a unique gene. Some are found only in a single trypanosome clone, whereas others are found in a side range of trypanosomes. A most striking result is the finding of an additional distinct copy of a VSG gene in the nuclear DNA of a cell population currently expressing that particular VSG. This has given rise to the hypothesis that each VSG gene exists as a basic copy that is not expressed. For expression to occur, this is duplicated and transferred to an expression site near a telomere of a chromosome. Silent VSG genes that already exist at telomeres can be activated without duplication. Antigenic variation can occur by replacement of the VSG gene at the current active telomeric expression site or by switching on an alternative telomeric expression site. Telomeric VSG genes may be activated *in situ,* by duplicative transposition, or by telomeric translocations. In this connection it is important to realize that separation of individual chromosomes of trypanosomes by ordinary methods is not possible, as the chromosomes of these organisms apparently do not undergo condensation in any part of the cell cycle.

A recently developed method, however, now permits the electrophoretic separation of chromosome-sized DNA molecules (see also Chapter 10). Application of this method of pulsed field gradient electrophoresis to the nuclear DNA of *T. brucei* has already revealed four fractions: (1) a minichromosomal fraction of about 100 DNA molecules ranging in size from 50 to 150 kilobases (kb) (by other methods the size of one minichromosome bearing a VSG gene had been estimated to be 80 kb); (2) a set of about six discrete chromosomes of 200–700 kb; (3) at least three chromosomes of about 2 megabases (Mb); (4) a large DNA fraction that stays close to the origin. The amount of this can be roughly calculated by subtracting from the kinetic complexity of *T. brucei* nuclear DNA of 43×10^6 bp the sum of 6 Mb in the 2-Mb group, 2.2 Mb in the 200- to 700-kb group, and 7 Mb in the minichromosome group. This leaves 27.8 Mb in the large DNA. It is of interest that two species of hemoflagellates that do not have antigenic variation, *Crithidia fasciculata* and *Leishmania tropica,* do not show any minichromosomal fraction. This is in keeping with the idea that the minichromosomes of *T. brucei* provide a way to expand the telomeric VSG gene repertoire. They hybridize strongly to VSG gene probes.

The method of pulsed field gradient electrophoresis has also shown that the basic copy genes for two particular VSGs (118 and 221) are in the large DNA whereas their expression-linked copies are in the 2-Mb DNA. One of these genes (221) can also be activated without duplication and when this occurs it is found in the large DNA. Thus, at least two sites (and maybe many more) exist for the expression of VSG genes. The mechanism regulating which of several putative telomeric expression sites is activated remains to be discovered.

The ordering of the sequence in which VATs occur may eventually provide clues as to how their expression is controlled. In rabbits infected with *T. equiperdum,* the VATs appear in a loosely defined order of early, middle, and late ones. Invariably, however, infection of a naive rabbit with any VAT results in a first relapse population expressing VSG-1. This expression is accompanied by duplication of the telomeric basic copy gene and its transposition to an expression site near a telomeric structure. Metacyclic VSGs and those expressed early in infection appear always to be derived from telomeric basic copies. There are indications for the presence, between the VSG-1 gene and the terminus, of an unidentified nucleoside (not methylcytidine) replacing a large proportion of deoxycytidine. Such a nucleoside could play a role in determining the specificity of enzymes that carry out the duplication–transposition event leading to the expression-linked copy.

It is of special interest that the VSG genes are totally suppressed with the beginning of the developmental cycle of the trypanosomes in the tsetse fly (see Diagram V). The so-called procyclic trypomastigotes of *T. brucei* that develop in the midgut of a fly or under appropriate culture conditions at 28°C, have no surface coat and no VSG. If the bloodstream form from which they were derived had a VSG dependent on an expression-linked copy, it has been possible to show that this copy is still present in the procyclic forms even after 2 months in culture. There is at this time, however, no mRNA for this VSG, as there would be in the bloodstream form. Transcripts with sequences homologous to the VSG mRNA spliced leader were equally abundant in the procyclic and bloodstream forms. The genomic organization homologous to the VSG leader, however, showed in each case examined a specific difference between the procyclic and bloodstream forms; this may be related to VSG suppression in the procyclic. With the appearance of metacyclic forms in the fly salivary glands, the VSGs are again expressed, but there is a special group of VSGs regularly associated with the metacyclics and bearing no relation to the VSG that characterized the bloodstream forms on which the fly had fed. Following cyclical transmission, however, the first patent parasitemia shows an anamnestic switch in which the VSG that went into the fly is again expressed. If this VSG was dependent on an expression-linked copy, and since this copy persists in the procyclic stage and perhaps through the entire cycle in the fly, this might account for its early expression in the blood of the new host. This would also fit well with the idea that expression of the metacyclic VSGs is controlled by a system distinct from that operating in the bloodstream VSGs.

The bearing of these ideas and observations on the whole problem of triggering of the different stages in the life cycle of a parasite is obvious. Parasites must be able to respond rapidly and appropriately to a change in environment (see also Chapters 8–10). When a *T. brucei* bloodstream form suddenly finds itself in the midgut of a tsetse fly, not only does it cease to express any VSG gene, but it also activates and suppresses batteries of other genes, such as those controlling its metabolism (see Chapter 9). Conversely, in the metacyclic forms, lying in wait in the salivary glands to be injected into a new vertebrate host, those genes are again expressed that equip the organism for survival in this host. Among these are the VSG genes.

The genes for two metacyclic variable antigens (MVATs) of *T. brucei rhodesiense* have been characterized and found to be located near chromosomal telomeres, like the genes for bloodstream-form variable antigens (BVATs). There was no evidence for gene rearrangements associated with their expression. Of great interest is the finding that at the time when *T. b. rhodesiense* are switching from MVATs to BVATs, up to about 3% of individual trypanosomes may show two variable surface antigens simultaneously, one an MVAT and the other any one of several BVATs (Fig. 18.4). This switching occurs 5–7 days after inoculation of a mouse with metacyclic trypanosomes from salivary glands of a tsetse fly. Living trypanosomes taken from a heavily infected mouse at this time were exposed to direct immunofluorescence using two different monoclonal antibodies simultaneously. One monoclonal antibody was specific for an MVAT and was conjugated with rhodamine; the other was specific for a BVAT and was conjugated with fluorescein. Only on days 4 through 6, and especially on day 5 of the infection, were there present individual organisms that showed both red and green fluorescence (Table 18.3). Each fluorescence of these trypanosomes was less intense than that of organisms that showed only one or the other, suggesting the gradual replacement of one VSG by another.

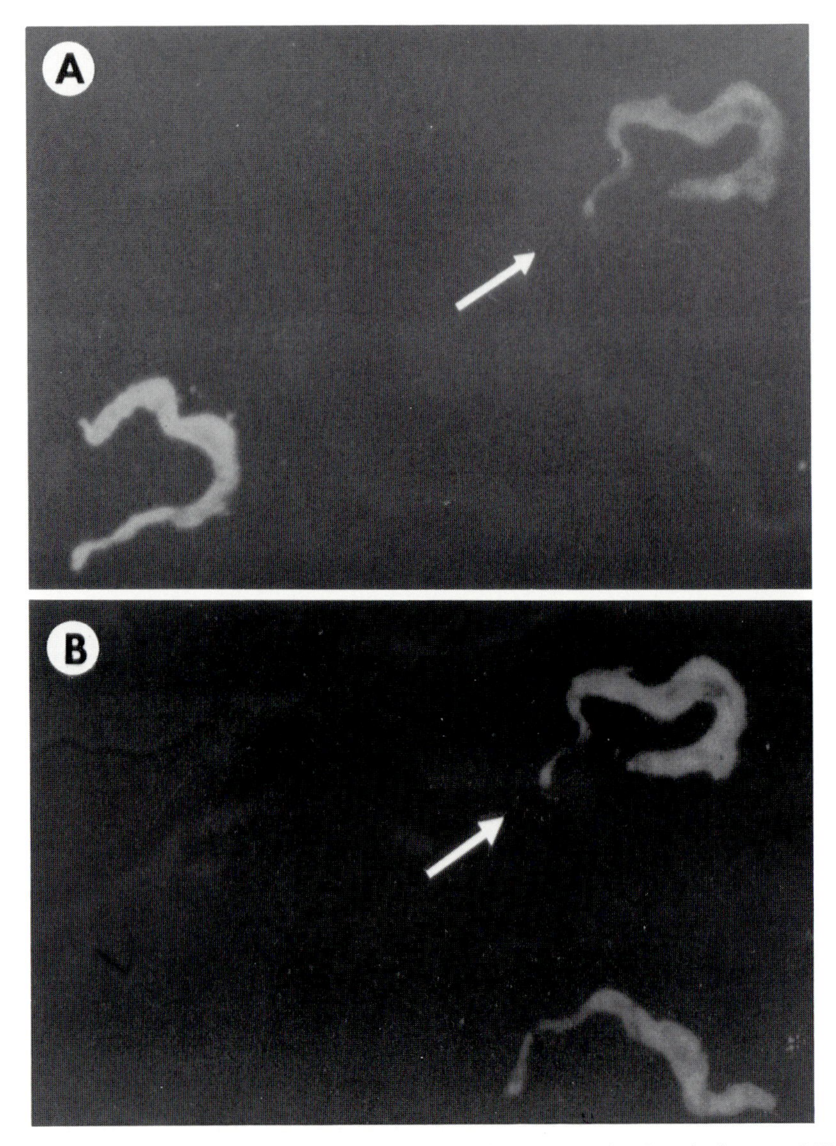

FIGURE 18.4. Reaction of individual trypanosomes with two monoclonal antibodies, one MVAT-specific (rhodamine-conjugated) and one BVAT-specific (fluorescein-conjugated). The panels are photographs of the same microscopic field taken through two different filter systems, one that detects rhodamine (A) and one that detects fluorescein (B). The trypanosome that appears only in A is an MVAT and fluoresces red (with MVAT-specific antibody 3.2C2.3); the trypanosome that appears only in B is a BVAT and fluoresces green (with BVAT-specific antibody 12.4F3.1). One trypanosome (marked by an arrow) in the process of switching from MVAT to BVAT is labeled with both antibodies and appears in both panels. Trypanosomes were isolated from blood collected from the tail vein of a mouse infected with metacyclic trypanosomes from WRATat 1.10-infected tsetse flies. Live trypanosomes obtained on day 5 of infection were diluted in phosphate-buffered saline (pH 7.2) containing 10% fetal bovine serum. Pairs of conjugated antibodies, one MVAT-specific and one BVAT-specific, were mixed with trypanosomes (on ice). After 5 min of incubation, wet mounts of this parasite–antibody mixture were examined for red or green fluorescence without prior washing. (From Esser and Schoenbechler, 1985. Original print courtesy of Dr. K. M. Esser.)

TABLE 18.3. *Trypanosoma brucei rhodesiense*: Simultaneous Presence of an MVAT and a BVAT on the Same Trypanosomes[a,b]

		BVAT specificity (%)[e]		
MVAT number[c]	Percentage of MVATs[d]	WRATat 1.1	WRATat 1.14	WRATat 1.10
		Day 4 of infection		
4	10	0	0	0
6	49	0.15	0.23	0
7	30	0.4	0.7	1.2
		Day 5 of infection		
4	5	0	3.7	0
6	33	1.5	1.7	0
7	16	3.9	5.0	0
		Day 6 of infection		
4	1	0	0	0
6	1	0	0	0
7	4	1.5	10.5	3.8

[a] Modified from Esser and Schoenbechler (1985).
[b] Percentage of MVAT-expressing trypanosomes that also expressed the indicated BVATs in a BALB/cJ mouse inoculated with metacyclic trypanosomes from tsetse flies infected with WRATat 1.10. Direct immunofluorescence reactions on live trypanosomes were carried out as described in the text for trypanosomes collected from one mouse on days 4 through 7 of infection. Similar results were obtained in other mice. No double-labeled trypanosomes were observed before day 4 or after day 6 of infection in any of the mice.
[c] All VATs are mutually exclusive, and each is defined by a monoclonal antibody: MVATs 4, 6, and 7 by antibodies 3.2C5.2, 3.2C2.2, and 3.103.1, respectively; and BVATs 1.1, 1.14, and 1.10 by antibodies 12.4F3.1, 21-14-146D, and 59-10-92J, respectively.
[d] Values indicate the percentage of trypanosomes expressing each MVAT at the time of assay for dual expression.
[e] Values are percentages of trypanosomes reactive with monoclonal antibodies specific for the indicated MVATs that also reacted with a monoclonal antibody specific for a BVAT. Percentages were calculated on the basis of single counts of 200 to 2000 trypanosomes. Values of zero indicate that the particular switch from MVAT to BVAT was not observed in samples of at least 2000 trypanosomes.

The interest in and utility of the VATs of the African trypanosomes in relation to the major biological problem of gene expression in eukaryotes is apparent. In addition, however, the VATs are of primary significance in relation to hopes for the development of vaccines against African sleeping sickness in humans and nagana and related trypanosomiases of cattle. If the trypanosomes can produce over 100 different antigenic types and if it is necessary to include all of them in the vaccine, the situation is clearly a difficult one. There are some indications for two possible ways around this difficulty. Since trypanosomes lose their surface coat during their early development in the tsetse fly, and then acquire a new one as metacyclic infective forms, it could be that the antigenic types of these forms have only a limited repertoire. Since these are the forms initially injected by the fly, a vaccine just against these might provide effective protection. In very recent work with *T. congolense* of cattle, made possible by the development of culture methods for the production of large numbers of metacyclic forms, Vickerman and his associates

TABLE 18.4. Percentage Labeling of *Trypanosoma congolense* Populations with Monoclonal Antibodies[a,b]

Hybridoma line	Trypanosome population							
	UM1 Ac.fix.	UM1 Living	CM1 Ac.fix.	CM2 Ac.fix.	CM3 Form.fix.	CM4 Ac.fix.	CFM Ac.fix.	UB1 Ac.fix.
GUPM 12.1	26	6	15	4	8	26	1/20	3
GUPM 12.2	15	18	2	12	5	12	3/15	26
GUPM 12.3	26	37	18	20	20	6	1/20	15
GUPM 12.4	10	5	2	12	20	4	4/20	18
GUPM 12.5	32	31	21	32	15	12	3/7	48
GUPM 12.6	12	2	20	21	8	6	3/20	2
GUPM 12.7	3	15	<1	<1	NL	<1	2/46	<1
GUPM 12.8	14	<1	2	10	NL	7	0/15	18
GUPM 12.9	<1	<1	NL	<1	NL	<1	5/30	<1
GUPM 12.10	45	20	22	30	31	52	1/15	26
GUPM 12.11	17	18	34	22	25	50	4/18	33
GUPM 12.12	20	8	10	18	10	10	1/17	8
Pooled GUPM 12.1–12.12	100	100	100	100	100	100	312/312	100

[a] Modified from Crowe *et al* (1983).
[b] All populations were acetone fixed (Ac.fix.) except trypanosomes from the CM3 metacyclic line which were formalin fixed (Form.fix.). Living metacyclics and an acetone-fixed preparation of the UM1 population were used. Results obtained from the small number of metacyclics extruded by infected tsetse flies (CFM) are presented as the number of metacyclics labeled by the mAb over the number counted. For all other populations, the percentage labeling was calculated on a count of at least 500 trypanosomes for each mAb, and a minimum of 2000 per population for the pool of mAbs. GUPM, Glasgow University Protozoology Monoclonal Antibody; NL, no labeling.

have shown that only 12 monoclonal antibodies will recognize all the mVATs from a single serodeme (A genotype) (see Table 18.4). Moreover, the same VATs present on the metacyclic forms were expressed in the initial populations of bloodstream trypanosomes at least through the 9th day in mice infected with these metacyclics. This suggests the possibility of a vaccine based on the limited number of mVATs. Such a vaccine, however, could only be locally effective since it has already been found that other serodemes show entirely different mVATs. The number of serodemes of *T. congolense* in nature is not known. Only if it is a small number would a vaccine based on a mixture of mVATs of all the serodemes be feasible. With *T. b. gambiense* and *rhodesiense,* the subspecies infective to humans, the number of serodemes does seem to be limited. Here, however, a different complication has appeared. The mVAT repertoire of *T. b. rhodesiense* has been found to be unstable. Isolates made in the same region of Africa over a 20-year period and shown to belong to the same serodeme expressed different mVATs. Furthermore, during ten sequential experimental fly transmissions, one mVAT was lost and two others were acquired.

The second possible way to circumvent the variant antigens is to try to immunize to a common surface antigen that is ordinarily obscured by the immunodominant variant glycoprotein. There are indications that such common accessible antigens do exist on the surface of *T. brucei* as minor components apparently ignored by the host. If these could be identified and purified, they could perhaps be used to produce an immunity not dependent on VAT. These antigens deserve at least as much attention as has been given to the VSGs.

Bibliography

Barry, J. D., Crowe, J. S., and Vickerman, K., 1983, Instability of the *Trypanosoma brucei rhodesiense* metacyclic variable antigen repertoire, *Nature* **306:**699–701.

Beat, D. A., Stanley, H. A., Choromanski, L., MacDonald, A. B., and Honigberg, B. M., 1984, Nonvariant antigens limited to bloodstream forms of *Trypanosoma brucei brucei* and *Trypanosoma brucei rhodesiense, J. Protozool.* **31:**541–548.

Brown, D. D., 1981, Gene expression in eukaryotes, *Science* **211:**667–674.

Buck, G. A., Longacre, S., Riband, A., Hibner, U., Giroud, C., Balty, T., Balty, D., and Eisen, H., 1984, Stability of expression-linked surface antigen gene in *Trypanosoma equiperdum, Nature* **307:**563–566.

Clarkson, A. B., Jr., and Mellow, G. H., 1981, Rheumatoid factor-like immunoglobulin M protects previously uninfected rat pups and dams from *Trypanosoma lewisi, Science* **214:**186–188.

Cohen, C., Reinhardt, B., Parry, D. A. D., Roelants, G. E., Hirsch, W., and Kanwé, B., 1984, α-Helical coiled-coil structures of *Trypanosoma brucei* variable surface glycoproteins, *Nature* **311:**169–171.

Cross, G. A. M., 1978, Antigenic variation in trypanosomes, *Proc. R. Soc. London Ser. B* **202:**55–72.

Cross, G. A. M., 1984, Structure of the variant glycoproteins and surface coat of *Trypanosoma brucei, Philos. Trans. R. Soc. London Ser. B* **307:**3–12.

Crowe, J. S., Barry, J. D., Luckins, A. G., Ross, C. A., and Vickerman, K., 1983, All metacyclic variable antigen types of *Trypanosoma congolense* identified using monoclonal antibodies, *Nature* **306:**389–391.

D'Alesandro, P. A., 1975, Ablastin: The phenomenon, *Exp. Parasitol.* **38:**303–308.

Delauw, M. -F., Pays, E., Steinert, M., Aerts, D., van Meirvenne, N., and Le Ray, D., 1985, Inactivation and reactivation of a variant-specific antigen gene in cyclically transmitted *Trypanosoma brucei, EMBO J.* **4:**989–993.

Doyle, J. J., Hirumi, H., Hirumi, K., Lupton, E. N., and Cross, G. A. M., 1980, Antigenic variation in clones of animal-infective *Trypanosoma brucei* derived and maintained *in vitro, Parasitology* **80:**359–369.

Esser, K. M., and Schoenbechler, M. J., 1985, Expression of two variant surface glycoproteins on individual African trypanosomes during antigen switching, *Science* **229:**190–193.

Ferguson, M. A. J., Haldar, K., and Cross, G. A. M., 1985, *Trypanosoma brucei* variant surface glycoprotein has a sn-1,2-dimyristyl glycerol membrane anchor at its COOH terminus, *J. Biol. Chem.* **260:**4963–4968.

Freymann, D. M., Metcalf, P., Turner, M., and Wiley, D. C., 1984, 6 Å-resolution X-ray structure of a variable surface glycoprotein from *Trypanosoma brucei, Nature* **311:**167–169.

Giannini, S. H., and D'Alesandro, P. A., 1982, Trypanostatic activity of rat IgG purified from the surface coat of *Trypanosoma lewisi, J. Parasitol.* **68:**765–773.

Giannini, S. H., and D'Alesandro, P. A., 1984, Isolation of protective antigens from *Trypanosoma lewisi* by using trypanostatic (ablastic) immunoglobulin G from the surface coat, *Infect. Immun.* **43:**617–621.

Gray, A. R., 1977, Antigenic variation in trypanosomes with particular reference to *Trypanosoma gambiense*, *Ann. Soc. Belge Med. Trop.* **57:**403–408.

Hajduk, S. L., 1984, Antigenic variation during the developmental cycle of *Trypanosoma brucei*, *J. Protozool.* **31:**41–47.

Hajduk, S. L., and Vickerman, K., 1981, Antigenic variation in cyclically transmitted *Trypanosoma brucei:* Variable antigen type composition of the first parasitaemia in mice bitten by trypanosome-infected *Glossina morsitans*, *Parasitology* **83:**609–621.

Hajduk, S. L., Cameron, C. R., Barry, J. D., and Vickerman, K., 1981, Antigenic variation in cyclically transmitted *Trypanosoma brucei:* Variable antigen type composition of metacyclic trypanosome populations from the salivary glands of *Glossina morsitans*, *Parasitology* **83:**595–607.

Hirumi, H., Hirumi, K., Doyle, J. J., and Cross, G. A. M., 1980, *In vitro* cloning of animal-infective bloodstream forms of *Trypanosoma brucei*, *Parasitology* **80:**371–382.

Jones, T. W., and Clarkson, M. J., 1974, The timing of antigenic variation in *Trypanosoma vivax*, *Ann. Trop. Med. Parisitol.* **68:**485–486.

Jones, T. W., Cunningham, I., Taylor, A. M., and Gray, A. R., 1981, The use of culture-derived metacyclic trypanosomes in studies on the serological relationships of stocks of *Trypanosoma brucei gambiense*, *Trans. R. Soc. Trop. Med. Hyg.* **75:**560–565.

Lenardo, M. J., Rice-Ficht, A. C., Kelly, G., Esser, K. M., and Donelson, J. E., 1984, Characterization of the genes specifying two metacyclic variable antigen types in *Trypanosoma brucei rhodesiense*, *Proc. Natl. Acad. Sci. USA* **81:**6642–6646.

Low, M., Ferguson, M. A. J., Futerman, A. H., and Silman, I., 1986, Covalently attached phosphatidyl inositol as a hydrophobic anchor for membrane proteins, *Trends in Biochem. Sci.*, in press.

Mellow, G. H., and Clarkson, A. B., Jr., 1982, *Trypanosoma lewisi:* Enhanced resistance in naive lactating rats and their suckling pups, *Exp. Parasitol.* **53:**217–228.

Murray, M., Grootenhus, J. G., Akol, G. W. O., Emergy, D. L., Shapiro, S. Z., Moloo, S. K., Dar, F., Bovell, D. L., and Paris, J., 1981, Potential application of research on African trypanosomiasis in wildlife and preliminary studies on animals exposed to tsetse infected with *Trypanosoma congolense*, in: *Wildlife Disease Research and Economic Development*, (L. Karstad, B. Nestel, and M. J. Graham, eds.), International Development Research Centre, Ottawa, Canada.

Olafson, R. W., Clarke, M. W., Kielland, S. L., Pearson, T. W., Barbet, A. F., and McGuire, T. C., 1984, Amino terminal sequence homology among variant surface glycoproteins of African trypanosomes, *Mol. Biochem. Parasitol.* **12:**287–298.

Parsons, M., Nelson, R. G., Stuart, K., and Agabian, N., 1983, Genomic organization of variant surface glycoprotein genes in *Trypanosoma brucei* procyclic culture forms, *J. Cell. Biochem.* **23:**27–33.

Parsons, M., Nelson, R. G., Stuart, K., and Agabian, N., 1984, Variant antigen genes of *Trypanosoma brucei:* Genomic alteration of a spliced leader orphon and retention of expression-linked copies during differentiation, *Proc. Natl. Acad. Sci. USA* **81:**684–688.

Raibaud, A., Gaillard, C., Longacre, S., Hibner, U., Buck, G., Bernardi, G., and Eisen, H., 1983, Genomic environment of variant surface antigen genes of *Trypanosoma equiperdum*, *Proc. Natl. Acad. Sci. USA* **80:**4306–4310.

Seed, J. R., 1984, Antigenic variation in trypanosomes, *J. Protozool.* **31:**41.

Sloof, P., Menke, H. H., Caspers, M. P. M., and Borst, P., 1983, Size fractionation of *Trypanosoma brucei* DNA: Localization of the 177-bp repeat satellite DNA and a variant surface glycoprotein gene in a mini-chromosomal DNA fraction, *Nucleic Acids Res.* **11:**3889–3901.

Strickler, J. E., Mancini, P. E., and Patton, C. L., 1978, *Trypanosoma brucei brucei:* Isolation of the major surface coat glycoprotein by lectin affinity chromatography, *Exp. Parasitol.* **46:**262–276.

Taliaferro, W. H., 1932, Trypanocidal and reproduction-inhibiting antibodies to *Trypanosoma lewisi* in rats and rabbits, *Am. J. Hyg.* **16:**32–84.

Turner, M. J., 1982, Biochemistry of the variant surface glycoproteins of salivarian trypanosomes, *Adv. Parasitol.* **21:**69–153.

Turner, M. J., 1984, Antigenic variation in its biological context, *Philos. Trans. R. Soc. London Ser. B* **307**:27–40.

Van der Ploeg, L. H. T., Cornelissen, A. W. C. A., Michels, P. A. M., and Borst, P., 1984, Chromosome rearrangements in *Trypanosoma brucei, Cell* **39**:213–221.

Van der Ploeg, L. H. T., Schwartz, D. C., Cantor, C. R., and Borst, P., 1984, Antigenic variation in *Trypanosoma brucei* analyzed by electrophoretic separation of chromosome-sized DNA molecules, *Cell* **37**:77–84.

Vickerman, K., and Barry, D., 1982, African trypanosomes, in: *Immunology of Parasitic Infections* (S. Cohen and K. Warren, eds.), Blackwell, Oxford, pp. 209–260.

CHAPTER 19

Immunity to Malaria and Related Intracellular Protozoa

Human malaria is a disease characterized by its long duration and chronicity. A typical infection with *Plasmodium vivax* lasts 2 to 3 years with periodic remissions and relapses. There is now good evidence to indicate that the relapses, whose frequency varies with the strain of parasite, originate from dormant forms, called hypnozoites. These are derived from inoculated sporozoites and lie in hepatic cells until somehow stimulated to develop into preerythrocytic schizonts, which then produce merozoites that invade red blood cells and initiate the relapse. Nothing is known as to whether the parasites of successive relapses differ antigenically. Immunity develops only slowly. The relapses, after a time, tend to be less severe clinically, even though more parasites may be present. Such a partially immune person may be walking around with a parasitemia of 50,000 per μl blood, whereas a nonimmune individual will be extremely ill with so few parasites (10 or even less per μl) that diagnosis by slide examination may be difficult. *P. ovale* infections resemble those with *P. vivax* and may have even longer latent periods. With the quartan parasite *P. malariae,* latent inapparent infection has been known to persist for 50 years. This has been shown by the accidental infection with this species of recipients of blood transfusions from donors who had left a malarious region 50 years earlier and had never been back. In such people the parasite and host are in a state of equilibrium. This may resemble the situation with experimental infection with *P. inui* in the rhesus monkey, where the spleen seems to play a positive role in maintaining the parasite (see Chapter 15).

P. falciparum is the species of human malaria that is most highly pathogenic and often fatal. It is responsible for a high infant mortality, about a million deaths of children each year in Africa alone. Even to this acute infection immunity develops only slowly. Under natural conditions, where exposure to reinfection is frequent and sustained, children who survive show an enlarged spleen and are likely to continue to show parasites until puberty or even beyond. The number of parasites present, the frequency of their presence, and the severity of their effects become progressively less. In such a holoendemic region, adults generally do not have a markedly enlarged spleen, rarely show any parasites, and are rarely ill with malaria. They are immune. This acquired immunity seems to be highly effective only against the local strains of falciparum malaria and to fade quite rapidly if the individual lives for some months in a nonmalarious region.

The natural history of malaria is thus very different from that of some viral diseases, such as smallpox or yellow fever, where recovery from the initial acute disease confers a long-lasting immunity and vaccination is dramatically effective. On the other hand, effective acquired immunity has been repeatedly demonstrated in a wide variety of experimental model systems using malarial parasites infective to convenient laboratory hosts: avian malaria in chickens, rodent malaria in mice and rats, and simian malaria in rhesus monkeys.

The model systems most intensively studied do not provide a close approximation to the situation in human malaria. There are available model systems that could more nearly do so. For example, *P. cynomolgi* in the rhesus monkey is very similar to *P. vivax* in humans, and *P. fragile* and *P. coatneyi* in the rhesus both resemble *P. falciparum* in humans. But these have been relatively little used for immunological studies. Most of the basic work of the past 25 years has been done with rodent malaria, mainly *P. berghei,* in mice and rats, and to a lesser extent with *P. knowlesi* in the rhesus *Macaca mulatta.* These are not natural hosts for these parasites. Various wild African rodents, such as those of the genus *Thamnomys,* are the natural hosts for *P. berghei.* The kra monkey or crab-eating macaque (*Macaca irus*) is the natural host for *P. knowlesi.* In this monkey, *P. knowlesi* produces relatively mild infections quite different from the fulminating, usually fatal infections it produces in the rhesus. Nevertheless, a large body of important information has been derived from these studies with *P. berghei* in mice and *P. knowlesi* in rhesus monkeys. In addition, two models have been developed for work directly with the human parasite *P. falciparum.* These are the New World monkeys *Aotus trivirgatus* (the owl or night monkey) and *Saimiri sciureus* (the squirrel monkey). Martin Young first showed that *Aotus* were susceptible to the human parasite *P. vivax* and soon after Quentin Geiman and Wasim Siddiqui demonstrated that *P. falciparum* could be adapted to produce intense infections in intact *Aotus.* In *Saimiri* on the other hand, infections tend to be lighter and somewhat more variable in intact monkeys, reaching high parasitemias only in splenectomized animals.

It is not surprising that the results from these varied model systems have not always been congruent. They do, however, permit certain generalizations to be made and they point to gaps in knowledge and to directions for further work. This has all acquired special importance and special relevance since it has become clear that existing methods for malaria control are inadequate and basis has developed for the hope that a vaccine for malaria could provide an additional effective weapon against it. I will try to summarize briefly where matters stand with regard both to acquired immunity to malaria and to the possibility for a successful vaccine.

In acquired immunity to malaria, both humoral and cell-mediated processes are involved, the relative importance of each varying with the particular parasite–host system under study. Antibodies are clearly important. Passive transfer of immunity with serum alone has been shown for many experimental malarias. It was also strikingly demonstrated for human falciparum malaria in the seminal experiment of Sidney Cohen, Ian McGregor, and S. P. Carrington, who in 1961 showed that gamma-globulin concentrates from adult West Africans would reduce the parasitemia when administered to children with *P. falciparum* infection (Fig. 19.1). Serum antibodies that confer such passive protection against the erythrocytic parasites seem to be directed against the merozoites, the invasive forms capable of only a short period of extracellular life. Many other kinds of antibodies to the erythrocytic stages are demonstrable by immunofluorescent reactions. In human serum these are found in the IgM and IgA fractions as well as in the IgG. Most of them probably have little relationship to the protective effect of a serum.

Antibodies also develop against malarial sporozoites. At least in certain experimental models they provide a completely effective immunity against a further inoculation with sporozoites of the same species. In mice an intra-

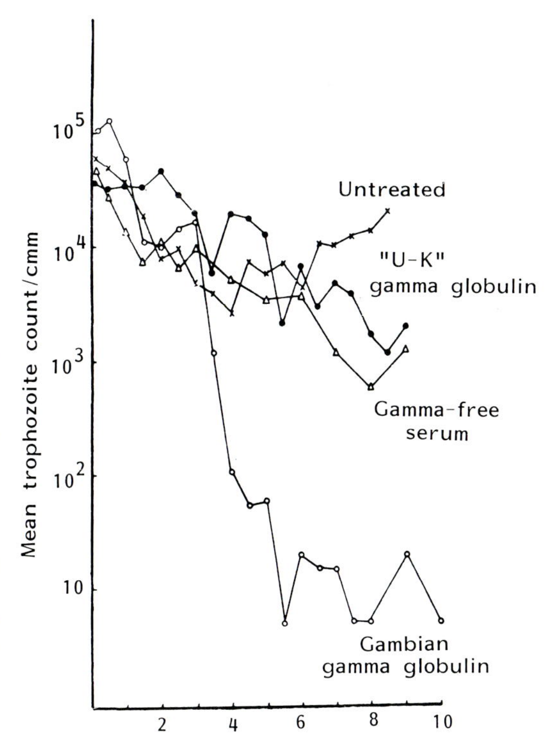

FIGURE 19.1. Passive transfer of immunity to *Plasmodium falciparum* in children. Mean trophozoite counts (logarithmic scale) in four groups of children with acute *P. falciparum* malaria: Untreated (2); "U-K" gamma globulin (4) given gamma globulins from sera of blood donors in Great Britain; Gamma-free serum (7) given Gambian serum after removal of the bulk of the 7 S gamma globulin (but still containing 7% of 7 S and 2% of 19 S gamma globulin); Gambian gamma globulin (12) given 7 S gamma globulin prepared from Gambian adults. All injections were given intramuscularly. (From Cohen *et al.*, 1961.)

venous injection of sporozoites of *P. berghei* inactivated with an appropriate dose of X-rays confers immunity to later injection of live sporozoites, but not to inoculation of erythrocytic forms of *P. berghei*. The immunity is stage specific. R. S. Nussenzweig and her colleagues furthermore have shown that the sporozoites have a single immunodominant surface antigen of 44,000 daltons (PB 44) and that a monoclonal antibody to this protein produced in mice will confer passive immunity to an otherwise lethal dose of sporozoites. With the recent development by M. Hollingdale of a tissue culture system in which sporozoites of *P. berghei* invade hepatic cells and develop through the preerythrocytic cycle into infective merozoites, it was possible to show that the anti-PB 44 antibody interferes with invasion by the sporozoites. Its action is thus analogous to the antibodies that interfere with merozoite invasion into erythrocytes. Antisporozoite antibodies involving surface proteins similar to PB 44 have also been demonstrated for the monkey malaria *P. knowlesi* and for the human parasites *P. falciparum* and *P. vivax*. The antisporozoite antibodies are demonstrable *in vitro* not only by their immunofluorescent reaction but also by the formation of a characteristic precipitate particularly around the posterior portion of the sporozoite, the so-called circumsporozoite precipitate (CSP) (Fig. 19.2). Such sporozoites are incapable of producing infection if injected into a susceptible host.

The development of these antisporozoite antibodies is thymus dependent, at least in mice. Homozygous nude mice (*nu/nu*) neither formed antisporozoite antibodies nor were protected by repeated intravenous injections of irradiated sporozoites of *P. berghei*, whereas their *nu/+* littermates showed both responses. Similarly, mice that had been thymectomized, irradiated, and bone marrow-reconstituted showed neither response. Further reconstitution

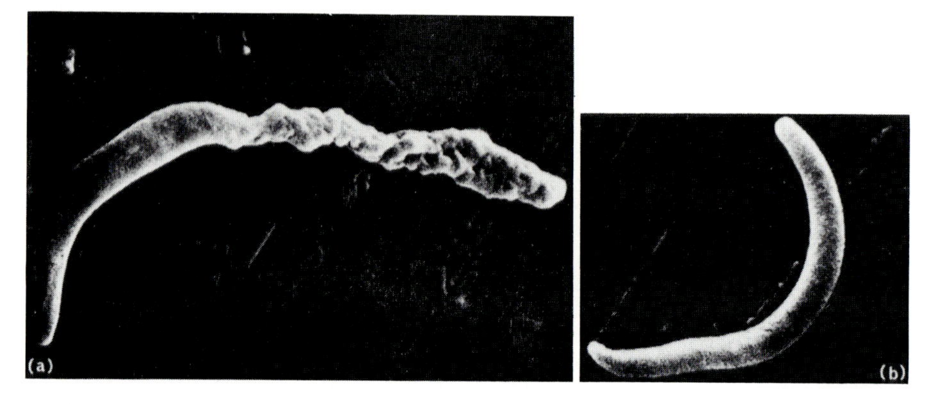

FIGURE 19.2. Alteration of sporozoites of *P. berghei* following immune serum incubation. (a) Sporozoite incubated in immune serum. The body of the parasite appears smooth. The irregular configuration of the CSP reaction extends a considerable distance posteriorly. (b) Sporozoite incubated in normal serum. Note the smooth surface of the parasite. The anterior end of the sporozoite is narrow, and can be clearly distinguished from the rounder posterior end. SEM. ×9000. (From Nussenzweig *et al.*, 1978. Original print courtesy of Dr. A. H. Cochrane.)

with thymocytes restored both the capacity to develop antisporozoite antibodies and to resist challenge with live infective sporozoites. Hence, while antibodies are here the principal effector mechanism against the sporozoites, T cells are essential to their formation.

Cell-mediated immunity plays many other roles in malaria. The major importance of cells of the lymphoid–macrophage system in both innate and acquired resistance has long been evident. One need only note the extensive hyperplasia of the liver and especially of the spleen (see Chapter 15). Macrophages perfused from the liver of mice infected with malaria are in a highly activated state. This probably results from the activities of T cells. In some model systems and particularly in rodent malaria, T-cell-dependent immunity against the erythrocytic stages is more important than antibody-dependent immunity. Thus, T-cell-deprived or nude mice cannot control and recover from infection with *P. yoelii*, a species of malaria not lethal to normal mice. That T cells are also important in primate malaria is indicated by the need for Freund's complete adjuvant in order to get successful vaccination against erythrocytic stages of *P. knowlesi* in rhesus monkeys or *P. falciparum* in owl monkeys. T cells could act, and probably do, by stimulating macrophages to ingest infected, and perhaps also uninfected, erythrocytes. They could increase the activity of NK cells. The NK cells in turn could exert antiparasitic effects through lymphokine mediators. Macrophages may secrete a number of products directly toxic to the parasites. Among these would be oxidative enzymes, such as glucose oxidase and polyamine oxidase. These would generate H_2O_2, superoxide and hydroxyl radicals, all shown to be toxic to malarial parasites *in vitro*. A macrophage product of special interest is the so-called tumor necrosis factor (TNF). This can be produced by endotoxin stimulation of animals previously injected with BCG (a tubercle bacillus preparation). TNF has been shown to be toxic to malarial parasites both *in vivo* and *in vitro* (Fig. 19.3). Although there is some evidence that the same polypeptide may be involved in the activity against tumor cells and against malaria, purified TNF has not been found to have any action on falciparum parasites. The interesting, and reasonable, suggestion has been made by J. H. L. Playfair that TNF evolved in the first place in connection with resistance to malaria rather than to tumors. Falciparum malaria, as a killer of children, would be much more likely to exert a selective effect to this end than would tumors. Whether this is really so remains to be seen.

In acute experimental malarial infections in animals, a crisis occurs which may be followed by either death or recovery. At this time abnormal forms of parasites are seen. These have been called "crisis forms" (Fig. 19.4) and are thought to be degenerating parasites that have been affected by some toxic conditions, perhaps resulting from the severe parasitemia. Moreover, it was early noted by W. H. Taliaferro that, in monkeys infected with *P. brasilianum*, as the infection progressed the average number of merozoites produced by a schizont decreased, decreasing the rate of reproduction (Fig. 19.5). Neither of these phenomena has been seen in human malaria *in vivo*, but it could

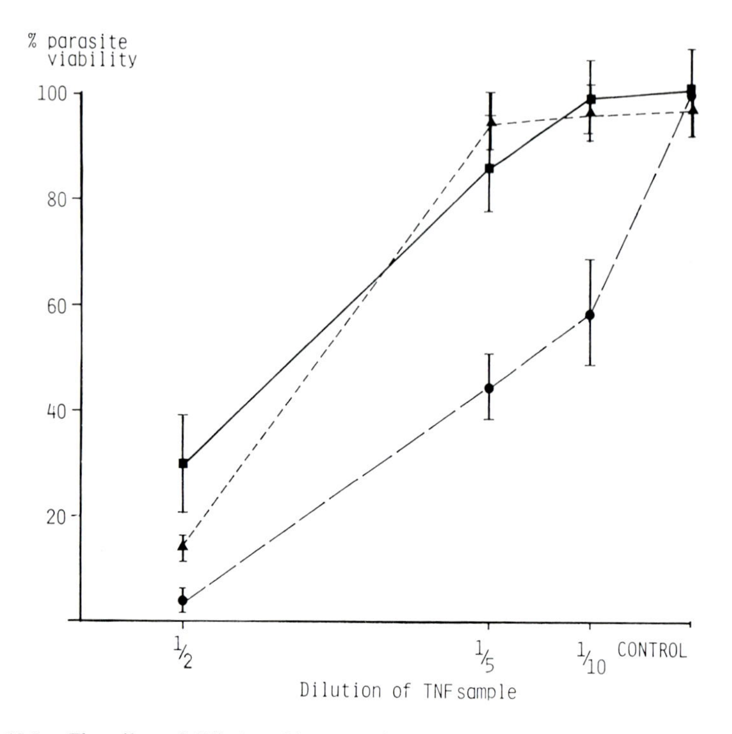

FIGURE 19.3. The effect of differing dilutions of TNF samples on the viability of *P. falciparum*, assessed morphologically after 24 hr. ●–●, human monocyte supernatant; ▲-----▲, mouse macrophage supernatant; ■—■, rabbit serum. 100% viability was equivalent to approximately 4–5% level of parasitemia. Each well contained a final volume of 50 μl at a hematocrit of 3%. (From Wozencraft *et al.*, 1984. Original print courtesy of Dr. A. O. Wozencraft.)

well be that appropriate material for their detection has not been examined. Indeed, in cultures *in vitro* certain human sera from a holoendemic region in the Sudan have been shown by J. B. Jensen to inhibit intracellular growth of *P. falciparum* and to produce abnormal forms resembling crisis forms (Table 19.1). There is no direct evidence that either these, or the crisis forms in experimental malarias, or the intracellular growth inhibition of *P. brasilianum*, are caused by TNF or by oxidative injury. Although preparations with TNF inhibit parasite growth *in vitro,* the inhibitory Sudanese sera were found not to exhibit TNF activity. There are a number of other possible mechanisms by which malarial parasites could be injured while in the red cell. For example, antibodies to transport proteins inserted in the red cell membrane but of parasite origin could interfere with movements of nutrients in or with escape of toxic wastes such as lactic acid. Oxidative mechanisms, as already noted (Chapter 14), almost surely play a role. Such oxidant-mediated cytotoxicity is well known for human monocytes activated by γ-interferon. In *in vitro* experiments with cultured *P. falciparum* the effect of the macrophages against

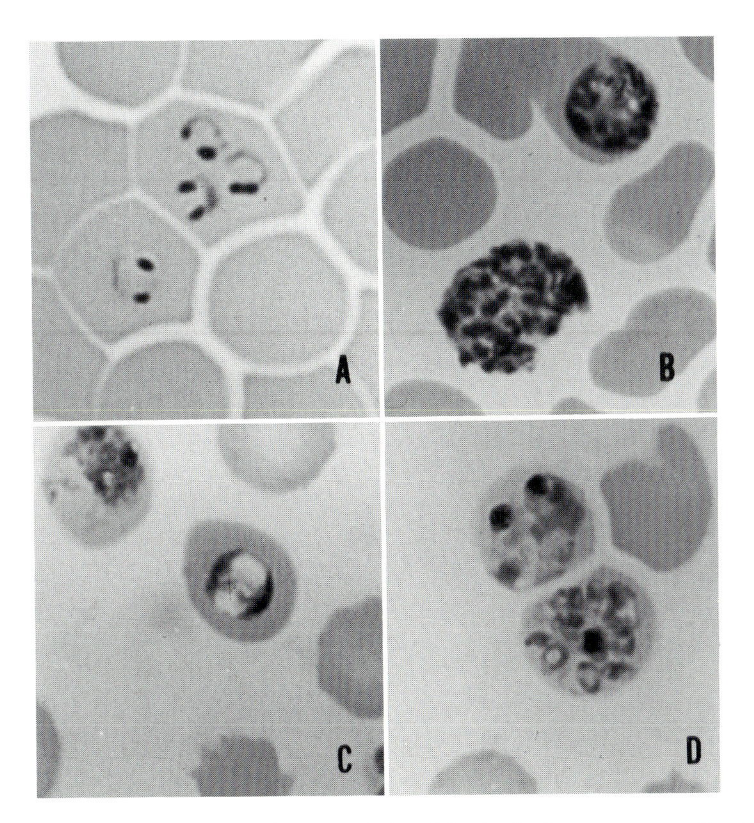

FIGURE 19.4. "Crisis forms" of *P. falciparum* induced *in vitro* by immune serum. Synchronized ring-stage cultures (A) are exposed to malaria-immune, or nonimmune serum in RPMI 1640 medium for 40–44 hr at 37°C. Parasites exposed to nonimmune serum mature to schizonts having 16–24 nuclei, or merozoites (B). Parasites exposed to immune serum either fail to mature, becoming pyknotic and karyorrhexic (C); or form schizonts having vacuolated cytoplasm and a greatly reduced number of nuclei, or merozoites (D). (Original prints courtesy of Dr. J. B. Jensen. See Jensen *et al.*, 1984.)

the parasites was closely connected with the release of H_2O_2 by the macrophages (Table 19.2). In these cultures, degenerate forms with pyknotic nuclei appeared that very much resembled crisis forms. The parasiticidal effect, however, was not wholly a result of oxidative mechanisms for activated macrophages from a patient with chronic granulomatous disease (CGD) and hence defective in oxidative metabolism, still had a partial inhibitory effect. Human neutrophils *in vitro* show similar inhibitory effects on parasite growth. Although these effects were enhanced by treatment of the neutrophils with phorbol myristate acetate, the inhibition was as great with neutrophils from two patients with CGD as it was with neutrophils from their mothers or from normal controls. Since the neutrophils from the CGD patients were totally

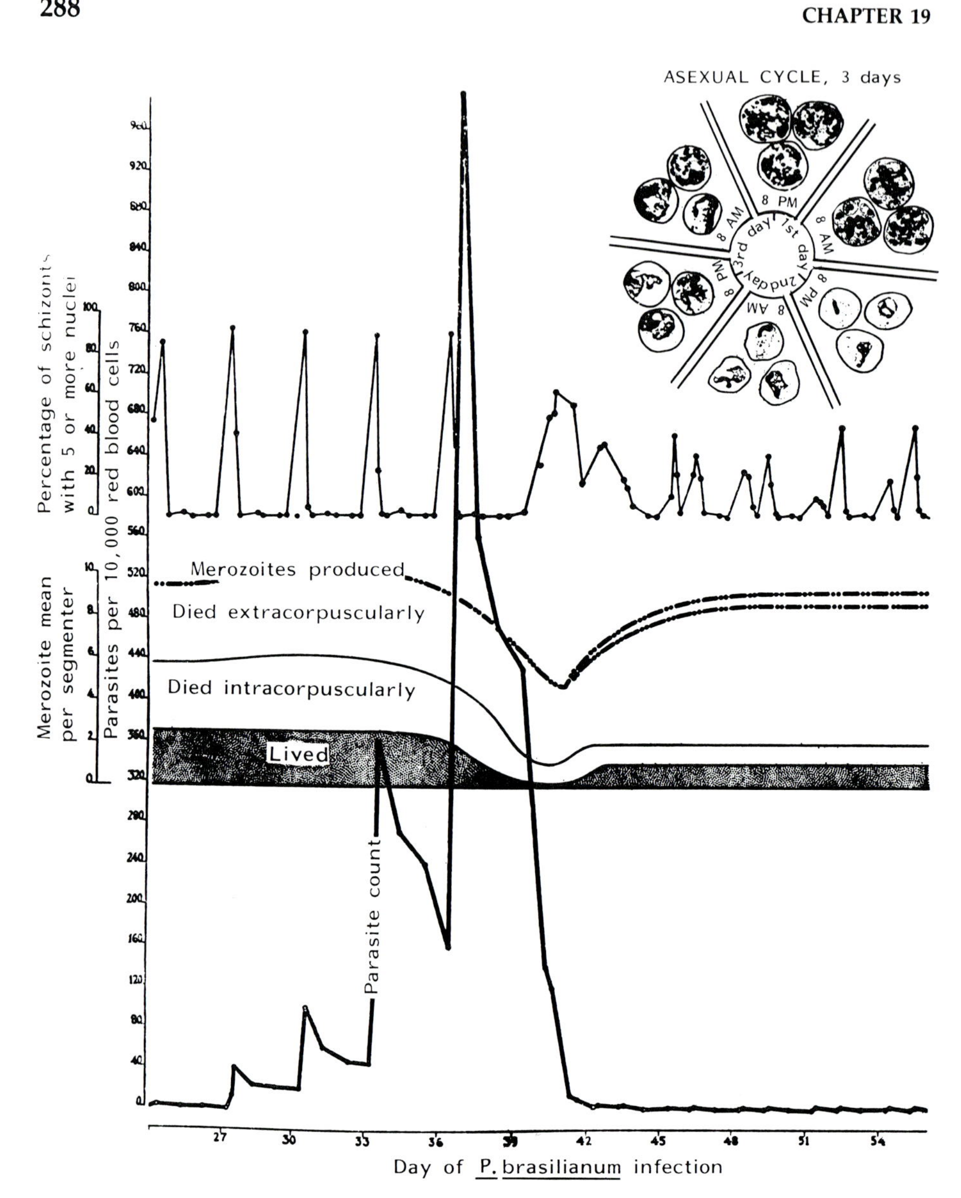

FIGURE 19.5. The rate of reproduction and death of parasites during the course of the parasitemia of *P. brasilianum* in a cebus monkey. The asexual cycle is shown in the upper right corner. The rate of reproduction is indicated by the occurrence of schizonts with five or more nuclei in conjuction with the merozoite mean per segmenter (merozoites produced). The number of merozoites which live and develop are computed from the net rate of rise in the parasitemia for each segmentation. The merozoites which die can be divided between those which die intra- and extra-corpuscularly.

Note that a drop in the parasitemia (crisis, 39th to 42nd day) occurs when less than one merozoite per segmenter lives and completes its development and that a static population occurs (after the 42nd day) when approximately one merozoite per segmenter lives and completes its development. (From Taliaferro, 1948.)

TABLE 19.1. Growth-Inhibiting Effect of Sera from a Highly Malarious Region of the Sudan on *P. falciparum in Vitro*[a–c]

Serum	1/IFA	% [^{3}H]-Hyp incorporation		% parasite stage distribution				% of control parasitemia
		48 hr	36 hr	AB	R	T	Sc	
Nonimmune	0	100	100	3	3	62	32	100
S-82-607	0	17	37	38	14	48	0	99
S-82-912	640	5	13	27	33	40	0	93
S-82-921	320	23	31	19	22	56	3	94
S-82-922	320	2	7	37	9	50	4	84
S-82-924	640	16	24	13	33	54	0	72
S-82-1000	80	8	15	14	16	70	0	100

[a] Modified from Jensen *et al.* (1984).
[b] All sera were dialyzed against culture medium to remove any possible drugs, None of the serum donors showed splenomegaly.
[c] The cultures were synchronized to within a 6-hr period and were started as rings for the 36-hr experiment or as schizonts for the 48-hr experiment. Note that the immunofluorescent antibody titers (1/IFA) were low to moderate and that there was relatively little effect on merozoite invasion as shown by the last column (% of control parasitemia) after 48 hr—schizont to schizont. The incorporation of hypoxanthine (%[^{3}H]-Hyp) was, however, markedly reduced in both the 36-hr and the 48-hr experiment. Also, note that the % parasite stage distribution was very different (at 44 hr) from the control for all the Sudanese sera. These cultures were started with late schizonts, the merozoites from which reinvaded cells successfully as shown by the % parasitemia. Their intracellular growth was then inhibited, as shown by the high proportion of abnormal forms (AB) and of rings (R) and trophozoites (T) and the small proportion of schizonts (Sc) as compared to the control.

defective in oxidative response to phorbol myristate acetate, other factors from the cells must have been responsible for the parasiticidal effect.

In sum, every resource of the immune system is brought to bear on the infecting malarial parasite. And yet, in human malaria, all of these are not adequate to eradicate existing infections or to prevent reinfection until after prolonged periods of fluctuating parasitemia. How do the parasites manage

TABLE 19.2. Correlation between Oxidative Activity and Killing of *P. falciparum* by Human Mononuclear Leukocytes[a]

Effector cells	% Parasitemia[b] ± S.E.	% macrophages reducing NBT	H_2O_2 release (nmoles/coverslip per 90 min)
—	16.7 ± 2.5%	—	—
Unseparated mononuclear leukocytes	12.5 ± 4.4%	18%	11 ± 3
Monocyte-enriched	6.3 ± 0.6%	68%	37 ± 7
T cell-enriched	15.0 ± 2.2%	0%	0
NK cells (large granular lymphocytes)	13.3 ± 2.7%	0%	0

[a] Modified from Ockenhouse *et al.* (1984).
[b] *In vitro* killing assay was performed with an initial parasitemia of 2%. The mean parasitemias in control and experimental cultures were determined after 48 hr on triplicate blood films ± S.E.

to do this, and in light of this situation, is there any real hope for a vaccine against malaria? The answer to the second question is, probably yes, providing we adequately take into consideration the answers to the first question.

The parasites escape the immune system in at least two general ways: They obfuscate it and subvert it, or they may undergo antigenic variation, much like the African trypanosomes. The latter has been clearly demonstrated for only one species, *P. knowlesi* in the rhesus monkey, but may well occur in others, as perhaps in the relapsing populations of *P. vivax* in humans. With *P. knowlesi*, W. D. Eaton had noted that serum from rhesus monkeys infected with this parasite and cured by drug treatment would agglutinate red cells containing late schizonts. Using this schizont-infected cell agglutination method (SICA), Neal Brown and his associates went on to show that antigenic variation occurs repeatedly in chronic infections with *P. knowlesi*. If *P. knowlesi* infections in rhesus monkeys are treated with subcurative doses of drug, one observes a series of high parasitemias before an effective immunity finally develops. Each of these populations shows a new variant by the SICA test. In the early populations only the agglutinating antibody is present and the infections have to be controlled by drug treatment. Eventually, however, an opsonizing antibody develops together with the SICA antibody (Fig. 19.6). Progressively lower parasitemias occur so that recovery ensues without drug treatment. As already noted (Chapter 13), the variant specific antigens are

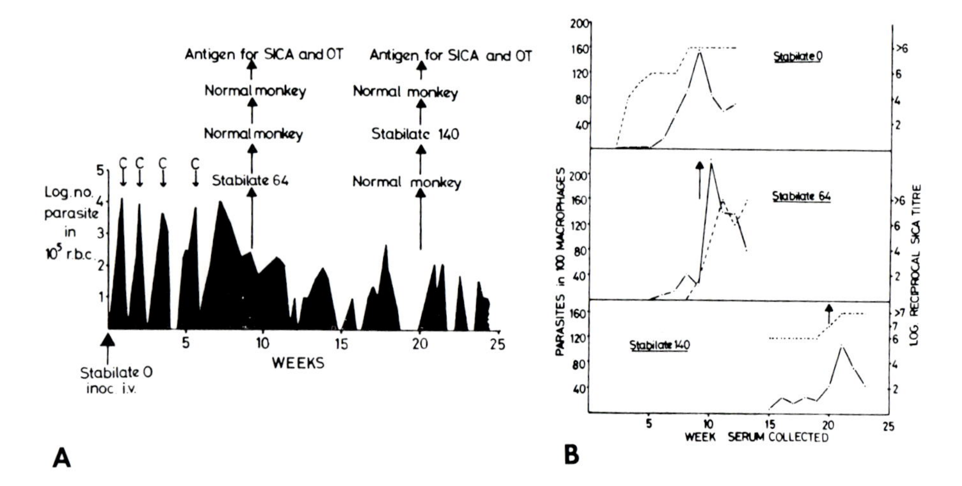

FIGURE 19.6. (A) History of *P. knowlesi* infection in a chronically infected rhesus monkey used as a serum and variant donor. The first four parasitemias had to be controlled with chloroquine at 20 mg base intramuscularly (C). SICA, schizont-infected cell agglutination; OT, opsonization test. (B) Levels of antibody in the same monkey against parasitized cells from stabilate 0, stabilate 64, and stabilate 140. Upward arrows indicate when variants and their stabilates were isolated from the chronic infection . . . ., SICA titer, –·–·–, opsonizing titer. Note that in the initial parasitemia (stabilate 0) the opsonizing antibodies did not appear until several weeks after appearance of the SICA antibodies, whereas in the later parasitemias both types of antibodies appeared at about the same time. (From Brown and Hills, 1974.)

formed by the parasite and come to be located on the surface of the schizont-infected erythrocyte. I have already discussed (Chapter 15) the peculiar relationship of the spleen to the expression of these variant specific antigens.

That antigenic variation of this type occurs with other malarial parasites is likely. That great antigenic diversity occurs with all species of malarial parasites is certain. This is now best documented for the major human malaria *P. falciparum.* The recent availability of a culture method for this species has made possible the isolation of many strains and clones and their detailed study by biochemical and immunological methods. These can be typed with monoclonal antibodies but so far there is no indication for the restriction of particular types to particular geographic regions. What is clear is that the numerous antigens of the parasites stimulate the production of antibodies that have no relevance to immunity. There is furthermore a polyclonal stimulation of B cells so that very high levels of gamma globulins are found in people repeatedly exposed to malaria. In Gambians unprotected from malaria, the rate of gamma-globulin synthesis was found to be seven times that in Europeans. This was considerably reduced in Gambians protected by antimalarial therapy or who had lived for some years in Europe. Whereas gamma-globulin levels in Europeans or Americans vary from 800 to 1600 mg/dl, in West Africans they are around 2500 mg/dl. Similarly, IgM levels of West Africans are about 250 mg/dl compared to only 80–90 mg/dl for Europeans. Observations in West Africans long resident in nonmalarious regions show that much of this increase is associated with malaria. Yet less then 5% of the globulin represents specific antimalarial antibody. The excess IgG includes a variety of autoantibodies which are important in some of the pathogenic effects of malaria. In tropical splenomegaly, which we have already seen is associated with a peculiar immunological response to malaria (Chapter 15), the IgM levels may range up to 30 times normal. It is presumed that this may result not only from direct stimulation of B lymphocytes by malarial mitogens (which have been demonstrated) but also from a malaria-induced defect in suppressive T cells permitting an unrestricted B-cell proliferation.

In addition to this kind of subversion of the immune system, direct immunosuppressive effects are well documented in both human and rodent malaria. A higher incidence of nonreactors to tetanus toxoid was observed among malarious children in Gambia than among nonmalarious children. Gambian children with falciparum malaria showed a reduced antibody response to the somatic "O" antigen of *Salmonella.* But the response to the flagellar "H" antigen of *Salmonella* was unchanged, illustrating how the suppression varies with the antigen involved. It is a selective, not a universal suppression. Similarly, in mice with a chronic *P. berghei* infection there was no response to pneumococcal polysaccharide SSS III but a normal response to dextran B 512. In cell-mediated immune responses also the effects of malaria depend on the parasite concerned, the particular antigen, and the particular response under study. In general, however, cell-mediated responses are less susceptible to suppression than antibody responses. The immunosuppressive effects probably result from a variety of mechanisms. These would include, in addition to polyclonal activation, defects in macrophage function, antigenic competition, and the activation of nonspecific suppressor T cells. A particularly interesting and important example is provided by Burkitt's lymphoma, a tumor of children that is common in Africa wherever malaria is holoendemic. The lymphoma has long been known to be associated with a herpesvirus, Epstein Barr virus (EBV). This virus infects B lymphocytes, which are stimulated to continuous division. *In vivo,* however, such B cells are under immunological control, and abnormal proliferation occurs only in immunosuppressed patients. EBV is ubiquitous, whereas Burkitt's lymphoma is largely restricted to malarious regions. It has recently been found that in an acute attack of *P. falciparum* malaria, T-cell populations are radically altered. Not only are T-cell numbers decreased and B cells increased, but the proportion of T-helper (T4) cells to T-suppressor (T8) cells is reduced. The low T4/T8 ratio resembles that in acquired immune deficiency syndrome. This may favor the unrestrained growth of B cells carrying EBV with development of the lymphoma, in this way accounting for the geographical correlation with holoendemic malaria. There may also be present suppressor cells that specifically depress the immune reactivity to the malarial parasite itself. This could be the basis for the prolonged chronic infection in intact rhesus monkeys with *P. inui* whereas splenectomized monkeys recover completely (see Chapter 15).

The picture at present is not a clear one. What is clear is that a complicated dynamic interplay occurs between the parasite with its many antigenic components and the whole immunological system of the host. The parasites produce substances that modulate and even suppress immune responses as well as other substances that stimulate effective immune responses. We are only at the beginning of our understanding of the antigenic complexity of malarial parasites. As various components of the parasites are isolated, we should begin to learn what does what in the host. Perhaps certain ones are involved with particular pathogenic effects, some with the anemia, others with the formation of immune complexes that lodge in the kidneys and are so important especially in pathology caused by *P. malariae,* others in still other ways. In *P. falciparum* there is a whole family of so-called S antigens (heat stable) that appear in the serum of infected people (and also in the fluid from cultures). These are likely to be different in different people in the same region. There is yet no reasonable hypothesis as to what they do for the parasite or to the host or whether any of them are concerned with immunity. All types of human malaria are characterized by the very high fever that tends to be intermittent and to occur at about the time of release of merozoites into the blood. Some factor must be liberated that causes monocytes to secrete an endogenous pyrogen, a small protein that in turn alters the activity of temperature sensitive neurons in the central nervous system, thereby raising the set temperature of the thermoregulatory system. Such a factor might be somewhat analogous to endotoxin or to other bacterial pyrogens, but nothing of this sort has yet been isolated from malarial parasites.

The possibility that we may be able to separate antigens that stimulate an effective immune response from all the others that modulate or suppress the immune response, or cause various pathogenic effects, provides the basis for the hope for development of a vaccine against malaria. Such a vaccine could be directed against the sporozoites or the erythrocytic stages. The first would prevent preerythrocytic development of the sporozoites injected by the bite of an infected mosquito. The second would suppress development of the erythrocytic parasites that produce the disease. Still a third type of vaccine has been envisaged, the so-called transmission-blocking vaccine. This is directed against the gametes which form in the midgut of a mosquito that has just fed on an infected person. Since this would be of no value to the vaccinated individual, it would have to be used together with one or both of the first two types of vaccine.

As noted earlier in this chapter, small numbers of X-irradiated sporozoites of *P. berghei* injected intravenously into mice will protect them against viable infective sporozoites of this species. The protection is strictly stage specific; the mice immune to sporozoites are fully susceptible to inoculation of the erythrocytic parasites of *P. berghei*. Moreover, it was found that three out of five human volunteers fed upon by a large number of mosquitoes infected with either *P. vivax* or *P. falciparum* whose sporozoites had been inactivated by appropriate exposure of the mosquitoes to X-rays were subsequently immune to challenge by bites of infective mosquitoes. The immunity to *P. vivax* lasted about 3 months, that to *P. falciparum* about 3–6 months. Since the sporogonic cycle of malaria has not been cultured and there is no way to obtain large numbers of sporozoites, a sporozoite vaccine would have been impractical had it not been for the methods of recombinant DNA. These, combined with the use of monoclonal antibodies, are being elegantly applied in attempts to develop a sporozoite vaccine. Sporozoites of the human parasites *P. vivax* and *P. falciparum*, like those of *P. berghei*, seem to have one main immunodominant surface antigen, CSP. This protein is characteristic for each species of parasite and has a molecular weight of 40,000 to 50,000. The CSPs of several species have been cloned and expressed in *E. coli;* that of the simian parasite *P. knowlesi* has also been cloned into yeast and into vaccinia virus. All these proteins show numerous repeat sequences.

Of special interest is the cloning of the complete gene for the CSP of *P. falciparum* in the expression vector λgt 11. This was done through the use of genomic DNA prepared from cultures of the erythrocytic stages. The DNA was treated with mung bean nuclease under conditions such that it was cut at positions before and after genes but not within gene coding regions. Colonies expressing the protein were selected by screening with a monoclonal antibody to the CSP of *P. falciparum*. The structure of the protein was then deduced from the nucleotide sequence of the cloned gene. This protein consists of a signal sequence, followed by a charged region, followed by 41 tandem repeats of a tetrapeptide, followed by two other charged regions, and finally an anchor sequence for a total of 412 amino acids (Fig. 19.7). As illustrated in the figure, the CSPs of *P. falciparum* and *P. knowlesi* show two regions of marked homology, one on either side of the repeat sequence. In the falciparum CS antigen, 37 of the 41 tandem repeat tetrapeptides are Asn-Ala-Asn-Pro whereas the other 4 are Asn-Val-Asp-Pro. Synthetic peptides of eight or more residues of these repeating units conjugated to a carrier protein have been shown to be immunogenic; they elicit antibodies in mice and rabbits that recognize the native CSP and that block the invasion of *P. falciparum* sporozoites into human hepatoma cells *in vitro* (see Chapter 8). Similar results have been obtained with the repeat units cloned and expressed in *E. coli,* and subsequently purified. These two kinds of preparations are about to be tested in human volunteers to determine whether they will call forth high levels of antibodies to the CSP of *P. falciparum.* If they do, they may be effective as a sporozoite vaccine.

With regard to a vaccine against the erythrocytic stages, it is important to note that the first protein of any sort to be purified from malarial parasites was obtained from a species of avian malaria *P. lophurae.* When this parasite develops in the red cells of ducks, it forms characteristic large granules that are not formed when it develops in two other susceptible hosts, chicks or turkeys. A. Kilejian found that the granules consist almost entirely of a single peculiar protein with an exceptionally high content of histidine (about 70%), relatively high proline, and little or no methionine. The purified protein was successfully used to immunize ducklings against *P. lophurae.* The function of the protein in the parasite is, however, unknown. Histidine-rich proteins have also been demonstrated in *P. falciparum;* one of them is associated with the knobs on the erythrocyte surface (see Chapter 13). Although serum containing antibodies to knob protein will interfere with sequestration of late-stage parasites in *Aotus* monkeys, there is no evidence that the knob itself is immunogenic.

A considerable number of antigens has now been identified from the erythrocytic stages of *P. falciparum.* This was greatly facilitated by the availability since 1976 of an *in vitro* culture method for the parasite (see Chapter 8). Indeed, this culture method has engendered a veritable explosion of studies on the molecular biology of this organism, with findings of significance not only to malaria but also to molecular biology in general (see Chapter 10). Special attention has been given to antigens appearing in the late schizont stage and on merozoites, since earlier work had shown that an effective immunity could be produced in experimental animals by vaccination with these stages together with Freund's complete adjuvant. Among numerous types of antigens so far identified from *P. falciparum* erythrocytic stages, the following are of special importance:

1. An antigen of about 195,000 relative molecular weight (M_r) present on the surface membrane of schizonts. With the formation of the merozoites this is broken down to products of lower molecular weight (83,000, 42,000, and

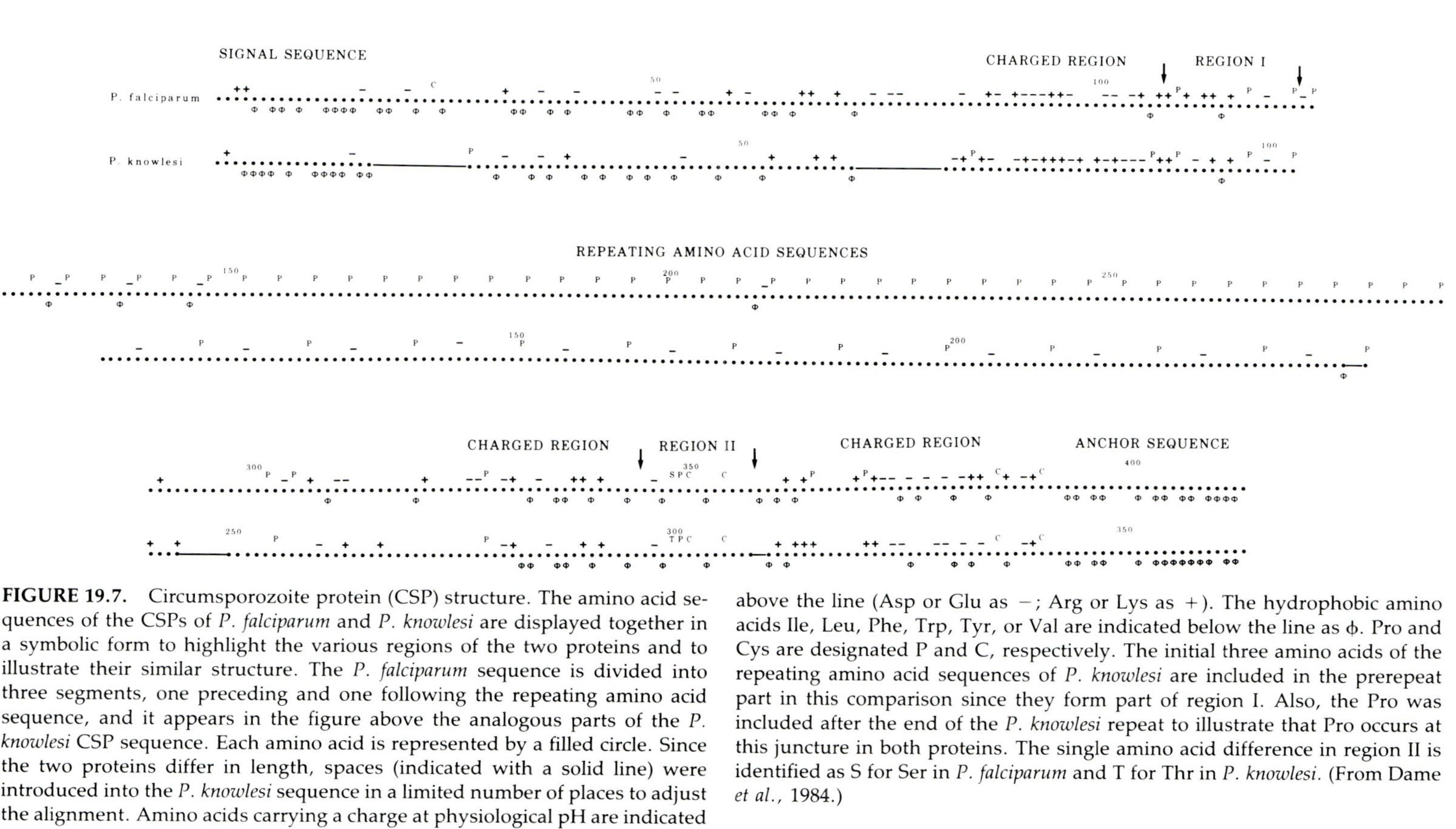

FIGURE 19.7. Circumsporozoite protein (CSP) structure. The amino acid sequences of the CSPs of *P. falciparum* and *P. knowlesi* are displayed together in a symbolic form to highlight the various regions of the two proteins and to illustrate their similar structure. The *P. falciparum* sequence is divided into three segments, one preceding and one following the repeating amino acid sequence, and it appears in the figure above the analogous parts of the *P. knowlesi* CSP sequence. Each amino acid is represented by a filled circle. Since the two proteins differ in length, spaces (indicated with a solid line) were introduced into the *P. knowlesi* sequence in a limited number of places to adjust the alignment. Amino acids carrying a charge at physiological pH are indicated above the line (Asp or Glu as −; Arg or Lys as +). The hydrophobic amino acids Ile, Leu, Phe, Trp, Tyr, or Val are indicated below the line as ϕ. Pro and Cys are designated P and C, respectively. The initial three amino acids of the repeating amino acid sequences of *P. knowlesi* are included in the prerepeat part in this comparison since they form part of region I. Also, the Pro was included after the end of the *P. knowlesi* repeat to illustrate that Pro occurs at this juncture in both proteins. The single amino acid difference in region II is identified as S for Ser in *P. falciparum* and T for Thr in *P. knowlesi*. (From Dame *et al.*, 1984.)

19,000) that are on the surface of the merozoites. Partial protection of vaccinated *Saimiri* monkeys has been obtained with a 200,000-M_r protein of this type (Fig. 19.8). This protein shows considerable antigenic diversity among different isolates from the same region as well as from different regions. Among clones it ranges in size from 190,000 to 200,000 and each parasite clone produces a protein distinct from the equivalent product in several other clones. These different clonal products are, however, immunologically interrelated and have determinants in common with all tested isolates of the parasite. They may eventually provide a useful system for antigenic typing of *P. falciparum*. The primary structure of this protein has been deduced from the nucleotide sequence of cDNA and a genomic clone covering the complete coding sequence of the gene for it.

2. A different protein, also from late schizonts and merozoites but with an M_r of 140,000, gave somewhat better immunization of *Saimiri* monkeys (Fig. 19.8).

3. In these experiments by L. Perrin, particularly good immunization of *Saimiri* monkeys was obtained with a 41,000-M_r protein apparently derived from the apical organelles of merozoites. When this protein was purified by affinity chromatography and used to vaccinate four monkeys according to the same protocol shown for Fig. 19.8, the infections showed peak parasitemias of only 1, 0.2, 0.08 and 0.05% reached on day 12, 34, 35, and 28, respectively, whereas the four controls had peaks of 11, 12, >20 and >20%,

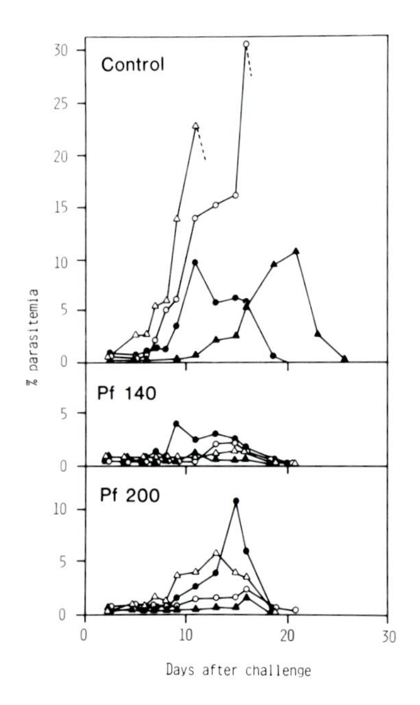

FIGURE 19.8. The course of infection with *P. falciparum* in three groups, of four *Saimiri* monkeys each, immunized with saline (Control), the 140,000-dalton polypeptide (Pf 140) or the 200,000-dalton polypeptide (Pf 200), both purified by affinity chromatography from cultures of an isolate from Zaire. With the first subcutaneous injection each monkey received Freund's complete adjuvant; with the second and third injection, also given subcutaneously 16 and 30 days respectively after the first, each monkey received Freund's incomplete adjuvant. On day 63 all the monkeys received an intravenous challenge inoculation of 2.5×10^7 parasites taken from a splenectomized *Saimiri* monkey infected with the Palo Alto FUP strain of *P. falciparum* from Uganda.

It will be noted that whereas two of the control monkeys showed parasitemia over 20% and were treated with chloroquine, all of the monkeys immunized with Pf 140 showed infections of only 1 to 5%, and those immunized with Pf 200 showed a definite though smaller degree of protection. (From Perrin *et al.*, 1984. Original print courtesy of Dr. L. Perrin.)

reached on day 10, 20, 10, and 16, respectively. This protein merits receiving much further attention.

4. Also of special interest are two proteins that bind to erythrocyte glycophorin and are on the surface of the merozoite (see also Chapter 3). These presumed receptors for attachment of merozoites have M_r of 155,000 and 130,000. Rabbit antiserum to them blocks invasion *in vitro.* No tests of immunization have yet been done.

5. Also of M_r 155,000 is a protein that has been localized to the micronemes of merozoites, rather than their surface. This protein can be demonstrated by immunofluorescence in merozoites. Immediately after invasion of the merozoite, it cannot be shown in the newly formed ring but it appears on the surface of the erythrocyte (Fig. 19.9). It has therefore been named the ring-infected erythrocyte surface antigen (RESA). Monoclonal antibody to this protein, as well as affinity-purified human immunoglobulin from malarial serum, blocks growth of the parasite *in vitro.* Again no tests of immunization have yet been done. The gene for this protein has been sequenced; there is a complex tandem repeat structure. Because of its presence in the micronemes of merozoites, this protein may be involved in the invasion processes that follow attachment. Once the ring stage has formed, the protein can be shown by immuno-electron microscopy only on the surface of the erythrocyte. How it gets there, and especially what it's doing there, are problems for the future. Perhaps it is somehow involved in the new permeability pathways with porelike properties that appear in infected red cells.

6. The last group of antigens I will consider has actually been known for the longest time. These are the S antigens first demonstrated by R. J. W. Wilson in the serum of malarious patients. These heat-stable proteins (to 5 min at 100°C) show great antigenic diversity and for this reason are not likely candidates for a vaccine. They range in molecular weight from 130,000 to 250,000. They are found in the culture fluid of *in vitro* cultures as well as in patients' plasma and are derived from the parasitophorous vacuole. The gene has been sequenced and again shows tandem repeat structures. It has been localized to chromosome 7 in pulsed field gradient gel electrophoresis (see Chapter 10), whereas the RESA gene (see above) is on chrommosome 1.

I turn now to a consideration of transmission-blocking immunity. It has been shown for several species of malarial parasites that antibodies induced in the vertebrate host to the gametes (which normally never occur in the vertebrate host) will act in the midgut of a mosquito that feeds on such a host to inactivate the malarial gametes so that the mosquito infection is greatly reduced or fully prevented. Such a vaccine would act like a gametocytocidal drug (see Chapter 25) to prevent mosquito infections and thereby break the chain of transmission. Here again the methods of monoclonal antibodies and recombinant DNA, together with the use of gametocyte-producing cultures of *P. falciparum,* are being effectively applied. The antigens from gametes, zygotes, and ookinetes so far identified are summarized in Table 19.3.

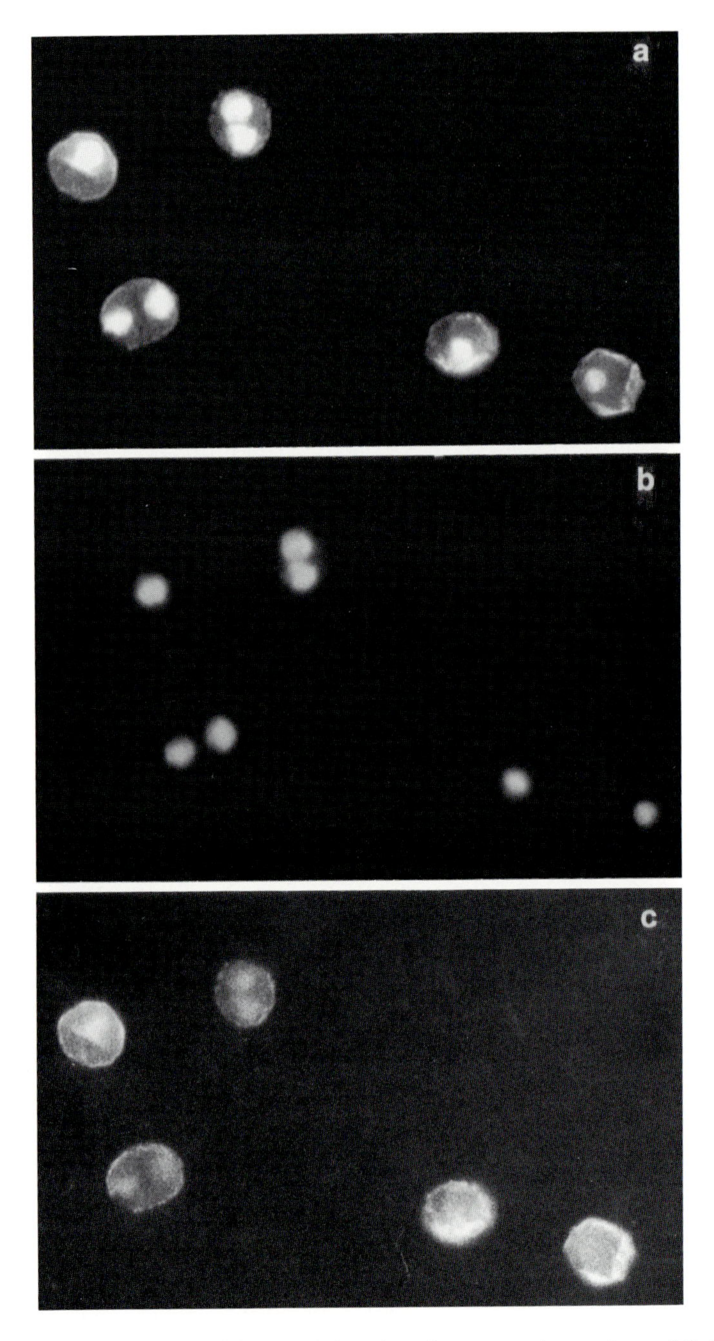

FIGURE 19.9. Demonstration of the ring-infected erythrocyte surface antigen of *P. falciparum*. Indirect immunofluorescence was performed on glutaraldehyde-fixed air-dried films of infected erythrocytes from culture using malarial serum from a Colombian donor. The parasite nuclei were then counterstained with ethidium bromide. (a) shows both the orange fluorescence of the nuclei and the green immunofluorescence of the surface of the infected erythrocytes. (b) shows only the fluorescence of the nuclei. (c) shows only the immunofluorescence of the surface of the erythrocytes. (From Perlmann *et al.*, 1984. Original prints courtesy of Dr. P. Perlmann.)

TABLE 19.3. Antigens of Gametes, Zygotes, and Ookinetes[a]

	M_r[b]	Location	Properties	*In vitro* activity of specific mAbs[c] or antisera
P. gallinaceum, *P. falciparum*	55,000, 60,000 260,000	Surface of male and female gametes Newly formed zygote surface	Synthesized in gametocytes 60K and 55K unlikely to be products of 260K precursor 60K and 55K are glycoproteins Species-shared epitopes for several *P. falciparum* isolates One epitope with *P. falciparum* exhibits antigenic diversity	mAbs synergize to block fertilization, thereby blocking mosquito transmission
P. gallinaceum	26,000	Surface of developing zygote and ookinete	Synthesized rapidly after fertilization Reaches zygote surface 2–3 hr postfertilization Not synthesized by mature ookinetes A glycoprotein	mAb blocks ookinete formation, thereby blocking mosquito transmission
P. yoelii	74,000	Surface of gametes?		Passivly transferred mAb blocks transmission Immunization with 74K blocks transmission

[a] Modified from World Health Organization (1984).
[b] Apparent molecular weight on SDS-polyacrylamide gel electrophoresis.
[c] mAb, monoclonal antibody.

In all work done so far, the induction of an active immunity to the erythrocytic stages with whole merozoites or with purified parasite proteins has required the use of an adjuvant. Freund's complete adjuvant has generally been used, but the much less toxic muramyl dipeptide is also effective. Even this, however, is not acceptable for use in humans. An integral part of vaccine development will therefore have to be its testing with those few adjuvants that are safe for human use.

Encouraging indications that effective vaccination against human malaria is a good possibility are provided by results obtained with bovine babesiosis, a disease of cattle with many similarities to human malaria. Parasites of the genus *Babesia,* like the plasmodia, multiply in erythrocytes of vertebrates. They differ from malarial parasites in several significant ways. The arthropod vector is a tick rather than an insect and the developmental cycle in the anthropod is quite different from that of malaria (see Diagram XI). The intraerythrocytic stages lie directly in the cytoplasm rather than within a parasitophorous membrane. At each cycle of division they form only two, or four, daughter parasites. A preerythrocytic cycle is not known. Nevertheless, the pathology and immunology of the diseases caused by *Babesia bovis* and *B. bigemina,* the two principal species in cattle, are quite like those of malaria. *B. bovis* in particular resembles *P. falciparum* in showing sequestration of infected cells in the capillaries of deep organs with blockage of capillaries and consequent pathological lesions. Cattle that recover may show prolonged low-grade infection, a state known as premunition. *B. bovis* has been grown in culture by methods very similar to those used for *P. falciparum.* Crude protein prepared from the culture fluid collected after a 24-hr growth period was tested in four cattle using Quil A, a saponin-type adjuvant that is acceptable for use in food animals. Two injections were given 2 weeks apart. Twelve weeks later, when the antibody titers to *Babesia* in these cattle had fallen to a plateau level, the cattle were each exposed to challenge by 1000 *Boophilus* tick larvae from a *B. bovis*-infected colony. At the same time, four control cattle that had been vaccinated with culture fluid from an uninfected erythrocyte suspension and four others that had recovered from infections with the related parasite *B. bigemina* were similarly exposed to challenge. All 12 animals became infected. The four that had received *B. bovis* culture fluid showed an accelerated antibody response, relatively mild clinical disease, and all survived, whereas two of the four controls and three of the four immune to *B. bigemina* died of the *B. bovis* infections.

Immunity to *Babesia,* like that to malaria, depends both on antibodies and on various cell-mediated factors. With the rodent parasites *B. microti* and *B. rodhaini,* nonspecific immunostimulants such as cord factor, zymosan (a yeast cell wall preparation), glucan, BCG, or *Corynebacterium parvum* will protect mice against subsequent infection. But such nonspecific factors did not protect cattle against their species of *Babesia.* In mice there is a cross-immunity between *B. microti* and the rodent malaria *Plasmodium vinckei.* This probably depends, like the immunity produced by nonspecific immunostimulants, on soluble nonantibody mediators that kill the parasite within the red blood cells. Although, as discussed earlier in this chapter, there is evidence for such mediators in primate malarias, there is little indication that nonspecific immunostimulants can elicit them in adequate amounts. For example, monkeys treated with Freund's complete or other adjuvants showed no resistance to malaria, resembling the situation with cattle and *Babesia.*

In bovine infections with *Theileria* (Diagram IX), the parasites causing East Coast fever, cytotoxicity against parasite-infected lymphoblasts mediated by T lymphocytes is a principal protective mechanism against reinfection. It will be remembered (see Chapter 13) that in the genus *Theileria,* infective forms introduced by an infected tick enter and transform lymphocytes so that they multiply extensively with the parasites simultaneously multiplying within them as macroschizonts. Vaccination against the two main species causing East Coast fever, *T. parva* and *T. lawrencei,* has been achieved by simultaneous inoculation of a stabilate of infective forms derived from ticks together with a single dose of a long-acting oxytetracycline. The drug permits development of a limited infection that provides immunity against field challenge with infected ticks of such severity as to result in death of all nonimmunized control cattle. The immunity is species specific; cattle vaccinated with *T. parva* were fully susceptible to *T. lawrencei.* It is very interesting that Russian workers have reported a large-scale field trial of a lymphoid cell culture vaccine against *T. annulata.* Over 100,000 cattle were vaccinated during the period 1976–1984. They were kept together with nonvaccinated calves in tick-infested pastures without treatment with acaricides. All the control animals became ill, with a death rate of 27 to 40%. There were no cases of theileriasis among the vaccinated animals. These animals showed a persistent immunity as long as they were exposed every year to *Theileria*-infected ticks.

Acquired immunity also develops with the Coccidia, the other major group of intracellular protozoan parasites of the subphylum Apicomplexa. This is of considerable economic importance in the production of domestic food animals and especially in the poultry industry. There is no doubt that chickens recovered from infection with *Eimeria tenella* (Diagram VI) or other species of *Eimeria* develop an effective immune response to further infection with the same species. This is in fact the basis for a commercially available live vaccine.

Little is known, however, as to the detailed mechanism of this immunity. Since chicken coccidia are parasites of epithelial cells in different regions of the alimentary tract, it is not surprising that IgA plays an important role. Circulating antibodies have also been demonstrated but their part in protection is uncertain, as are the roles of macrophages and other cells. It is interesting that splenectomy has no effect on coccidial infections, in sharp contrast to its dramatic effects on infection with malarial and other protozoan parasites of the blood. This is of course in keeping with the fact that the spleen plays a minor role in antibody formation to antigens given by routes other than the intravenous one.

The most important coccidian parasite of humans, *Toxoplasma gondii* (see Diagram VII and several earlier chapters), presents a diverse and fascinating immunological picture, as might be expected from its complex life cycle, the diversity of its hosts, and its several modes of transmission. Most infections in most species are essentially asymptomatic, as shown by the large proportion of animals and humans with antibody. In all of these an effective immunity develops before any significant lesions can be produced. The initial asymptomatic infection may be followed by prolonged chronic infection, also usually asymptomatic. On the other hand, acute infections may be very severe, even fatal, and all possible gradations in between are seen. Initial asymptomatic infection in a pregnant woman is extremely dangerous to the fetus, in whom lesions damage the brain and eyes. In chronic toxoplasmosis, cysts develop containing organisms that grow very slowly but are protected from the host's immune system by the cyst wall. Such cysts are common in the brain and retina, and in people are an important cause of chorioretinitis. Toxoplasmic infection, even more than the other intracellular protozoa I have considered, is characterized by prolonged chronicity; in most animals that can develop immunity, it becomes chronic. Such chronically infected rodents with viable organisms in cysts in the brain serve to infect cats, the only known definitive host in which the sexual cycle and formation of oocysts occur. It is interesting to reflect that this mechanism of transmission serves to bypass the sanitary habits of cats. Whereas chickens always ingest feces with their food so that direct transmission via oocysts is simple, cats are much less likely to do so.

Cats develop a very effective immunity. If infected as adolescents or adults, they do not shed oocysts when reinoculated. The extraintestinal stages are also affected by the host immune response and tend to disappear toward the end of the primary infection in the intestine, when oocyst production has almost ceased.

In intermediate hosts, such as humans or rodents, the immunity which develops in natural infections is accompanied by low-grade chronic infection; it is again a state of premunition. In experimental animals such chronic infections persisting up to 5 years have been documented. Although specific antibody to *Toxoplasma* is found (and is the basis for a diagnostic test), it does not act on intracellular toxoplasms. All indications are that the immunity, or the premunition, that controls spread of the infection, is largely based on cell-mediated factors. Lymphokines have been shown to exert both growth-inhibiting and killing effects on toxoplasms contained with macrophages. The killing effects probably result from the release of toxic oxygen intermediates, such as H_2O_2 and superoxide radical. The static effects are less well understood.

One lymphokine of defined nature that inhibits intracellular growth of *Toxoplasma* is type II or γ-interferon. Very recently, E. R. Pfefferkorn has disclosed a unique mechanism whereby γ-interferon stops the growth of *Toxoplasma*. He observed that this inhibiting effect was strongly dependent on the tryptophan concentration of the medium in which the host fibroblasts were maintained (Fig. 19.10); with 50 μg/ml tryptophan, four times more γ-interferon was required to produce the same degree of inhibition as with 10

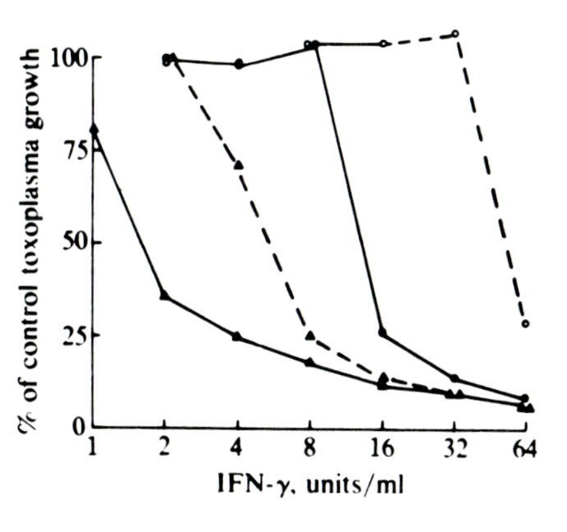

FIGURE 19.10. Assay of antitoxoplasma activity of γ-interferon in the presence of four different tryptophan concentrations. Cultures of human fibroblasts were treated with twofold dilutions of γ-interferon in media differing only in their tryptophan concentrations. Twenty-four hours later all cultures were infected with 3×10^4 plaque-forming units of *Toxoplasma gondii;* 48 hr after infection all cultures received 5 μCi of [^{3}H]uracil and incorporation into acid-precipitable material was measured 2 hr later. This gave a measure of the growth of the toxoplasma as the host cells lack uracil phosphoribosyl transferase and do not incorporate uracil. Results are expressed as a percentage of incorporation in control cultures not treated with interferon. The concentrations of tryptophan were in μg/ml: 0, ▲; 2, △; 10, ●; 50, ○. (From Pfefferkorn, 1984.)

μg/ml tryptophan. Further work showed that the γ-interferon induced the host cells to produce an indole-amine dioxygenase that degraded tryptophan to kynurenine, *N*-formylkynurenine, and other unidentified products. The parasite would thus be starved of tryptophan, an essential amino acid it cannot synthesize.

The overriding importance of the immune system in protection of people from toxoplasmosis and certain other protozoa has been dramatically underscored ever since the clinical use of toxic immunosuppressive agents in cancer chemotherapy and in organ transplantation. In individuals under treatment with these agents, chronic, previously inapparent infection with *Toxoplasma* is revealed as acute disease. Even more serious are the outbreaks of disease with *Pneumocystis carinii* and *Cryptosporidium*, two obscure species of protozoa that exist in nature as parasites of various animals. In humans they develop only in immunologically compromised hosts, and are among the serious complications in patients suffering from acquired immune deficiency.

Bibliography

Allison, A. C., and Eugui, E. M., 1981, Theileriosis—Cell-mediated and humoral immunity, *Am. J. Pathol.* **102:**114–120.

Allison, A. C., and Eugui, E. M., 1983, The role of cell-mediated immune responses in resistance to malaria, with special reference to oxidant stress, *Annu. Rev. Immunol.* **1:**361–392.

Ballou, W. R., Rothbard, J., Wirtz, R. A., Gordon, D. M., Williams, J. S., Gore, R. W., Schneider, I., Hollingdale, M. R., Beaudoin, R. L., Maloy, W. L., Miller, L. H., and Hockmeyer, W. T., 1985, Immunogenicity of synthetic peptides from circumsporozoite protein of *Plasmodium falciparum*, *Science* **228:**996–999.

Bautista, C. R., and Kreier, J. P., 1980, The action of macrophages and immune serum on growth of *Babesia microti* in short-term cultures, *Tropenmed. Parasitol.* **31:**313–324.

Boyle, D. B., Newbold, C. I., Wilson, R. J. M., and Brown, K. N., 1983, Intraerythrocytic development and antigenicity of *Plasmodium falciparum* and comparison with simian and rodent malaria parasites, *Mol. Biochem. Parasitol.* **9:**227–240.

Brown, G. V., Gulvenor, J. G., Crewther, P. E., Bianco, A. E., Coppel, R. R., Saint, R. B., Stahl, H.-D., Kemp, D. J., and Anders, R. F., 1985, Localization of the ring-infected erythrocyte surface antigen (Resa) of *Plasmodium falciparum* in merozoites and ring-infected erythrocytes, *J. Exp. Med.* **162:**774–779.

Brown, K. N., and Hills, L. A., 1974, Antigenic variation and immunity to *Plasmodium knowlesi:* Antibodies which induce antigenic variation and antibodies which destroy parasites, *Trans. R. Soc. Trop. Med. Hyg.* **68:**139–142.

Carlin, J. M., Jensen, J. B., and Geary, T. G., 1985, Comparison of inducers of crisis forms in *Plasmodium falciparum in vitro, Am. J. Trop. Med. Hyg.* **34:**668–674.

Clark, I. A., 1978, Does endotoxin cause both the disease and parasite death in acute malaria and babesiosis?, *Lancet* **2:**75–77.

Clark, I. A., 1979, Protection of mice against *Babesia microti* with cord factor, COAM, zymosan, glucan, Salmonella and Listeria, *Parasite Immunol.* **1:**179–196.

Clyde, D. F., McCarthy, V. C., Miller, R. M., and Woodward, W. E., 1975, Immunization of man against falciparum and vivax malaria by the use of attenuated sporozoites, *Am. J. Trop. Med. Hyg.* **24:**397–401.

Cohen, S., 1979, Review lecture—Immunity to malaria, *Proc. R. Soc. London Ser. B* **203:**323–345.

Cohen, S., and Lambert, P. H., 1982, Malaria, in: *Immunology of Parasitic Infections* (2nd edition) (S. Cohen and K. S. Warren, eds.), Blackwell, Oxford, pp. 422–474.

Cohen, S., McGregor, I. A., and Carrington, S., 1961, Gamma-globulin and acquired immunity to human malaria, *Nature* **192:**733–737.

Coppel, R. L., Cowman, A. F., Anders, R. F., Bianco, A. E., Saint, R. B., Lingelbach, K. R., Kemp, D. J., and Brown, G. V., 1984, Immune sera recognize on erythrocytes a *Plasmodium falciparum* antigen composed of repeated amino acid sequences, *Nature* **310:**789–792.

Cox, F. E. G., 1978, Heterologous immunity between piroplasms and malaria parasites: The simultaneous elimination of *Plasmodium vinckei* and *Babesia microti* from the blood of doubly infected mice, *Parasitology* **76:**55–60.

Cox, F. E. G., 1980, Non-specific immunization against babesiosis, in: *Isotope and Radiation Research on Animal Diseases and Their Vectors,* Proceedings of the International Symposium 7–11 May 1979, International Atomic Energy Agency, Vienna, pp. 95–104.

Cunningham, M. P., 1977, Immunization of cattle against *Theileria parva,* in: *Theileriosis* (J. B. Henson and M. Campbell, eds.), International Development Research Centre, Ottawa, pp. 66–75.

Dame, J. B., Williams, J. L., McCutchan, T. F., Weber, J. L., Wirtz, R. A., Hockmeyer, W. T., Maloy, W. L., Haynes, J. D., Schneider, I., Roberts, D., Sanders, G. S., Reddy, E. P., Diggs, C. L., and Miller, L. H., 1984, Structure of the gene encoding the immunodominant surface antigen on the sporozoite of the human malaria parasite *Plasmodium falciparum, Science* **225:**593–599.

Davis, P. J., Parry, S. H., and Porter, P., 1977, The role of secretory IgA in anticoccidial immunity in the chicken, *Immunology* **34:**879–888.

Doyle, J. J., 1977, Antigenic variation in babesia, *Adv. Exp. Med. Biol.* **93:**27–29.

Enea, V., Arnot, D., Schmidt, E. C., Cochrane, A., Gwadz, R., and Nussenzweig, R. S., 1984, Circumsporozoite gene of *Plasmodium cynomolgi* (Gombak); cDNA cloning and expression of the repetitive circumsporozoite epitope, *Proc. Natl. Acad. Sci. USA* **81:**7520–7524.

Enea, V., Zavala, J. E. F., Arnot, D. E., Asavanich, A., Masuda, A., Quakyi, I., and Nussenzweig, R. S., 1984, DNA cloning of *Plasmodium falciparum* circumsporozoite gene: Amino acid sequence of repetitive epitope, *Science* **225:**628–630.

Ferrante, A., Rzepczyk, C. M., and Allison, A. C., 1983, Polyamine oxidase mediates intraerythrocytic death of *Plasmodium falciparum, Trans. R. Soc. Trop. Med. Hyg.* **77:**789–791.

Freeman, R. R., and Holder, A. A., 1983, Surface antigens of malaria merozoites: A high molecular weight precursor to an 83,000 mol wt form expressed on the surface of *Plasmodium falciparum* merozoites, *J. Exp. Med.* **158:**1647–1653.

Ginsburg, H., Krugliak, M., Eidelman, O., and Cabantchik, Z. I., 1983, New permeability pathways induced in membranes of *Plasmodium falciparum* infected erythrocytes, *Mol. Biochem. Parasitol.* **8:**177–190.

Greenwood, B. M., Stratton, D., Williamson, W. A., and Mohammed, I., 1978, A study of the role of immunological factors in the pathogenesis of the anaemia of acute malaria, *Trans. R. Soc. Trop. Med. Hyg.* **72**:378–385.

Gysin, J., Barnwell, J., Schlesinger, D. H., Nussenzweig, V., and Nussenzweig, R. S., 1984, Neutralization of the infectivity of sporozoites of *Plasmodium knowlesi* by antibodies to a synthetic peptide, *J. Exp. Med.* **160**:935–940.

Handman, E., Goding, J. W., and Remington, J. S., 1980, Detection and characterization of membrane antigens of *Toxoplasma gondii, J. Immunol.* **124**:2578–2583.

Holder, A. A., and Freeman, R. R., 1981, Immunization against blood-stage rodent malaria using purified parasite antigens, *Nature* **294**:361–364.

Holder, A. A., and Freeman, R. R., 1982, Biosynthesis and processing of a *Plasmodium falciparum* schizont antigen recognized by immune serum and a monoclonal antibody, *J. Exp. Med.* **156**:1528–1538.

Holder, A. A., Freeman, R. R., and Newbold, C. I., 1983, Serological cross-reaction between high molecular weight proteins synthesized in blood schizonts of *Plasmodium yoelii, Plasmodium chabaudi* and *Plasmodium falciparum, Mol. Biochem. Parasitol.* **9**:191–196.

Holder, A. A., Lockyer, M. J., Odink, K. G., Sandhu, J. S., Riveros-Moreno, V., Nicholls, S. C., Hillman, Y., Davey, L. S., Tizard, M. L. V., Schwarz, R. T., and Freeman, R. R., 1985, Primary structure of the precursor to the three major surface antigens of *Plasmodium falciparum* merozoites, *Nature* **317**:270–273.

Hollingdale, M. R., Nardin, E. H., Tharavanij, S., Schwartz, A. L., and Nussenzweig, R. S., 1984, Inhibition of entry of *Plasmodium falciparum* and *P. vivax* sporozoites into cultured cells: An *in vitro* assay of protective antibodies, *J. Immunol.* **132**:909–913.

Hope, I. A., Hall, R., Simmons, D. L., Hyde, J. E., and Scaife, J. G., 1984, Evidence for immunological cross-reaction between sporozoites and blood stages of a human malaria parasite, *Nature* **308**:191–194.

Hope, I. A., Mackay, M., Hyde, J. E., Goman, M., and Scaife, J., 1985, The gene for an exported antigen of the malaria parasite *Plasmodium falciparum* cloned and expressed in *Escherichia coli, Nucleic Acids Res.* **13**:369–379.

Jayawardena, A. N., Mogil, R., Murphy, D. B., Burger, D., and Gershon, R. K., 1983, Enhanced expression of H-2K and H-2D antigens on reticulocytes infected with *Plasmodium yoelii, Nature* **302**:623–626.

Jensen, J. B., Hoffman, S. L., Boland, M. T., Akood, M. A. S., Laughlin, L. W., Kurniawan, L., and Marwoto, H. A., 1984, Comparison of immunity to malaria in Sudan and Indonesia: Crisis-form versus merozoite-invasion inhibition, *Proc. Natl. Acad. Sci. USA* **81**:922–925.

Jones, T. C., 1977, Control of intracellular parasitism by macrophages, in: *Immune Effector Mechanisms in Diseases* (M. E. Weksler, S. D. Litwin, R. R. Riggio, and G. W. Siskin, eds.), Grune & Stratton, New York, pp. 29–48.

Jones, T. C., and Len, L., 1976, Pinocytic rates of macrophages from mice immunized against *Toxoplasma gondii* and macrophages stimulated to inhibit toxoplasma *in vitro, Infect. Immun.* **14**:1011–1013.

Kan, S.-C., Yamaga, K. M., Kramer, K. J., Case, S. E., and Siddiqui, W. A., 1984, *Plasmodium falciparum:* Protein antigens identified by analysis of serum samples from vaccinated *Aotus* monkeys, *Infect. Immun.* **43**:276–282.

Kemp, D. J., Corcoran, L. M., Coppel, R. L., Stahl, H. D., Bianco, A. E., Brown, G. V., and Anders, R. F., 1985, Size variation in chromosomes from independent cultured isolates of *Plasmodium falciparum, Nature* **315**:347–350.

Kharazmi, A., and Jepsen, S., 1984, Enhanced inhibition of *in vitro* multiplication of *Plasmodium falciparum* by stimulated human polymorphonuclear leucocytes, *Clin. Exp. Immunol.* **57**:287–292.

Kharazmi, A., Jepsen, S., and Valerius, N. H., 1984, Polymorphonuclear leucocytes defective in oxidative metabolism inhibit *in vitro* growth of *Plasmodium falciparum, Scand. J. Immunol.* **20**:93–96.

Krotoski, W. A., Bray, R. S., Garnham, P. C. C., Gwadz, R. W., Killick-Kendrick, R., Draper, C. C., Targett, G. A. T., Krotoski, D. M., Guy, M. W., Koontz, L. C., and Cogswell, F. B., 1982, Observations on early and late post-sporozoite tissue stages in primate malaria. II.

The hypnozoite of *Plasmodium cynomolgi bastianellii* from 3 to 105 days after infection, and detection of 35 to 40 hour pre-erythrocytic forms, *Am. J. Trop. Med. Hyg.* **31**:211–225.

Krotoski, W. A., Collins, W. E., Bray, R. S., Garnham, P. C. C., Cogswell, F. B., Gwadz, R. W., Killick-Kendrick, R., Wolf, R., Sinden, R., Koontz, L. C., and Stanfill, P. S., 1982, Demonstration of hypnozoites in sporozoite-transmitted *Plasmodium vivax* infection, *Am. J. Trop. Med. Hyg.* **31**:1291–1293.

Krotoski, W. A., Garnham, P. C. C., Bray, R. S., Krotoski, D. M., Killick-Kendrick, R., Targett, G. A. T., and Guy, M. W., 1982, Observations on early and late post-sporozoite tissue stages in primate malaria. I. Discovery of a new latent form of *Plasmodium cynomolgi* (the hypnozoite) and failure to detect hepatic forms within the first 24 hours after infection, *Am. J. Trop. Med. Hyg.* **31**:24–35.

Kuttler, K. L., and Johnson, L. W., 1980, Immunization of cattle with a *Babesia bigemina* antigen in Freund's complete adjuvant, *Am. J. Vet. Res.* **41**:536–538.

Levy, M. G., Clabaugh, G., and Ristic, M., 1982, Age resistance in bovine babesiosis: Role of blood factors in resistance to *Babesia bovis, Infect. Immun.* **37**:1127–1131.

Lunn, J. S., Chin, W., Contacos, P. G., and Coatney, G. R., 1966, Changes in antibody titers and serum protein fractions during the course of prolonged infections with vivax or with falciparum malaria, *Am. J. Trop. Med. Hyg.* **15**:3–10.

McBride, J. S., Newbold, C. I., and Anand, R., 1985, Polymorphism of a high molecular weight schizont antigen of the human malaria parasite *Plasmodium falciparum, J. Exp. Med.* **161**:160–180.

McCutchan, T. F., Hansen, J. L., Dame, J. B., and Mullins, J. A., 1984, Mung bean nuclease cleaves *Plasmodium* genomic DNA at sites before and after genes, *Science* **225**:625–628.

McGarvey, M. J., Sheybani, E., Loche, M. P., Perrin, L., and Mach, B., 1984, Identification and expression in *Escherichia coli* of merozoite stage-specific genes of the human malarial parasite *Plasmodium falciparum, Proc. Natl. Acad. Sci. USA* **81**:3690–3694.

McGregor, I. A., 1964, Studies in the acquisition of immunity to *Plasmodium falciparum* infections in West Africa, *Trans. R. Soc. Trop. Med. Hyg.* **58**:80–92.

Murphy, J. R., 1981, Host defenses in murine malaria: Analysis of plasmodial infection-caused defects in macrophage microbicidal capacities, *Infect. Immun.* **31**:396–407.

Nardin, E. H., Nussenzweig, V., Nussenzweig, R. S., Collins, W. E., Harinasuta, K. T., Tapchaisri, P., and Chomcharn, Y., 1982, Circumsporozoite proteins of human malaria parasites *Plasmodium falciparum* and *Plasmodium vivax, J. Exp. Med.* **156**:20–30.

Nussenzweig, R. S., Cochrane, A. H., and Lustig,, H. J., 1978, Immunological responses, in: *Rodent Malaria* (R. Killick-Kendrick and W. Peters, eds.), Academic Press, New York, pp. 248–307.

Ockenhouse, C. F., Schulman, S., and Shear, H. L., 1984, Induction of crisis forms in the human malaria parasite *P. falciparum* by gamma-interferon-activated, monocyte-derived macrophages, *J. Immunol.* **133**:1601–1608.

Odink, K. G., Lockyer, M. J., Nicholls, S. C., Hillman, Y., Freeman, R. R., and Holder, A. A., 1984, Expression of cloned cDNA for a major surface antigen of *Plasmodium falciparum* merozoites, *FEBS Lett.* **173**:108–112.

Oka, M., Aikawa, M., Freeman, R. R., Holder, A. A., and Fine, E., 1984, Ultrastructural localization of protective antigens of *Plasmodium yoelii* merozoites by the use of monoclonal antibodies and ultrathin cryomicrotomy, *Am. J. Trop. Med. Hyg.* **33**:342–346.

Perkins, M. E., 1984, Surface proteins of *Plasmodium falciparum* merozoites binding to the erythrocyte receptor, glycophorin, *J. Exp. Med.* **160**:788–798.

Perlmann, H., Berzins, K., Wahlgren, M., Carlsson, J., Björkman, A., Patarroyo, M. E., and Perlmann, P., 1984, Antibodies in malarial sera to parasite antigens in the membrane of erythrocytes infected with early asexual stages of *Plasmodium falciparum, J. Exp. Med.* **159**:1686–1704.

Perrin, L. H., Merkli, B., Loche, M. Chizzolini, C., Smart, J., and Richle, R., 1984, Antimalarial immunity in Saimiri monkeys: Immunization with surface components of asexual blood stages, *J. Exp. Med.* **160**:441–451.

Pfefferkorn, E. R., 1984, Interferon γ blocks the growth of *Toxoplasma gondii* in human fibroblasts by inducing the host cells to degrade tryptophan, *Proc. Natl. Acad. Sci. USA* **81**:908–912.

Pfefferkorn, E. R., and Guyre, P. M., 1984, Inhibition of growth of *Toxoplasma gondii* in cultured fibroblasts by human recombinant gamma interferon, *Infect. Immun.* **44**:211–216.

Pinder, M., and Hewett, R. S., 1980, Monoclonal antibodies detect antigenic diversity in *Theileria parva* parasites, *J. Immunol.* **124**:1000–1001.

Potocnjak, P., Yoshida, N., Nussenzweig, R. S., and Nussenzweig, V., 1980, Monovalent fragments (Fab) of monoclonal antibodies to a sporozoite surface antigen (Pb44) protect mice against malarial infection. *J. Exp. Med.* **151**:1504–1513.

Ravetch, J. V., Feder, R., Pavlovec, A., and Blobel, G., 1984, Primary structure and genomic organization of the histidine-rich protein of the malaria parasite *Plasmodium lophurae*, *Nature* **312**:616–620.

Richards, W. H. G., Mitchell, G. H., Butcher, G. A., and Cohen, S., 1977, Merozoite vaccination of rhesus monkeys against *Plasmodium knowlesi* malaria; immunity to sporozoite (mosquito-transmitted) challenge, *Parasitology* **74**:191–198.

Sakuri, H., Takei, Y., Omata, Y., and Suzuki, N., 1981, Production and properties of toxoplasma growth inhibitory factor (Toxo-GIF) and interferon (IFN) in the lymphokines and the circulation of toxoplasma immune mice, *Zentralbl. Bakteriol. Parasitenkd. Infektionskr. Hyg. Abt. 1 Orig. Reihe A* **251**:134–143.

Samantaray, S. N., Bhattacharyulu, Y., and Gill, B. S., 1980, Immunization of calves against bovine tropical theileriosis (*Theileria annulata*) with graded doses of sporozoites and irradiated sporozoites, *Int. J. Parasitol.* **10**:355–358.

Saul, A., Myler, P., Schofield, L., and Kidson, C., 1984, A high molecular weight antigen in *Plasmodium falciparum* recognized by inhibitory monoclonal antibodies, *Parasite Immunol.* **6**:39–50.

Sharma, S., and Godson, G. N., 1985, Expression of the major surface antigen of *Plasmodium knowlesi* sporozoites in yeast, *Science* **228**:879–882.

Skamene, E., Stevenson, M. M., and Lemieux, S., 1983, Murine malaria: Dissociation of natural killer (NK) cell activity and resistance to *Plasmodium chabaudi*, *Parasite Immunol.* **5**:557–565.

Smith, G. L., Godson, G. N., Nussenzweig, V., Nussenzweig, R. S., Barnwell, J., and Moss, B., 1984, *Plasmodium knowlesi* sporozoite antigen: Expression by infectious recombinant vaccinia virus, *Science* **224**:397–399.

Smith, R. D., Carpenter, J., Cabrera, A., Gravely, S. M., Erp, E. E., Osorno, M., and Ristic, M., 1979, Bovine babesiosis: Vaccination against tick-borne challenge exposure with culture-derived *Babesia bovis* immunogens, *Am. J. Vet. Res.* **40**:1678–1682.

Smith, R. D., James, M. A., Ristic, M., Aikawa, M., and Vega y Murguia, C. A., 1981, Bovine babesiosis: Protection of cattle with culture-derived soluble *Babesia bovis* antigen, *Science* **212**:335–338.

Stahl, H.-D., Crewther, P. E., Anders, R. F., Brown, G. V., Coppel, R. L., Bianco, A. E., Mitchell, G. F., and Kemp, D. J., 1984, Interspersed blocks of repetitive and charged amino acids in a dominant immunogen of *Plasmodium falciparum*, *Proc. Natl. Acad. Sci. USA* **82**:543–547.

Stepanova, N., Zablotsky, V., and Mutuzkina, Z., 1985, Specific prevention of bovine theileriasis in the Soviet Union, in: *Proceedings of the VII International Congress of Protozoology*, Nairobi, Abstract 168, p. 95.

Stocker, R., Hunt, N. H., Buffinton, G. D., Weidemann, M. J., Lewis-Hughes, P. H., and Clark, I. A., 1985, Oxidative stress and protective mechanisms in erythrocytes in relation to *Plasmodium vinckei* load, *Proc. Natl. Acad. Sci. USA* **82**:548–551.

Taliaferro, W. H., 1948, The inhibition of reproduction of parasites by immune factors, *Bacteriol. Rev.* **12**:1–17.

Taliaferro, W. H., and Taliaferro,, L. G., 1944, The effect of immunity on the asexual reproduction of *Plasmodium brasilianum*, *J. Infect. Dis.* **75**:1–32.

Taverne, J., Dockrell, H. M., and Playfair, J. H. L., 1981, Endotoxin-induced serum factor kills malarial parasites *in vitro*, *Infect. Immun.* **33**:83–89.

Taverne, J., Matthews, N., Depledge, P., and Playfair, J. H. L., 1984, Malarial parasites and tumor cells are killed by the same component of tumor necrosis serum, *Clin. Exp. Immunol.* **57**:293–300.

Thomas, A. W., Deans, J. A., Mitchell, G. H., Alderson, T., and Cohen, S., 1984, The Fab fragments of monoclonal IgG to a merozoite surface antigen inhibit *Plasmodium knowlesi* invasion of erythrocytes, *Mol. Biochem. Parasitol.* **13:**187–199.

Trager, W., Lanners, H. N., Stanley, H. A., and Langreth, S. G., 1983, Immunization of owl monkeys to *Plasmodium falciparum* with merozoites from cultures of a knobless clone, *Parasite Immunol.* **5:**225–236.

Vande Waa, J. A., Jensen, J. B., Akood, M. A. S., and Bayoumi, R., 1984, Longitudinal study on the *in vitro* immune response to *Plasmodium falciparum* in Sudan, *Infect. Immun.* **45:**505–510.

Wåhlin, B., Wahlgren, M., Perlmann, H., Berzins, K., Björkman, A., Patarroyo, M. E., and Perlmann, P., 1984, Human antibodies to a M_r 155,000 *Plasmodium falciparum* antigen efficiently inhibit merozoite invasion, *Proc. Natl. Acad. Sci. USA* **81:**7912–7916.

Wedderburn, N., and Dracott, B. N., 1977, The immune response to type III pneumococcal polysaccharide in mice with malaria, *Clin. Immunol.* **28:**130–137.

Whittle, H. C., Brown, J., March, K., Greenwood, B. M., Seidelin, P., Tighe, H., and Wedderburn, L., 1984, T-cell control of Epstein–Barr virus-infected B cells is lost during *P. falciparum* malaria, *Nature* **312:**449–450.

Williamson, W. A., and Greenwood, B. M., 1978, Impairment of the immune response to vaccination after acute malaria, *Lancet* **1:**1328–1329.

World Health Organization Scientific Working Group on Immunology of Malaria, 1984, Recent progress in the development of malaria vaccines: Memorandum from a WHO meeting, *Bull. WHO* **62:**715–727.

Wozencraft, A. O., Dockrell, H. M., Taverne, J., Targett, G. A. T., and Playfair, J. H. L., 1984, Killing of human malaria parasites by macrophage secretory products, *Infect. Immun.* **43:**664–669.

Wright, I. G., Mirre, G. B., Rode-Bramanis, K., Chamberlain, M., Goodger, B. V., and Waltisbuhl, D. J., 1985, Protective vaccination against virulent *Babesia bovis* with a low molecular weight antigen, *Infect. Immun.* **48:**109–113.

Wyler, D. J., Oppenheim, J. J., and Koontz, L. C., 1979, Influence of malaria infection on the elaboration of soluble mediators by adherent mononuclear cells, *Infect. Immun.* **24:**151–159.

Yamaga, K. M., Kan, S. C., Kramer, K. J., and Siddiqui, W. A., 1984, *Plasmodium falciparum:* Comparative analysis of antigens from continuous *in vitro* cultured and *in vivo* derived malarial parasites, *Exp. Parasitol.* **58:**138–146.

Yoshida, N., Nussenzweig, R., Potocnjak, P., Nussenzweig, V., and Aikawa, M., 1980, Hybridoma produces protective antibodies directed against the sporozoite stage of malaria parasite, *Science* **207:**71–73.

Young, J. F., Hockmeyer, W. T., Gross, M., Ballou, W. R., Wirtz, R. A., Trosper, J. H., Beaudoin, R. L., Hollingdale, M. R., Miller, L. H., Diggs, C. L., and Rosenberg, M., 1985, Expression of *Plasmodium falciparum* circumsporozoite proteins in *Escherichia coli* for potential use in a human malaria vaccine, *Science* **228:**958–962.

CHAPTER 20

Immunology of Leishmaniasis and American Trypanosomiasis (Chagas' Disease)

These two important protozoan diseases, leishmaniasis and Chagas' disease, will be considered together not only because both are caused by species of hemoflagellates that develop intracellularly, but also because the manifestations of both depend to a great extent on the immunological reactions of the host. This is particularly true of the leishmaniases, a group of diseases caused by various species and subspecies of the genus *Leishmania* all transmitted by sand flies of the genera *Phlebotomus, Lutzomyia,* and others. All are zoonoses in the sense that a variety of mammals, wild or domesticated, such as rodents, opossums, or dogs, serve as reservoir hosts. As already discussed (Chapter 3), the most remarkable characteristic of the *Leishmania* is that their habitat in the vertebrate host is within the macrophage, the very cell that normally constitutes the first line of defense against invading organisms. Furthermore, they actually multiply within the phagolysosomes. Somehow they are impervious to the low pH (around 5) and the powerful hydrolases of the phagolysosomal vacuole. Indeed, they have on their surface at least one powerful hydrolase of their own, an acid phosphatase. This enzyme has been found to inhibit superoxide anion production by human neutrophils activated with the chemoattractant peptide f-Met-Leu-Phe. The phosphatase apparently acts on the receptor for the formylated peptide activator. One leishmania contains enough acid phosphatase to reduce by 50% the superoxide anion production of three neutrophils. Since the enzyme has the same effect on macrophages, this may help to explain the reduced respiratory bursts elicited by amastigotes of *Leishmania.* No doubt other enzymes will be found that help *Leishmania* to live and grow in the adverse conditions of a phagolysosome.

Immunity to these parasites develops only slowly. It may be completely effective as in some forms of dermal leishmaniasis, notably that caused by *L. tropica* in the Middle East. Here a local skin lesion develops, presumably at the site of introduction of parasites by the bite of an infected sand fly. After several months to a year or two the lesion heals, and the person is then immune to reinfection. In the serious forms of leishmaniasis, however, nonhealing metastatic foci develop either in the skin (the diffuse cutaneous form), in the mucocutaneous tissues (as in South American espundia), or in the viscera (as in kala-azar). These are generally associated with suppressed cell-mediated immunity, but it is difficult to say which is cause and which effect. In experimental animals, such as guinea pigs with the specific guinea pig parasite *L. enriettii*, nonhealing metastatic foci may develop even in the presence of a protective immunological response that subsequently wanes. I have previously discussed the genetic basis for resistance and susceptibility of mice to *L. donovani*, the parasite causing visceral leishmaniasis. Here the susceptible animals could be grouped into those that would recover spontaneously, i.e., would develop an effective acquired immunity, and those that would not. This difference is evidently controlled by at least two genes within or closely adjacent to the major histocompatibility complex (*H-2*) of the mouse. The results of an experiment demonstrating this are shown in Fig. 20.1. Note that the mice of haplotype $H\text{-}2^{d/d}$ and $H\text{-}2^{b/d}$ both failed to reduce their parasite load whereas those of haplotype $H\text{-}2^{b/b}$ did so, especially by the 50th day and thereafter. Similar results were obtained with four other *H-2* haplotypes tested on the B10 genetic background and with the same *b* and *d* haplotypes on a different genetic background (BALB). An additional gene, not in the *H-2* complex, seems to be involved in the final clearing of parasites.

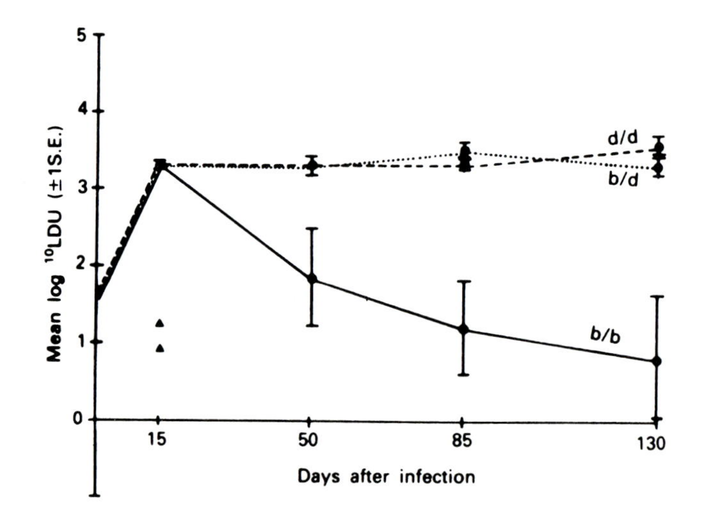

FIGURE 20.1. *Leishmania donovani* infection in the liver of mice with homozygous *b/b*, heterozygous *b/d*, and homozygous *d/d H-2* haplotypes. Groups of ten mice were infected intravenously with 10^7 amastigotes. Two mice from each group were killed on days 15, 50, and 85 and four on day 130. Their livers were used to prepare Giemsa-stained impression smears, from which the number of parasites per 500 liver cell nuclei was counted and expressed as *L. donovani* units (LDU). The triangles represent individual counts from control homozygous *Lsh'* (innately resistant) mice. (From Blackwell, 1982.)

In cutaneous leishmaniasis also certain strains of mice provide excellent models for the self-healing and the diffuse proliferative forms of the disease. In C3H/HeJ mice inoculated with either *L. tropica major* or *L. mexicana amazonensis* subcutaneously at the base of the tail, a local lesion develops, reaches a maximum diameter of 4–5 mm in about 5 to 8 weeks, and then spontaneously regresses and heals completely. In these mice the depletion of B cells and the complete suppression of antibody formation by the injection of goat anti-mouse IgM had no effect, indicating that this immunity is almost entirely cell-mediated. In BALB/c mice, however, the lesion grows rapidly, reaching 5-mm diameter in only 2 weeks, and then continues to increase progressively in size and produces in 8 to 12 weeks a disseminated cutaneous infection as well as visceralization of parasites to the liver and spleen. Many of these mice die. Most interesting were results obtained in such BALB/c mice that were treated throughout with the goat anti-mouse IgM. In these mice the development of the lesion stopped suddenly at 2 weeks (Fig. 20.2) and no disseminated disease developed up to 12 weeks as long as this treatment was continued. Development of antileishmanial antibodies was almost completely suppressed in these animals, and they showed marked delayed-type hypersensitivity to leishmanial extracts, not shown by untreated controls. Further work revealed that the antibodies were not directly suppressing cell-mediated

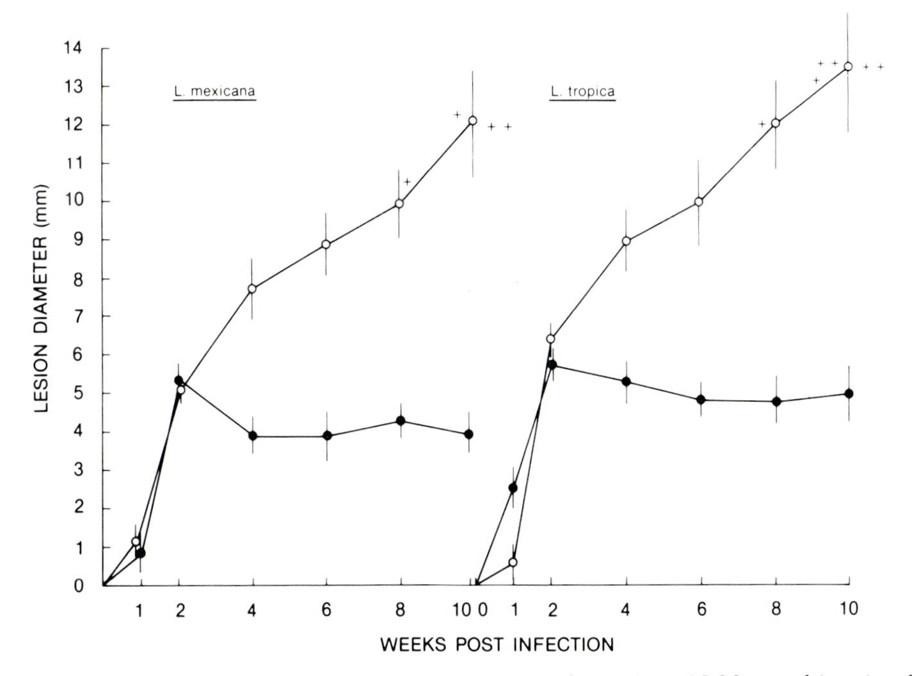

FIGURE 20.2. The course of *L. tropica* and *L. mexicana* infection in anti-IgM-treated (●—●) and untreated, control (○—○) BALB/c mice. The mean lesion size ± S.E. ($n = 15$) is shown. (Pluses indicate deaths due to disseminated leishmaniasis.) (From Sacks *et al.*, 1984.)

immunity by blocking induction or expression of antigen-reactive T cells. The transfer of splenic T cells from normal BALB/c mice into BALB/c mice that had been rendered resistant by sublethal irradiation abrogated their resistance, whereas transfer of T cells from mice depleted of B cells by the anti-mouse IgM treatment did not have this effect. Thus, it would appear that B cells are required for the generation of suppressor T cells, a homeostatic control network. In the BALB/c mouse (and perhaps also in humans of certain genetic constitutions) these T cells function in a counterproductive way to suppress the cell-mediated immunity, such as develops in C3H/HeJ mice (and in humans of different genetic constitution).

It is of interest that very high levels of serum IgG and IgM, even higher than in malaria, characterize kala-azar, the human visceral leishmaniasis caused by *L. donovani.* These antibodies are not only useless to the host but probably serve to help protect the parasites by interfering with the establishment of an effective cell-mediated immunity.

An effective immunity, whether in human or mouse, is cell-mediated and is always accompanied by delayed-type hypersensitivity. There is evidence from experimental work with mice and *L. tropica* for parasite-specific T cells that mediate the following functions: (1) helper activity for antibody responses; (2) transfer of antigen-specific delayed-type hypersensitivity responses to normal mice; (3) specific activation of parasitized macrophages resulting in destruction of intracellular parasites. In addition, macrophages may be activated in such a way as to render them resistant to infection by the amastigotes. Lymphokines having such effects may be induced specifically or nonspecifically, as in spleen cells stimulated with a mitogen or taken from mice previously injected with BCG. It may be that the lymphokines merely induce enhanced levels of the same microbicidal mechanisms already present but at ineffective levels, as for example the mechanism of oxidative killing (see Chapter 17). Or it could well be that the lymphokines induce entirely different additional killing mechanisms. The starvation of *Toxoplasma* for tryptophan, produced by γ-interferon (see Chapter 19), serves to illustrate the interesting possibilities that await the original investigator.

Cutaneous leishmaniasis of the Old World, caused by *L. tropica major* (also known as *L. major*), is ordinarily a self-curing disease. Furthermore, immunity has been induced by exposure to living promastigotes. These two facts suggest that this widespread and sometimes disfiguring infection might be amenable to control by vaccination. In keeping with this is the recent finding that mice could be protected against *L. major* infection by vaccination with a purified glycolipid prepared from cultures of promastigotes of this species. The glycolipid was purified by affinity chromatography using a monoclonal antibody. Mice of the highly susceptible BALB/c line vaccinated with only 2 μg of the glycolipid together with Freund's complete adjuvant and then challenged with 1×10^4 promastigotes, showed no lesions until after 60 days, and even at 75 days only 10 of the 16 animals had lesions. The 8 controls that received adjuvant alone all showed lesions at 50 days, and at

75 days their lesion score was much higher than that of the vaccinated ones. This glycolipid has been shown to be the leishmanial receptor for macrophages; it also appears on the surface of infected macrophages. Hence, immunity induced by it might operate both by interfering with uptake of promastigotes by macrophages and also by effects on macrophages that did become infected. No immunity could be induced with the water-soluble carbohydrate portion of the glycolipid, which is one of the "excreted factors" found in the medium in which promastigotes have been grown.

In Chagas' disease, as in leishmaniasis, serious consequences of the infection may be as much a result of immunological reactions of the host as of activities of the parasite. It is probable that only about one-third of individuals infected with *Trypanosoma cruzi* show an acute febrile disease coinciding with the original infection. In both individuals with and without initial symptoms, after a more or less prolonged latent period, the symptoms of chronic Chagas' disease may appear. Despite the early appearance of antibodies to *T. cruzi*, the parasites persist throughout much of the lifetime of the infected person. Mortality in chronic *T. cruzi* infections reaches almost 13 per 1000 with about 60% dying as a direct result of the disease, most of them from cardiac insufficiency. Pseudocysts of clusters of the intracellular amastigote forms of the parasite occur in heart muscle, as well as in other striated muscle. I have already noted (Chapter 3) that although *T. cruzi* can infect and develop in many types of vertebrate cells, the infective trypomastigotes of many strains have a special affinity for muscle cells. Some strains, however, are reticulotrophic in mice. The intracellular clusters of amastigotes presumably are protected from the action of antibodies by their intracellular location. It is a curious fact that clinical manifestations in chronic Chagas' disease are not related to the level of parasitemia, as far as this can be determined. It must be noted that this is a difficult determination to make. From living patients parasites can be demonstrated (as by culture or by allowing triatomid bugs to feed on them) in about half the cases, but this does not correlate with severity of the symptoms. In postmortem examinations, where tissue sections of heart muscle for example may reveal no parasites despite much evidence of pathology, one cannot be certain but that further examination would have revealed a pseudocyst of amastigotes close by. In recent years this situation has stimulated much work and speculation concerning the possibility that the effects of chronic Chagas' disease may result from autoimmune reactions of the host.

T. cruzi is highly immunogenic. During the acute phase, high titers of antibodies appear, mainly of the IgM type. Lower levels persist during the latent and chronic phases and these are mainly IgG and IgA. These antibodies, in the presence of complement, readily lyse the metacyclic trypomastigote forms found in the feces of infected triatomids. Hence, these antibodies provide protection from superinfections. They are not, however, very effective against bloodstream trypomastigotes. In the presence of complement these are lysed only with excess antibody. This fact, together with the protected intracellular position of the amastigotes, permits the infection to smolder on for years despite the development of antibodies and of cell-mediated immunity. At the same time there are often present antibodies that react against heart muscle and other striated muscle cells, as well as against neural cells. They can be removed by absorption of the serum with *T. cruzi*. It has been claimed that the pathological changes in cardiac muscle result from cell-mediated autoimmune reactions, but this is not fully accepted. The best evidence for cross-reactive antigens has been supplied by recent work in which a monoclonal antibody, CE5, raised against dorsal root ganglia of normal rats, was shown to bind to the surface of viable amastigotes of *T. cruzi* (Fig. 20.3). This monoclonal also bound to human and rat neurons known to degenerate in Chagas' disease. Moreover, the CE5 antibody immunoprecipitated an 87,500-dalton polypeptide obtained from *in vitro* translation of mRNA from *T. cruzi*,

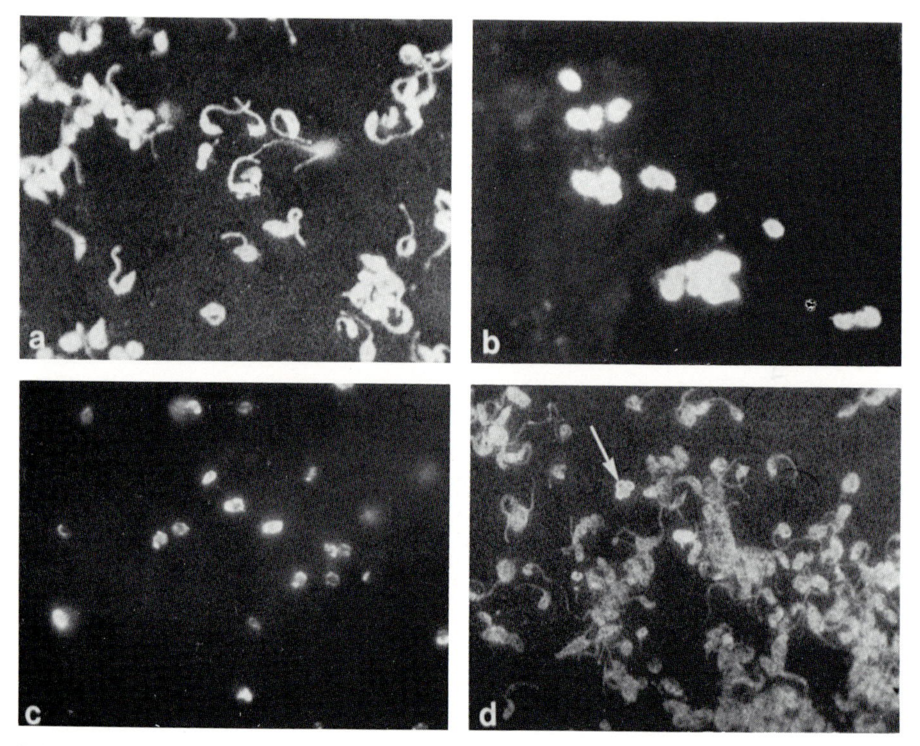

FIGURE 20.3. Staining of *T. cruzi* (Y strain) organisms with CE5 monoclonal antibody. (a) Formaldehyde-fixed epimastigotes showing homogeneous cytoplasmic fluorescence. (b) Formaldehyde-fixed amastigotes showing bright fluorescence associated with the surface membrane. (c) Viable amastigotes stained in suspension; living epimastigotes gave no reaction with CE5. (d) Batch culture of epimastigotes containing a low percentage of amastigotes, after formaldehyde fixation. Epimastigotes gave typical cytoplasmic fluorescence whereas in amastigotes (arrowed) the antigen detected by CE5 is limited to the cell-surface membrane. As a specificity control, parasites were also treated with an IgM monoclonal antibody against dinitrophenol or with ascites fluid containing MOPC 104E myeloma IgM. Positive staining was only observed with CE5. (From Wood *et al.*, 1982.)

showing that an antigen or at least an epitope, corresponding to this antibody is specified by the parasite genome and not passively acquired from the host. Neuronal damage is a major aspect of the pathology of chronic Chagas' disease. Indeed, F. Köberle has argued cogently that this is the principal factor and that most of the other pathology, such as enlarged heart, aneurysms of the heart, megaesophagus, and megacolon, follows from this. At one time it was thought this neuronal damage might be caused by a toxin, but no toxin has been demonstrated. Autoimmune reactions seem more likely agents of these effects.

This complex immunological situation is perhaps the main reason why so little progress has been made toward a vaccine against Chagas' disease. There is no doubt that a primary infection with *T. cruzi* in animals induces a significant immunity to reinfection. Such immunity has also been produced by various attenuated live vaccines; some protection against virulent *T. cruzi* has even resulted from inoculation of mice with the insect hemoflagellate *Leptomonas pessoai*. Immunization has also been obtained through injection of killed *T. cruzi* and various crude fractions of it. With all these methods the specter of autoimmunization cannot be ignored. A safer approach would seem to be the use of purified antigens of the parasite that would induce a protective

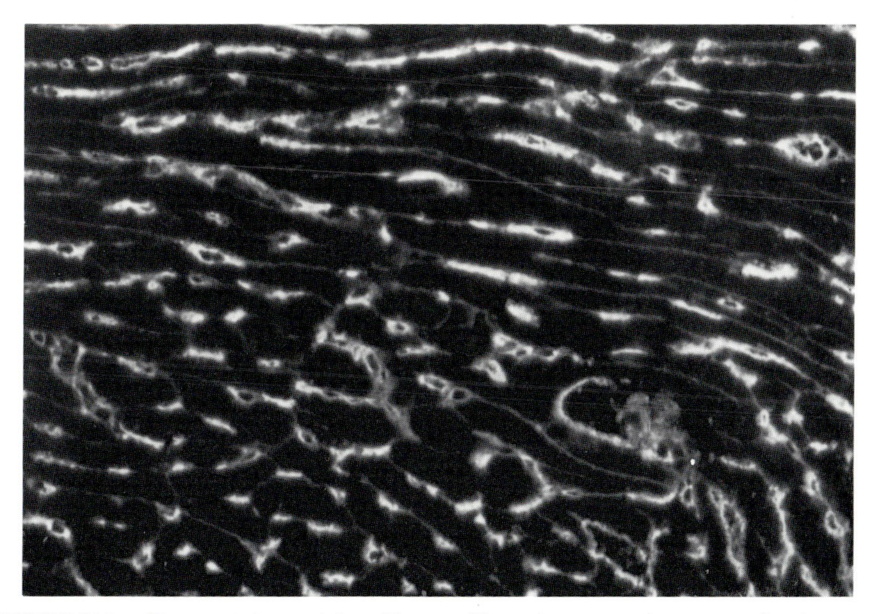

FIGURE 20.4. Characteristic reactivity of human Chagas' serum against mammalian heart tissue. Cryostat sections of mouse heart (5 μm thick) were prepared and stored at −20°C. They were incubated with serial dilutions of test serum for 20 min at room temperature, washed, developed with a 1 in 20 dilution of FITC-labeled rabbit anti-human IgG for 20 min at room temperature, and, after a final wash, examined using a fluorescence microscope. Normal human serum in identical conditions gave no detectable staining. (From Scott and Snary, 1979.)

immunity without cross-reacting with any host protein. An encouraging result has been obtained with a surface glycoprotein of *T. cruzi*. Mice injected with this material together with an adjuvant, were protected against an otherwise lethal challenge with a virulent strain of *T. cruzi*. The anti-heart tissue reactivity of a human Chagas' serum, as revealed by an immunofluorescent reaction (Fig. 20.4), was not removed by this glycoprotein, as it was by whole *T. cruzi* (Table 20.1). Unfortunately, this protein was not effective by itself; it produced immunity only when injected together with an adjuvant, such as Freund's complete adjuvant or Quillaya saponin. A safer adjuvant would have to be sought.

Just as lymphokines affect the uptake of leishmanial amastigotes by macrophages, so they also exert several kinds of effects on *T. cruzi*. In a comparison of two mouse strains—one (C3H/HeJ) exhibiting high parasitemia, and the other (C57BL/6) low parasitemia when infected with *T. cruzi*—it was found that the spontaneous release of lymphokine *in vitro* by spleen cells from infected mice was significantly higher for the cells from the C57BL/6 mice. Furthermore, supernatants from infected C57BL/6 spleen cells, after fivefold concentration by ultrafiltration, inhibited both the motility and the infectivity of bloodstream trypomastigotes, whereas similar preparations from C3H/HeJ splenocytes had little effect on motility and delayed but did not prevent infection (Table 20.2). Similarly, infection *in vitro* of macrophages or rat heart myoblasts by the bloodstream forms of *T. cruzi* was inhibited by β-interferon. Interferons are known to appear early in experimental infections with *T. cruzi* and it is likely that they play a mitigating role in initial resistance as well as in later immunity. It would be of interest to know what happens in this regard in the many wild animal hosts of *T. cruzi* where the infection seems to be nonpathogenic.

TABLE 20.1. Absorption of Autoreactive Human Chagas' Serum with Whole *T. cruzi* or Glycoprotein Fraction[a]

Serum	IFA titer against	
	Heart[b]	*T. cruzi*[c]
Chaga's serum unabsorbed	1 : 512	1 : 4096
Chagas' serum absorbed *T. cruzi*[d]	1 : 64	1 : 8
Chagas' serum absorbed glycoprotein[e]	1 : 512	1 : 1024
Normal human serum	1 : 8	1 : 16

[a] Modified from Scott and Snary (1979)
[b] IFA (indirect fluorescence assay) for antiheart reactivity as for Fig. 20.4.
[c] Air-dried smears of *T. cruzi* epimastigotes from *in vitro* culture were substituted for heart in IFA.
[d] Absorbed with 45 mg lysophilized epimastigotes (≃ 4.5 × 10^9 organisms)/ml serum; 1 hr at 37°C and then overnight at 4°C.
[e] Absorbed with glycoprotein (500 μg protein, ≃ 10^{10} organisms/ml serum); 1 hr at 37°C and then overnight at 4°C.

TABLE 20.2. Direct Effect of Naturally Released, Fivefold Concentrated Supernatants from Spleen Lymphocytes on Infectivity of Bloodstream Trypomastigotes of *T. cruzi* for the Human Diploid Cell Line W138 (ATCC No. CCL75)[a]

Lymphocytes	Time of observation							
	Day 1		Day 2		Day 3		Day 4	
	No. INF cells/50 fields	% inhib.	No. INF cells/50 fields	% inhib.	No. INF cells/50 fields	% inhib.	No. INF cells/50 fields	% inhib.
Medium control	76.5 ± 6.4	—	74.0 ± 1.4	—	120.0 ± 7.0	—	112.0 ± 24.7	—
Uninfected C3H/HeJ lymphocytes	Not done	—	50.5 ± 0.7	32	136.4 ± 60.0	(190)[b]	202.0 ± 44.8	(180)[b]
Infected C3H/HeJ lymphocytes	6.0 ± 0.0	93	16.8 ± 1.7	78	300.0 ± 21.2	(250)[b]	240.0 ± 28.3	(214)[b]
Uninfected C57BL/6 lymphocytes	Not done	—	39.0 ± 4.2	48	Not done	—	Not done	Not done
Infected C57BL/6 lymphocytes	17.0 ± 1.4	78	7.7 ± 6.6	90	8.25 ± 8.46	94	2.5 ± 3.5	98

[a] Modified from Krassner *et al.* (1982).
[b] Percent infected (INF) cells above medium control.

Bibliography

Barral, A., Petersen, E. A., Sacks, D. L., and Neva, F. A., 1983, Late metastatic leishmaniasis in the mouse: A model for mucocutaneous disease, *Am. J. Trop. Med. Hyg.* **32:**277–285.

Blackwell, J. M., 1982, Genetic control of recovery from visceral leishmaniasis, *Trans. R. Soc. Trop. Med. Hyg.* **76:**147–151.

Blackwell, J., Freeman, J., and Bradley, D., 1980, Influence of H-2 complex on acquired resistance to *Leishmania donovani* infection in mice, *Nature* **283:**72–74.

Brener, Z., 1980, Immunity to *Trypanosoma cruzi, Adv. Parasitol.* **18:**247–292.

Chang, K.-P., 1983, Cellular and molecular mechanisms of intracellular symbiosis in leishmaniasis, *Int. Rev. Cytol.* (Suppl.) **14:**267–305.

Channon, J. Y., Roberts, M. B., and Blackwell, J. M., 1984, A study of the differential respiratory burst activity elicited by promastigotes and amastigotes of *Leishmania donovani* in murine resident peritoneal macrophages, *Immunology* **53:**345–354.

De Souza, W., 1984, Cell biology of *Trypanosoma cruzi, Int. Rev. Cytol.* **86:**197–283.

Garnham, P. C. C., and Humphrey, J. H., 1969, Problems in leishmaniasis related to immunology, *Curr. Top. Microbiol. Immunol.* **48:**29–42.

Grimaud, J. A., and Andrade, S. A., 1984, *Trypanosoma cruzi:* Intracellular host–parasite relationship in murine infection, *Cell. Mol. Biol.* **30:**59–65.

Handman, E., and Mitchell, G. F., 1985, Immunization with Leishmania receptor for macrophages protects mice against cutaneous leishmaniasis, *Proc. Natl. Acad. Sci. USA* **82:**5910–5914.

Howard, J. G., and Liew, F. Y., 1984, Mechanisms of acquired immunity in leishmaniasis, *Philos. Trans. R. Soc. London Ser. B* **307:**87–98.

Hudson, L., 1983, *Trypanosoma cruzi:* The immunological consequences of infection, *J. Cell. Biochem.* **21:**299–304.

Hudson, L., 1983, Immunopathogenesis of experimental Chagas' disease in mice: Damage to the autonomic nervous system, *Ciba Found. Symp.* **99:**234–246.

Jones, T. C., 1981, Interactions between murine macrophages and obligate intracellular protozoa, *Am. J. Pathol.* **102:**127–132.

Kierszenbaum, F., and Sonnenfeld, G., 1984, β-Interferon inhibits cell infection by *Trypanosoma cruzi, J. Immunol.* **132:**905–908.

Köberle, F., 1974, Pathogenesis of Chagas' disease, *Ciba Found. Symp.* **20:**137–158.

Krassner, S. M., Granger, B., Morrow, C., and Granger, G., 1982, *In vitro* release of lymphotoxin by spleen cells from C3H/HeJ and C57BL/6 mice infected with *Trypanosoma cruzi, Am. J. Trop. Med. Hyg.* **31:**1080–1089.

Louis, J. A., Zubler, R. H., Coutinko, S. G., Lima, G., Behin, R., Mouel, J., and Engers, H. D., 1982, The *in vitro* generation and functional analysis of murine T cell populations and clones specific for a protozoan parasite *Leishmania tropica, Immunol. Rev.* **61:**215–243.

Mauel, J., 1982, *In vitro* induction of intracellular killing of parasitic protozoa by macrophages, *Immunobiology* **161:**392–400.

Mauel, J., 1984, Mechanisms of survival of protozoan parasites in mononuclear phagocytes, *Parasitology* **88:**579–592.

Nguyen, T. V., Mauel, J., Etges, R. J., Bouvier, J., and Bordier, C., 1984, Immunogenic proteins of *Leishmania major* during mouse infection, *Parasite Immunol.* **6:**265–273.

Nogueira, N., and Coura, J. R., 1984, American trypanosomiasis (Chagas' disease), in: *Tropical and Geographical Medicine* (K. S. Warren and A. A. F. Mahmoud, eds.), McGraw–Hill, New York, pp. 253–269.

Oster, C. N., and Nagy, C. A., 1984, Macrophage activation to kill *Leishmania tropica:* Kinetics of macrophage response to lymphokines that induce antimicrobial activities against amastigotes, *J. Immunol.* **132:**1494–1500.

Plata, F., Wietzerbin, J., Pons, F. G., Falcoff, E., and Elsen, H., 1984, Synergistic protection by specific antibodies and interferon against infection by *Trypanosoma cruzi, Eur. J. Immunol.* **14:**930–935.

Poulter, L. W., 1979, The kinetics and quality of acquired resistance in self-healing and metastatic leishmaniasis, *Clin. Exp. Immunol.* **36:**30.

Poulter, L. W., 1980, Mechanisms of immunity to leishmaniasis. II. Significance of the intramacrophage localization of the parasite, *Clin. Exp. Immunol.* **40:**25–35.

Ramaley, A. T., Kuhns, D. B., Basford, R. E., Glew, R. H., and Kaplan, S. S., 1984, Leishmanial phosphatase blocks neutrophil O^{-2} production, *J. Biol. Chem.* **259:**11173–11175.

Ribeiro dos Santos, R., and Hudson, L., 1980, *Trypanosoma cruzi:* Immunological consequences of parasite modification of host cells, *Clin. Exp. Immunol.* **40:**36–41.

Rowin, K. S., Tanowitz, H. B., Wittner, M., Nguyen, H. T., and Nadel-Ginard, B., 1983, Inhibition of muscle differentiation by *Trypanosoma cruzi, Proc. Natl. Acad. Sci. USA* **80:**6390–6394.

Sacks, D. L., Scott, P. A., Asofsky, R., and Sher, F. A., 1984, Cutaneous leishmaniasis in anti-IgM-treated mice: Enhanced resistance due to functional depletion of a B cell-dependent T cell involved in the suppressor pathway, *J. Immunol.* **132:**2072–2077.

Scott, M. T., and Snary, D., 1979, Protective immunization of mice using cell surface glycoprotein from *Trypanosoma cruzi, Nature* **282:**73–74.

Slutzky, G. M., El-On, J., and Greenblatt, C. L., 1979, Leishmanial excreted factor: Protein-bound and free forms from promastigote cultures of *Leishmania tropica* and *Leishmania donovani, Infect. Immun.* **26:**916–924.

Teixeira, A. R. L., 1979, Chagas' disease: Trends in immunological research and prospects for immunoprophylaxis, *Bull. WHO* **57:**697–710.

Turco, S. J., Wilkerson, M. A., and Clawson, D. R., 1984, Expression of an unusual acidic glycoconjugate in *Leishmania donovani, J. Biol. Chem.* **259:**3883–3889.

Villalta, F., and Kierszenbaum, F., 1983, Immunization against a challenge with insect vector, metacyclic forms of *Trypanosoma cruzi* simulating a natural infection, *Am. J. Trop. Med. Hyg.* **32:**273–276.

Wood, J. N., Hudson, L., Jessell, T. M., and Yamamoto, M., 1982, A monoclonal antibody defining antigenic determinants on subpopulations of mammalian neurons and *Trypanosoma cruzi* parasites, *Nature* **296:**34–38.

Wrightsman, R., Krassner, S., and Watson, J., 1982, Genetic control of responses to *Trypanosoma cruzi* in mice: Multiple genes influencing parasitemia and survival, *Infect. Immun.* **36:**637–644.

Zehavi, U., Abrahams, J. C., Granoth, R., Greenblatt, C. L., Slutzky, G. M., and El-On, J., 1983, Leishmanial excreted factors (EFs): Purification by affinity chromatography, *Z. Parasitenkd.* **69:**695–701.

Zuckerman, A., and Lainson, R., 1977, Leishmania, in: *Parasitic Protozoa,* Volume I (J. P. Kreier, ed.), Academic Press, New York, pp. 57–133.

CHAPTER 21

Schistosomiasis and Concomitant Immunity

Most helminthic parasites do not multiply within a single host. The intensity of infection depends on the amount of exposure to invasive forms and on the development of an acquired immunity to these invasive forms. In some worm–host combinations, acquired immunity extends also to the adult forms resulting in "self-cure" (see Chapter 23), but in others the worms are beautifully adapted to continue living and mating and laying eggs in a host that is immune to reinfection. The blood flukes, including the three main species parasitic in humans—*Schistosoma mansoni, S. japonicum,* and *S. haematobium*—provide excellent examples of this phenomenon of "concomitant immunity." In these worms the sexes are separate but live together in pairs, the narrow rounded body of the female enclosed in a gynecophoric canal formed by the curling of the flat body of the male around it. Such pairs of *S. mansoni* can live for years in the mesenteric veins of a person and continue to produce eggs. Yet the indications are that people in endemic areas eventually become immune to superinfection by additional cercariae of *S. mansoni.* This is clearly a situation that is advantageous to the parasite as well as the host. Heavy infection with *S. mansoni* could result in death of the host and the accompanying death of the worms within it. As it is, both live on to seed heavily the environment with the worm's eggs excreted with the person's feces, thereby infecting many snails and disseminating the infection to many new hosts (see Diagram XIII). There are here two questions of great immunological interest. First, what are the immune mechanisms that prevent superinfection? Second, how can the adult worms evade this acquired immunity to their earlier stages and live on in the presence of immune reactions which they themselves have provoked? Closely related to these questions is the problem of pathogenesis of schistosomiasis, one of the most widespread and serious of parasitic diseases.

That these questions are not yet fully answered is not surprising if we consider the complex developmental and migratory path of the worm within its definitive host, a human or a mouse. First, the free-swimming cercariae

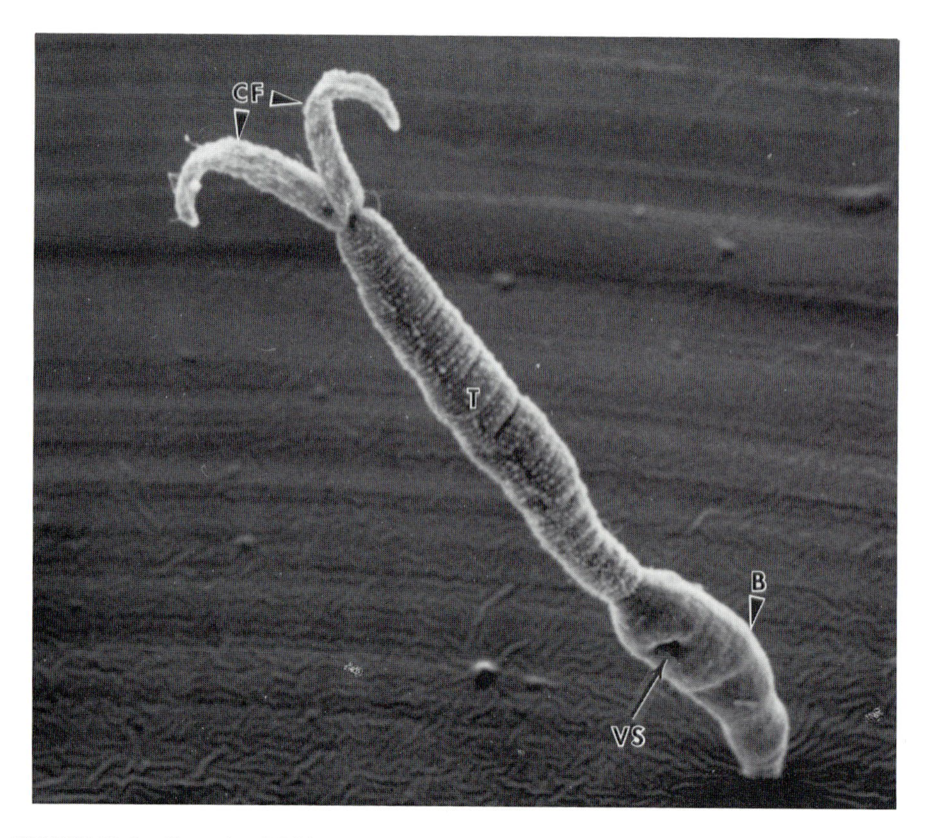

FIGURE 21.1. Cercaria of *Schistosoma mansoni* attached to the substrate by its oral sucker. B, body; VS, ventral sucker; T, tail, CF, caudal furcae. SEM. ×200. (From McLaren, 1980. Original print courtesy of Dr. D. J. McLaren.)

(Fig. 21.1) penetrate the skin and within minutes transform to schistosomula with very different physiological characteristics (see Chapter 2). The schistosomula are carried via the blood vessels and heart to the lungs in one or several cycles while they develop for a week or so. Again via the circulation they are carried to the small portal veins of the liver where, by about 3–4 weeks after infection, the now nearly mature worms start pairing. The male worm then carries the female into mesenteric veins. Here, in the embrace of the male, the female completes maturation, is fertilized, and begins laying increasing numbers of eggs. The spined eggs are designed to pass through the walls of the venules into the gut, from which many are shed with the feces. But a great excess of eggs is produced so that many others are carried by the blood flow mostly back to the liver (remember that the portal flow is from the alimentary tract to the liver) and also to other organs. Thus, the host is exposed to antigens first from the penetrating cercariae, then from the migrating schistosomula, then from the developing worms in the lungs and liver, and finally from the adult worms and their eggs. The most likely targets of protective immunity to superinfection could be either the cercariae, the migrating schistosomula, or the young worms in the lungs.

Much evidence, largely from *in vitro* studies, indicates that the newly transformed schistosomula are among the targets of effective immune reactions. Of special interest is the surface of the organism. This undergoes drastic modification just after penetration of a cercaria through the skin of a host. The surface of the cercaria bounded by a single trilaminate outer membrane surrounded by a glycocalyx (Fig. 21.2), changes into a double outer membrane without a glycocalyx. The glycocalyx is lost quickly upon skin penetration and it is probable that this loss is responsible for the acquired sensitivity to water; within 30 min after skin penetration, 90% of transforming schistosomula are killed by exposure to water. Whereas the cercariae were very much at home in water and could not survive if placed directly in saline, the schistosomula derived from them now survive in saline and indeed will develop *in vitro* in appropriate media (see Chapter 8). The changes can also be induced mechanically by centrifuging and vortexing the cercariae, followed by incubation at 37°C in an appropriate culture medium with serum, a procedure that has greatly facilitated experimental work with these organisms. At 30 min after skin penetration, schistosomula still have a single trilaminate lim-

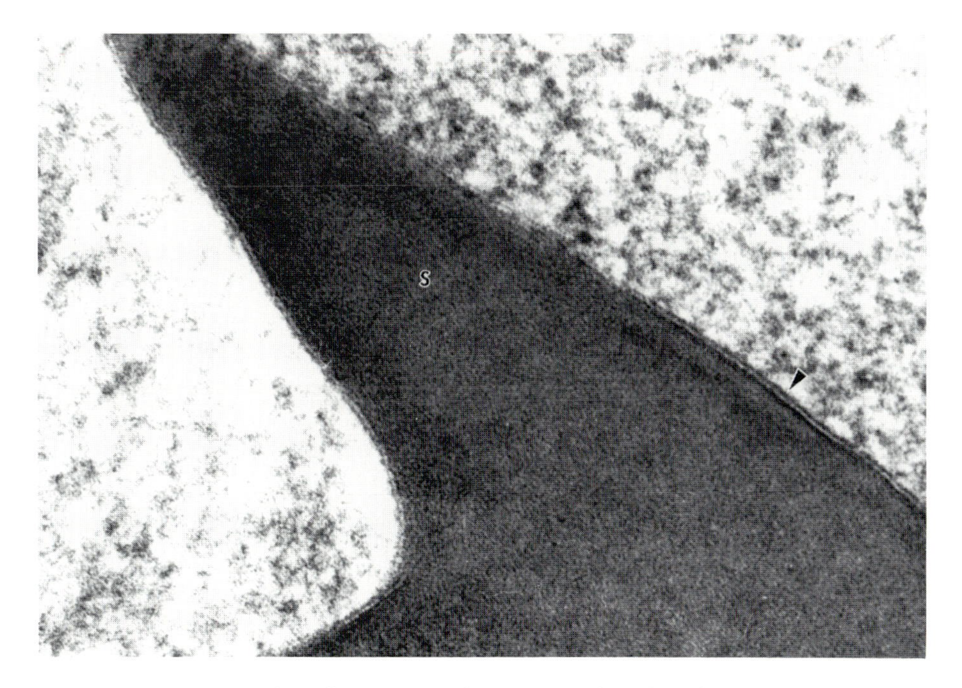

FIGURE 21.2. A spine (S) on the tegument of a cercaria to show the trilaminate outer membrane (arrowhead) and the fibrillar nature of the surrounding glycocalyx. TEM. ×97,000. (From McLaren, 1980. Original print courtesy of Dr. D. J. McLaren.)

iting membrane but have lost most of their surface coat. During the next hour, membranous inclusion bodies from the tegmental cells move to the surface (Fig. 21.3); their double membranes fuse with the tegmental outer membrane and their contents spread over the surface. Numerous microvilli are formed and then shed, so that the host is exposed to a large amount of parasite antigen. By 3 hr after penetration, the schistosomula have acquired the double outer membrane (Fig. 21.4) that they will maintain into adult life and that characterizes all flukes that live in blood. They are no longer sensitive to complement as were the cercariae.

Of the various experimental models used for the study of immunity and pathogenesis in *S. mansoni* infection, the mouse model seems to be closest to the situation in humans. In mice as in humans, immunity to reinfection is incomplete and coincides with continued life of the adult worms, i.e., we have a concomitant immunity. Some immunity can be produced by the adult worms alone if they are artificially transplanted into the mesenteric veins of a naive mouse, showing that the several migrating stages are not essential to its development. It can also be produced by infecting mice with irradiated cercariae that do not develop beyond the schistosomulum stage, showing that adult worms are not essential. Moreover, there is evidence that egg production

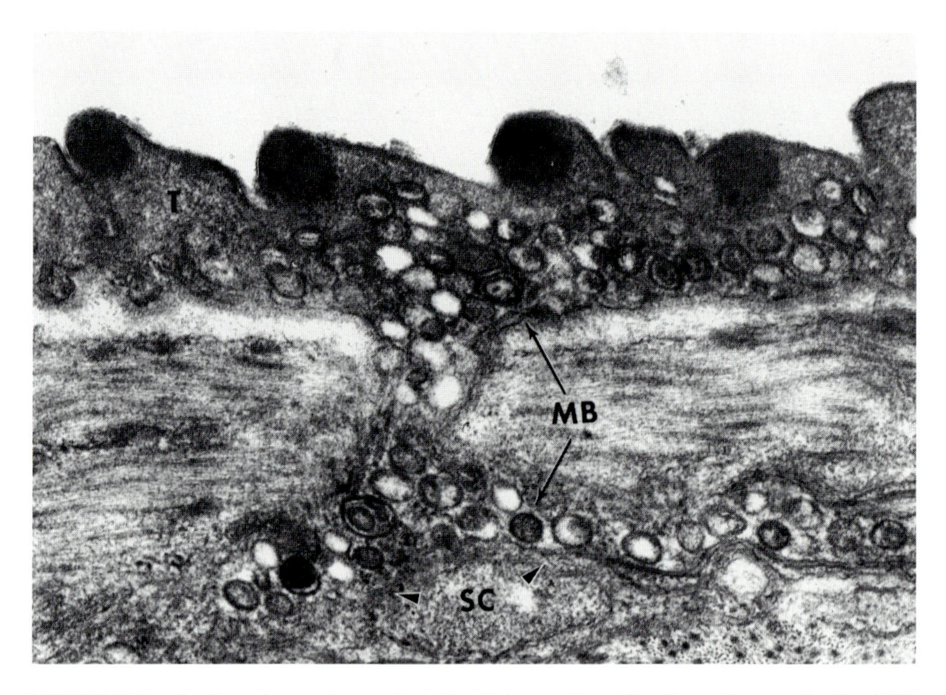

FIGURE 21.3. Surface of a newly penetrated schistosomulum. Small membranous bodies (MB) are passing from a subtegmental cell (SC) through a connection to the tegument (T). TEM. ×36,000. (From McLaren, 1980. Original print courtesy of Dr. D. J. McLaren.)

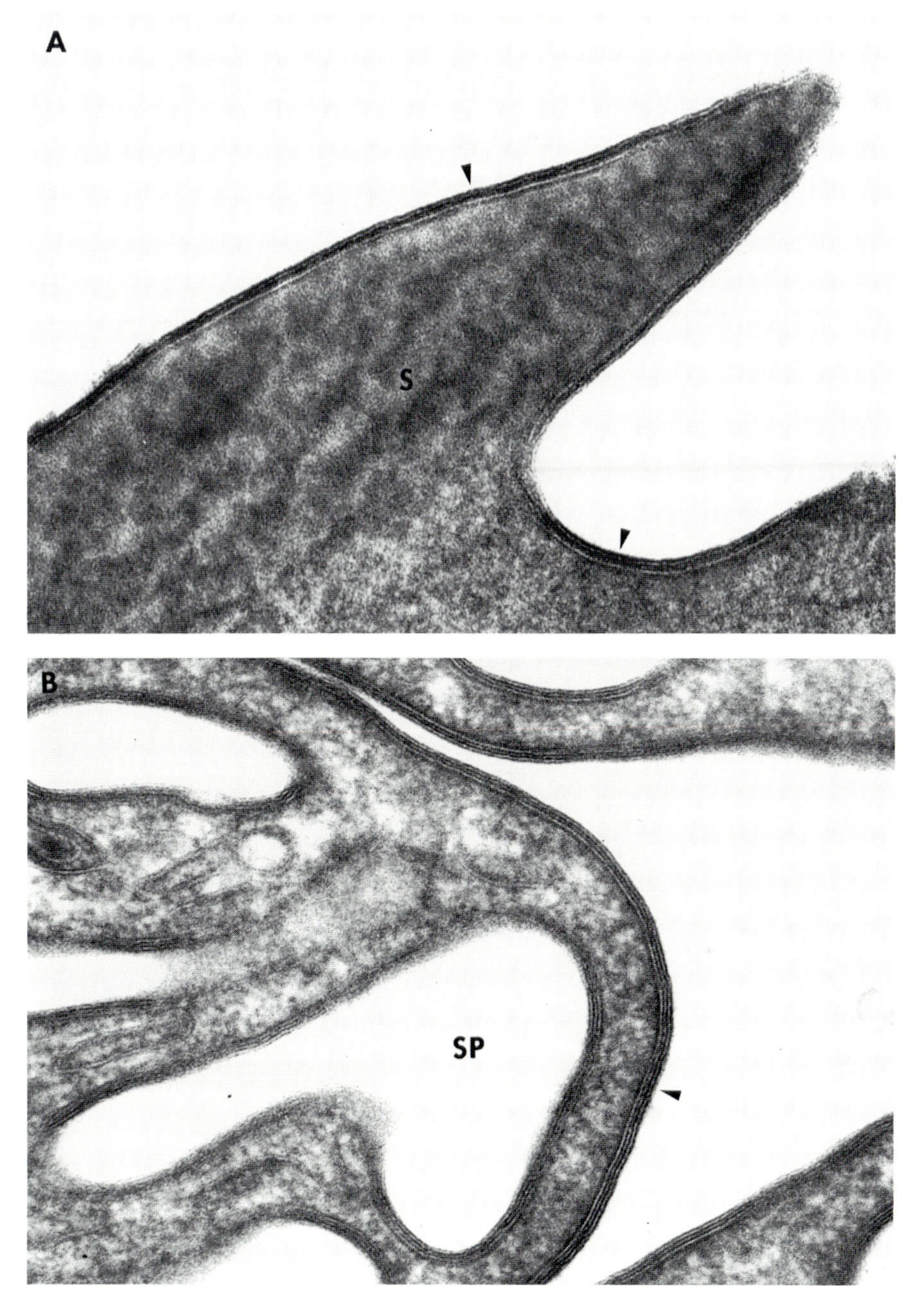

FIGURE 21.4. (A) Newly formed double outer membrane (arrowheads) of a schistosomulum 3 hr after skin penetration. S, spine. TEM. ×124,000. (B) Double outer membrane (arrowhead) of adult *S. mansoni*. This tegumental membrane also lines the surface pits (SP). TEM. ×112,000. (From McLaren, 1980. Original prints courtesy of Dr. D. J. McLaren.)

can also play an important role in the acquisition of an effective immunity. In mice, unlike the situation in less suitable hosts, a unisexual infection with either male or female *S. mansoni* confers less resistance to a challenge infection than does a bisexual infection.

Furthermore, in mice infected with worms of both sexes and challenged 6–8 weeks later, and hence at a time when the first worms were adult but had not yet begun laying eggs, the worms of the second infection were found to be completely eliminated if the count was based on adult worms present 7 weeks after giving the challenge. If, however, the mice were killed at only 6 days after giving the challenge and the schistosomula in the lungs were counted, there was no reduction in their numbers as compared with controls infected for the first time. Hence, in these mice the immunity was exerted on worms beyond the lung stage. On the other hand, if mice were exposed to a challenge infection 12 or more weeks after the initial infection, the number of living schistosomula present 6 days later was about 70% less than in controls (Fig. 21.5). In such mice, schistosome eggs were present in the lungs at the time of challenge. This strongly suggests that the formation of granulomas around the eggs in the lungs, and in other tissues, plays a significant role in this immunity. Indeed, under these conditions a significant portion of the apparent immunity may be a result of the pathological hemodynamic and inflammatory changes produced by the egg granulomas. If such effects were eliminated by immunizing mice with irradiated cercariae, it was found that in the immunized mice the additional elimination of challenge worms over

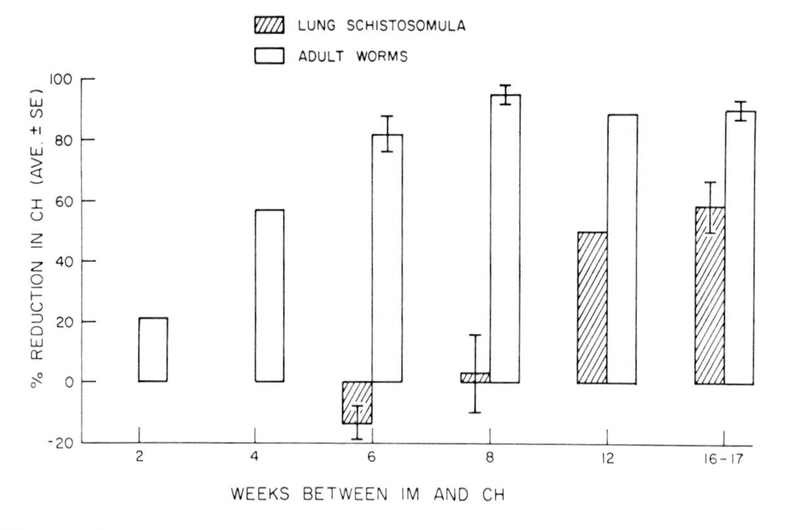

FIGURE 21.5. *S. mansoni* in mice. Comparison of resistance based on lung schistomulum (6 day) and adult worm (7 week) counts. Shown are averages from all experiments from all mice challenged at 6, 8, 12, and 16–17 weeks after initial infection. IM, immunization; CH, challenge. See text. (From Dean *et al.*, 1978b.)

TABLE 21.1. Passive Transfer of Protection to *Schistosoma mansoni* Using Monoclonal Antibody EG1C4B1[a,b]

Experiment no.		Mean no. of worms	Percent protection
1	Normal sera	44 ± 15	—
	Chronic sera	16 ± 11	66
	Monoclonal Ab	19 ± 8	57
2	Normal sera	40 ± 20	—
	Chronic sera	12 ± 9	70
	Monoclonal Ab	5 ± 6	88
3	Normal sera	63 ± 16	—
	Chronic sera	19 ± 16	60
	Monoclonal Ab	16 ± 11	75

[a] Modified from Harn *et al.* (1984).
[b] 500 mechanical somula injected i.v. Monoclonal antibody (*culture supernatant*) given 3–12 hr earlier. Lung worms harvested on day 4. Experiments 2 and 3 used fivefold concentrated *culture supernatant*.

that seen in controls occurred after the lung stage, so that reduced numbers reached the liver. The schistosomula in the lungs were not reduced in number but their migration from the skin was delayed by several days. Some workers, however, have also observed an early effect.

K. S. Warren and his associates have shown the importance of the egg granulomas in the immunopathogenesis of schistosomiasis. Mice infected with *S. mansoni* have provided an excellent model for some features of the human disease. For example, within 3 months after heavy infection, they develop hepatomegaly, splenomegaly, portal hypertension, and esophageal varices. The granulomas in mice are T cell dependent and are accelerated and potentiated by delayed hypersensitivity. Study of the antigens present in the eggs and responsible for the sensitization resulting in granuloma formation has revealed at least three soluble proteins, of which two are glycoproteins.

Attempts to immunize mice with egg antigens or whole egg extracts have given generally negative results. Recently, however, an antiegg monoclonal antibody has been obtained (using eggs of *S. mansoni*) that protected mice against *in vivo* challenge with cercariae as effectively as serum from chronically infected mice (Table 21.1). This antibody reacts with the surface of schistosomula (Fig. 21.6) even 96 hr after their transformation. It binds to egg antigens with molecular weights of 111,000, 200,000, and over 200,000, and also to schistosomulum antigens with molecular weights of 130,000 and 160,000. Like serum from infected mice, the monoclonal antibody had a direct killing effect in the presence of complement (Fig. 21.7). A 38,000-molecular-weight glycoprotein (GP 38) on the surface of the schistosomula has also been characterized with a monoclonal antibody. This is present during the first 3 days of parasitic life but is no longer detectable by day 5 in the lung stage. The same monoclonal antibody recognizes a GP 115 present in the metabolic

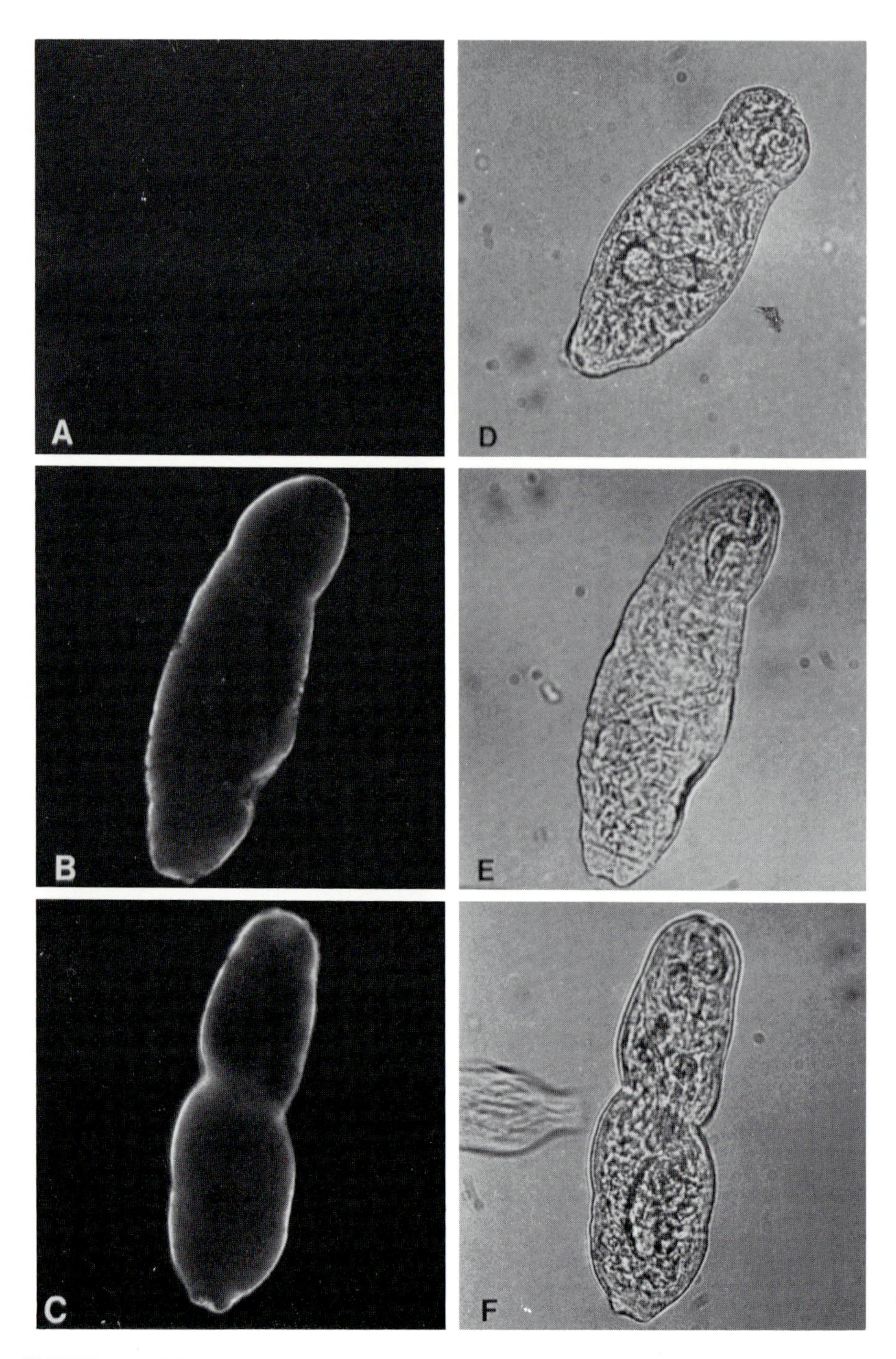

FIGURE 21.6. Schistosomula exposed to normal mouse sera (A, D), monoclonal antibody EG1C4B1 (B, E), and infected mouse sera (C, F) as seen with immunofluorescence (A–C) and light microscopy (D–F). (From Harn *et al.*, 1984. Original prints courtesy of Dr. D. A. Harn.)

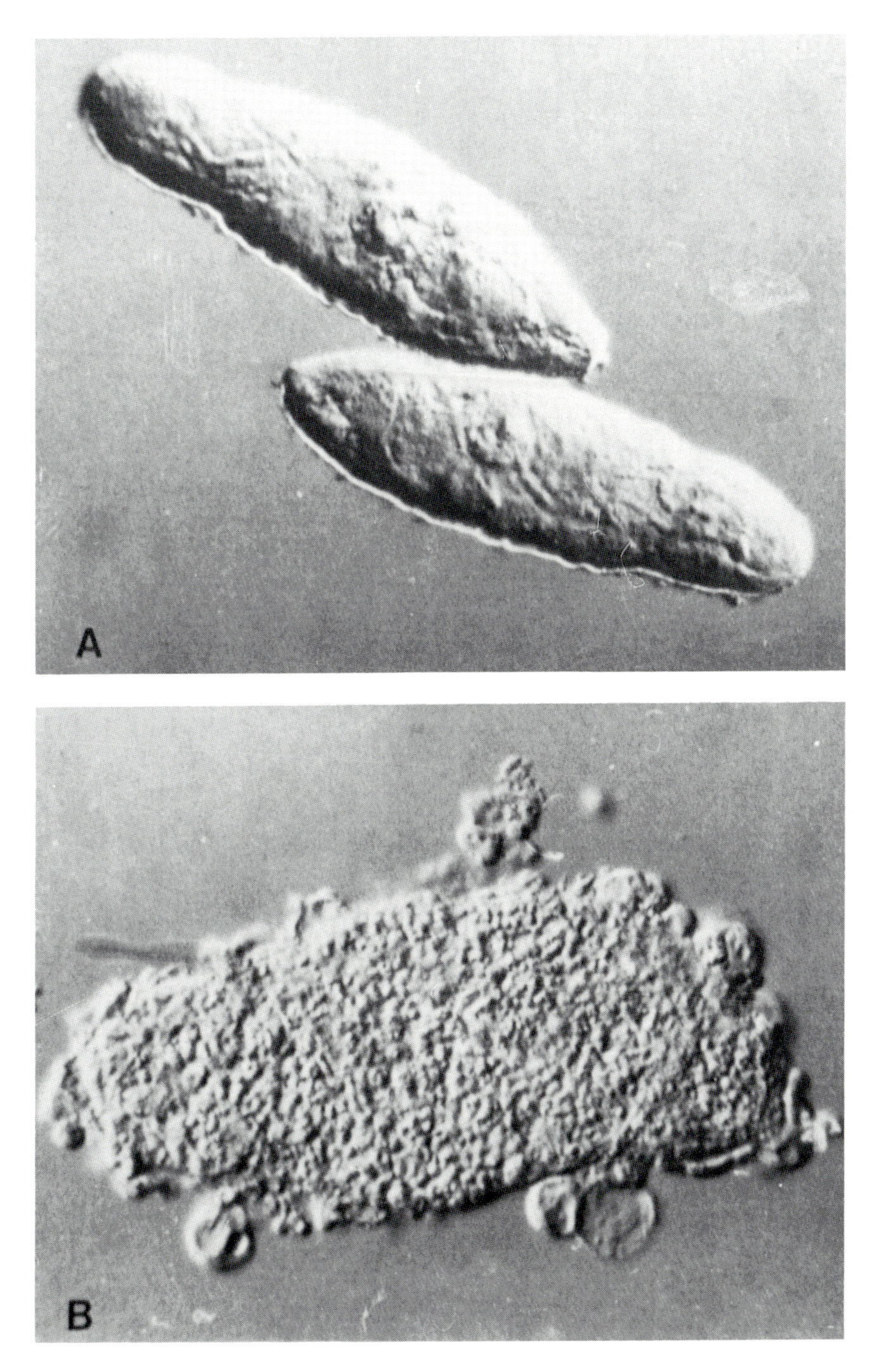

FIGURE 21.7. Schistosomula exposed to (A) normal mouse serum plus fresh guinea pig serum; (B) monoclonal antibody EG1C4B1 plus fresh guinea pig serum, showing extensive complement-mediated damage. (From Harn *et al.*, 1984. Original prints courtesy of Dr. D. A. Harn.)

products of the adult worm, but not on its surface. Immunization of rats with GP 115 induces IgG2a antibody that binds to GP 38 on the schistosomula.

The serum of infected humans has similar killing activity in the presence of complement. Perhaps of greater relevance to the actual *in vivo* situation is the action of such serum against schistosomula in conjunction with white cells, and here eosinophils seem to play a special role. If schistosomula are placed with serum from a schistosomiasis patient and eosinophils from a normal person, or from the patient, the worms are soon killed (Table 21.2). Complement is not required, although it does enhance the effect. The detailed mechanism of killing is of great interest. In the presence of specific antibody of the IgG type, the eosinophils attach tightly to the surface of the schistosomula. They then degranulate, releasing electron-dense deposits onto the surface of the worm. Not only are peroxidase and a variety of other enzymes released, but in addition, and probably more important, the major basic protein of the eosinophil granule is released onto the surface of the schistosomulum (Fig. 21.8) as well as into the medium. This and a second cationic protein from the granules of eosinophils are toxic to schistosomula at low concentrations. Similar effects have been observed with rat eosinophils and serum from immune rats. In either case the tegument of the schistosomulum is disrupted, permitting eosinophils actually to penetrate inside the worm (Fig. 21.9).

The role of eosinophils acquires special significance in relation to the

TABLE 21.2. Preferential Capacity of Human Eosinophils to Cause Antibody-Dependent Damage to Schistosomula[a]

Cells	Antibody dilution	% dead schistosomula[b]	
		Skin-transformed	Mechanically prepared
Eosinophils	1/6	90 ± 2[c]	35 ± 1
	1/30	91 ± 4	63 ± 4
	0	3 ± 4	2 ± 1
Neutrophils	1/6	2 ± 2	3 ± 3
	1/30	5 ± 6	1 ± 3
	0	4 ± 3	2 ± 2
Medium	1/6	3 ± 1	6 ± 2
	1/30	3 ± 2	1 ± 1
	0	1 ± 1	1 ± 1

[a] Modified from Butterworth *et al.* (1982).

[b] Schistosomula, prepared either by penetration of rat skin or mechanically, were cultured in flat-bottomed microtiter wells in the presence of antischistosomulum serum and eosinophils (94% pure) or neutrophils (96% pure) at a ratio of 4000 : 1. Damage was assessed microscopically after 40 hr of incubation.

[c] Mean ± S.D. of triplicate determinations.

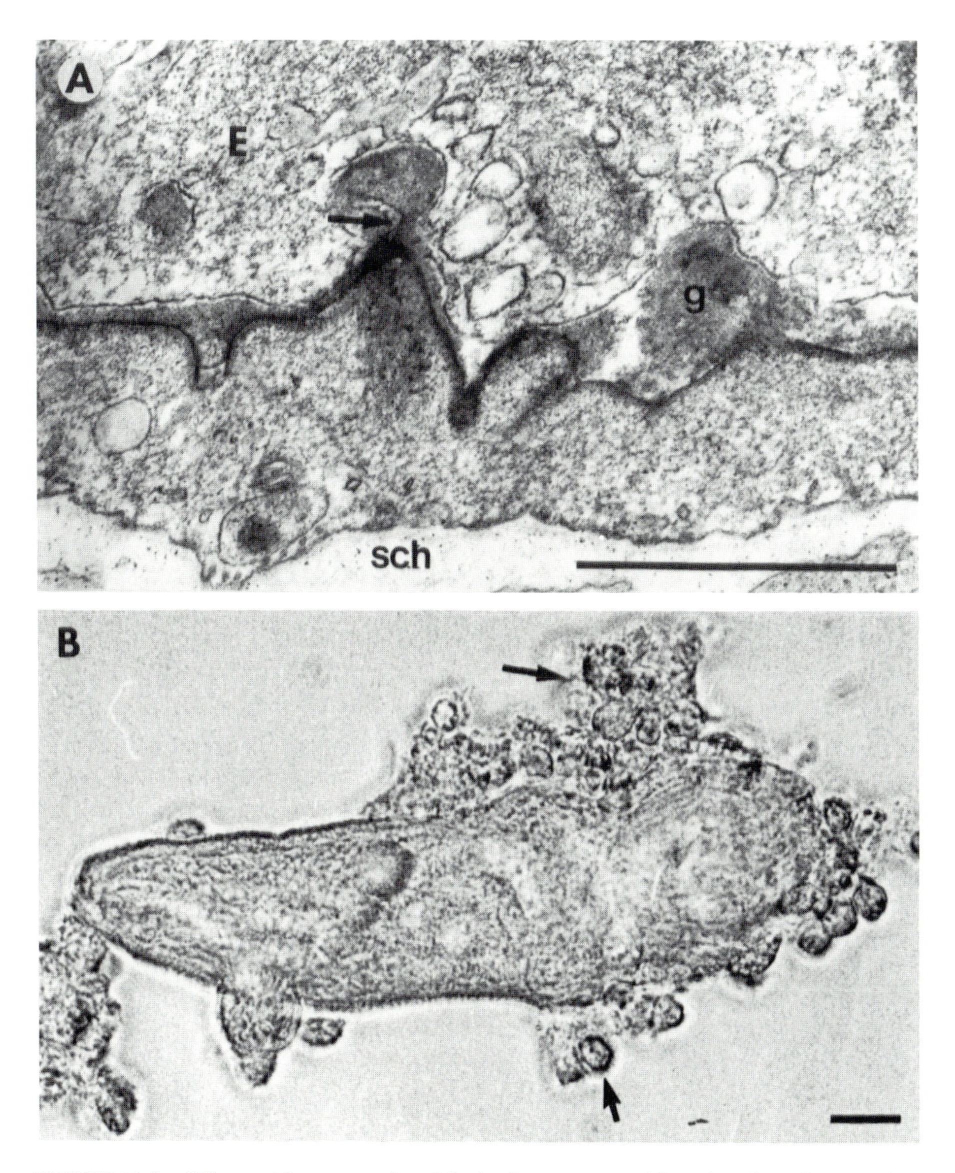

FIGURE 21.8. Effects of human eosinophils in the presence of heat-inactivated serum from patients with active *S. mansoni* infection on schistosomula *in vitro*. (A) The membrane of a dense cytoplasmic granule in an eosinophil (E) has fused with the plasma membrane in a region of adherence to a schistosomulum (sch), and the contents of the granule appear to be in the process of secretion onto the surface of the worm (arrow). Dense material, as at *g*, similar to the contents of the cytoplasmic granules of the eosinophil, is present between the cell and the parasite. TEM. Bar = 1 μm. (B) A schistosomulum of *S. mansoni* after 4 hr incubation with human eosinophils and antibody. Note the localized regions where material has been released from within the parasite and the eosinophils attached to this material (arrows). Phase contrast. Bar = 10 μm. (From Glauert *et al.*, 1978. Original prints courtesy of Dr. A. M. Glauert.)

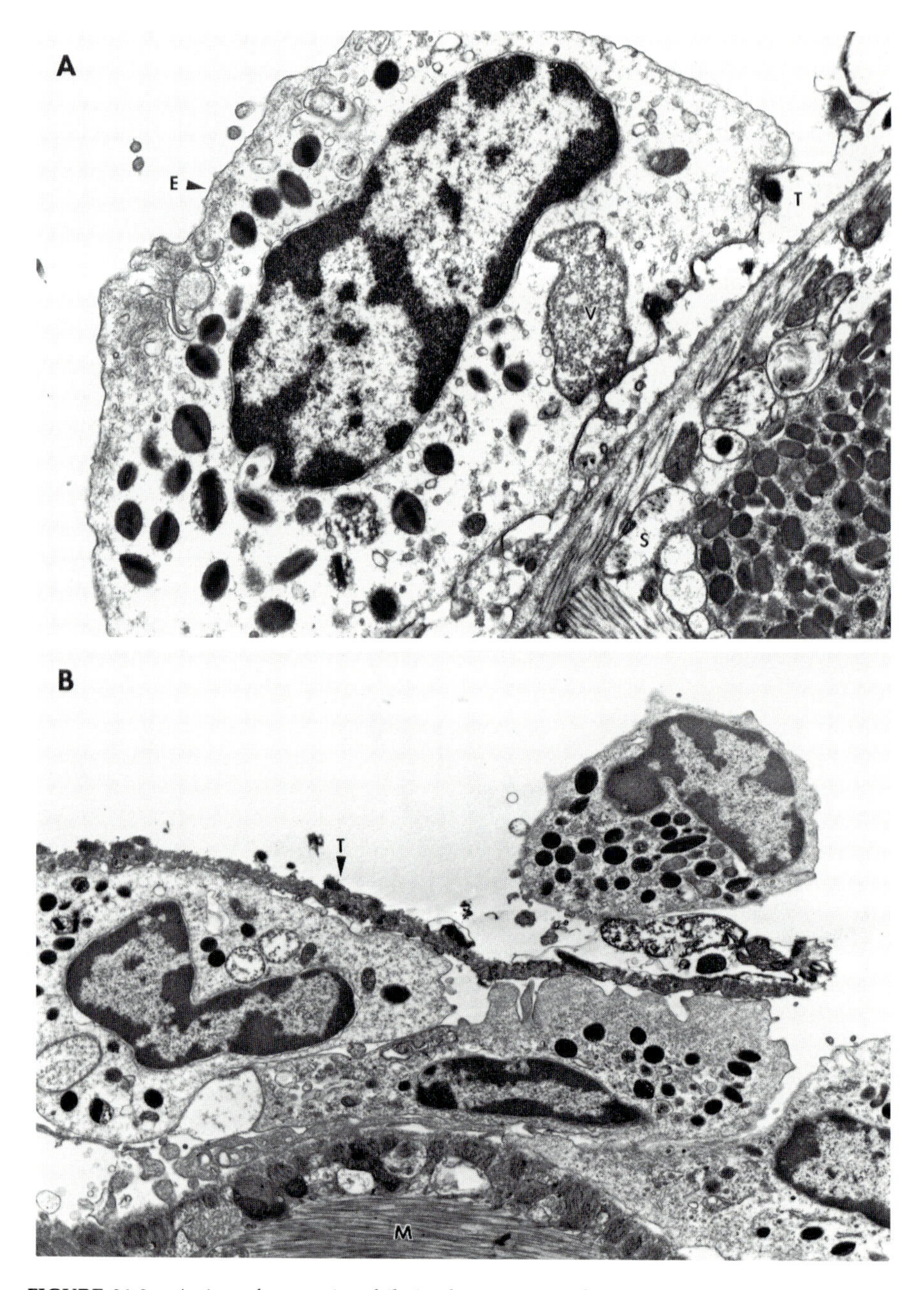

FIGURE 21.9. Action of rat eosinophils in the presence of immune serum with complement on schistosomulum *in vitro*. (A) Tegument (T) of the parasite (S) becomes vacuolated and disrupted in the vicinity of the secreting eosinophil (E). V, vacuole. TEM. ×83,000. (B) Flattened eosinophils situated between the tegument (T) and the muscle layers (M) of a damaged parasite. The tegument is being stripped off the worm's body. TEM. ×5500. (From McLaren, 1980. Original prints courtesy of Dr. D. J. McLaren.)

eosinophilia long known to be associated with many helminthic infections, including those with schistosomes. Eosinophils taken from such individuals having a moderate eosinophilia, associated either with schistosomiasis or with some other helminthic infection, were found to be more active in killing schistosomula in the presence of suboptimal concentrations of antischistosomulum serum than were eosinophils from normal individuals. This enhancement probably results in part from agents secreted by mononuclear cells. There is also evidence that the degranulation of mast cells releases products that are not directly toxic to schistosomula but enhance eosinophil-mediated killing. Since mast cells degranulate in reactions of immediate hypersensitivity, mediated by IgE or IgG2a, this implies that in the immediate hypersensitivity reaction to invading schistosomula there may be not only selective localization of eosinophils at the site of invasion but also enhancement of their functional activity.

The identity of the antigens that react with the antibodies involved in antibody-dependent killing by eosinophils has not been determined. Presumably they are expressed on the surface of the young skin-stage schistosomula, and extensive efforts at their identification are in progress. Whatever they are, their reactivity on the surface of the worm becomes much reduced within the first 2–3 days of schistosomulum development, as is evidenced by a decline in the extent of binding by the worm of antibodies present in immune serum. This is a result of at least two processes, both complex and neither fully understood. One is masking of worm antigens by the acquisition of host macromolecules. If a worm is transferred surgically from a host of species A, in which it has developed, to a host of another susceptible species B, it will survive and become established. If, however, the species B host has first been immunized against species A (as by injection of red cells of species A), the worm is destroyed. Its surface contains species A macromolecules. One mechanism for their acquisition is a fusion observed to occur between the outer membrane of schistosomula and the membrane of attached host cells, especially neutrophils. Host molecules then become demonstrable in the outer membrane as intramembranous particles that will form characteristic aggregates in the presence of antihost antibodies. Antigens of the human ABO blood groups were shown to be specifically acquired when schistosomula were grown *in vitro;* those grown with type A human red cells acquired type A; those with type B cells acquired type B; and those with AB cells expressed both A and B antigens. Similarly, there is acquisition of host-specific histocompatibility antigens on the parasite surface. In addition to such masking, there are changes in expression of worm antigens on the surface. Certain antigens seem to be internalized and certain new ones appear. This occurs *in vitro* in schistosomula prepared mechanically (rather than by skin penetration) and cultured in a defined serum-free culture medium. Hence, it occurs in the complete absence of host macromolecules. It is furthermore a process sensitive to puromycin. These two mechanisms and also others, such as the secretion of proteases by the worms and in particular certain intrinsic changes in the worm surface not directly related to its antigenicity, together permit the worm to survive in a host immune to reinfection.

That such an immunity can exist, together with some indications that it does exist in human schistosomiasis as well as in animal models, has encouraged the hope that vaccination against the human schistosome infections might be possible. Immunization of experimental animals with either cercariae or schistosmula, still living but attenuated by irradiation (as exposure to 50 kR of γ-irradiation from a ^{60}Co source), has been repeatedly accomplished for all three species of human schistosomes as well as for *S. matthei* and *S. bovis*, parasites of ruminants. In this way Sudanese cattle were successfully protected under field conditions from morbidity and mortality caused by *S. bovis*. Moreover, it was shown that vaccination of American cattle with *S. japonicum* resulted in a significant reduction in challenge infections. This has potential public health importance since in the regions where *S. japonicum* is endemic, cattle are among the major reservoir hosts of this parasite. The safety of a live attenuated schistosomulum vaccine is being explored by recent studies with mice and primates. Thus far, it has been found that 1000 such schistosomula injected intramuscularly gave rise to only reversible local pathology, like that elicited by bacterial vaccines but differing in the kinds of cells involved in the inflammatory reactions. The lesions healed completely in 4–5 weeks. Furthermore, in mice vaccinated when already infected, or given a second vaccination, the lesions were no more severe than in naive mice. However, local inflammation has been more substantial and longer lasting in vaccinated baboons.

A vaccine based on one or several purified antigens would of course be preferable but much work with extracts and fractions of worms or worm eggs has not given encouraging results. However, more recently, partial (40–50% effective) passive immunization has been obtained with several monoclonal antibodies to schistosomes, including the one against schistosome eggs already discussed above. If such antibodies can be used to characterize the relevant antigens, as has already been done with the antiegg monoclonal and several others, these antigens would certainly be among the candidates for a vaccine against schistosomiasis.

There is also strong evidence that cell-mediated, rather than antibody-mediated immunity may be of primary importance. Thus, inbred mice of strains (P/J or P/N) defective in delayed-type hypersensitivity reaction could develop only minimal immunity after vaccination with irradiated cercariae. It was shown that this resulted from failure of these mice to develop an activated macrophage response. In mice of strains that exhibit the usual immunity after vaccination with irradiated cercariae, lymphokine-activated macrophages kill schistosomula both *in vivo* and *in vitro* at least as effectively as do eosinophils. This approach, based on the importance of cell-mediated immunity, has yielded successful vaccinations of mice with a crude fraction of an adult worm extract given intradermally together with the adjuvant BCG. Surface proteins seem not to be involved; rather, host protection is correlated with footpad hypersensitivity to a 97,000-dalton antigen of somatic origin.

Whatever the purified antigens are that may eventually be used in a schistosome vaccine, it is likely that an effective and safe immunopotentiator will have to be included. It may also be necessary to take into account immunosuppressive factors. Thus, adult schistosomes have been shown to release low-molecular-weight inhibitory factors. One of these strongly depresses T-lymphocyte proliferation *in vivo* as well as *in vitro*. There are also indications for blocking antibody modulating the efficiency of immune effectors. It is clear that many questions about immunity to schistosomiasis are still unresolved or controversial, but there is every reason to expect continued progress.

Bibliography

Barral-Netto, M., Cheever, A. W., Lawley, T. J., and Ottesen, E. A., 1983, Cell-mediated and humoral immune responses in capuchin monkeys infected with *Schistosoma japonicum* or *Schistosoma mansoni*, *Am. J. Trop. Med. Hyg.* **32:**1335–1343.

Bickle, Q. D., Taylor, M. G., Doenhoff, M. J., and Nelson, G. S., 1979, Immunization of mice with gamma-irradiated intramuscularly injected schistosomula of *Schistosoma mansoni*, *Parasitology* **79:**209–222.

Butterworth, A. E., Taylor, D. W., Veith, M. C., Vades, M. A., Dessein, A., Stunock, R. F., and Wall, E., 1982, Studies on the mechanisms of immunity in human schistosomiasis, *Immunol. Rev.* **61:**5–39.

Capron, A., Dessaint, J.-P., Capron, M., Joseph, M., and Torpier, G., 1982, Effector mechanisms of immunity to schistosomes and their regulation, *Immunol. Rev.* **61:**41–66.

Damian, R. T., 1979, Molecular mimicry in biological adaptation, in: *Host–Parasite Interfaces* (B. B. Nichol, ed.), Academic Press, New York, pp. 103–126.

Dean, D. A., 1983, Schistosoma and related genera—acquired resistance in mice—a review, *Exp. Parasitol.*, **55:**1–104.

Dean, D. A, Minard, P., Stirewalt, M. A., Vannier, W. E., and Murrell, K. D., 1978a, Resistance of mice to secondary infections with *Schistosoma mansoni*. I. Comparison of bisexual and unisexual initial infections, *Am. J. Trop. Med. Hyg.* **27:**951–956.

Dean, D. A., Minard, P., Murrell, K. D., and Vannier, W. E., 1978b, Resistance of mice to secondary infections with *Schistosoma mansoni*. II. Evidence for a correlation between egg deposition and worm elimination, *Am. J. Trop. Med. Hyg.* **27:**957–965.

Dean, D. A., Mangold, B. L., Georgi, J. R., and Jacobson, R. H., 1984, Comparison of *Schistosoma mansoni* migration patterns in normal and irradiated cercaria-immunized mice by means of autoradiographic analysis, *Am. J. Trop. Med. Hyg.* **33:**89–96.

Dissous, C., and Capron, A., 1983, *Schistosoma mansoni:* Antigen community between schistosomula surface and adult worm incubation products as a support for concomitant immunity, *FEBS Lett.* **162:**355–359.

Glauert, A. M., Butterworth, A. E., Sturrock, R. F., and Houba, V., 1978, The mechanism of antibody-dependent, eosinophil-mediated damage to schistosomula of *Schistosoma mansoni in vitro:* A study by phase-contrast and electron microscopy, *J. Cell Sci.* **34:**173–192.

Harn, D. A., Mitsuyama, M., and David, J. R., 1984, *Schistosoma mansoni:* Anti-egg monoclonal antibodies protect against cercarial challenge *in vivo*, *J. Exp. Med.* **159:**1371–1387.

Hsü, S. Y. L., Hsü, H. F., Penick, G. D., Lust, G. L., Osborne, J. W., and Cheng, H. F., 1975, Mechanisms of immunity to schistosomiasis: Histopathologic study of lesions elicited in rhesus monkeys during immunizations and challenge with cercariae of *Schistosoma japonicum*, *J. Reticuloendothel. Soc.* **18:**167–185.

Hsü, S. Y. L., Hsü, H. F., Xu, S. T., Shi, F. H., He, Y. X., Clarke, W. R., and Johnson, S. C., 1983, Vaccination against bovine schistosomiasis japonica with highly X-irradiated schistosomula, *Am. J. Trop. Med. Hyg.* **32:**367–370.

Hussain, R., Hofstetter, M., Goldstone, A., Knight, W. B., and Ottesen, E. A., 1983, IgE responses in human schistosomiasis. I. Quantitation of specific IgE by radioimmunoassay and correlation of results with skin test and basophil histamine release, *Am. J. Trop. Med. Hyg.* **32**:1347–1355.

James, S. L., Correa-Oliveira, R., and Leonard, E. J., 1985, Defective vaccine-induced immunity to *Schistosoma mansoni* in P strain mice. II. Analysis of cellular responses, *J. Immunol.* **133**:1587–1593.

James, S. L., Pearce, E. J., and Sher, A., 1985, Induction of protective immunity against *Schistosoma mansoni* by a non-living vaccine. I. Partial characterization of antigens recognized by antibodies from mice immunized with soluble schistosome extract, *J. Immunol.* **134**:3432–3438.

Lewis, F. A., Stirewalt, M., and Leef, J. L., 1984, *Schistosoma mansoni:* Radiation dose and morphologic integrity of schistosomules as factors for an effective cryopreserved live vaccine, *Am. J. Trop. Med. Hyg.* **33**:125–131.

McLaren, D. J., 1980, *Schistosoma mansoni:* The parasite surface in relation to host immunity, in: *Tropical Medicine Research Studies* (K. H. Brown, ed.), Research Studies Press, Chichester.

McLaren, D. J., and Boros, D. L., 1983, *Schistosoma mansoni:* Schistosomulicidal activity of macrophages isolated from liver granulomas of infected mice, *Exp. Parasitol.* **56**:346–357.

Mahmoud, A. A. F., 1982, Nonspecific acquired resistance to parasitic infection, in: *Immunology of Parasitic Infections* (S. Cohen, ed.), Blackwell, Oxford, pp. 99–115.

Mahmoud, A. A. F., 1982, The ecology of eosinophils in schistosomiasis, *J. Infect. Dis.* **145**:613–622.

Mahmoud, A. A. F., and Woodruff, A. W., 1975, Renal lesions caused by immune complex deposition in schistosomiasis, *Trans. R. Soc. Trop. Med. Hyg.* **69**:187–188.

Olds, G. R., and Ellner, J. J., 1984, Modulation of macrophage activation and resistance by suppressor T lymphocytes in chronic murine *Schistosoma mansoni* infection, *J. Immunol.* **133**:2720–2724.

Sher, A., and Moser, G., 1981, Schistosomiasis—Immunological properties of developing schistosomula, *Am. J. Pathol.* **102**:121–126.

Simpson, A. J. G., Payares, G., Walker, T., Knight, M., and Smithers, S. R., 1984, The modulation of expression of polypeptide surface antigens on developing schistosomula of *Schistosoma mansoni*, *J. Immunol.* **133**:2725–2730.

Smith, M. A., Clegg, J. A., Snary, D., and Trejdosiewicz, M., 1982, Passive immunization of mice against *Schistosoma mansoni* with an IgM monoclonal antibody, *Parasitology* **84**:83–91.

Smithers, S. R., and Doenhoff, M. J., 1982, Schistosomiasis, in: *Immunology of Parasitic Infections* (S. Cohen and K. S. Warren, eds.), Blackwell, Oxford, pp. 527–607.

Smithers, S. R., McLaren, D. J., and Remalko-Pinto, F. J., 1977, Immunity to schistosomes—the target, *Am. J. Trop. Med. Hyg.* **26**(Suppl.):11–19.

Stirewalt, M. A., Cousin, C. E., and Dorsey, C. H., 1983, *Schistosoma mansoni:* Stimulus and transformation of cercariae into schistosomules, *Exp. Parasitol.* **56**:358–368.

von Lichtenberg, F., 1977, Experimental approaches to human schistosomiasis, *Am. J. Trop. Med. Hyg.* **26**(Suppl.):79–87.

von Lichtenberg, F., 1985, Conference on contended issues of immunity to schistosomes, *Am. J. Trop. Med. Hyg.* **34**:78–85.

Warren, K. S., 1982, The secret of the immunopathogenesis of schistosomiasis: *In vivo* models, *Immunol. Rev.* **61**:189–213.

Wilson, R. A., Coulson, P. S., and McHugh, S. M., 1983, A significant part of the "concomitant immunity" of mice to *Schistosoma mansoni* is the consequence of a leaky hepatic portal system, not immune killing, *Parasite Immunol.* **5**:595–601.

Zodda, D. M., and Phillips, S. M., 1982, Monoclonal antibody-mediated protection against *Schistosoma mansoni* infection in mice, *J. Immunol.* **129**:2326–2328.

CHAPTER 22

Entamoeba histolytica and Other Intestinal Protozoa. Pathogenesis and Immunology

---

Entamoeba histolytica provides a striking example of how the infliction of harm to its host is altogether counterproductive for the normal propagation of the parasite. Paradoxical as it may appear, people with amebic dysentery and an ulcerated colon are noninfective to others (except under special circumstances to be considered later). True enough, the feces of such a person swarm with amebae, highly motile and full of red blood cells and tissue debris. But such amebae die quickly in the external environment. Even if swallowed immediately by another person, or if fed to a susceptible animal, such as a dog or a kitten, they cannot establish infection since they are killed by the acidity of the stomach. Only the cysts can survive externally for some days in a moist environment at ordinary temperatures, and only the cysts are able to pass through the stomach without injury and to excyst in the intestine (see Chapter 2). For reasons that are not understood, no cysts are formed when there is active amebic disease of the colon. Similarly, the other main manifestation of amebic disease, abscess of the liver, again constitutes a blind alley for the parasite. Cysts are not formed, and even if formed they would have no outlet from the host. It is only in healthy carriers, when relatively smaller amebae live mainly as harmless commensals in the lumen of the colon, feeding largely on bacteria, that large numbers of the cysts essential to infection of other hosts are formed. Over a million cysts per gram of feces may be excreted daily by such a person. What triggers cyst formation remains one of the unsolved major problems in the biology of *E. histolytica*. As already noted (Chapter 8), in axenic culture encystation has never been induced. The only way to get cysts *in vitro* is still the method developed years ago by that pioneer protozoologist C. Dobell, requiring cultivation with bacteria without rice starch and the subsequent addition of rice starch. This experimental difficulty in obtaining cysts remains as a hindrance in work with animal models of amebiasis.

Such symptomless cyst-producing carriers constitute 80–90% of individuals infected with *E. histolytica*. They are responsible for the wide distribution of the parasite, which is found in most of the world. Both the incidence of such carriers and the proportion of infected individuals who show amebic disease are higher in regions where sanitation is poor. In India and Mexico, for example, various recent surveys show 25 to 50% of asymptomatic people to be harboring the parasite. In path-breaking studies in 1913, E. L. Walker and A. W. Sellards experimentally infected 18 volunteers with *E. histolytica*. Only 4 developed amebic dysentery and of these 2 had very mild disease. As one might expect there is a whole range of gradations between the symptomless carrier and the person suffering from fulminant dysentery and/or amebic abscess of the liver. A person who has recovered from amebic disease may become and remain for years a symptomless carrier. A carrier may suddenly develop acute disease. What determines these events is not fully understood. It must result from interplay of the invasive properties of the parasite, i.e., its virulence, and the immune reactions of the host.

From what has already been said, it is clear that effective immunity to infection of the lumen of the colon, the habitat of *E. histolytica*, never develops. Carriers remain infective for years. Furthermore, in endemic regions the incidence of such infection increases with age as do both morbidity and mortality from amebic disease (Fig. 22.1). Antibodies are formed against amebic antigens but their role is not clear. Among infected individuals who remain asymptomatic, only about 8 to 15% show significant antibody levels in serum. Hence, a serological immunity is not important in preventing invasive disease. Once colonic invasion occurs, however, 81 to 98% show antibody titers of 1:128 or higher by an indirect hemagglutination test. These antibodies are of the IgG class. Together with complement, they will kill amebae *in vitro* but

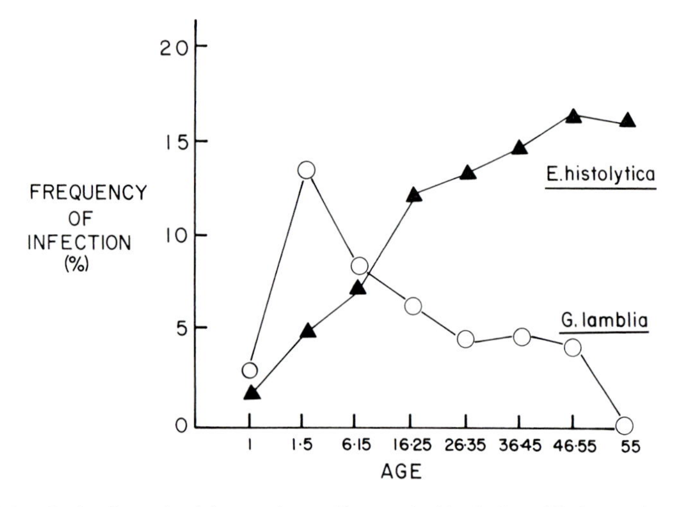

FIGURE 22.1. Lack of acquired immunity to *Entamoeba histolytica* with increasing age. This is contrasted with the very different situation for another intestinal protozoan parasite (*Giardia lamblia*) in the same African population. (From Ravdin and Guerrant, 1982, adapted from Oyerinde *et al.*, 1977).

not *in vivo.* In their presence, reinfection is possible and may even be enhanced. Coproantibodies of the IgA class increase during amebic disease. Antiamebic IgE is sometimes present in the serum, but there is no evidence that this plays any protective role. Nevertheless, it is clear that an intact immune system is essential to prevent tissue invasion. Fulminating amebiasis occurs in patients given corticosteroids or who are otherwise immunologically incompetent. Pregnancy may exacerbate amebic disease and malnutrition, particularly protein deficiency, regularly does so. T cells appear to be of principal importance and lymphocytes from active or recently cured cases are cytotoxic to the amebae. In most individuals, where the amebic infection remains a luminal one without tissue invasion, we have again a situation where the immune system of the host provides for the prolonged coexistence of parasite and host without the latter being overwhelmed and destroyed. In schistosome infection (Chapter 21), concomitant immunity prevents superinfection while allowing worms already present to live and lay eggs for many years. In amebic infections, the immune system in most infected people, by preventing tissue invasion, likewise permits a prolonged period of formation of cysts and the propagation of the species.

There is unfortunately no good animal model for the situation where oral administration of *E. histolytica* cysts would produce in most animals a chronic infection without disease but with excretion of cysts in the feces. Animals have generally been infected by the intrarectal installation of trophozoites with the object of producing an ulcerative colitis. Rats, mice, and guinea pigs, as well as dogs and kittens, have been used in this way. Such infections ordinarily do not result in cyst production. Hamsters have been particularly useful for the study of hepatic amebiasis by the direct inoculation of amebae into the liver. Although immunization of such experimental animals can be achieved, this does not seem highly relevant to amebic infections in people. The animal models have, however, provided important information with regard to the other side of the coin—the invasive properties of *E. histolytica.* When the amebae produce pathogenic effects, these are associated mainly with their direct invasive activities. These have been extensively investigated in animal models and also *in vitro* in tissue cultures. Some significant correlations are appearing.

Of special importance is the demonstration that some strains of *E. histolytica* retain their virulence after prolonged axenic culture *in vitro.* This can be shown by inoculating such amebae either directly or via the portal vein into hamster liver, where they form abscesses. Bacteria do, however, aid in intestinal colonization, perhaps by lowering oxidation reduction potential, by providing nutrients, and by assisting in adherence to the mucosa. This explains early observations in which axenic amebae introduced into germfree guinea pigs did not establish infection, whereas the same amebae did produce infections in guinea pigs with their usual intestinal bacteria. Some strains, however, after prolonged axenic culture, have greatly reduced virulence as measured by the ability to form abscesses in hamster liver. A few passages of such strains through hamster liver will restore their virulence. Virulence for hamster liver has also been restored by growing the amebae in a medium supplemented with cholesterol (Table 22.1). Note that of the two strains tested, only the one (HK-9) that had essentially lost invasiveness for the hamster liver was affected by the cholesterol. The already high invasiveness of strain HB-301 was not further enhanced. It is of interest that cholesterol is an essential nutrient for *E. histolytica* as it is for many other protozoa and other invertebrates (see Chapter 7). It seems reasonable to assume that some adequate level of cholesterol in the surface membrane may play a role in the phagocytic and tissue-destroying activities that constitute the invasive process.

A remarkable correlation between phagocytic activity, as measured by uptake of erythrocytes or latex beads, and virulence, as measured by the percent of inoculated hamsters showing liver abscesses, has been demonstrated in recent experiments of such elegance as to deserve detailed description. Amebae of a pathogenic and highly phagocytic strain grown axenically were allowed to feed on a suspension of *Escherichia coli* bacteria strain CR34 Thy⁻ that had been grown with 5-bromo-2′-deoxyuridine (BUdR). Such amebae incorporated BUdR into their DNA and were rendered sensitive to near-visible irradiation (peak at 310 μm). Consequently, only the nonphagocytic amebae survived such irradiation. From these, clones were prepared that showed a dramatic loss of virulence. Clones were also prepared from the untreated virulent strain. The erythrophagocytic activity of the original strain HMI:IMSS, a clone (A) prepared from it, and of a clone (L-6) prepared from the selected nonphagocytic population are shown in Fig. 22.2, and their invasiveness for hamster liver in Table 22.2. In addition, virulence revertants were obtained from clone L-6 by four passages in hamster liver followed each time by a short period of axenic culture. These amebae that had recovered

TABLE 22.1. Effect of Cholesterol and Epicholesterol on the Virulence of *E. histolytica* HK-9 and HB-301[a]

Strain	Treatment[b]	No. of experiments	Size of inoculum (× 10^6)[c]	No. of hamsters with abscesses per No. of hamsters inoculated	
HK-9	Untreated	5	1.1–4.8	0/12	(0%)
	Cholesterol	5	1.2–5.8	6/15	(40%)
	Epicholesterol	2	1.5–2.0	4/8	(50%)
HB-301	Untreated	2	1.7–3.6	4/6	(67%)
	Cholesterol	2	1.5	5/8	(63%)

[a] Modified from Bos and van de Griend (1977).
[b] Cholesterol and epicholesterol were dissolved in chloroform and evaporated in culture tubes (0.35 mg/tube) before adding the TP-S-1 medium.
[c] Amebae in 75–150 μl. The largest inoculum tested consisted of 75 μl packed cell volume and 75 μl TP-S-1 medium. One week after infection, lesions were usually more or less confluent granulomatous necrotic areas. Amebae were demonstrated by microscopy or culture of smears, or in histological cross sections.

FIGURE 22.2. Erythrophagocytosis by a strain of *E. histolytica* (HM1:IMSS, ○) and two clones derived from it, clone A (●) and a clone from the selected nonphagocytic population (L-6, △). Trophozoites were mixed with human red blood cells (RBC) in a ratio of 1 trophozoite to 100 RBC and incubated at 37°C for 2, 5, and 10 min. The erythrophagocytosis indices were obtained by multiplying the mean number of RBC per ameba by the number of trophozoites that showed at least one ingested RBC. (From Orozco *et al.*, 1983. Original print courtesy of Dr. E. Orozco.)

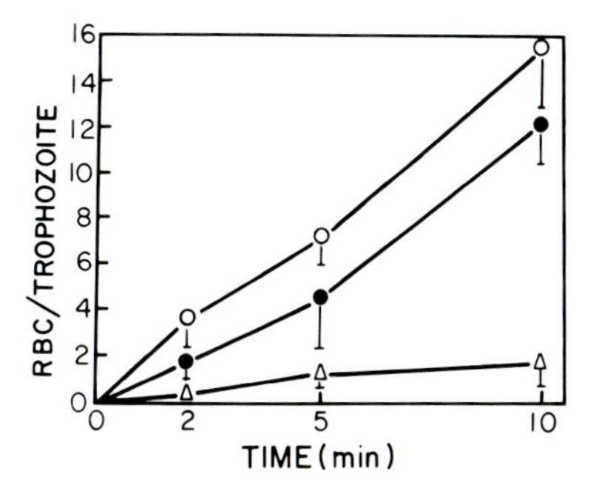

the ability to produce abscesses in hamster liver simultaneously recovered high rates of erythrophagocytosis.

Possibly related to phagocytic activity is the agglutination of *E. histolytica* trophozoites by concanavalin A, and this again is related to the degree of pathogenicity. Whereas two still highly pathogenic strains of *E. histolytica* were agglutinated by as little as 2.5 μg concanavalin A/ml, a third strain that had lost its invasiveness after more than 50 years in culture was not agglutinated by even 10 μg/ml. Virulence as measured by *in vitro* destructive effect of amebae on a monolayer culture of baby hamster kidney cells has also been correlated with the content of proteolytic enzyme activity on Azocoll, an insoluble collagen. It seems reasonable to suppose that such enzymes able to digest attachment proteins of cells would play a role in the invasion of tissue by amebae.

The cytolethal action of *E. histolytica* trophozoites has been studied in various tissue culture cell lines (Fig. 22.3). It involves three steps: (1) adherence of amebae to target cells; (2) cytolysis of the adherent cells; (3) phagocytosis of viable or killed target cells (Fig. 22.4). Note that I have already provided examples of how all three of these activities are correlated with virulence.

TABLE 22.2. Virulence of *E. histolytica* (HM1 : IMSS and Derivative Clones)[a]

Strain	No. of trophozoites	No. of animals[b]	Percent of hamsters with abscesses
HM1 : IMSS	2×10^4	20	95
HM1 : IMSS[c]	1.5×10^4	20	50
Clone A	2×10^4	20	90
Clone A[c]	1.8×10^4	20	50
Clone L-6	2×10^5	30	0
Clone L-6	5×10^5	20	0

[a] Modified from Orozco *et al.* (1983).
[b] Newborn hamsters were intrahepatically inoculated. At 8 or 16 d postinoculation, animals were sacrificed.
[c] AD_{50} was defined as the number of trophozoites required to produce abscesses in 50% of the animals.

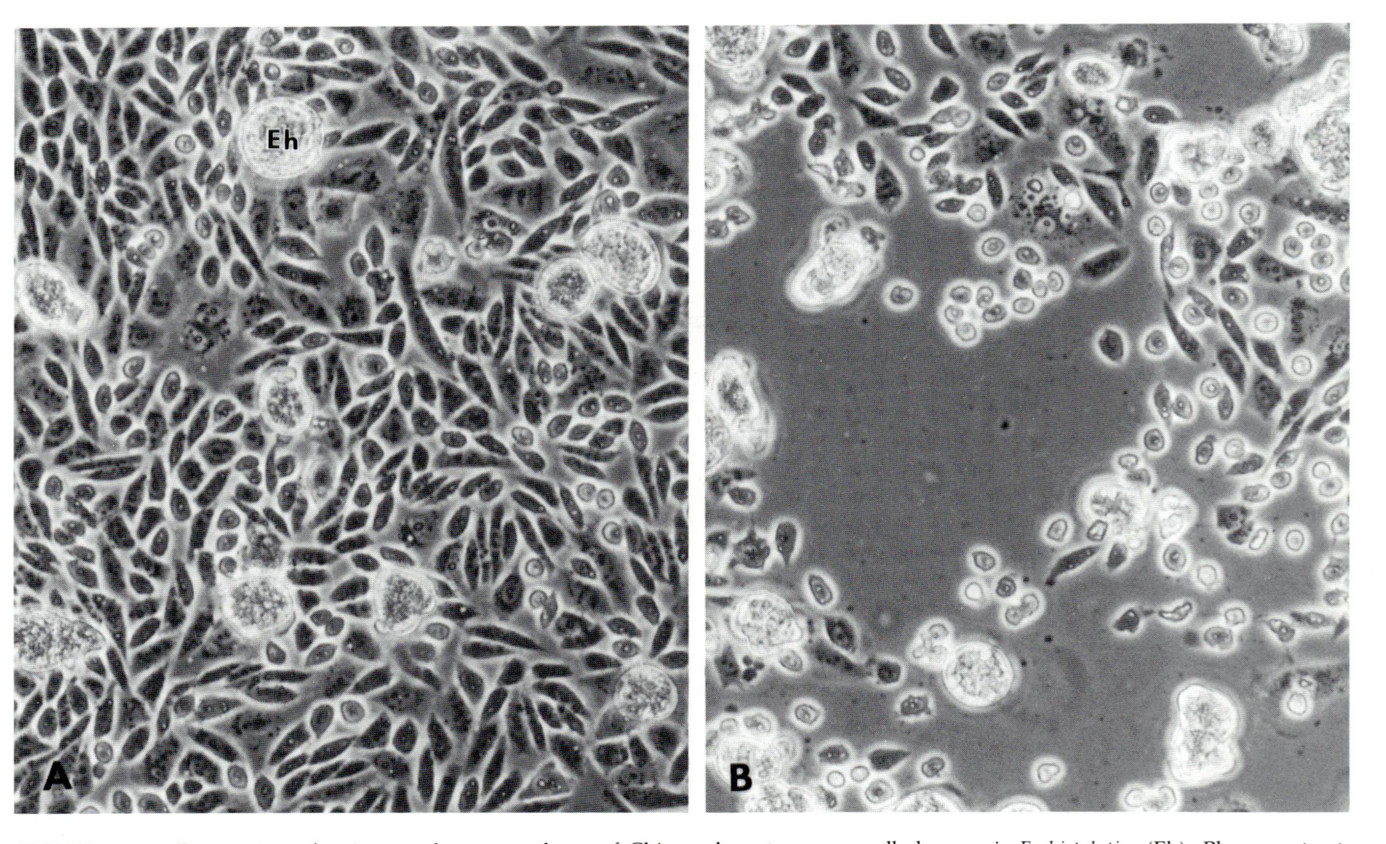

FIGURE 22.3. Destruction of a tissue culture monolayer of Chinese hamster ovary cells by axenic *E. histolytica* (Eh). Phase-contrast photographs taken (A) just after addition of amebae and (B) after 3 hr incubation. (From Ravdin and Guerrant, 1982. Original prints courtesy of Dr. J. I. Ravdin.)

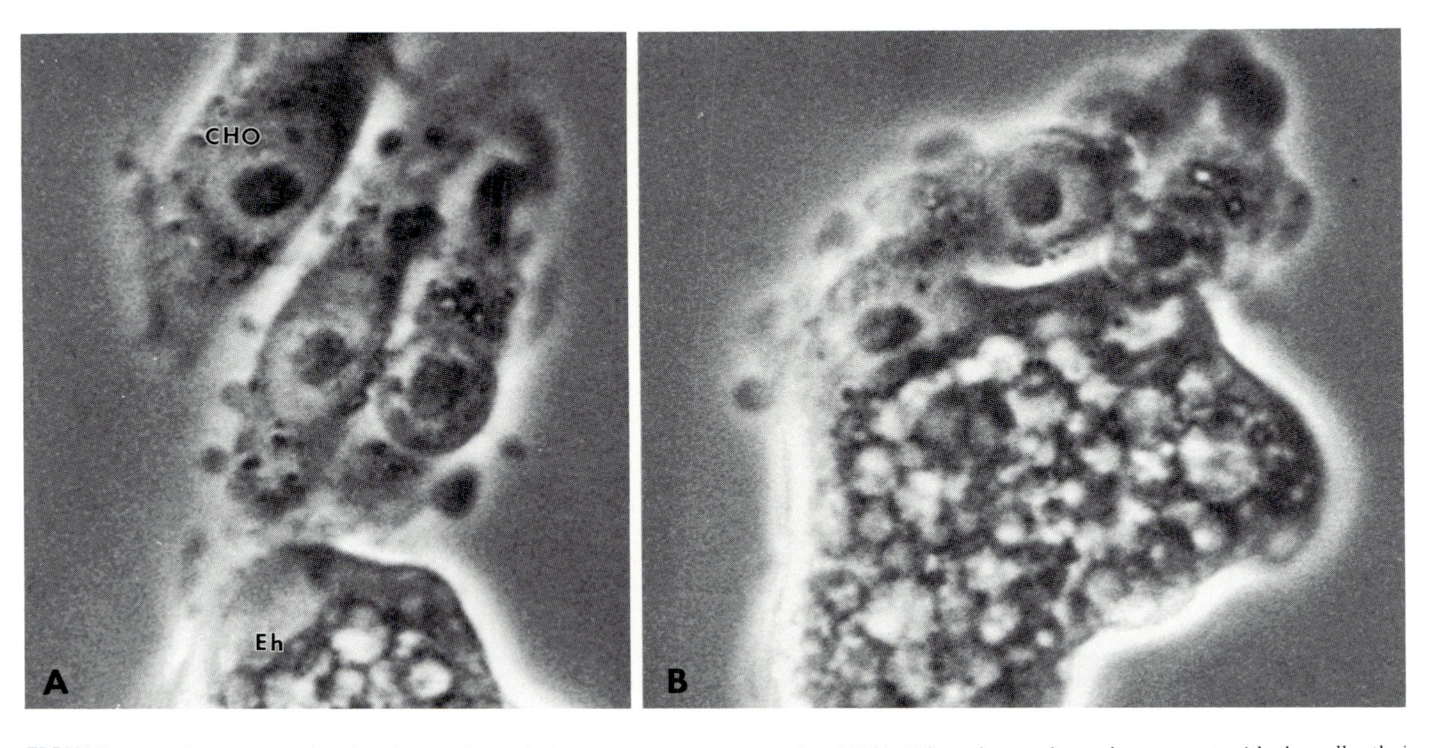

FIGURE 22.4. Interaction of *E. histolytica* (Eh) with Chinese hamster ovary cells (CHO). When the ameba makes contact with the cells, their membrane blebs and releases from the surface (A). Amebic phagocytosis then begins (B). From cinemicrography. (From Ravdin and Guerrant, 1982. Original prints courtesy of Dr. J. I. Ravdin.)

Amebae adhere to human red cells more than to red cells of other species, and this adherence is mediated by lectins. Such lectin-mediated direct contact with target cells is essential to the cytolethal effect. There is evidence that calcium is involved in the killing action and that ion flux changes probably occur in the target cell membrane and the membrane of the amebae. Some special properties of the ameba plasma membrane have already been noted (see Chapter 2). The roles of amebic microfilaments and of lectins are indicated by the blocking effects of cytochalasin and *N*-acetyl-D-galactosamine, respectively (Fig. 22.5). Target cells that adhere at 4°C are not killed unless warmed to 37°C, when they are killed before phagocytosis occurs. These activities are responsible for the typical pathology of an amebic ulcer of the colon (Fig. 22.6) and a hepatic abscess where the amebae, feeding as they do only on living cells, are found at the periphery of the lesion with necrotic debris in its interior portion. There is ordinarily little or no white cell response, and the amebae will kill leukocytes on contact, as they will other types of target cells. In some active or recently cured cases, however, lymphocytes have been found to be cytotoxic to the amebae. Such an amebicidal effect of gut mucosal lymphocytes has also been shown in infected guinea pigs. It may be concerned in the normal resistance to invasion of tissue by amebae, but there is no evidence for this.

Under exceptional circumstances, as in homosexual men engaging in sodomy and probably having reduced immunological competence, direct in-

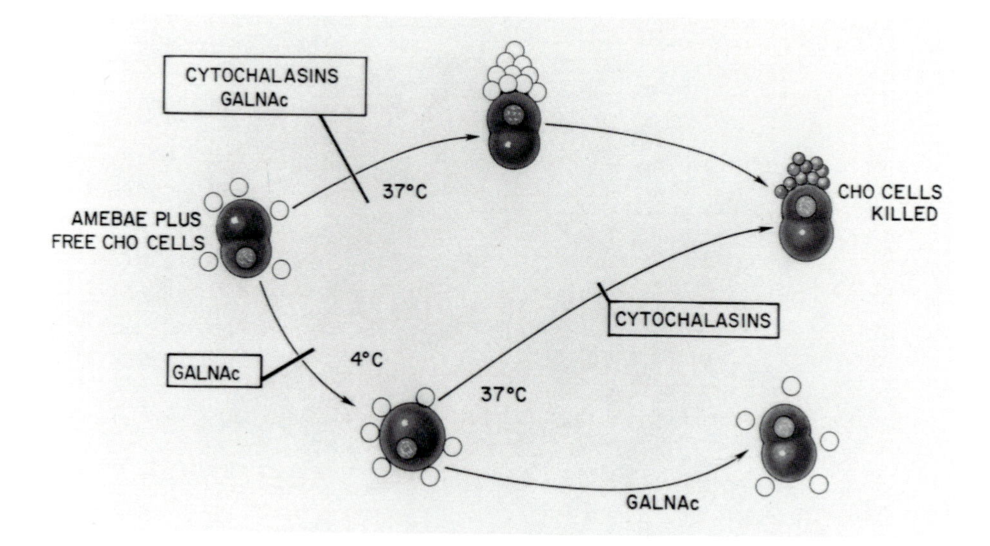

FIGURE 22.5. A representation of amebic adherence (occurring at 4 or 37°C) and amebic killing of previously adherent target cells (only at 37°C). Cytochalasins block both adherence (37°C) and amebic killing of adherent target cells. *N*-Acetyl-D-galactosamine (GALNAc) prevents amebic adherence at 37 or 4°C and elutes previously adherent target cells. (From Ravdin and Guerrant, 1982. Original print courtesy of Dr. J. I. Ravdin.)

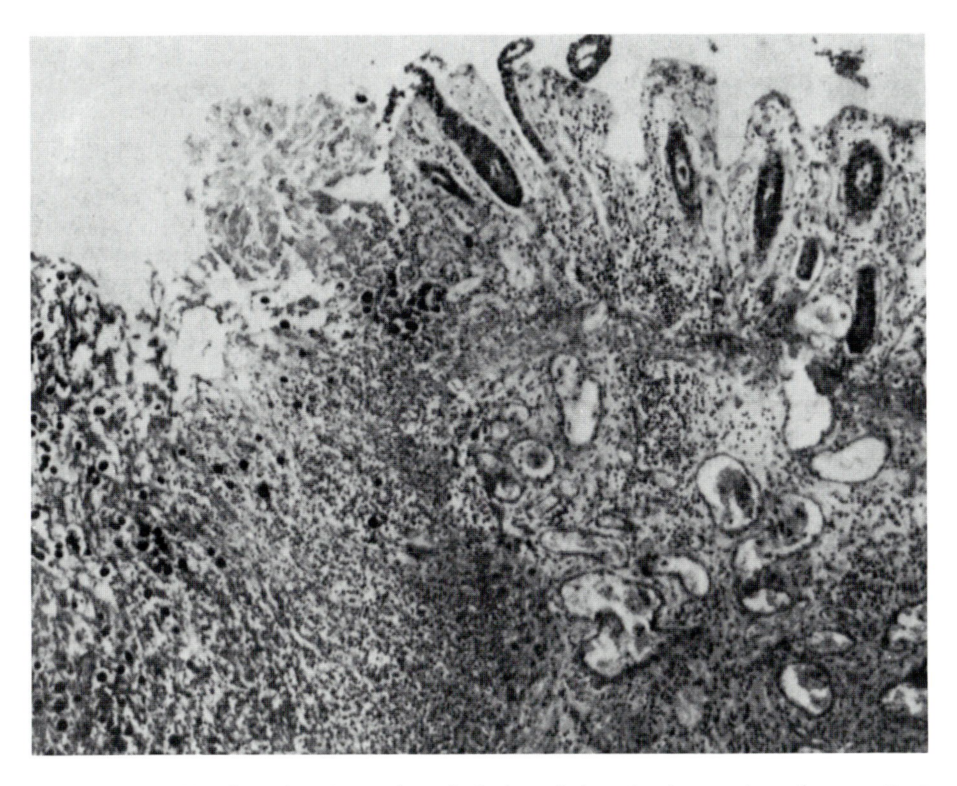

FIGURE 22.6. Histological section to show flask-shaped ulceration in a specimen from a patient with severe colonic amebiasis. Periodic acid–Schiff stain. ×160. (From Ravdin and Guerrant, 1982.)

vasion of the genitalia by trophozoites may occur. Under these conditions, amebiasis, like some other enteric infections, becomes a sexually transmitted disease.

The great significance of further work on the cell biology of *E. histolytica,* and elucidation of the mechanisms involved in the blocking of invasive disease in most infected individuals by strains of amebae that are invasive in other individuals, is apparent.

There are a number of species of colon-dwelling amebae of humans and other animals that are never invasive. Those found in people are of importance in relation to the differential diagnosis of *E. histolytica* infections. Particular attention has to be paid to *E. hartmanni,* a species so similar to *E. histolytica* that it was formerly considered as a small nonpathogenic strain of this species. Like those nonpathogenic amebae, there are a number of commensal flagellates that inhabit the lumen of the large intestine.

One genus of flagellate, however, lives on the mucosal surface of the small intestine, and the species infecting people, *Giardia lamblia,* has definite pathological effects. As already noted (Chapter 4), the villi of the small intestine are often flattened, presumably as a result of adherence of numerous *Giardia* trophozoites to it. Unlike the situation with *E. histolytica*, pathogenesis and excretion of cysts exist together, though symptomless infections are common as well as infections with such vague symptoms as to make diagnosis difficult. Effective immunity does develop (Fig. 22.1) and depends on secretory IgA antibodies and on T cells.

Bibliography

Bos, H. J., and van de Griend, R. J., 1977, Virulence and toxicity of axenic *Entamoeba histolytica*, *Nature* **265**:341–343.

Dobell, C., 1919, *The Amoebae Living in Man*, John Bale, Sons & Danielson, Ltd., London.

Dutta, G. P., 1981, *Experimental and Clinical Studies in Amoebiasis*, Tata McGraw–Hill, New Delhi.

Elsdon-Dew, R., 1968, The epidemiology of amoebiasis, *Adv. Parasitol.* **6**:1–62.

Erlandson, S. L., and Meyer, E. A. (eds.), 1984, *Giardia and Giardiasis*, Plenum Press, New York.

Gadasi, H., and Kobiler, D., 1983, *Entamoeba histolytica:* Correlation between virulence and content of proteolytic enzymes, *Exp. Parasitol.* **55**:105–110.

Ghadirian, E., and Meerovitch, E., 1984, Lectin-induced agglutination of trophozoites of different species and strains of Entamoeba, *Z. Parasitenkd.* **70**:147–152.

Martinez-Palomo, A., 1982, *The Biology of Entamoeba Histolytica*, Research Studies Press and Wiley, New York.

Meerovitch, E., and Ghadirian, E., 1978, Restoration of virulence of axenically cultivated *Entamoeba histolytica* by cholesterol, *Can. J. Microbiol.* **24**:63–65.

Mirelman, D., Bracha, R., and Sargeaunt, P. G., 1984, *Entamoeba histolytica:* Virulence enhancement of isoenzyme-stable parasites, *Exp. Parasitol.* **57**:172–177.

Orozco, E., Guarneros, G., Martinez-Palomo, A., and Sanchez, T., 1983, *Entamoeba histolytica*—Phagocytosis as a virulence factor, *J. Exp. Med.* **158**:1511–1521.

Oyerinde, J. P. O., Ogunbi, O., and Alonge, A. A., 1977, Age and sex distribution of infections with *Entamoeba histolytica* and *Giardia intestinalis* in the Lagos population, *Int. J. Epidemiol.* **6**:231–234.

Ravdin, J. I., and Guerrant, R. L., 1982, A review of the parasite cellular mechanisms involved in the pathogenesis of amebiasis, *Rev. Infect. Dis.* **4**:1185–1207.

Sepulveda, B., 1982, Amebiasis: Host–pathogen biology, *Rev. Infect. Dis.* **4**:1247–1253.

Sepulveda, B., and Diamond, L. S. (eds.), 1976, *Proceedings of the International Conference on Amoebiasis*, Instituto Mexicano del Sequero Social, Mexico.

Smith, P. D., Keister, D. B., and Elson, C. O., 1983, Human host response to *Giardia lamblia*. II. Antibody-dependent killing *in vitro*, *Cell. Immunol.* **82**:308–315.

Trissl, D., 1982, Immunology of *Entamoeba histolytica* in human and animal hosts, *Rev. Infect. Dis.* **4**:1154–1184.

Tsutsumi, V., Mena-Lopez, R., Anaya-Velazquez, F., and Martinez-Paloma, A., 1984, Cellular bases of experimental amebic liver abscess formation, *Am. J. Pathol.* **117**:81–91.

Walker, E. L., and Sellards, A. W., 1913, Experimental entamoebic dysentery, *Philipp. J. Sci. (B Trop. Med.)* **8**:253.

CHAPTER 23

Acquired Immunity to Intestinal Nematodes and to Ticks

In 1947 Norman R. Stoll, in "This Wormy World," estimated that, out of the then world population of 2000 million people, 1700 million were parasitized by nematodes. Although the world population has increased, this proportion has not changed. Especially prevalent and associated with poor sanitation are the intestinal nematodes *Ascaris lumbricoides* and the hookworms *Ancylostoma duodenale* and *Necator americanus*. There are indications that the immune system serves to limit the worm burden, but little direct information is available regarding acquired immunity to these human parasites. A great deal has been learned, however, about acquired immunity to related nematode parasites of domestic animals and about another type of nematode parasite of humans, the ubiquitous *Trichinella spiralis*.

T. spiralis which will infect many kinds of mammals, depends on predation and the carnivorous habit for its dissemination (see Diagram XVIII). Viviparous female worms living in an intramulticellular niche within the enterocytes of the small intestine produce large numbers of larvae. These do not make their way to the outside but instead travel via the lymphatics and the bloodstream to invade striated muscle. Here they develop intracellularly and form resting cysts called nurse cells (see Chapter 5). When infected meat is eaten by another suitable host, the cysts are digested in the stomach, liberating the larvae which are moved to the small intestine. They penetrate the columnar epithelium and develop rapidly, undergoing four molts, to become adult males and females within about 30 hr. The females begin to deposit larvae within 5 days and continue to do so for another 5–10 days, each female producing about 1000. The exact number of offspring is largely dependent on the capacity of the host to respond immunologically to the infection. During the 8th to 10th day of the infection in an adult rat, the worms in the intestine begin to be expelled and all are gone within about 2 weeks. This expulsion is the result of immunological reactions. In rats and some inbred strains of mice, if a second *T. spiralis* infection is given, most of the worms are expelled from the intestine within 48 hr (Table 23.1). Such a

TABLE 23.1. Immunity to Reinfection in *Trichinella spiralis*[a,b]

Group	Mean intestinal worm count at 24 hr	48 hr
Controls, primary infection	1086 ± 96	1105 ± 117
Secondary infection	80 ± 60	41 ± 14

[a] Modified from Love *et al.* (1976).
[b] A group of female rats was infected with 4000 larvae and 8 weeks later was reinfected with 1000 larvae. At this time 1000 larvae were given to a second previously uninfected control group. Rats from each group were killed at 24 and 48 hr after infection to determine the number of intestinal worms.

TABLE 23.2. Immunization of Mice with Purified Antigens[a,b]

Antigen and amount (μg)		Number of larvae recovered (mean ± S.E.)	% reduction from control	Significance (p values from Student's t test)	
None (controls		71,400 ± 4200	—		
S_3—crude fraction	50	22,100 ± 2800	69		
from which	10	44,900 ± 3000	37	0.001	
antigens were	1	54,600 ± 5200	17	0.1	(n.s.)
isolated	0.1	73,800 ± 3700	—		
48K	50	22,500 ± 3000	69		
	10	13,000 ± 3000	81		
	1	27,600 ± 2700	61		
	0.1	39,300 ± 3000	45	0.001	
	0.01	72,700 ± 5000	—	0.1	(n.s.)
50/55K	50	35,000 ± 3400	51		
	10	29,200 ± 1600	59		
	1	45,000 ± 3700	37		
	0.1	47,700 ± 3800	31	0.001	
	0.01	70,700 ± 4300	—	0.1	(n.s.)
37K	50	39,500 ± 4000	44	0.001	
	10	62,100 ± 3000	13	0.1	(n.s.)
	1	61,200 ± 4200	14		
	0.1	69,200 ± 3000	3		

[a] Modified from Silberstein and Despommier (1984).
[b] A control group of seven mice and treatment groups of five mice each received an oral challenge of 360 infective larvae. The treatment groups had been injected with the indicated antigen and amount emulsified in Freund's complete adjuvant, one-third of the total dose being given at each of three injections at intervals of 1 week. The challenge infection was administered 7 days after the last injection. Thirty-five days after challenge the mice were killed and the numbers of muscle larvae determined.

sharp reduction in numbers of adult worms means an equally sharp or even greater reduction in resultant encysted muscle larvae. The encysted forms already present from the first infection, and even some of the larvae produced by the few worms of the second infection, are not affected by the immunity. Rarely is there a complete elimination of the worms of a challenge infection.

Such a protective immunity has now been produced in mice with highly purified antigens (Table 23.2). These were prepared by antibody affinity chromatography using three monoclonal antibodies. Of special interest is the 48,000-dalton antigen (48K). This antigen in itself could account for the entire immunizing effect of the crude fraction S_3 from which it was obtained. It represents about 1% of the protein content of S_3 and 0.1 μg of it gave about the same protection as 10 μg of S_3 (Table 23.2). Immunization of mice with this antigen protects them from otherwise lethal infection with *Trichinella* (Fig. 23.1). The 48K antigen is present in the granules of the β stichocytes, the lining of the gut, and the cuticle of infective larvae, as shown by immunoperoxidase staining. It is present in the secretions of the stichosome, a special organ composed of large cells, the stichocytes, that surround the narrow esophagus of the worm (Figs. 23.2, 23.3). The secretion granules of these cells have been shown to be important in induction of immunity.

Immunity can be induced by the intestinal phase alone of the infection, as shown experimentally with infections abrogated by drug treatment before production of larvae had begun. The immunity can also be fully accounted for by the early expulsion of most of the intestinal worms of a reinfection. It

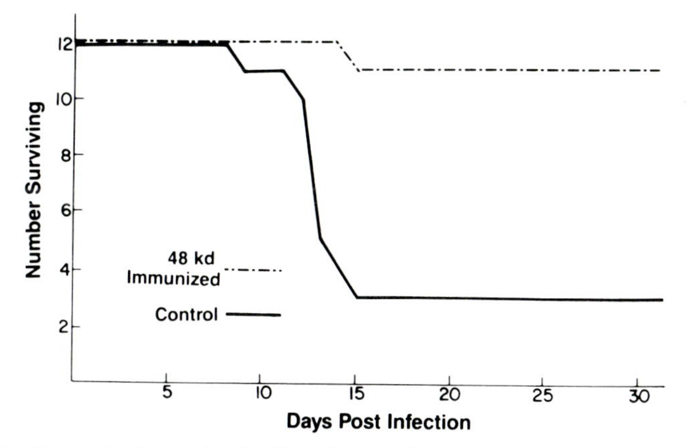

FIGURE 23.1. Immunization to *Trichinella* with a purified antigen. Survival of control mice and mice immunized with the 48K antigen after an oral dose of 1200 infective L_1 larvae. The immunized mice received three intraperitoneal injections at weekly intervals of 3.3 μg antigen with Freund's complete adjuvant, whereas the control mice received buffer with adjuvant. One week after the last injection all mice were infected with an oral dose of larvae. The 11 immunized mice that survived at 30 days had about 40,000 muscle larvae per mouse as compared to about 124,000 in the surviving controls. (From Silberstein and Despommier, 1985b.)

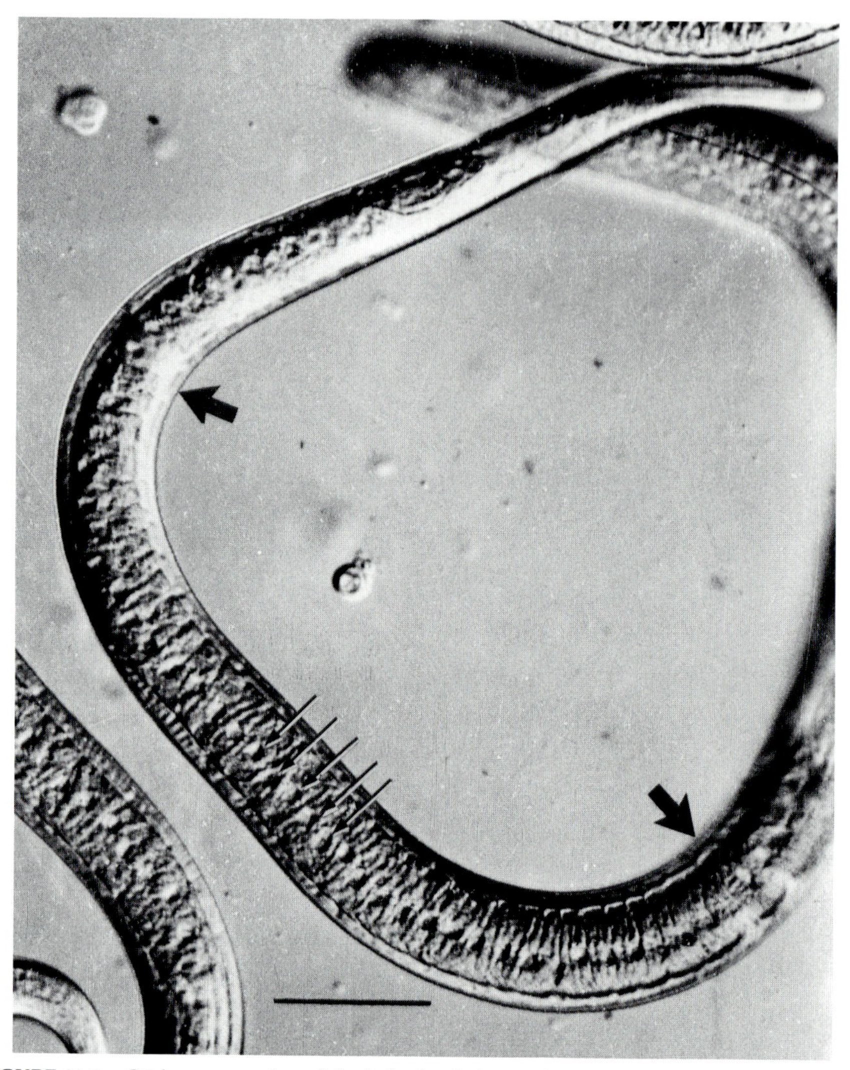

FIGURE 23.2. Stichosome region of the infective L_1 larva of *Trichinella*. The large arrows mark its beginning and end. Each stichosome contains 50–55 stichocytes (small arrows). Nomarski interference microscopy. Bar = 50 μm. (From Despommier, 1983. Original print courtesy of Dr. D. Despommier.)

is probable that secretions from the newly invading intestinal worms of a first infection induce both a humoral and a cell-mediated immunity. Immunity has been passively transferred either with antiserum alone from a previously infected animal or with living spleen or lymph node cells alone, and the effect of the two together is additive. Based on these findings, the following hypothesis is put forth as the mechanism(s) which act to expel the infection.

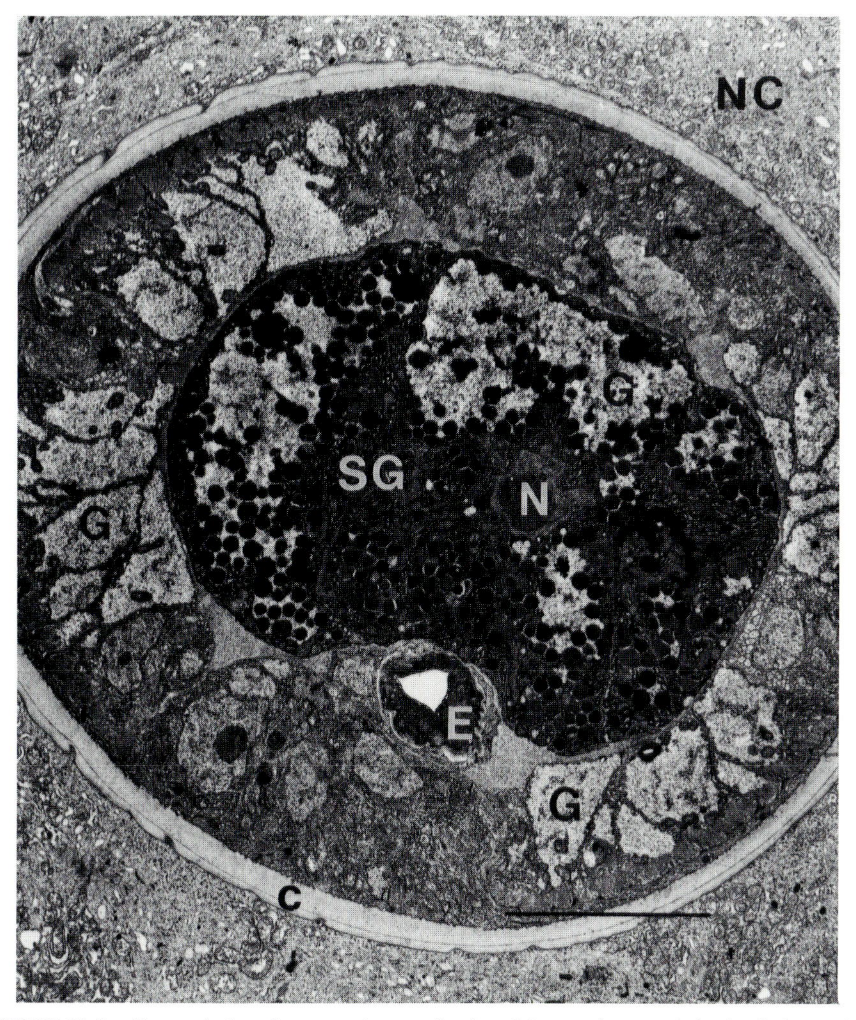

FIGURE 23.3. Transmission electron micrograph of a stichocyte from an infective L_1 larva. SG, β secretory granule; N, nucleus; G, glycogen; E, esophagus; C, cuticle; NC, nurse cell. Bar = 8 μm. (From Despommier, 1983. Original print courtesy of Dr. D. Despommier.)

Upon challenge, antibodies damage most of the invading worms in the intestine. This is folowed by an intense inflammatory reaction that is ultimately responsible for expulsion of the worms. The same two events occur in an initial infection, but at a longer interval after the entry of the parasites.

In addition to this protective mechanism, a complete infection also results in antibodies directed against the surface components of newborn larvae. Such larvae can be obtained by appropriate *in vitro* incubation of gravid female worms. When injected intravenously they are fully infective and establish themselves in the muscles to form normal encysted muscle larvae. A monoclonal antibody, prepared from spleen cells of mice orally infected with 200 muscle larvae, was found to react with a surface labeled antigen of 64,000 daltons on newborn larvae. This antibody (designated NIM-M5) promoted adherence of granulocytes, chiefly eosinophils, to newborn larvae, with damage to their surface (Fig. 23.4). Both of these effects were much augmented

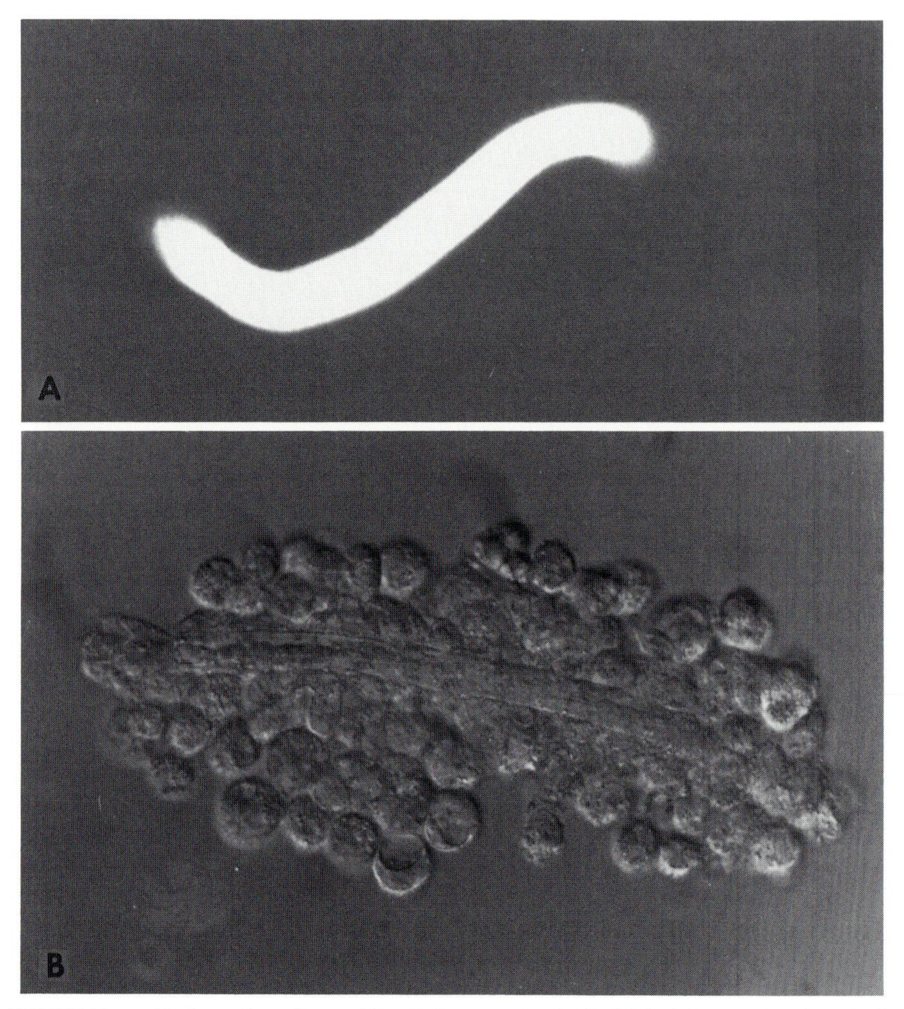

FIGURE 23.4. (A) A newborn larva of *T. spiralis* stained with FITC-labeled monoclonal antibody NIM-M5 (0.16 mg/ml) followed by FITC goat anti-mouse immunoglobulin. (B) A newborn larva following culture with NIM-M5 and rat eosinophil-enriched cell suspension for 4 hr. The parasite is covered with many adherent granulocytes and the internal structure of the worms is seen to be degenerating. ×1000. (From Ortega-Pierres *et al.*, 1984. Original prints courtesy of Dr. C. Mackenzie.)

by complement. Moreover, incubation of newborn larvae with immune mouse serum or with NIM-M5 reduced their infectivity by about 40%. We have here a situation altogether analogous to the effects of antibody with eosinophils on the schistosomulum of *Schistosoma mansoni* (see Chapter 21). In both cases a tissue-invading larval form is attacked by eosinophils through the mediation of antibodies. In both cases increase in the worm burden is prevented. With the schistosomes, adult worms of the first infection persist unaffected by the immunity to newly invading larval forms. With *Trichinella* the muscle larvae from the first infection likewise persist, unaffected by antibodies directed against the newborn larvae. Indeed, NIM-M5 does not react with any surface component of the muscle larvae. In the case of *Trichinella* there is an additional and probably even more important protective mechanism—the early expulsion of most of the intestinal worms of a reinfection.

Such worm expulsion is the principal mechanism of acquired immunity in many economically important intestinal nematode infections of domestic animals. It was N. R. Stoll who in 1929 called attention to the "self-cure" of sheep from infection with the stomach worm *Haemonchus contortus*. He showed that early stages in development of a challenge infection triggered an immune reaction blocking egg production by the worms of an initial infection and causing many of these to be expelled. After sensitization by several infections, further challenge results in a local hypersensitivity reaction in the abomasum, the habitat of adult worms, causing most of them to be rejected. The immune reactions occur when larvae of the challenge infection are molting from the ensheathed third larval stage (the free-living infective stage, see Diagram XXI) to the fourth larval stage, a molt that occurs in the abomasum. This suggests a role of antigens secreted in the molting process, and immunization of sheep has been effected with the use of materials secreted into the medium of short-term cultures of third- or fourth-stage larvae. Young lambs, unfortunately, are immunologically unresponsive and are the principal casualties from severe *Haemonchus* infections.

The economic losses among domestic ruminants caused by *Haemonchus* and related trichostrongylid nematodes cannot be overemphasized. This has stimulated much experimental work directed toward development of effective vaccines, but so far with relatively little success.

One species, *Trichostrongylus colubriformis*, a parasite inhabiting the intestine of sheep and goats, lends itself well to laboratory work because it will develop to egg-laying adults in guinea pigs. These animals can be effectively vaccinated with as little as 10 μg protein/100 g body wt of partially purified antigens from homogenates of fourth-stage larvae (Table 23.3). These results incidentially also show that worm acetylcholinesterase, an enzyme suspected to have an immunoprotective role, does not here have this. If similar results could be obtained in sheep, especially in the very young animals at risk, a practical vaccine could be developed. With the identification of antigens responsible for protective immunity, the methods of recombinant DNA might be applicable for large-scale production of a vaccine. Meanwhile the only

TABLE 23.3. Worm Counts in Guinea Pigs Vaccinated with 10 μg Protein/100 g Body wt of Fractions Obtained from *Trichostrongylus colubriformis* Fourth-stage Larvae by DEAE-Sephadex Chromatography[a]

Group	Dose of AChE[b]	No. of animals per group	Geometric mean worm count	Mean log worm count[c]
Controls	—	13	410	2.61*
Vaccinated				
Before fractionation	12.0	9	225	2.35*
Fraction I	0.07	9	64	1.81†
Fraction II	5.2	9	305	2.48*
Fraction III	71.8	6	607	2.78*

[a] Modified from Rothwell and Merritt (1975).
[b] Micromoles acetylcholine hydrolyzed per hour.
[c] Means with the same superscripts are not significantly different; those with different superscripts are significantly different at the 5% level by Duncan's new multiple range test.

commercially available vaccine against a nematode infection remains a preparation of irradiated larvae of *Dictyocaulus viviparus* against this lungworm of cattle. Third-stage infective larvae of this worm (of the family Metastrongylidae) are inactivated by exposure to 40,000 R and administered to calves in two injections to give a high degree of immunity to the field infection (Table 23.4). A very similar vaccine against the dog hookworm *Ancylostoma caninum* has also been produced but has not been commercially successful.

A particularly favorable model system for studies on immunity to intestinal nematodes is provided by the rat parasite *Nippostrongylus brasiliensis.* This organism has a life cycle very similar to that of the human hookworms (Diagram XX). Eggs passed with feces hatch and develop in soil through three larval instars to infective forms. These enter a new host through the skin, migrate via the heart to the lungs, where they molt to stage four. These fourth-stage larvae are carried up the bronchi and trachea and then down the pharnyx to the intestine, where they arrive about 2–3 days after infection. Here the final molt to the adult stage occurs, followed soon after by mating, with egg production beginning on about day 6. If the initial infection is moderate to heavy (several hundred or more worms), the worms are expelled after about 2 weeks and the rats show a strong immunity to reinfection. With smaller infections (e.g., 50 larvae), worm loss is more gradual, extending over a 30-day period, but as few as 10 adult female worms will stimulate resistance to reinfection. Just as with *Trichinella,* worm expulsion depends first on damage to the worms produced by antibodies and then on an acute inflammatory reaction. Passive transfer of immune serum to normal rats will damage worms at an early stage of infection and they will also be prematurely expelled by a nonspecific inflammatory reaction. In both cases degranulation of mast cells

TABLE 23.4. Immunization of Calves with Irradiated *Dictyocaulus viviparus* Larvae[a,b]

	Immunizing phase					Challenge phase		
Group	Calf Nos.	X-ray dose (R)	No. of irradiated larvae, day 0	No. of worms, day 35 (mean ± S.E.)	Lesion score, day 35 (a.v.)	No. of normal infective larvae given, day 50	No. of worms, day 85 (mean ± S.E.)	Lesion score, day 85 (av.)
B_1	11–15	40,000	4000	2.0 ± 1.7	1.4			
B_2	16–20	40,000	4000			4000	13.6 ± 18	1.6
C_1	21–25	60,000	4000	0	0			
C_2	26–30	60,000	4000			4000	919.8 ± 619	6.0
E	41–45	0	0			4000	1188.4 ± 375	8.8

[a] Modified from Jarrett *et al.* (1960).
[b] Note that with 40,000 R a few worms developed and caused slight lung pathology. This gave effective protection (compare groups B_2 and the controls E). With 60,000 R no worms developed in the immunizing infection and there was no pathology. There was no immunity either (group C_2).

occurs and amines are thought to be among the effector mechanisms. Extensive experimental work has clearly shown that the worms must be damaged by antibodies before expulsion can occur. Precipitates form at the worms' orifices. There is strong indication that antibodies to the acetylcholinesterase of the worms may be of special importance, but this has not been unequivocally proven. Damaged worms seem to lose their hold in the intestinal mucosa and are more likely to be found free in the lumen. In the second step in worm expulsion, reaginic antibodies formed by IgE-producing cells show a striking increase. Such increase is characteristic of all helminthic infections. Thus, the IgE of children in regions with endemic *Ascaris* is 25 times higher than normal.

In a second infection of rats with *Nippostrongylus* the fourth-stage larvae in the lungs are somewhat affected, but loss of worms does not occur until they reach the intestines. Here the worms that remain are stunted and are soon expelled. If such stunted worms present 7 days after reinfection were transferred to a fresh rat, they recovered and laid normal numbers of eggs for 6 days. They actually lived longer in fresh rats than normal worms from an initial infection similarly transferred to fresh rats. Hence the survivors in the reinfection show a partial "adaptation" of some sort. This is, however, insufficient to overcome the host's immune response, for if left in the original reinfected rat the worms would have been soon expelled.

Perhaps a more efficient "adaptation" occurs in what have been called "trickle" infections. If rats are exposed daily to only a few infective larvae, they build up a high worm burden, and no dramatic elimination of the worms occurs. This is probably analogous to the usual situation in nature. It may also be analagous to the situation in people where hookworm infection persists indefinitely. It must be stressed that this is not a case of failure to develop an acquired immunity. An immunity does develop and serves to prevent the host from being overwhelmed by the parasite. At the same time, however, limited numbers of parasites are able to continue to live and reproduce in such a partially immune host. Just how this comes about is among the many fascinating problems that remain to be explored. An additional model that may be useful in such exploration is provided by the infection of dogs with *Ancylostoma ceylonicum*. This is a third species of hookworm that infects humans, but it is much less common than the other two, *A. duodenale* and *Necator americanus*. The latter will not infect any experimental animal. The *A. ceylonicum* infection in dogs, on the other hand, shows the same acute and chronic infections and the same kind of hookworm disease as seen in humans infected with any of the three species.

Immunity to Ectoparasites

Analagous in several ways to the mechanisms involved in expulsion of nematodes from the intestinal mucosa are the reactions in the skin that confer an acquired immunity to ticks. As with the nematodes, serum antibodies play a role, but of special importance are local cell-mediated inflammatory reactions at the site of attachment of the tick. As with the intestinal nematodes the immunity shows a wide range of effectiveness depending on the species of host and the species and stage of the tick. With a model system of guinea pig as host and any of several species of ixodid (so-called hard-bodied) ticks, a small initial infestation with larval ticks will produce virtually complete immunity to a second such infestation. This has been well shown with *Dermacentor variabilis,* the American dog tick, and with *Amblyomma americanum.* A similar immunity to larval ticks was demonstrated in the natural host–parasite association of the rabbit and the rabbit tick *Haemaphysalis leporis-palustris.* But when *D. variabilis* larvae were studied on their natural host, the deer mouse *Peromyscus,* these animals became only relatively resistant after two or three infestations. With the tick *Ixodes triangulicipes* on its natural host, the field mouse *Apodemus sylvaticus,* no acquired resistance could be seen even after four infestations with small numbers of larvae, whereas with the same infestations on laboratory mice some resistance was already apparent after the first. It is likely that these are quantitative differences rather than qualitative ones, and that with larger early infestation resistance would become apparent.

With the guinea pig model with either *Dermacentor* or *Amblyomma,* acquired immunity to the larger nymphs and adult stages is expressed more in a reduction in engorged weight of the nymphs and adult females than in a reduction in the number that are able to attach and feed. This reduction in amount of blood taken secondarily results in a reduction in successful molting by nymphs to the adult stage and a reduction in the number of eggs laid by the adult females. With the larval stage, where there may be 80% or better reduction in the number of larvae able to feed at all on an immune guinea pig, those few larvae that have succeeded in engorging are small, do not have the usual dark gray nearly black color, and fail to molt to the nymphal stage.

TABLE 23.5. Feeding of *Amblyomma americanum* Larvae on Guinea Pig Recipients of Peritoneal Exudate Cells or Serum from Actively Sensitized Animals or Naive Controls[a]

Treatment (No. of animals)	Tick yield (%)	Tick rejection (%)	Tick weight (mg)	% weight reduction[b]
Naive control (18)	83 ± 4[c]	—	0.91 ± 0.04	—
Normal serum (15)	83 ± 4	0.0	0.88 ± 0.04	3.3
Immune serum (18)	59 ± 4[d]	28.9	0.78 ± 0.04[e]	14.3
Normal PEC (3)	83 ± 5	0.0	0.91 ± 0.01	0.0
Immune PEC (6)	38 ± 3[d]	54.2	0.65 ± 0.05[d]	28.6
Actively sensitized controls (9)	21 ± 4[d]	74.7	0.58 ± 0.04[d]	36.3

[a] Modified from Brown and Askenase (1983).
[b] Percent rejection or percent weight reduction = 100 × [1 − (tick yield or weight from treated/tick yield or weight from controls)].
[c] Mean ± S.E.M.
[d] $p < 0.001$ (versus naives).
[e] $p < 0.05$ (versus naives).

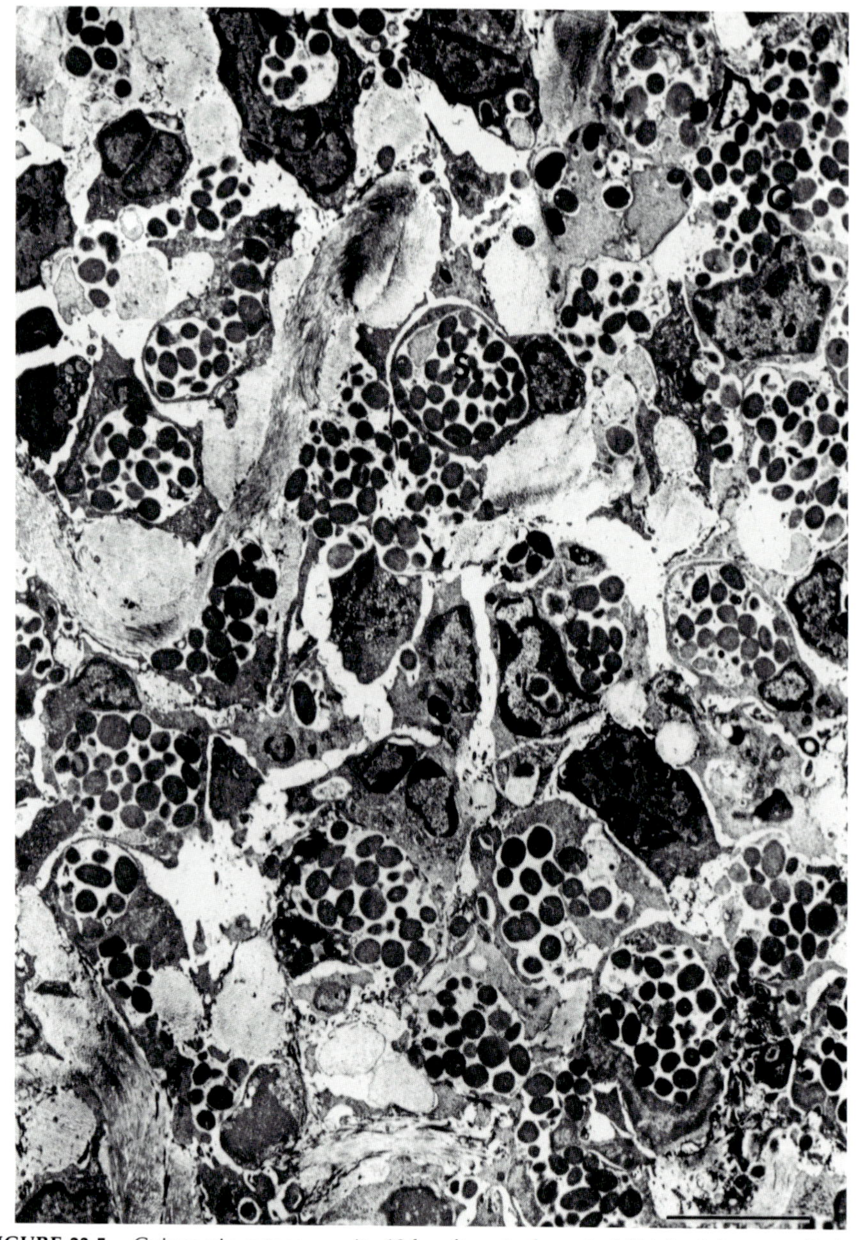

FIGURE 23.5. Guinea pig cutaneous site 18 hr after attachment of *Rhipicephalus appendiculatus* larvae in a host immunized by prior infestation with tick larvae. The space between bundles of dermal collagen is completely filled by a cellular infiltrate dominated by basophils exhibiting degranulation sacs (S) characteristic of compound exocytosis, in which membranes around individual granules fuse with each other and with the plasma membrane causing the granules to lie in a common sac exposed to the extracellular milieu. Membrane-free granules (G) have been liberated to the exterior of the cells. Bar = 50 μm. (From McLaren *et al.*, 1983. Original print courtesy of Dr. P. W. Askenase.)

Whitish larvae of this type appear to have fed on a serous exudate rather than on blood. The relatively few bright red ones appear unable to digest the ingested blood.

It has now been clearly shown that this immune rejection results from the combined action of IgG serum antibodies and sensitized T cells (Table 23.5) that recruit basophils and eosinophils to the site of attachment of the ticks (Fig. 23.5). The basophils are of prime importance as shown by the greater effect of antibasophil serum than of antieosinophil serum in abrogating the immune response (Table 23.6). Whereas in initial infestation there is little cellular reaction around the point of attachment of a larval tick even on day 4 when the larvae have completed engorgement, extensive basophil accumulation occurs around the site of attachment on an immune guinea pig (Fig. 23.6).

Again as with nematodes, the actual mechanisms by which basophils adversely affect tick feeding are not known. When these cells degranulate there may be a number of toxic effects. Moreover, some direct damage by antibodies may occur. Recent *in vitro* feeding experiments have shown inhibition of feeding by histamine plus serotonin at concentrations (about 10 μM) likely to be present in an anaphylactic basophil reaction such as occurs in immune animals. In these experiments adult female *Dermacentor andersoni* were induced to feed through a treated mouse skin membrane on a mixture of defibrinated bovine blood with the phagostimulants ATP and reduced glutathione, each at 10 μM. These *in vitro* results are consonant with *in vivo*

TABLE 23.6. Immune Resistance and Cutaneous Cellular Response to Tick (*A. americanum*) Challenge in Guinea Pigs Treated with Antibasophil or Antieosinophil Serum[a]

Sensitized	Treatment[b] (No. of hosts)	Tick rejection (%)[c]	Cells per 0.032 mm² of dermis at tick feeding sites: Basophils	Eosinophils	Mononuclears
−	NIL, SAL or NRS (15)	—	23 ± 8 [d]	12 ± 3	25 ± 4
+	NIL or SAL (12)	48	81 ± 9	87 ± 16	54 ± 9
+	NRS (12)	47	67 ± 21	91 ± 20	58 ± 8
+	AES (9)	23[e]	58 ± 9	19 ± 9[e]	72 ± 10
+	ABS (6)	7[e]	1 ± 1[e]	34 ± 7[e]	93 ± 19

[a] Modified from Brown and Askenase (1983).
[b] Animals were treated before tick challenge with one of the following: NIL, no treatment; SAL, saline; NRS, normal rabbit serum; AES, antieosinophil serum; ABS, antibasophil serum.
[c] Percent rejection = 100 × [1 − (tick yield from sensitized/tick yield from controls)].
[d] Mean cell counts ± S.E.M. at day 4 after tick challenge.
[e] Significantly different ($p < 0.05$) from sensitized and challenged controls (NIL, SAL, or NRS).

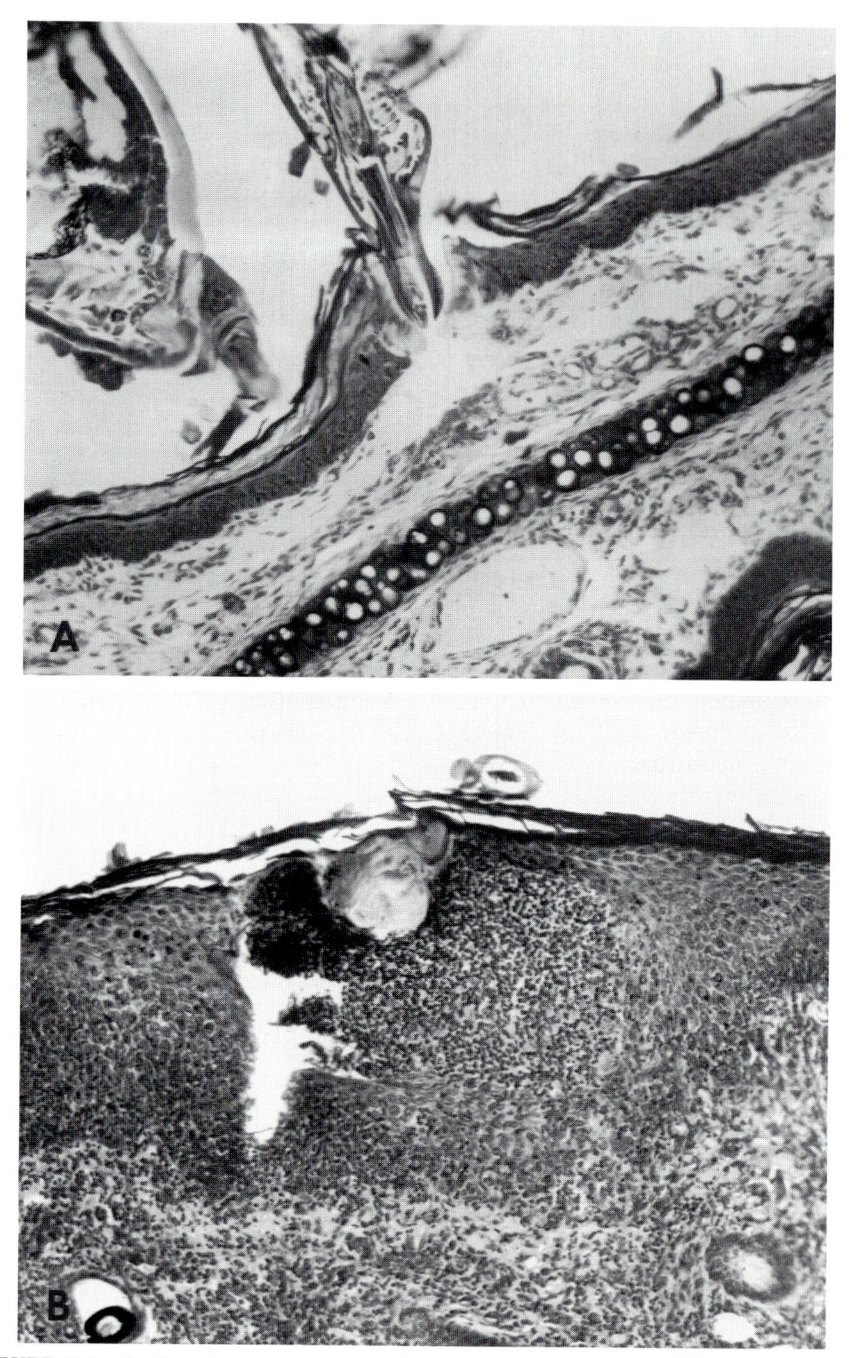

FIGURE 23.6. Sections through the point of attachment of a larva of *Dermacentor variabilis* on the ear of a susceptible guinea pig (A) and an immune guinea pig (B). Both preparations fixed on day 4 after application of the ticks. (From Trager, 1939.)

observations indicating effects of histamine in reducing attachment of ticks on cattle as well as guinea pigs. Such a reaction could account for the observed greater degree of wandering of larval ticks on an immune host. When they insert their mouthparts, they might elicit a rapid anaphylactic reaction causing them to withdraw and move elsewhere.

In guinea pigs at least, active immunity has been produced by injection of tick tissue extracts, particularly salivary gland extracts, suggesting the possibility of an effective vaccine. That such a vaccine might have practical application in two different ways is indicated by the following observations. In controlled trials that simulated field conditions, rabbits that had been immunized against the vector tick (*D. andersoni*) were protected from tularemia, a bacterial disease transmitted by ticks (Table 23.7). Hence, immunization to ticks could be of material value in protection against tick-transmitted diseases of cattle, such as *Theileria parva*, cause of East Coast fever, transmitted by three-host ticks such as *Rhipicephalus*. In addition, vaccination of cattle against the cattle tick *Boophilus microplus*, a one-host tick and the vector of *Babesia* (see Chapter 19), might protect them both against babesiosis and against the considerable direct damage that results from heavy infestation with ticks. Cattle have been shown to become relatively resistant to further infestations with *Boophilus* after a primary infestation. Many years ago observant stock raisers in Australia noted that some cattle after heavy primary infestation with *Boophilus* became relatively resistant. The resistance manifested itself first with a high proportion of "sickly, yellow, infertile" ticks accompanied by pinpricklike sores and an exudate. Infestation with *Boophilus* occurs only by the larvae; these subseqently molt twice while on the host and again engorge, with only the engorged females dropping off (Diagram XI). Accordingly, the immunity would need only to be effective against the larvae, the stage most susceptible in other tick–host associations. Resistant cattle often show greater grooming activity, and this is more effective in removing ticks, presumably because fewer of the larvae are able to attach properly. The extent of this acquired resistance varies greatly, however, not only with the breed but also among

TABLE 23.7. Effect of Host Resistance to Tick Infestation on the Transmission of *Francisella tularensis*[a]

	Tick-resistant group	Control group
Percent of animals dying when infested with infected ticks[b]	36.4%[c]	100%
Percent of animals dying which yielded *F. tularensis* on necropsy	100%	100%

[a] Modified from Wikel (1980).
[b] Eleven tick-resistant and ten control animals were infested with 25 infected nymphs for 14 days.
[c] $p < 0.02$

TABLE 23.8. Number of Female Ticks Engorged on Previously Unexposed Brahman (*Bos indicus*) and Shorthorn (*Bos taurus*) Cattle Infested Four Times with 20,000 *Boophilus microplus* Larvae[a–c]

Breed	Animal No.	Primary a–d	Second e	Third f	Fourth g
Brahman	276	5258[a]	956	295	920
	340	5478[b]	3440	3628	3329
	342	4643[c]	1587	1085	1064
	389	4350[d]	2123	703	742
Shorthorn	8	3787[a]	7182	6074	6039
	9	5726[b]	3023	3864	4532
	10	3708[c]	4569	4016	4599
	5	6160[d]	5209	4767	4839

[a] Modified from Wagland (1975).
[b] Dates of infestation: a, 6/4/71; b, 7/27/71; c, 9/14/71; d, 11/9/71; e, all 1/10/72, f, all 2/16/72; g, all 3/21/72.
[c] Note that Brahman and Shorthorn did not differ in innate resistance (primary infestation) but only in acquisition of resistance. In other experiments, however, *Bos taurus* has also been seen to acquire resistance.

individual animals (Table 23.8). This suggests that a combination of selective breeding with an appropriate vaccine could lead to the desired result.

It is of considerable interest that infection of cattle with *Babesia bovis* has been observed to reduce their acquired resistance to the tick *Boophilus microplus*, the natural vector of *B. bovis*. The nature of this suppression remains to

TABLE 23.9. Comparative Responses of the Host to Intermittent Exposure to Resident or Free-Living Ectoparasites[a]

Type	Parasite		Process	Feeding or population
Free-living	Insects	Mosquitoes	Desensitization	Not affected
		Fleas	Desensitization	Not affected
Resident		Keds		
		Melophagus ovinus	Resistance	Inhibited
		Lice		
		Polyplax serrata	Resistance	Inhibited
		Haematopinus eurysternus	Resistance	Inhibited
Free-living	Mites		Desensitization	Not affected
Resident		*Sarcoptes scabei*	Desensitization	Inhibited
		Demodex canis	Desensitization	Inhibited
Free-living	Ticks	Argasidae, adults	Desensitization	Not affected
Resident		Argasidae, larvae	Resistance	Inhibited
		Ixodidae	Resistance	Inhibited

[a] Modified from Nelson *et al.* (1977).

be determined. On the other hand, tick infestation itself likewise has immunosuppressive effects. These may be mediated by prostaglandin E_2, shown to be present in saliva of the tick *Ixodes dammini* in significant amounts, as well as by other constituents of the tick saliva. Thus, the ectoparasitic ticks, like internal parasites, are equipped to counteract in part the immune reactions of their host.

The ixodid ticks are ectoparasites that require several days of attachment on a host to engorge fully. Other kinds of hematophagous arthropods can engorge in minutes and it is not surprising that an effective acquired immunity to these, if it exists at all, would have to be of a different nature. In general, resident ectoparasites induce a resistance manifested by reduced feeding and reduced population as a result of either reduced fecundity or death of females. Nonresident ectoparasites, such as mosquitoes and fleas, bring about desensitization of the host rather than actual resistance (Table 23.9).

Bibliography

Adams, D. B., and Beh, K. J., 1981, Immunity acquired by sheep from an experimental infection with *Haemonchus contortus*, *Int. J. Parasitol.* **11**:381–386.

Askenase, P. W., 1983, Effector and regulatory functions of cells in the immune response to parasites, *Fed. Proc.* **42**:1740–1743.

Brown, S. J., and Askenase, P. W., 1983, Immune rejection of ectoparasites (ticks) by T cell and IgG_1 antibody recruitment of basophils and eosinophils, *Fed. Proc.* **42**:1744–1749.

Brown, S. J., and Knapp, F. W., 1981, Response of hypersensitized guinea pigs to the feeding of *Amblyomma americanum* ticks, *Parasitology* **83**:213–223.

Callow, L. L., and Stewart, N. P., 1978, Immunosuppression by *Babesia bovis* against its tick vector, *Boophilus microplus*, *Nature* **272**:818–819.

Campbell, W. C. (ed.), 1983, *Trichinella and Trichinosis*, Plenum Press, New York.

Carroll, S. M., and Grove, D. I., 1984, Parasitological, hematologic, and immunologic responses in acute and chronic infections of dogs with *Ancylostoma ceylanicum:* A model of human hookworm infection, *J. Infect. Dis.* **150**:284–294.

Clegg, J. A., and Smith, M. A., 1978, Prospects for the development of dead vaccines against helminths. *Adv. Parasitol.* **16**:165–218.

Despommier, D. D., 1981, Partial purification and characterization of protection-inducing antigens from the muscle larva of *Trichinella spiralis* by molecular sizing chromatography and preparative flatbed isoelectric focusing, *Parasite Immunol.* **3**:261–272.

Despommier, D. D., 1983, Biology, in: *Trichinella and Trichinosis* (W. C. Campbell, ed.), Plenum Press, New York, pp. 75–151.

Jarrett, W. F. H., Jennings, F. W. McIntyre, W. I. M., Mulligan, W., and Urquhart, G. M., 1960, Immunological studies on *Dictyocaulus viviparus* infection: Immunity produced by the administration of irradiated larvae, *Immunology* **3**:145–151.

Johnston, T. H., and Bancroft, H. J., 1918, A tick-resistant condition in cattle, *Proc. R. Soc. Queensl.* **30**:219–317.

Koudstaal, D., Kemp, D. H., and Kerr, J. D., 1978, *Boophilus microplus:* Rejection of larvae from British breed cattle, *Parasitology* **76**:379–386.

Larsh, J. E., and Race, G. J., 1975, Allergic inflammation as a hypothesis for the expulsion of worms from tissues: A review, *Exp. Parasitol.* **37**:251–266.

Larsh, J. E., Jr., and Weatherly, A. F., 1975, Cell-mediated immunity against certain parasitic worms, *Adv. Parasitol.* **13**:183–222.

Lloyd, S., 1981, Progress in immunization against parasitic helminths, *Parasitology* **83**:225–242.
Love, R. J., Ogilvie, B. M., and McLaren, D. J., 1976, The immune mechanism which expels the intestinal stage of *Trichinella spiralis* from rats, *Immunology* **30**:7–15.
McGowan, M. J., and Barker, R. W., 1980, A selected bibliography of tick-host resistance and immunological relationships, *Bull. Entomol. Soc. Am.* **26**:17–25.
McLaren, D., Worms, M. J., and Askenase, P. W., 1983, Cutaneous basophil associated resistance to ectoparasites (ticks). III. Electron microscopy of *Rhipicephalus appendiculatus* larval feeding sites in actively sensitized guinea pigs and recipients of immune serum, *J. Pathol.* **139**:291–308.
Miller, T. A., 1978, Industrial development and field use of the canine hookworm vaccine, *Adv. Parasitol.* **16**:333–342.
Miller, T. A., 1979, Hookworm infections in man, *Adv. Parasitol.* **17**:315–384.
Nelson, W. A., Bell, J. F., Clifford, C. M., and Keirans, J. E., 1977, Review article: Interaction of ectoparasites and their hosts, *J. Med. Entomol.* **13**:389–428.
Ogilvie, B. M., and de Savigney, D., 1982, Immune response to nematodes, in: *Immunology of Parasitic Infections* (S. Cohen and K. S. Warren, eds.), Academic Press, New York, pp. 715–757.
Ogilvie, B. M., and James, V. E., 1971, *Nippostrongylus brasiliensis:* A review of immunity and the host/parasite relationship in the rat, *Exp. Parasitol.* **29**:138–177.
Ortega-Pierres, G., Mackenzie, C. D., and Parkhouse, R. M. E., 1984, Protection against *Trichinella spiralis* induced by a monoclonal antibody that promotes killing of newborn larvae by granulocytes, *Parasite Immunol.* **6**:275–284.
Paine, S. H., Kemp, D. H., and Allen, J. R., 1983, *In vitro* feeding of *Dermacentor andersoni* (Stiles): Effects of histamine and other mediators, *Parasitology* **86**:419–428.
Randolph, S. E., 1979, Population regulation in ticks: The role of acquired resistance in natural and unnatural hosts, *Parasitology* **79**:141–156.
Ribeiro, J. M. C., Makoul, G. T., Levine J., Robinson, D. R., and Spielman, A., 1985, Antihemostatic, anti-inflammatory and immunosuppressive properties of the saliva of a tick, *Ixodes dammini, J. Exp. Med.* **161**:332–344.
Roberts, J. A., 1968, Acquisition by the host of resistance to the cattle tick, *Boophilus microplus* (Canestrini), *J. Parasitol.* **54**:657–662.
Roberts, J. A., 1968, Resistance of cattle to the tick *Boophilus microplus* (Canestrini). II. Stages of the life cycle of the parasite against which resistance is manifest, *J. Parasitol.* **54**:667–673.
Rogers, W. P., 1982, Enzymes in the exsheathing fluid of nematodes and their biological significance, *Int. J. Parasitol.* **12**:495–502.
Rothwell, T. L. W., and Merritt, G. C., 1975, Vaccination against the nematode *Trichostrongylus colubriformis.* II. Attempts to protect guinea pigs with worm acetylcholinesterase, *Int. J. Parasitol.* **5**:453–460.
Silberstein, D. S., and Despommier, D. D., 1984, Antigens from *Trichinella spiralis* that induce a protective response in the mouse, *J. Immunol.* **132**:898–904.
Silberstein, D. S., and Despommier, D. D., 1985a, Effects on *Trichinella spiralis* of host responses to purified antigens, *Science* **227**:948–950.
Silberstein, D. S., and Despommier, D. D., 1985b, Immunization with purified antigens protects mice from lethal infection with *Trichinella spiralis, J. Parasitol.* **71**:516–517.
Soulsby, E. J. L., 1966, The mechanisms of immunity to gastro-intestinal nematodes, in: *Biology of Parasites* (E. J. Soulsby, ed.), Academic Press, New York, pp. 255–276.
Stoll, N. R., 1947, This wormy world, *J. Parasitol.* **33**:1–18.
Stoll, N. R., 1961, The worms: Can we vaccinate against them?, *Am. J. Trop. Med. Hyg.* **10**:293–303.
Trager, W., 1939, Acquired immunity to ticks, *J. Parasitol.* **25**:57–81.
Utech, K. B. W., Seifert, G. W., and Wharton, R. H., 1978, Breeding Australian Illawarra Shorthorn cattle for resistance to *Boophilus microplus.* I. Factors affecting resistance, *Aust. J. Agric. Res.* **29**:411–422.
Wagland, B. M., 1975, Host-resistance to the cattle tick (*Boophilus microplus*) in Brahman (*Bos indicus*) cattle. I. Responses of previously unexposed cattle to farm infestations with 20,000 larvae, *Aust. J. Agric. Res.* **26**:1073–1080.

Wakelin, D., and Denham, D. A., 1983, The immune response, in: *Trichinella and Trichinosis* (W. C. Campbell, ed.), Plenum Press, New York, pp. 265–308.
Wikel, S. K., 1980, Host resistance to tick-borne pathogens by virtue of resistance to tick infestation, *Ann. Trop. Med. Parasitol.* **74:**103–104.
Wikel, S. K., and Osburn, R. L., 1982, Immune responsiveness of the bovine host to repeated low-level infestations with *Dermacentor andersoni, Ann. Trop. Med. Parasitol.* **76:**405–414.
Wikel, S. K., Graham, J. E., and Allen, J. R., 1978, Acquired resistance to ticks. IV. Skin reactivity and *in vitro* lymphocyte responsiveness to salivary gland antigen, *Immunology* **34:**257–263.
Willadsen, P., and Williams, P. G., 1976, Isolation and partial characterization of an antigen from the cattle tick, *Boophilus microplus, Immunochemistry* **13:**591–597.

CHAPTER 24

Symbiosis

As we have seen in the preceding chapters, a state of truce often develops between a parasite and its host. With the animal parasites of vertebrates, this is generally based on immunological reactions of the host and their partial evasion by the parasite. The numbers of the parasite are reduced to or held at a minimal level. The host lives on with little injury while the relatively few parasites within it continue to disseminate their species to new hosts, a situation clearly advantageous to the parasite as well as to the host. Thus, in most people infected with *Entamoeba histolytica,* the amebae are prevented from invading the tissues but live mainly in the lumen of the colon. Here they form large numbers of cysts that are excreted with the feces and are infective to new hosts. Such infections are asymptomatic. Similarly, a person infected with a small number of *Schistosoma mansoni* will be unaware of the infection, will be protected from further infection, and will continue to excrete large numbers of schistosome eggs for years.

Such associations may be considered examples of symbiosis in the broader meaning of the word: an intimate living together of two different species of organisms. Through common usage, however, *symbiosis* has come to imply an intimate association of two different species in which each is beneficial to the other and indeed may be essential to the life of the other. It is in this sense of mutualism that I use the word for the discussion in this chapter.

Symbiosis, or mutualism, is a widespread phenomenon among both plants and animals. One thinks immediately of the root nodule bacteria that fix nitrogen for leguminous plants, the green hydra and many other kinds of invertebrates with their endocellular algae whose photosynthetic products feed the host, the chimeric plants known as lichens consisting of a fungus and an alga in intimate obligate association, the cellulose-digesting rumen bacteria of cattle, and the cellulose-digesting protozoa of termites. Less well known are the obligate intracellular bacteria essential to the life of many kinds of invertebrates from protozoa to arthropods and mollusks. Here are included also the recently found chemolithotrophic bacteria associated with tube worms and mollusks living in deep-sea hydrothermal vents. The bacteria comprise up to 60% of the net weight of the worm *Riftia pachyptila,* which has no ingestive or digestive organs, suggesting that it is nutritionally dependent on its symbiotic bacteria.

There are few examples of mutualism between parasitic protozoa or helminths and their vertebrate hosts, the parasite–host combinations with which this book is mainly concerned. Nevertheless, the symbiotic associations display so many of the same characteristics found in parasitic associations that they must be discussed in relation to the biology of parasitism. Especially striking are the analogies between endocellular symbionts and intracellular parasites. I have already noted that intracellular parasitic protozoa modify their host cell and that the parasite–host cell combination must be regarded as a new, integrated unit rather than the sum of parasite plus host cell. The endosymbiotic association would then represent a further integration. If parts of the symbiont genome were then to become incorporated into the nuclear genome of the host, the ultimate integration would be achieved. This has not been shown for any endosymbiotic association but this is exactly the situation for chloroplasts and mitochondria. These organelles have genomes and certain enzymes like those of blue-green algae and bacteria, supporting the widely accepted hypothesis that they arose from such endosymbiotic organisms. Many of the proteins of chloroplasts and even more of mitochondria are coded for by the nuclear genome, not by the genome of the organelle, and recent work indicates that genes or parts of them can indeed move from one organelle to another. In the chloroplasts of photosynthetic eukaryotes the important CO_2-fixing enzyme ribulose-1,5-bisphosphate carboxylase occurs as two subunits, a shorter one coded for in the nucleus and a longer one produced by the chloroplast genome. A different situation has been found in recent work with the flagellate *Cyanophora paradoxa.* This organism always bears a pair of endosymbiotic blue-green algae called cyanelles. Here both subunits of ribulose-1,5-bisphosphate carboxylase are coded for in the genome of the cyanelle, supporting other evidence that this is still a symbiont, not yet a chloroplast. Whereas the chloroplasts of land plants and most groups of eukaryotic algae appear to have been derived from the prokaryotic blue-green algae, there are several groups of algae whose chloroplasts seem to have come from other eukaryotic cells that in turn had endosymbiotic blue-green algae. Such instances of endosymbionts within endosymbionts are not uncommon. The plasmids of bacteria often occur in bacteria that are themselves symbionts, as well as in all kinds of free-living bacteria.

Plasmids, defined originally as extrachromosomal hereditary determinants, occur in the cytoplasm of bacteria as double-stranded DNA circles ranging in size from about 1.5 to 100 million daltons. They may be regarded as the ultimate in endosymbiotic specialization, a group of extranuclear genes dependent for their replication on the host cell and conferring to that host a number of additional specific characters. Plasmids are involved in conjugational transfer (F plasmids) and in resistance to many kinds of antibiotics and other antibacterial drugs (R plasmids). Whether a symbiont is a plasmid, a bacterial or algal cell, or a higher eukaryote, it must in all cases effect a balance with its host, producing numbers high enough to maintain itself, but not so much as to injure the host. With endocellular symbionts the number per host cell varies greatly with the species of symbiont. If the number is relatively large, division of the host cell can easily effect approximately equal distribution to the daughter cells. But if the symbionts are relatively few, some special mechanisms must exist for their distribution to the daughter cells. In addition, many endocellular symbionts have characters permitting their extracellular transfer to new host cells. Thus, some plasmids of bacteria can even cross intergeneric boundaries. Of special interest and importance is the control of replication. This may also have particular relevance to the control of replication of intracellular parasites that is seen as latency in some kinds of parasitic infections.

For most endosymbionts little is known concerning the control of replication. For certain types of plasmids there is now strong evidence that the plasmids themselves control their own replication rate. This requires some way for the plasmid to "sense" its own numbers and adjust replication accordingly. A negative control loop could be based on an inhibitor produced by the plasmid. Just such a replication inhibitor has been demonstrated for the plasmid R1; it is a small untranslated RNA molecule synthesized in proportion to the gene dosage. It may well be that inhibitors produced by the symbionts also operate in other symbiotic associations. It is equally possible that control is effected by host products or by nutritional deprivation.

Nutritional interdependence is certainly at the basis of many kinds of symbiotic associations. It was L. R. Cleveland who in the 1920s first clearly demonstrated a nutritional symbiosis. He showed that wood-feeding termites, if deprived of the teeming intestinal flagellates that inhabit their midgut, were unable to live on a cellulose diet and died of starvation. Later work with termites and with the wood-feeding roach *Cryptocercus punctulatus,* which also contains in its enlarged hindgut a remarkable assemblage of highly specialized flagellate protozoa, showed beyond doubt that certain of these protozoa are the source of the cellulase that enables the host insect to utilize cellulose as its sole energy source. Several species of the protozoa have been studied in axenic culture (Figs. 24.1, 24.2). They are free not only from external bacteria but also from intracellular ones, so that their cellulase cannot be derived from bacteria. Furthermore, defaunated termites (*Zootermopsis*) if reinfected with a pure axenic culture of one species of their protozoa (*Trichomitopsis termopsidis*) were again able to live on a cellulose diet (Fig. 24.3), proving that it is the protozoa, not any bacteria, that are here the essential symbiotic organisms. Spread of the infection to newly hatched larval termites and the reinfection of termites that have lost their protozoa after molting are readily effected by the habit of proctodeal feeding. Termites solicit and imbibe drops of anal fluid from other members of the colony; these drops contain large numbers of the gut flagellates. The termites in turn contribute to the life of the protozoa not only by providing them with an appropriate sheltered environment and probably certain nitrogenous nutrients, but also by furnishing them with finely

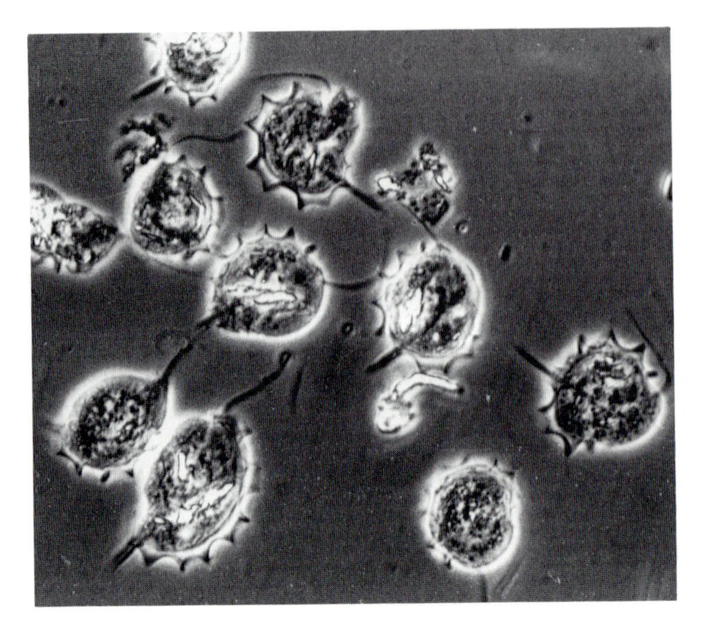

FIGURE 24.1. *Trichomitopsis termopsidis*, a cellulose-digesting symbiotic flagellate from the hindgut of the termite *Zootermopsis*, in axenic culture. Phase contrast after glutaraldehyde fixation. ×320. (From Yamin, 1978. Original print courtesy of Dr. M. A. Yamin.)

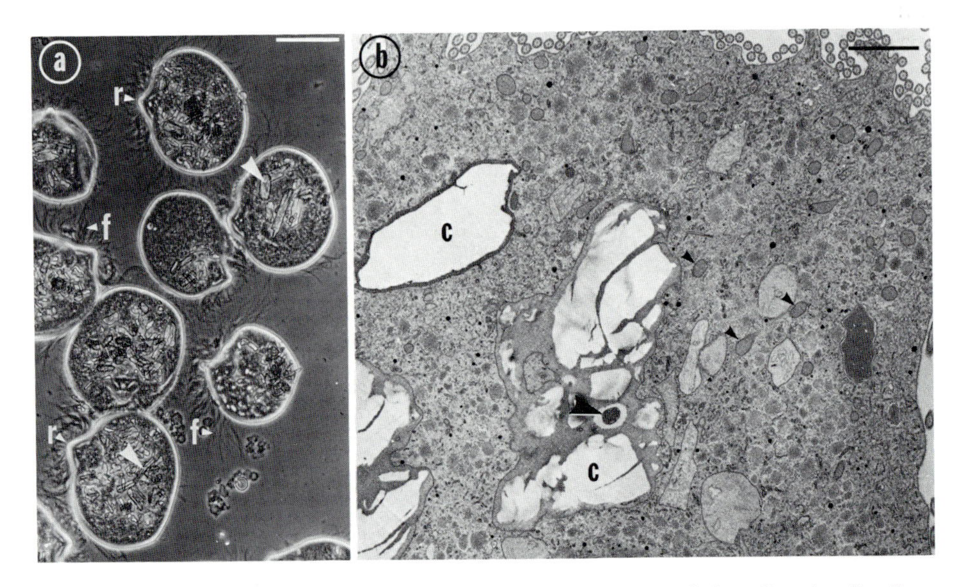

FIGURE 24.2. *Trichonympha sphaerica*, a hypermastigote symbiotic cellulose-digesting flagellate from the hindgut of the termite *Zootermopsis* in axenic culture. (a) Phase-contrast micrograph of cultured cells after glutaraldehyde fixation. Note rostrum (r), flagella (f), and ingested cellulose particles (arrowheads). Bar = 100 μm. (b) Transmission electron micrograph of part of a cell to show absence of endosymbiotic bacteria. Note microbodies (small arrowheads), food vacuoles with cellulose particles (C), and a heat-killed food bacterium (large arrowhead). Bar = 2 μm. (From Yamin, 1981. Original prints courtesy of Dr. M. A. Yamin.)

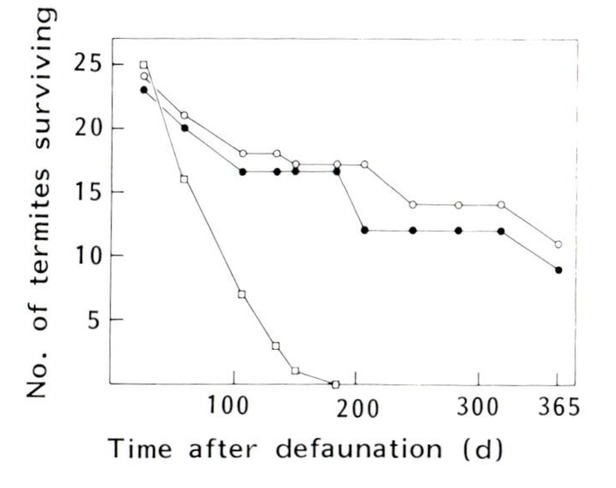

FIGURE 24.3. Survival of defaunated *Zootermopsis* after feeding with (○) *Trichomitopsis termopsidis* from axenic culture (24 termites), (●) untreated *Zootermopsis* intestinal contents (23 termites), and (□) heat-killed *T. termopsidis* (25 termites). Termites were fed 25 days after defauntation. (From Yamin and Trager, 1979.)

comminuted particles of cellulose, the only energy source they are able to utilize.

Many other groups of insects exhibit such nutritional symbiosis. This is especially prominent in insects that have restricted diets, such as those living on plant sap or blood. There are now a number of instances where experimental work has shown that it is the symbionts that make possible development on these restricted diets. It is also symbiotic bacteria that enable an omnivorous insect like the common cockroach (*Blatella germanica*) to thrive on meager diets low in growth factors. The symbiotic microorganisms are usually intracellular in specialized cells, the mycetocytes, often forming a conspicuous organ, the mycetome (Fig. 24.4). For such symbionts transovarial transmission is the rule (Fig. 24.5). Of special interest, and of special relevance to the biology of animal parasitism, are the symbiotic bacteria of certain bloodsucking insects.

V.B. Wigglesworth first pointed out that regardless of their taxonomic position, those insects that were completely dependent on blood as food throughout their life cycle harbored symbiotic bacteria, whereas those that fed on blood during only part of the life cycle did not (Table 24.1). Thus, fleas, mosquitoes, sand flies, blackflies, and horseflies all feed as larvae on diets rich in microorganisms and do not have symbiotic bacteria, whereas the bloodsucking Hemiptera and Anoplura feed only on blood and have symbiotic bacteria. A remarkable contrast is provided by the two closely related muscid flies, the stable fly *Stomoxys* and the tsetse fly *Glossina*. The larvae of *Stomoxys* are free-living and feed on manure, rich in microorganisms. The larvae of *Glossina* develop within the mother fly, where they are nourished by special glands (Diagram XXIV). Only *Glossina* has symbiotic bacteria. These occur in a mycetome forming a portion of the midgut of the fly (Fig. 24.6). They are transmitted transovarially, probably via the nurse cells, and perhaps also by means of the secretion of the milk glands to the feeding larva. Treatment of female tsetse flies with bactericidal agents destroys the symbionts and at the same time interferes with the ability of the flies to reproduce. Most likely the symbiotic bacteria supply essential growth factors. Similarly, if a female louse

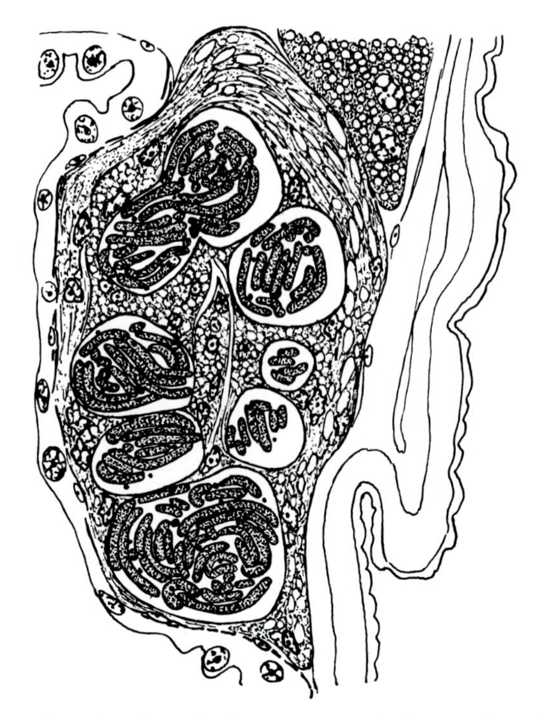

FIGURE 24.4. Sagittal section through the mycetome of a human louse (*Pediculus capitis*) to show the large intracellular bacteria (at left the gut epithelium, at right the hypodermis). ×770. (After Ries, from Buchner, 1965.)

(*Pediculus capitis*) was surgically deprived of its mycetome before the symbionts had migrated to the oviducts (see Fig. 24.5), it failed to reproduce, whereas it would reproduce in a normal way if the mycetome was extirpated after infection of the oviducts had occurred. Larval lice deprived of their mycetome would die suddenly 5 or 6 days later, but their life could be prolonged by intrarectal feeding of yeast extract. Again the indications are that the symbionts supply various growth factors. This has been best demonstrated in work with the hemipteran bug *Rhodnius prolixus*. In this insect the symbiotic bacteria are a species of Actinomyces (*Nocardia rhodnii*) that lives extracellularly in the insect's gut. The bacteria are not transmitted transovarially. Since the bugs live gregariously, young bugs get infected by simple fecal contamination from older bugs. Hence, bacteria-free bugs could be easily obtained by isolating freshly laid eggs in a sterile container. When such bugs were allowed to take blood meals on rabbits, they grew as well as infected ones only through the first two of their five larval stages. None developed to adults. They could, however, be made to grow and develop normally if reinfected with the specific bacteria. Their growth could also be improved if the rabbit or mouse on which they were fed was injected with B vitamins just before, or if they were fed through a membrane on blood with various supplements. It was shown that

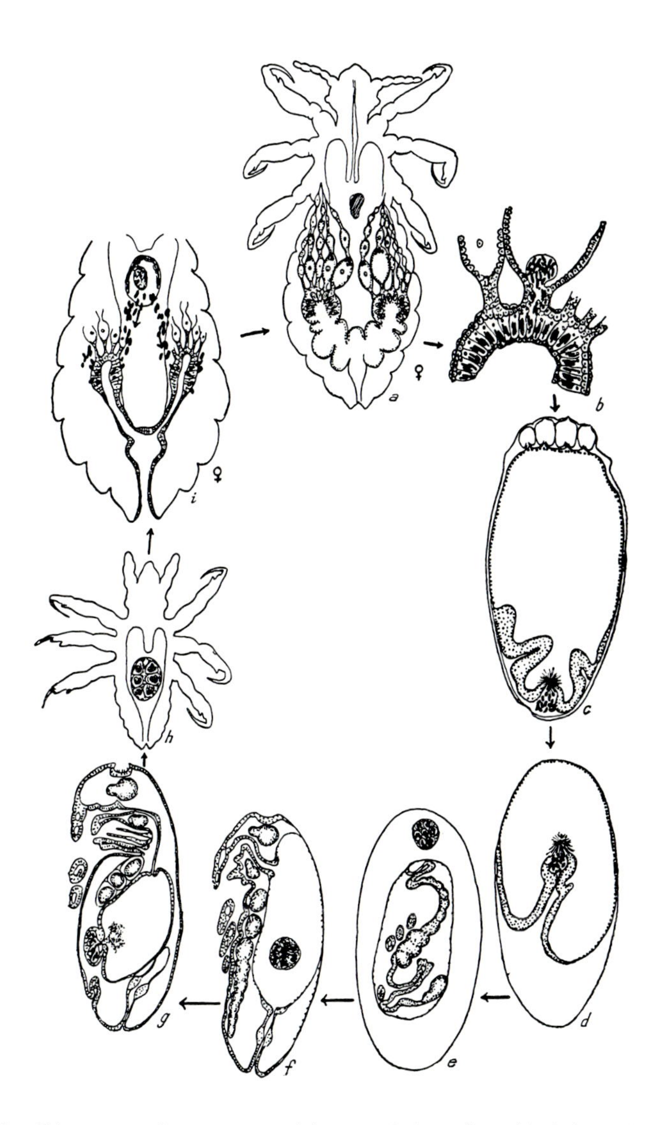

FIGURE 24.5. Diagram to show transovarial transmission of symbiotic bacteria in the human louse *P. capitis*. (a) Adult female with evacuated mycetome and infected ovarial ampullae. (b) Infection of oocytes from ampullae. (c, d) Invagination of germ band; symbionts carried into interior of embryo. (e) Primary mycetome in the yolk. (f) Transitory mycetome in embryonal gut. (g) Abscission of the stomach disk to form the definitive mycetome. (h) First nymphal stage with mycetome. (i) Transfer of symbionts from mycetome into ovarial ampullae in adult female. (After Ries, from Buchner, 1965.)

TABLE 24.1. The Main Groups of Insects Feeding on Vertebrate Blood[a]

Order	Family	Examples	Type of metamorphosis	Food of		Symbionts
				Young	Adults	
Hemiptera	Cimicidae	Bedbugs	Incomplete	Blood	Blood	+
	Reduviidae	Conenose bugs (*Rhodnius*)	"	"	"	+
Anoplura	Pediculidae	Lice	"	"	"	+
Siphonaptera	Pulicidae	Fleas	Complete	Detritus	"	–
Diptera	Culicidae	Mosquitoes	"	Micro organisms	Blood ♀	–
	Psychodidae	Sandflies	"	"	Blood	–
	Simuliidae	Blackflies	"	"	"	–
	Tabanidae	Horseflies	"	"	"	–
	Muscidae (subfamily Stomoxydinae)	*Stomoxys* (stable fly)	"	"	"	–
		Glossina (tsetse)	"	Secretion of uterine glands	"	+
	Hippoboscidae	Sheep ked (*Melophagus*)	"	"	"	+
		Pigeon fly (*Pseudolynchia*)	"	"	"	+

[a] Modified from Trager (1970).

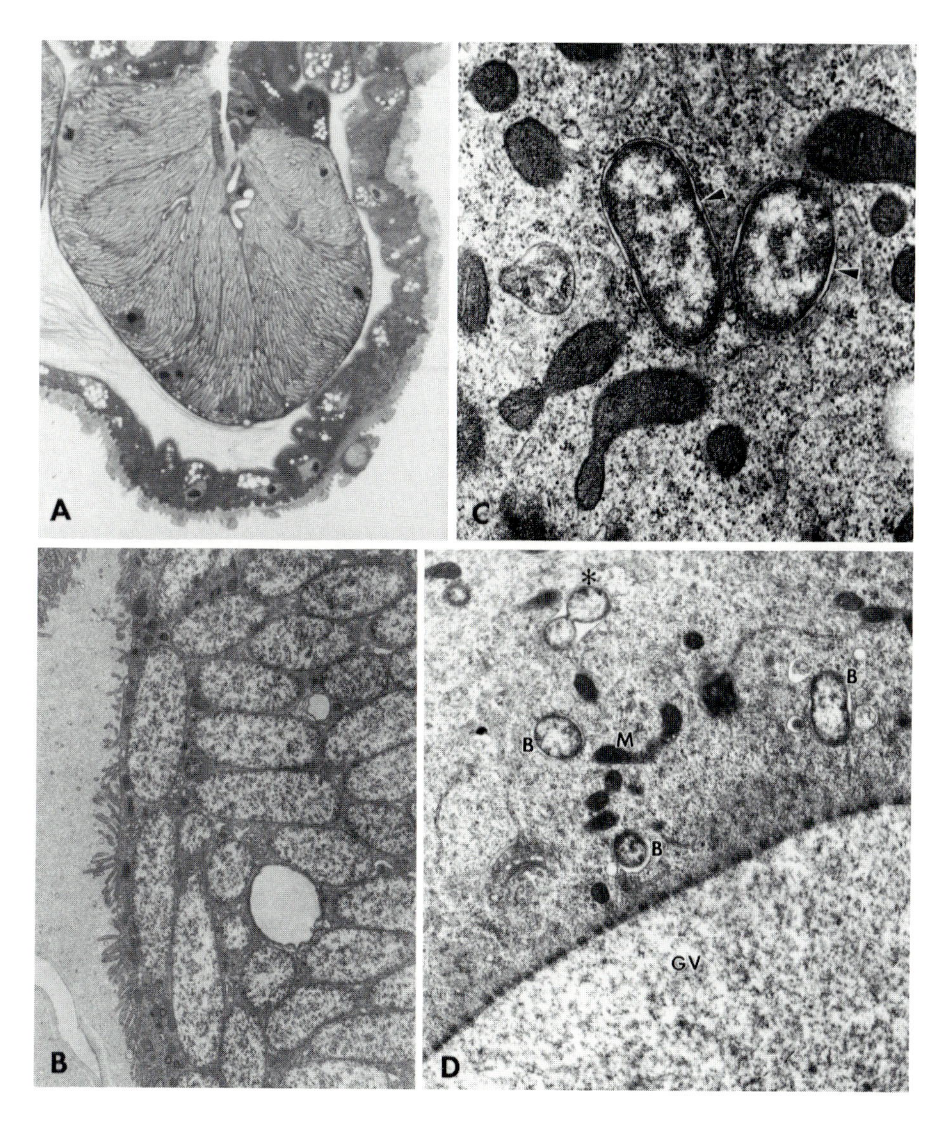

FIGURE 24.6. Symbiotic bacteroids of the tsetse fly *Glossina austeni*. (A) Light micrograph of a section through a portion of the alimentary tract to show the mycetome filled with the bacteroids. ×400. (B) Bacteroids in the apical cytoplasm of the mycetome of an adult female. TEM. ×7000. (C) bacteroids in the cytoplasm of a nurse cell in a newly emerged female. Note the host membrane (arrowheads) around the bacteroid. TEM. ×39,000. (D) Part of a young oocyte in a newly emerged fly. Note the bacteroids (B) in the ooplasm, clearly distinguishable from the mitochondria (M). A dividing bacteroid (*) is also visible. GV, germinal vesicle. TEM. ×14,000. (From Huebner and Davey, 1974. Original prints courtesy of Dr. E. Huebner.)

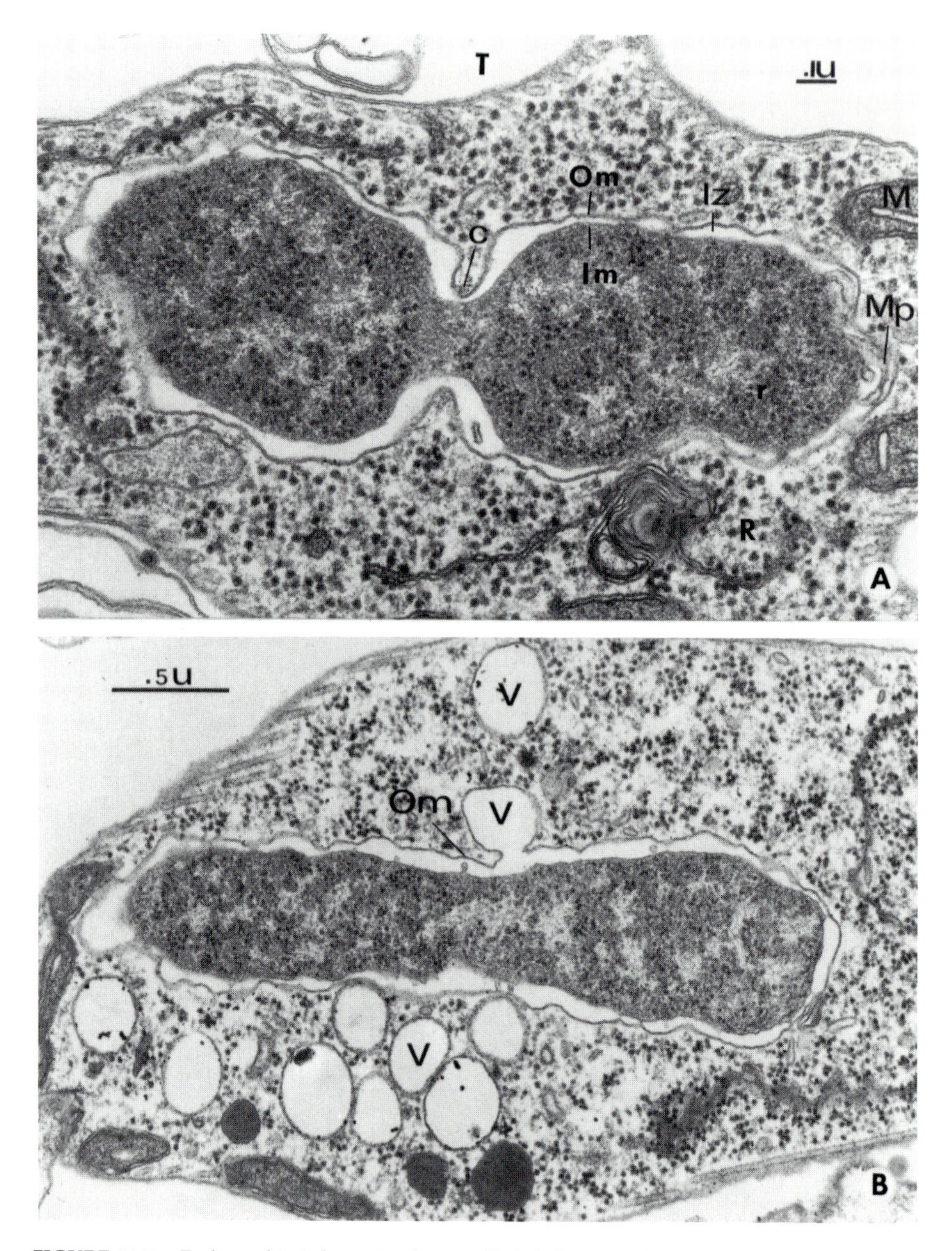

FIGURE 24.7. Endosymbiotic bacteria, the so-called diplosome of the trypanosomatid flagellate *Blastocrithidia culicis* from the gut of mosquitoes. (A) A dividing diplosome surrounded by two membranes, an inner (Im) and an outer (Om) separated by a clear space (Iz). Note the constriction (C), the absence of a septum between the two portions, the ribosomes (r) which are smaller than the host ribosomes (R), and the membrane protrusions (Mp) in the intermediate zone. Also shown are a portion of the flagellate's mitochondrion (M) and its subpellicular microtubules (T). (B) A diplosome with its outer membrane (Om) protruding into a structure very similar to the adjacent host vacuoles (V). TEM. (From Chang, 1974. Original prints courtesy of Dr. K.-P. Chang.)

Rhodnius depends on its bacteria for adequate supplies of pyridoxine, pantothenate, nicotinamide, thiamine, folic acid, and biotin, factors in which blood is relatively poor.

A nutritional role of endosymbiotic bacteria is especially well shown for certain trypanosomatid flagellates of insects (see also Chapter 7). I have already noted that three species which could be grown in very simple defined media all turned out to contain endosymbiotic bacteria (Fig. 24.7). When *Crithidia oncopelti*, *C. deanei*, and *Blastocrithidia culicis* were made aposymbiotic by treatment with antibiotics, they showed all of the nutritional requirements of related forms not having symbionts, such as *C. fasciculata*, and even some additional requirements. Aposymbiotic flagellates also lacked threonine deaminase, enzymes of the urea cycle, and those for heme biosynthesis, all of which could be easily demonstrated in the wild-type symbiont-bearing organisms. Since hemin is such an important essential nutrient for every species of hemoflagellate so far examined, except *C. oncopelti* and *B. culicis*, the biosynthesis of heme in these two species is of special interest. These two flagellates as they occur in nature, *C. oncopelti* in the gut of a plant-feeding bug and *B. culicis* in the gut of mosquitoes, always contain a diploid body whose multiplication keeps pace with that of the flagellate. It resembles bacteria in its ultrastructure, genome complexity, and sensitivity to antibiotics. The wild-type flagellates containing these symbiotic bacteria grow well in a defined medium without heme. If freed from their bacteria, as by treatment with chloramphenicol, they require a medium with hemin at a minimum concentration of 0.1 μg/ml. Like the closely related but naturally symbiont-free *C. fasciculata*, they can utilize protoporphyrin IX. Accordingly they have a ferrochelatase, the last enzyme in the heme biosynthetic pathway, catalyzing the insertion of iron into protoporphyrin IX. But they lack earlier enzymes in the pathway to heme. Uroporphyrinogen synthase, though present in high activity in the symbiont-bearing flagellates was essentially absent from the aposymbiotic flagellates, as it was also from two other species of hemoflagellate naturally without symbionts (Fig. 24.8A). Even more significant was the finding that the specific activity of uroporphyrinogen synthase was much higher in the isolated symbiotic bacteria than in the symbiont-bearing flagellate (Fig. 24.8B). Clearly the symbiotic bacteria provide their host with an essential nutrient which it would otherwise have to take up from its environment. Presumably this gives the symbiont–host cell combination a selective advantage in environments where hemin, and various other essential factors, are low.

It is extremely interesting to note that in one entirely different kind of symbiotic association, that of the *Rhizobia* in the root nodules of leguminous plants, the heme portion of the leghemoglobin that forms an essential part of the nitrogen-fixing system is synthesized by the bacteria. The host plant cells form the apoprotein, but they do this only if heme is being supplied by the bacteria. In this way an integration is effected between the two partners. It would be important to know the further utilization of heme in the flagellates

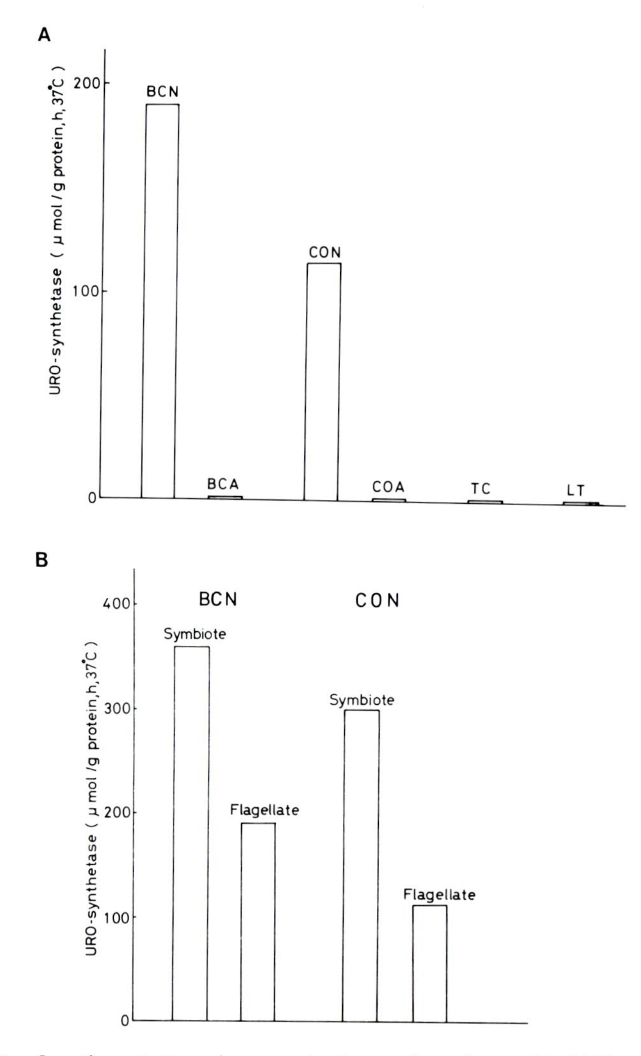

FIGURE 24.8. Specific activities of uroporphyrinogen I synthase. (A) Of *Blastocrithidia culicis* with and without symbionts (BCN and BCA, respectively); of *Crithidia oncopelti* with and without symbionts (CON and COA, respectively); and of two related flagellates naturally devoid of symbionts, *Trypanosoma conorhini* (TC) and *Leishmania tarentolae* (LT). (B) Of the symbiont-containing flagellates (flagellate) and of the isolated symbiont fraction (symbiote) of *B. culicis* (BCN) and *C. oncopelti* (CON). (From Chang *et al.*, 1975.)

and the quantitative relationships when it is supplied by the symbionts and when it is taken from the environment by symbiont-free organisms.

The role of endosymbiotic microorganisms is not always so clear as it is in these hemoflagellates, or legume root nodules, or the blood-feeding insects. Many kinds of free-living ciliates contain endocellular bacteria. The first to be studied in detail was observed as a cytoplasmic hereditary factor in *Paramecium aurelia*. T. M. Sonneborn discovered that certain lines of *P. aurelia* exerted a killing effect on others. The killer lines were shown to contain an extranuclear hereditary factor that was called kappa. Kappa could exist only in lines of suitable genetic character. Kappa is now known to be a rather large endosymbiotic bacterium that forms a refractile body, the R body. Ultrastructural studies have shown that the R body is a tightly wound ribbon that unwinds under certain conditions to form a filament up to 15 μm long. This sudden elongation may be responsible for the killing action. Paramecia that contain kappa are not susceptible to its killing action. If artificially freed from kappa, as under certain growth conditions, they become as susceptible to it as are lines lacking the genetic factor that enables them to harbor kappa. The biochemical bases for these actions remain unknown.

Many different kinds of endosymbiotic bacteria in ciliates are now known. Some exert a killing action like that of kappa. Others are so-called mate killers, like mu of *Paramecium.* A. T. Soldo has proposed the term *xenosomes* for the endosymbiotic bacteria he has described from certain marine ciliates, notably *Parauronema acutum.* Here it has been shown that infective forms are released from intact cells so that the symbiont does not depend solely on cell division of its host for its propagation. All of these symbionts have a genome size around 0.5×10^9 daltons, hence like that of *Mycoplasma* and considerably smaller than that of *Escherichia coli* (2.5×10^9). The killing action in lines not infected with the symbiont may or may not confer some selective advantage. Certainly, ciliates lacking these symbionts are quite able to multiply and propagate successfully under their normal environmental conditions.

On the other hand, in some species of the hypotrich ciliate *Euplotes,* there are present symbiotic bacteria essential to the life of their host. In *E. aediculatis* these symbionts, called omikron, occur at about 1000 per cell. If the *Euplotes* are freed from them, as with an appropriate X-ray dosage, or by exposure to penicillin at 100–150 U/ml for 5–6 days, they swim about, feed, and undergo one or two divisions. Then they stop feeding, grow smaller, and die in about 2–3 weeks. If the X-ray or penicillin treatment failed to eliminate all the symbionts, the *Euplotes* would start to grow and divide after a lag period of up to 15 days. Such organisms again had the full complement of about 1000 omikron per cell. Similarly, *Euplotes* fully freed from omikron could be rescued if reinfected either by adding a symbiont-containing homogenate to the medium or by injecting them with symbiont-containing cytoplasm. In either case active growth would begin when the number of omikron reached about 900 to 1000 per cell. There can be no doubt that omikron supplies its host with some essential factor, but at present we have no inkling as to what it is or why some species of *Euplotes* should require it. It must be noted that all species of *Euplotes* feed on algae and bacteria and that axenic cultures of this organisms have not been obtained, unlike the situation with *Paramecium* and *Parauronema.*

Intracellular symbiotic associations appear so highly specialized that we tend to think of them as being long established, and probably this is generally so. That they may be still arising, however, is indicated by a series of remarkable observations made by K. W. and M. S. Jeon. They studied a strain (designated strain D) of the large freshwater ameba *Amoeba proteus.* Certain

TABLE 24.2. Summary of Injection Experiments: Resuscitation of xD_nD_c Cells[a] by Injection of Whole or Fractionated xD Cytoplasm[b]

Expt	Material injected	No. of cells	No. of cells dividing	Viable clones[c]	
				No.	%
I	Whole xD cytoplasm	41	22	14	34
II	D cytoplasm (injection control)	28	4	0	0
III	Supernatant of homogenized and centrifuged xD cytoplasm	31	4	0	0
IV	Resuspended sediment of III	39	30	13	33
V	Separated and washed xD endosymbionts	27	22	8	30
VI	Killed xD endosymbionts[d]	43	7	0	0

[a] xD_nD_c = nucleus from xD ameba in enucleated D cell. These would all die if left without further treatment.
[b] Modified from Jeon and Jeon (1976).
[c] All these viable clones contained many endosymbionts.
[d] Killed by heat or osmotic shock.

of their cultures became spontaneously infected with large numbers of an unidentified species of bacteria. These are first were somewhat pathogenic to the ameba. A few years later, however, it became apparent that these amebae, now designated strain xD, could not survive without the bacteria. Although *A. proteus* has to be fed on other protozoa and again has not been grown axenically, it does provide special experimentally favorable conditions in that it can withstand nuclear transplantation. It was found that if the nucleus of an infected xD ameba was transferred carefully into the enucleated cytoplasm of an uninfected strain D ameba, the cell was nonviable. If, however, the recipient cell was also given a small volume of infected cytoplasm, it was viable. This effect is a result of the presence of the bacteria (Table 24.2). These bacteria accordingly must be considered endosymbionts. That this dependence on the endosymbionts could be rapidly acquired was shown by the following experiment. Nuclei from uninfected D amebae were placed in enucleated cytoplasm of infected xD amebae, producing a perfectly viable combination. These nuclei were subsequently transferred back into D amebae. After as few as four divisions in the infected cytoplasm, these nuclei already gave 74% nonviable combinations in the absence of the endosymbiont. The presence of the bacteria must somehow affect the nuclear genome, rendering it dependent on a bacterial product. What this product might be is not known. The bacteria themselves are not digested by the host cell; like long-established intracellular symbionts and like intracellular parasites such as *Toxoplasma*, they are contained in membrane-lined vacuoles with which lysosomes do not fuse (Fig. 24.9). These bacteria, like most established endosymbiotic bacteria, have not been grown in culture. Where they came from remains as unknown as for the many cases of endosymbionts found in nature.

From the preceding discussion a number of analogies will be evident between endosymbiotic bacteria and intracellular parasites, including the intracellular protozoa. In either case the symbiont or parasite may lie directly in the host cytoplasm but more usually it lies in a membrane-bound vacuole. With most of the intracellular protozoa and with all endosymbionts so far studied in this respect, the membrane of this vacuole does not fuse with secondary lysosomes. This parasitophorous membrane, as I have called it for the intracellular protozoa (see Chapters 5, 6, 13), although it may originate in a manner resembling an endocytotic vacuole, rapidly changes its character. It seems to come under the control of the parasite or symbiont. Noteworthy in this regard is the division of the vacuole that accompanies division of the organism within it, so that most of the time each parasite or symbiont lies within its own vacuole. This phenomenon occurs with associations as diverse as chlorella in *Paramecium bursaria, Leishmania donovani* in a hamster spleen cell, or the endosymbiotic bacteria in *Blastocrithidia* (see Fig. 24.7). In trying to find out how this happens, we will have to learn much more about intracellular membrane systems and their turnover than we known at present.

At first thought one might consider endosymbionts and intracellular parasites to differ sharply with regard to the control of their replication and their ultimate effects on the host cell. It is true that an erythrocytic malarial parasite destroys its host cell as it completes its single cycle of multiplication within it. But this is not the case for many other kinds of intracellular parasites. I

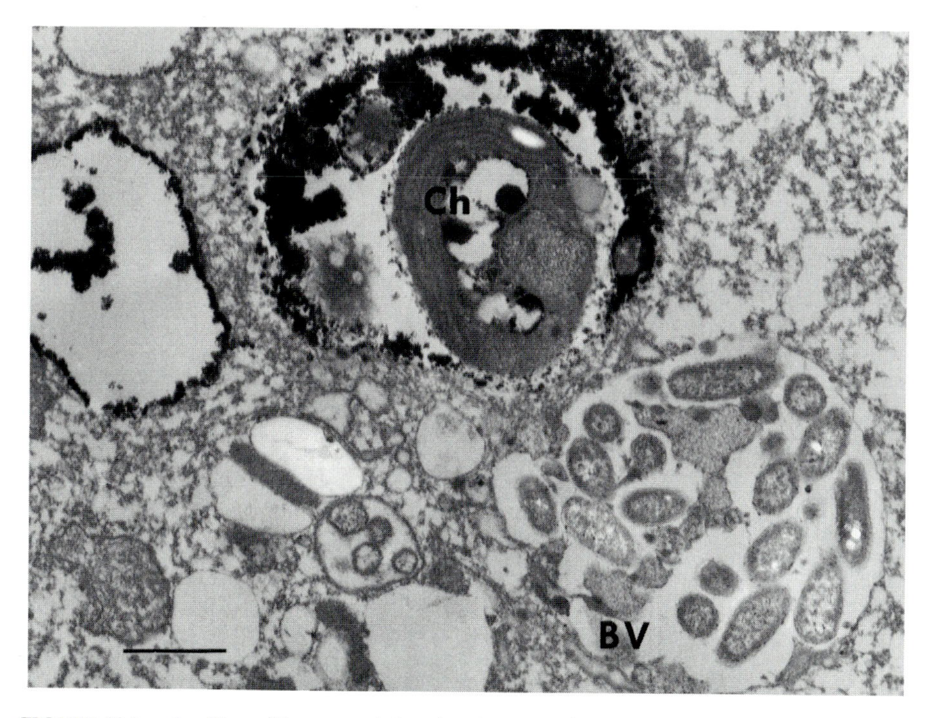

FIGURE 24.9. An xD symbiont-containing *Amoeba proteus* fixed 16 hr after induced phagocytosis of chlorella and prepared to demonstrate acid phosphatase activity. As shown by the dark reaction product, this activity was present in a phagosome containing an ingested chlorella (Ch) but not in a vesicle with symbiotic bacteria (BV). Bar = 1 μm. (From Jeon and Jeon, 1976. Original print courtesy of Dr. K. W. Jeon.)

have noted in earlier chapters how some, such as certain microsporidia, not only permit the continued life of their host cell but actually bring about its hypertrophy with the formation of a polyploid nucleus. Such hypertrophy and polyploidy are also brought about by endosymbionts as diverse as *Rhizobia* in root nodule cells and bacteroids in the specialized mycetocytes of roaches. Even with the parasites there is a certain amount of control of their replication. This is much more pronounced with the endosymbionts. Interestingly enough, under certain conditions the control may break down. For example, kappa may multiply and become pathogenic to its host *Paramecium*. Knowing more about these controls might have not only great academic interest but also perhaps much practical significance. At present we do not know whether similar principles operate in all the different kinds of endosymbiotic associations. Perhaps the hemoflagellates bearing endosymbionts provide exceptionally favorable material for further study. Both the wild-type symbiont-bearing and the aposymbiotic forms can be grown in defined media and the heme biosynthesis by the symbionts may provide a key to unlocking the nature of the mutual interactions between symbiont and host cell.

This system might also be useful in the study of how intracellular symbionts or parasites exert effects on the host cell surface. It was found for *Crithidia oncopelti* that symbiont-containing and symbiont-free organisms differ in agglutinability by lectins. Symbiont-free flagellates had a three-fold higher agglutination titer with concanavalin A than symbiont-containing ones. Correspondingly the symbiont-free organisms had 24 to 27 $\times$ 10^4 binding sites per cell for concanavalin A whereas the symbiont-containing ones had only 6 to 7 $\times$ 10^4. Conversely, symbiont-containing flagellates had two- to three-fold greater agglutination titers with fucose-binding lectin (from *Lotus*). It would be good to know how the endosymbiotic bacteria bring about this change in saccharide ligands in the surface membrane of *C. oncopelti*. Knowing this might even be useful in trying to explain the much more extensive changes produced by malarial parasites on the surface of their host erythrocytes.

Intracellular parasites and endosymbiotic bacteria also have in common the property that few of them have been grown axenically, i.e., in a nonliving medium. The reasons for this difficulty in extracellular cultivation remain obscure. In the present state of ignorance, speculation is of little value. The endosymbionts of the tsetse fly *Glossina* have been maintained in a cell-free medium apparently viable up to about 3 months with very limited elongation and division during the first 3 weeks. Essential to this development were succinate (10 mM) and pyruvate (20 mM), shown to be the principal substrates metabolized by the symbionts, plus the nucleotides ADP, CTP, UTP, TMP, GTP, and NAD (nicotinamide adenine dinucleotide) each at 1 mM, in a buffered PPLO broth. I have already noted the essential role of exogenous ATP and pyruvate for the limited extracellular development of the erythrocytic stages of the malarial parasites *Plasmodium lophurae* and *P. falciparum* (see Chapters 7–9). For *P. lophurae* it has been shown that the erythrocytic stage lacks enzymes for CoA biosynthesis. I have also ready noted that intracellular parasites of the groups of rickettsiae and chlamydiae utilize the ATP of their host cells. It would not be surprising if at least some of the endosymbiotic bacteria likewise took advantage of the abundant supplies of ATP and perhaps other cofactors available in the cytoplasm of their host cells. In this way the symbionts could conserve energy and use it in the biosynthesis of other metabolites essential both to themselves and to their host cells, as for example heme in the symbiosis of bacteria with the flagellate *Blastocrithidia culicis.*

Bibliography

Alfieri, S. C., and Camargo, E. P., 1982, Trypanosomatidae: Isoleucine requirement and threonine deaminase with and without endosymbionts, *Exp. Parasitol.* **53**:371–380.

Broda, P., 1979, *Plasmids,* Freeman, San Francisco.

Buchner, P., 1965, *Endosymbiosis of Animals with Plant Microorganisms* (revised English version), Interscience, New York.

Chang, K.-P., 1974, Ultrastructure of symbiotic bacteria in normal and antibiotic-treated *Blastocrithidia culicis* and *Crithidia oncopelti, J. Protozool.* **21**:699–707.

Chang, K.-P., Chang, C. S., and Sassa, S., 1975, Heme biosynthesis in bacterium–protozoan symbioses: Enzymatic defects of host hemoflagellates and the complemental role of their intracellular symbiotes, *Proc. Natl. Acad. Sci. USA* **72**:2979–2983.

Cleveland, L. R., 1934, The wood-feeding roach *Cryptocercus,* its protozoa, and the symbiosis between protozoa and roach, *Mem. Am. Acad. Arts Sci.* **17**:185–342.

Cook, C. B., Pappas, P. W., and Rudolph, E. D. (eds.), 1980, *Cellular Interactions in Symbiosis and Parasitism,* Ohio State University Press, Columbus.

Dwyer, D. M., and Chang, K.-P., 1976, Surface membrane carbohydrate alterations of a flagellated protozoan mediated by bacterial endosymbiotes, *Proc. Natl. Acad. Sci. USA* **73**:852–856.

Galinari, S., and Camargo, E. P., 1979, Urea cycle enzymes in wild and aposymbiotic strains of *Blastocrithidia culicis, J. Parasitol.* **65**:88.

Huebner, E., and Davey, K. G., 1974, Bacteroids in the ovaries of a tsetse fly, *Nature* **249**:260–261.

Jannasch, H. W., and Mottl, M. J., 1985, Geomicrobiology of deep-sea hydrothermal vents, *Science* **229**:717–725.

Jeon, K. W. (ed.), 1983, *Intracellular Symbiosis,* Academic Press, New York.

Joen, K. W., and Jeon, M. S., 1976, Endosymbiosis in amoebae: Recently established endosymbionts have become required cytoplasmic components, *J. Cell Physiol.* **89**:337–344.

Jolley, E., and Smith, D. C., 1980, The green hydra symbiosis. 2. The biology of the establishment of the association, *Proc. R. Soc. London* **207**:311.

Jordan, A. M., and Trewern, M. A., 1976, Sulphaquinoxaline in host diet as the cause of reproductive abnormalities in the tsetse fly (*Glossina* spp.), *Entomol. Exp. Appl.* **19**:115–129.

Lederberg, J., 1952, Cell genetics and hereditary symbiosis, *Physiol. Rev.* **32**:403–430.

Lee, J. J., 1983, Perspective on algal endosymbionts in larger *Foraminifera, Int. Rev. Cytol.* (Suppl.) **14**:49–77.

McLaughlin, G. L., Wood, D. L., and Cain, G. D., 1983, Lipids and carbohydrates in symbiotic and aposymbiotic *Crithidia oncopelti* and *Blastocrithidia culicis, Comp. Biochem. Physiol. B* **76**:143–152.

Muscatine, L., and Greene, R. W., 1973, Chloroplasts and algae as symbionts in molluscs, *Int. Rev. Cytol.* **36**:137–169.

Nogge, G., 1976, Sterility in tsetse flies (*Glossina morsitans* Westwood) caused by loss of symbionts, *Experientia* **32**:995–996.

Nordström, K., Molin, S., and Light, J., 1984, Control of replication of bacterial plasmids: Genetics, molecular biology, and physiology of the plasmid R1 system, *Plasmid* **12**:71–90.

Odelson, D. A., and Breznak, J. A., 1985, Cellulase and other polymer-hydrolyzing activities of *Trichomitopsis termopsidis,* a symbiotic protozoan from termites, *Appl. Environ. Microbiol.* **49**:622–626.

Odelson, D. A., and Breznak, J. A., 1985, Nutrition and growth characteristics of *Trichomitopsis termopsidis,* a cellulolytic protozoan from termites, *Appl. Environ. Microbiol.* **49**:614–621.

Trager, W., 1970, *Symbiosis,* Van Nostrand Reinhold, New York.
Verma, D. P. S., and Long, S., 1983, The molecular biology of *Rhizobium*–legume symbiosis, *Int. Rev. Cytol.* (Suppl.) **14:**211–245.
Wink, M., 1979, The endosymbionts of *Glossina morsitans* and *G. palpalis:* Cultivation experiments and some physiological properties, *Acta Trop.* **36:**215–222.
Yami, M. A., 1978, Axenic cultivation of the cellulolytic flagellate *Trichomitopsis termopsidis* (Cleveland) from the termite *Zootermopsis, J. Protozool.* **25:**535–538.
Yamin, M. A., 1981, Cellulose metabolism by the flagellate *Trichonympha* from a termite is independent of endosymbiotic bacteria, *Science* **211:**58–59.
Yamin, M. A., and Trager, W., 1979, Cellulolytic activity of an axenically-cultivated termite flagellate, *Trichomitopsis termopsidis, J. Gen. Microbiol.* **113:**417–420.

CHAPTER 25

Chemotherapy

Introduction

Seven chapters (17–23) have been devoted to discussions of acquired immunity in vertebrate hosts to parasitic infections. This is not out of proportion to the great fundamental significance of immunological reactions in animal parasitism. It also is not out of proportion to the potentially immense practical value of vaccines against some of these animal parasites. At the same time it has to be admitted that such vaccines do not yet exist. At present, prevention of these diseases must rest as in the past mainly on ecological measures, such as sanitation and vector control, and to a small extent on antiparasitic drugs. Once the disease has been acquired, we have to rely on specific chemotherapeutic agents. The importance of these cannot be exaggerated, since prevention is often difficult. Furthermore, studies of mode of action of effective drugs provide insights into the physiology of the parasite. Conversely, adequate understanding of the physiology of the parasite could lead to the rational design of new and more effective drugs. Most new antiparasitic agents still continue to be found mainly by intelligent screening procedures rather than on the basis of knowledge of the biochemistry of the parasite. Enough is now known, however, about the mechanisms of action of a number of different classes of drugs to permit a discussion based on their targets and in this way to provide an approach to the design of new agents. New agents must be sought continuously, since the parasites often develop drug-resistant mutants.

The practice of chemotherapy is surely as old as human social organization. Plant extracts in particular have been used throughout the ages for the treatment of various ailments, including diseases now known to be caused by infectious agents. As early as 430 A.D. a Chinese "Handbook of Prescriptions for Emergency Treatments" advised the use of an extract of qing hao (*Artemisia annua* L., sweet or annual wormwood; Fig. 25.1A,B) to reduce fever. Recent work, principally in China, has shown that this extract contains the

FIGURE 25.1. *Artemisia annua* growing in the Washington, D.C. region as it appears in early August (A) and with its pale yellow flowers (B). This plant is a source of qinghaosu or artemisinin (C). (From Klayman, 1985. Original print courtesy of Dr. D. L. Klayman.)

active antimalarial agent now identified as quinghaosu or artemisinin (Fig. 25.1C). A "Compendium of Materia Medica" produced in China in 1596 recommended qing hao specifically for the treatment of chills and fever of malaria. It was at about this same time that Europeans learned from natives of Peru about the antimalarial effectiveness of the bark of the cinchona tree (Fig. 25.2A). As is well known, quinine (Fig. 25.2B) was later isolated by Pelletier in 1834, from cinchona bark. This was for many years the world's principal chemotherapeutic agent for treatment of an infectious disease.

That an antiparasitic agent must injure the parasite without any important harm to the host seems obvious. Ehrlich was the first to clearly formulate the concept of specific receptors in the parasite with higher avidity for the drug than those present in the host, or ideally, altogether lacking in the host. Penicillin is such a remarkable drug because it interferes with bacterial cell wall synthesis, a biochemical pathway of no significance in the vertebrate host. Similarly, the sulfa drugs interfere with the synthesis of pteroic acid from *p*-aminobenzoic acid, an essential step in the formation of the folates. Vertebrates must obtain preformed folic acid in their diet; they cannot synthesize folates and accordingly sulfa drugs have no target in the host. As already noted (Chapter 9), many parasites also require a folate source in their diet and these are not susceptible to sulfa drugs. On the other hand, the intracellular protozoa of the groups of coccidia and the malarial parasites apparently synthesize folates and are adversely affected by analogues of *p*-aminobenzoic acid, such as the sulfa drugs. These competitively inhibit the parasites' 7,8-dihydropteroate synthase. The structural resemblance of these drugs to the natural metabolite *p*-aminobenzoic acid led to the development of a general theory of antimetabolite action, formulated with special clarity by D. W. Woolley. It was evident that compounds structurally related to important metabolites and with a sufficient affinity for the enzymes or receptors concerned could replace the metabolites, interfering in this way with the synthesis of essential substances, or with transport of the metabolite. These concepts led to a number of successful antiparasitic drugs, including the antimalarial pyrimethamine (Fig. 25.3a) and the anticoccidial agent amprolium (Fig. 25.3b). Pyrimethamine, a pyrimidine antimetabolite, binds avidly to the dihydrofolate reductase of malarial parasites, interfering with pyrimidine synthesis and hence with DNA synthesis. Although this biosynthetic pathway is essential to the host as well as the parasite, the parasite's enzyme is very different from that of the host (molecular weight 100,000 to 200,000 as against about 25,000 for the host dihydrofolate reductase) and the drug has a much higher binding affinity for the parasite enzyme. Moreover, malarial parasites lack a pyrimidine salvage pathway and cannot utilize preformed pyrimidines from the host. Amprolium, on the other hand, is an analogue of thiamine, an essential cofactor in carbohydrate metabolism, and its effect is reversible by thiamine. Amprolium evidently acts by interfering with the transport of thiamine. Its K_i for transport of thiamine in the coccidian parasite *Eimeria tenella* is 50 times lower than for thiamine transport in the

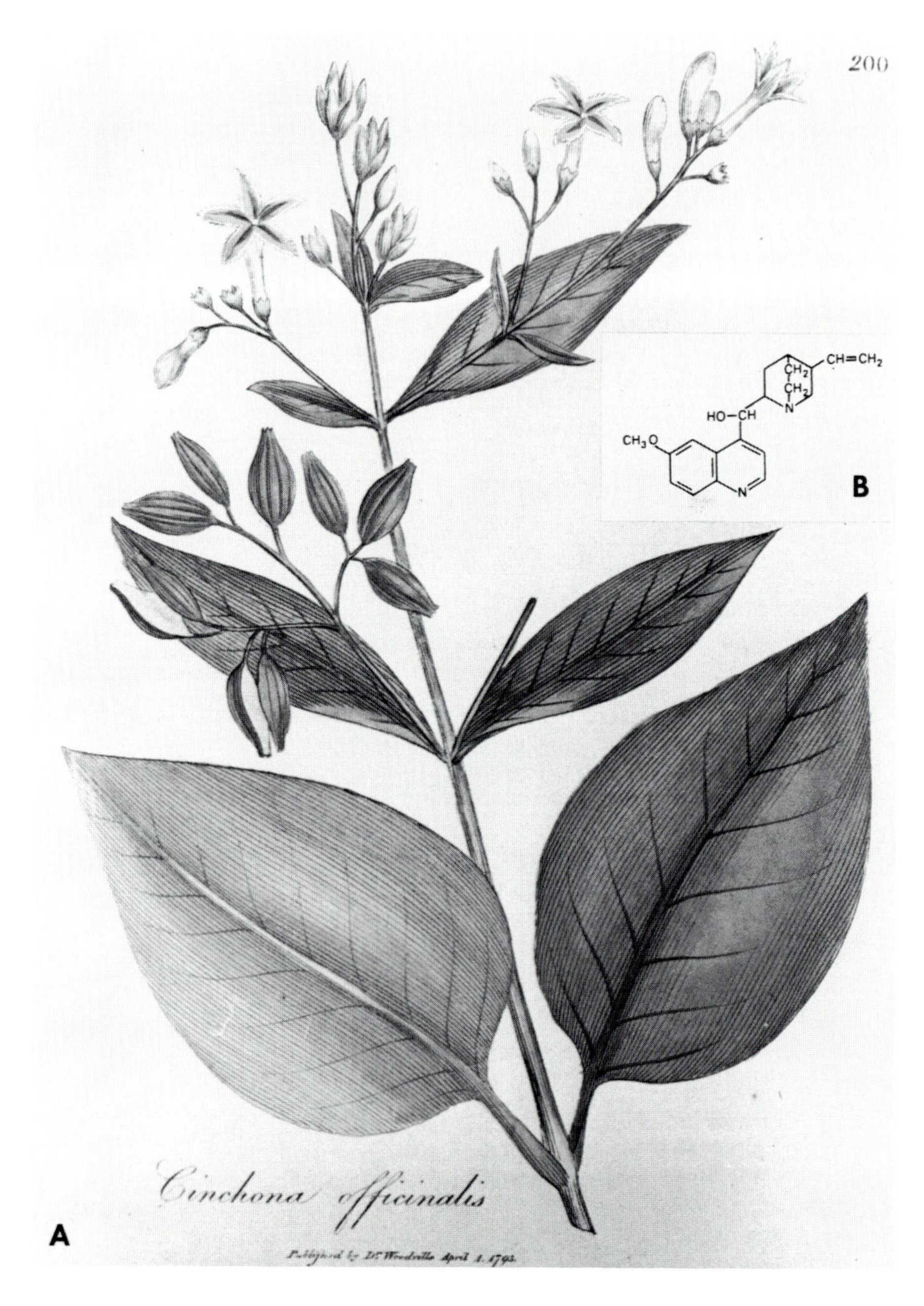

FIGURE 25.2. *Cinchona officinalis,* twig with flowers (A), a source of quinine (B). (A reproduced from a color plate in *Medical Botany* by William Woodville, published in London in 1793. Photograph obtained through the kind cooperation of Mrs. L. Lynas, Head Reference Librarian, The New York Botanical Garden, Bronx, N.Y.)

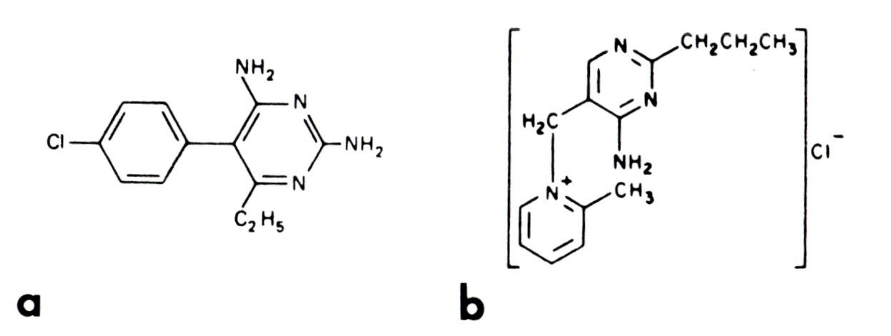

FIGURE 25.3. Two kinds of antimetabolites: (a) the antimalarial pyrimethamine; (b) the anticoccidial amprolium.

chicken intestine. Both of these drugs provide examples of agents that affect systems present in both the parasite and the host but in a different way so that the parasite is injured at levels of drug far below those injurious to the host.

I will now proceed to consider in a systematic way the probable targets and modes of action of some important antiparasitic drugs. This will be followed by a discussion of important drugs whose mode of action is not understood. I will then consider the problem of drug resistance and its relation to the immune response. The use of drugs in prevention and community control of parasitic diseases will be treated in Chapter 26.

Probable Modes of Action of Certain Antiparasitic Drugs

Formation of Toxic Free Radicals

It must be emphasized at the outset that the demonstration that a drug affects a particular enzyme or biological activity does not constitute proof that this is the mode of action of the drug. This may be a secondary effect and may be of little consequence relative to other effects of the agent. Moreover, a single agent might affect more than one metabolic activity. The latter statement is particularly true of those agents that exert their effects through the formation of toxic radicals.

Metronidazole, a 5-nitroimidazole derivative (Fig. 25.4) very useful against *Trichomonas vaginalis, Giardia lamblia,* and *Entamoeba histolytica,* provides a particularly instructive example of such an agent. Metronidazole itself is relatively nontoxic, but reduction of its nitro group results in the formation of short-lived cytotoxic intermediates. Organisms with metabolic pathways having a low redox potential and linked to ferredoxin or flavodoxinlike electron transport components are especially efficient in reducing nitroimidazoles (Fig. 25.4),

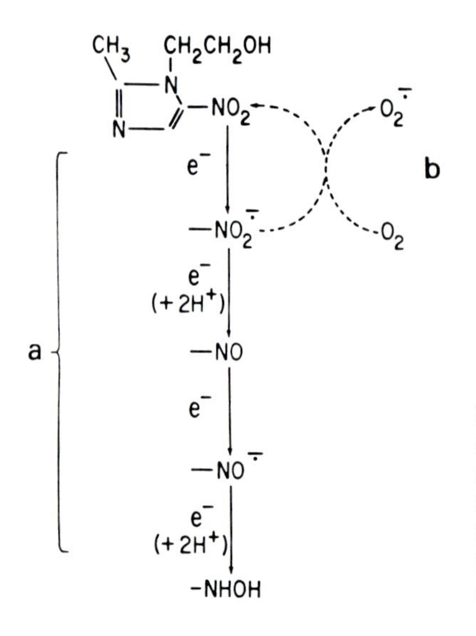

FIGURE 25.4. Metronidazole. Formula (at top) and scheme of its reductive activation: (a) reduction of the nitro group in one-electron steps, giving successively the nitro–free radical, nitroso, nitroso–free radical, and hydroxylamine derivatives; (b) reoxidation of the nitro–free radical by O_2, giving superoxide anion and the original drug. (From Müller, 1983.)

and it is against such organisms that these compounds are most effective. Such pathways are found in anaerobic bacteria and in certain protozoa with an anaerobic metabolism. In the trichomonads the reactions occur in the hydrogenosome (see Chapter 9) with the conversion of pyruvate to acetyl-CoA. In *Giardia* and *Entamoeba* they occur in the cytosol. In either case the toxic intermediates exert their deleterious effects on the cell by damaging its DNA and possibly other targets (Fig. 25.5). The details of this action are not understood since the intermediate reduction products have not been isolated; their existence has been inferred indirectly. The final products of reduction are inactive.

The enzyme responsible for the reduction of metronidazole is a pyruvate-ferredoxin oxidoreductase. This has no counterpart in the aerobic mammalian cell, accounting for the low toxicity of metronidazole to the host. The redox potential of ferredoxin is −470 mV, just under that of metronidazole at −486 mV. Under aerobic conditions the toxicity of metronidazole to facultative anaerobic protozoa is less than under anaerobic conditions, indicating that oxygen is the main competitor with metronidazole for the available electrons.

Nifurtimox (Fig. 25.6), a nitrofuran derivative, is one of the few drugs effective against *Trypanosoma cruzi.* It again is a generator of free radicals. In the presence of reduced nicotinamide adenine dinucleotide (NADH) or of NADPH, homogenates of, or intact *T. cruzi* convert nifurtimox to the nitro-amino radical with increase in production of superoxide anion and hydrogen peroxide. It is highly significant that the concentration of nifurtimox of 10–20 μM required to inhibit *T. cruzi* growth *in vitro* is about the same as the serum concentration in humans after one oral dose of 15 mg/kg, and is also the same

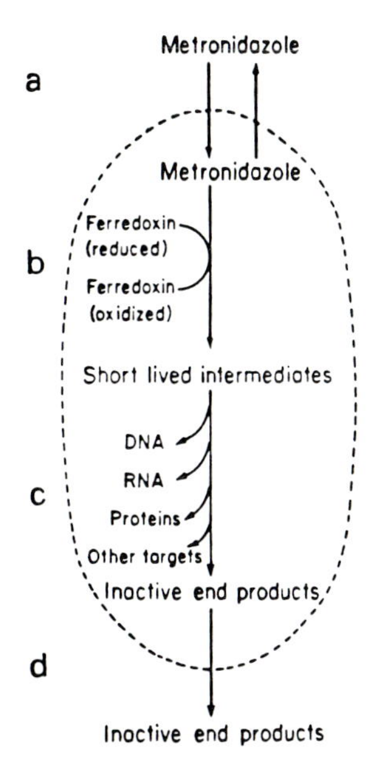

FIGURE 25.5. Scheme for action of metronidazole on an anaerobic microorganism. (a) Passage through the cell membrane; (b) reductive activation; (c) interaction with intracellular targets; (d) release of inactive end products. (From Müller, 1983.)

as that giving maximal stimulation of O_2^- formation by the parasite mitochondrial fraction. This is consistent with the view that the free radical metabolites lead to lipid peroxidation and hence to damage to the cell. Although *T. cruzi* has superoxide dismutases, it lacks catalase and is low in peroxidases. The generation of free radicals is probably also the basis of nifurtimox toxicity to the mammalian host.

A number of other types of antiparasitic agents may owe their efficacy at least in part to the generation of free radicals. Among these would be the naphthoquinones such as menoctone, a drug very useful in the treatment of cattle with East Coast fever (*Theileria parva*). (But see below under purine and pyrimidine metabolism.) It may well be that some oxidative mechanism is involved in the action of qinghaosu since this sesquiterpene lactone is inactivated if its peroxide grouping is removed (see later section for further discussion of qinghaosu). As already noted in several previous chapters, the antiparasitic action of phagocytic cells depends in part on an oxidative burst

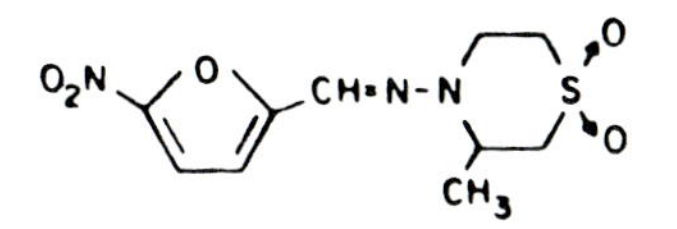

FIGURE 25.6. Nifurtimox, a nitrofuran derivative effective against *Trypanosoma cruzi*.

releasing oxygen radicals and hydrogen peroxide. This is without doubt a possible mode of action that must always be kept in mind.

Interference with Folate Biosynthesis and Metabolism

Here we find some very active antimalarial and anticoccidial drugs. I have noted (Chapter 9) that the intracellular sporozoa synthesize their pyrimidine *de novo*, a process in which folate coenzymes are involved. Vertebrate hosts synthesize pyrimidine but require an exogenous source of folic acid. In the introductory discussion of this chapter I have already called attention to the activity of sulfa drugs, as analogues of *p*-aminobenzoic acid, in inhibiting the reaction of 2-amino-4-hydroxy-6-(hydroxymethyl) dihydropteridine diphosphate with *p*-aminobenzoate to form 7,8-dihydropteroate, which then forms pteroylglutamic acid (folic acid). How the 2-amino-4-hydroxy-6-(hydroxymethyl) dihydropteridine is formed in malarial parasites is unknown. Furthermore, the 7,8-dihydropterate synthases of plasmodia and coccidia are probably quite different from the corresponding bacterial enzymes. For example, metachloridine and 2-ethoxy-*p*-aminobenzoate, two drugs ineffective against sulfa-sensitive bacteria, are active, the first against malarial parasites and the second against a chicken coccidia. The action of both is antagonized by *p*-aminobenzoic acid.

In the further biosynthesis of folate cofactors, dihydrofolate must be reduced to tetrahydrofolate, a step catalyzed by the enzyme dihydrofolate reductase. This enzyme in malarial parasites and coccidia is quite different from the corresponding enzyme in vertebrates. In addition, it contains thymidylate synthase activity, catalyzing the formation of deoxythymidine phosphate from deoxyuridine phosphate. The dihydrofolate reductase and the thymidylate synthase, together with serine hydroxymethyl transferase, catalyzing formation of N^5,N^{10}-methylene tetrahydrofolate, are involved in the biosynthesis of pyrimidines via a "thymidylate synthesis cycle" (Fig. 25.7). Accordingly, pyrimethamine (Fig. 25.3a) and certain related pyrimidine analogues with a high affinity for the dihydrofolate reductase of malarial parasites are very effective antimalarials. It is worth noting that G. H. Hitchings developed these compounds as antifolates interfering with the growth of certain lactobacilli that require exogenous folic acid. Malarial parasites are generally considered to be unable to use exogenous folate and the parasite dihydrofolate reductase will not reduce folate. No dihydrofolate synthase activity has been demonstrated. There are also a number of observations suggesting that we do not yet fully understand folate metabolism in malarial parasites. For example, when *Plasmodium falciparum* is grown *in vitro* its sensitivity to pyrimethamine is affected by the folic acid content of the medium as well as by the *p*-aminobenzoic acid level. With both avian and rodent malaria, parasites made resistant to a sulfonamide were cross-resistant to pyrimethamine.

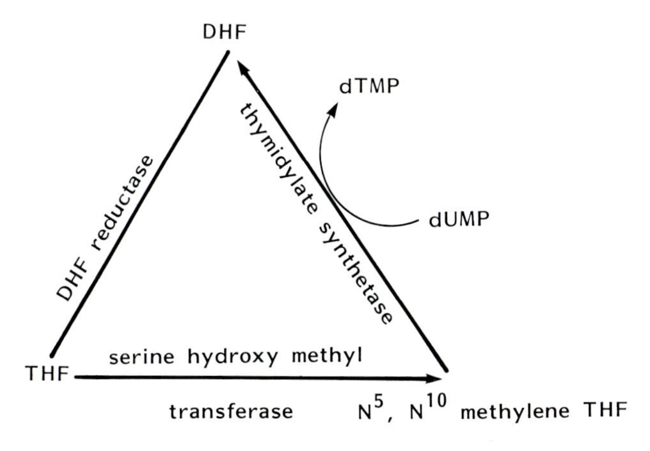

FIGURE 25.7. Thymidylate synthesis cycle. DHF, dihydrofolate; THF, tetrahydrofolate; dUMP, uridine monophosphate; dTMP, thymidine monophosphate.

As might be expected from the fact that sulfonamides and pyrimethamine attack the folate pathway at two different points, these drugs given together show a synergistic action. This is the basis for a valuable antimalarial, Fansidar, which combines pyrimethamine with the long-acting sulfa drug sulfadoxine (Fig. 25.8). Even to this combination the parasites have developed resistance, a subject to which I return later in this chapter.

Most parasitic eukaryotes require preformed folic acid in their diet (see Chapter 7) so that sulfa drugs could not be expected to have any activity. The antifolates like pyrimethamine are also relatively ineffective against parasites other than malaria perhaps because these other species have a pyrimidine salvage pathway permitting them to utilize exogenous thymidine. Malarial parasites, however, lack any such pathway and are entirely dependent on *de novo* synthesis of pyrimidine. I turn now to consideration of another point of attack on this synthetic pathway.

Interference with Pyrimidine Biosynthesis

Although hydroxynaphthoquinones have been studied for over 40 years as antiprotozoal agents and have yielded a few compounds of some practical value, it was only recently that investigation of their effects on *P. falciparum*

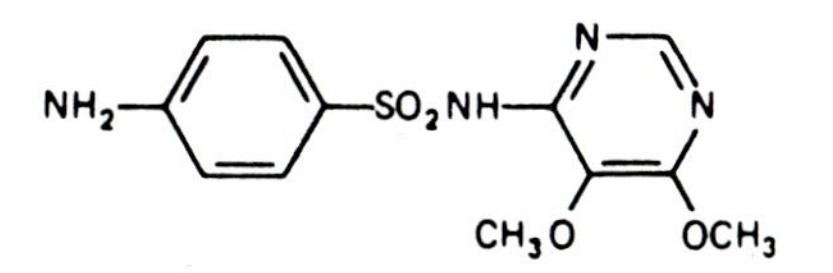

FIGURE 25.8. Sulfadoxine, used with pyrimethamine as an antimalarial (Fansidar).

growing *in vitro* has led to potentially very effective drugs and to an understanding of their mode of action.

The structures of menoctone and parvaquone, the two members of this group previously found highly effective in cattle against *Theileria parva,* are shown in Fig. 25.9, together with the structures of the newly developed compounds BW 58C and two closely related compounds, BW 568C and BW 720C. Against *P. falciparum in vitro* BW 58C was over 5600 times as active as menoctone and 650 times more active than chloroquine. Against *P. yoelii* in mice it was 4 times as active as chloroquine. It also appears to have causal prophylactic activity (see also later in this chapter) against sporozoite-induced *P. yoelii nigeriense* infections in mice. It was also found active both *in vitro* and *in vivo* against the chicken coccidian *Eimeria tenella.* For this latter parasite, BW 568C was even more effective, being more active than monensin, the current drug of choice. BW 720C was especially effective against *T. parva* with a therapeutic dose in cattle of only 2.5 mg/kg as compared to 20 mg/kg for parvaquone.

For *P. falciparum* it was shown that the quinones specifically inhibit synthesis of the pyrimidine nucleotides (see Fig. 9.6). At only 0.1 mM, BW 58C caused within 1 hr a decrease in UTP content from 2.2 nmoles in controls to

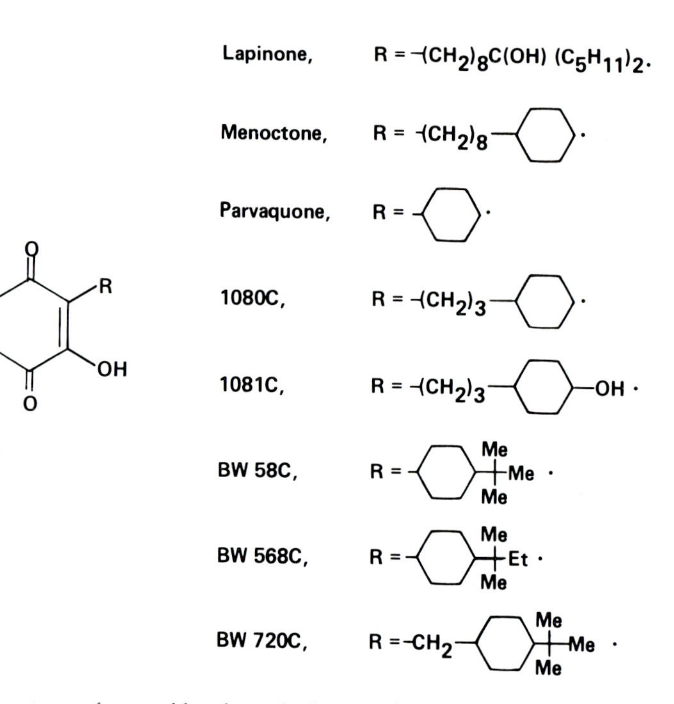

FIGURE 25.9. Structures of some old and new hydroxynaphthoquinones. See text. (From Hudson *et al.*, 1985.)

only 0.8 nmole and up to a tenfold increase in concentration of ^{14}C-labeled carbamoyl aspartate, formed from $H^{14}CO_3$, from 0.37 nmole in controls to 3.8 nmoles. The site of inhibition is probably at the dihydroorotate dehydrogenase rather than the dihydroorotase (see Fig. 9.6). Dihydroorotate did not increase to high levels since the equilibrium constant of dihydroorotase greatly favors formation of carbamoyl aspartate.

In view of these results it will certainly be important to look also for agents that would interfere at later stages in pyrimidine synthesis, such as at orotate phosphoribosyl transferase and orotidylate decarboxylase, two enzymes already shown to have different properties in *P. falciparum* from the corresponding enzymes in vertebrate cells.

Interference with Purine Metabolism

All protozoan parasites so far studied require an exogenous source of purines (see Chapters 7, 8); none can synthesize purines *de novo.* The same is true of *Schistosoma mansoni.* Less is known about other parasitic helminths; most of them probably resemble other invertebrates so far investigated in requiring a purine in their diet. Thus, the parasites, unlike vertebrate hosts, depend on various purine salvage pathways. These accordingly offer possible targets for chemotherapeutic agents, particularly if they involve unique enzymes not found in the host.

One such unique nucleotide phosphotransferase has been found in species of *Leishmania.* It transfers the phosphate group from a variety of monophosphate esters to the 5′-position of purine nucleosides. It also phosphorylates analogues of purine nucleosides such as allopurinol riboside, formycin B, and thiopurinol riboside (Fig. 25.10) to form the corresponding nucleotides. These may then be converted to triphosphates and incorporated into the nucleic acids or they may become inhibitors of other essential enzymes in purine metabolism. All three of these agents are indeed potent leishmanicidal agents, both *in vivo* and *in vitro,* and have little toxicity for the mammalian host.

The effects of allopurinol [4-hydroxypyrazolo(3,4-*d*)pyrimidine, HPP] and of thiopurinol [4-thiopyrazolo(3,4,-*d*)pyrimidine, TPP] and their corresponding ribosides (HPPR and TPPR, respectively) on growth of promastigotes of three species of *Leishmania* are shown in Table 25.1. These agents are also active against amastigotes growing in macrophages *in vitro.* Allopurinol, a drug commonly used for the treatment of gout (where it inhibits xanthine oxidase to reduce blood uric acid levels), has already been found useful in the treatment of patients with visceral leishmaniasis resistant to the antimony compound pentostam, and further human trials with this compound and its riboside are in progress.

These hypoxanthine and inosine analogues are also effective against some strains of *Trypanosoma cruzi* and under appropriate conditions against the

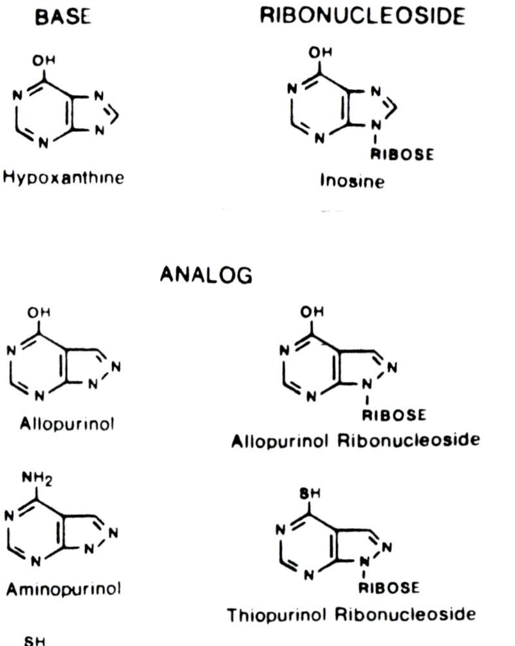

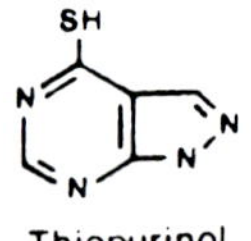

FIGURE 25.10. Structures of certain purines and purine analogues. (From Fish *et al.*, 1985.)

bloodstream forms of the African trypanosome *T. brucei* and its subspecies *gambiense* and *rhodesiense*. In all these organisms the compounds seem to be metabolized in a similar way with final incorporation into the RNA. 9-Deazainosine was especially active against *T. brucei* and seems to be nontoxic to mammalian cells.

Another target in purine metabolism is provided in the malarial parasite *P. falciparum*. This organism utilizes hypoxanthine as its purine source for synthesis of both its adenylates and guanylates. The antileukemia drug bredinin (4-carbamoyl-1-β-D-ribofuranosyl-imidazolium-5-olate) which inhibits conversion of inosine monophosphate (IMP) to guanosine monophosphate

TABLE 25.1. Activities of Pyrazolo(3,4-*d*)pyrimidines against *Leishmania*[a]

Species	Concentration (μM) of agent giving 90% inhibition of growth of promastigotes			
	TPP[b]	HPP	TPPR	HPPR
L. brasiliensis	20	50	4	3
L. donovani	100	300	2	0.3
L. mexicana	110	40	30	40

[a] Modified from Marr *et al.* (1982)
[b] See text.

(GMP) was found to inhibit growth of *P. falciparum* in culture. It apparently acts on the IMP dehydrogenase of the parasite.

Interference with Methylation Reactions

Sinefungin, an adenosine analogue (Fig. 25.11) known to inhibit methyl transferases, completely inhibited growth of *P. falciparum in vitro* at 0.3 μM. It was much more active in this regard than other methylation inhibitors tested: deazaadenosine, *S*-isobutyladenosine, and 5′-deoxy-5′-(isobutylthio)-3-deazaadenosine. Sinefungin has also been found active against some other protozoan parasites. In no case, however, has it been actually demonstrated that its activity in these organisms is a result of methyl transfer inhibition. In view of the demonstration that inhibitors of methyl transfer are highly specific for particular methylation reactions, and in view of the importance of methyl transfer in RNA synthesis and in other metabolic reactions, this seems an appropriate field for further work.

Interference with Polyamine Synthesis

Polyamines are present in all living organisms and seem to be especially important in cellular proliferation and differentiation. The first and rate-limiting reaction in biosynthesis of polyamines is the formation of putrescine from ornithine, a reaction catalyzed by ornithine decarboxylase. The other polyamines, spermidine and spermine, are then seccessively formed from putrescine in reactions involving decarboxylated *S*-adenosylmethionine. An antimetabolite of ornithine, α-(difluoromethyl)ornithine (DFMO) (Fig. 25.12), specifically inhibits the rate-limiting enzyme ornithine decarboxylase. This compound inhibits the growth of a number of parasitic protozoa including *P. falciparum* in culture and African trypanosomes *in vivo*.

The activity of DFMO against African trypanosomes is of special importance since it has been found effective against central nervous system infection with either *Trypanosoma gambiense* or *T. rhodesiense* in animal models. Because

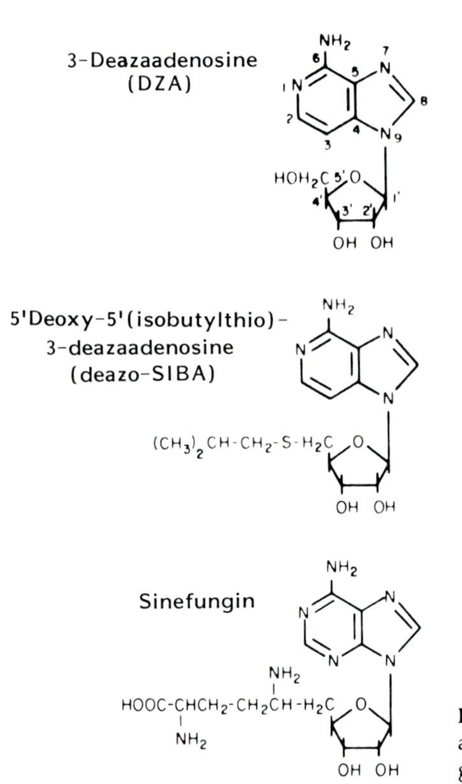

FIGURE 25.11. Methylation inhibitors active against *Plasmodium falciparum in vitro*. (From Trager *et al.*, 1980.)

many drugs fail to pass the blood–brain barrier, there has been no useful drug for trypanosomiasis of the central nervous system other than melarsoprol or Mel B, drugs with frequent deleterious side effects. In limited clinical trials, DFMO has been found effective against central nervous system trypanosomiasis in humans. But there is much room for improvement with regard to reduction in the dose and the period of treatment, which was 6–8 weeks. The search is on for more active analogues and for possible synergistic agents. In an animal model the effect of DFMO was greatly potentiated by low doses of suramin, an antitrypanosomal drug that by itself is useless against the central nervous system infection. Also of considerable interest is a very recent finding by C. J. Bacchi and his colleagues that sinefungin acts synergistically with DFMO to cure acute as well as central nervous system infections in the animal model. Perhaps sinefungin interferes with the spermidine and spermine synthesis that requires *S*-adenosylmethionine.

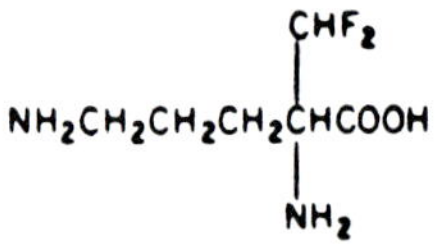

CHF_2
$NH_2CH_2CH_2CH_2CHCOOH$
NH_2

FIGURE 25.12. α-(Difluoromethyl)ornithine, an inhibitor of ornithine decarboxylase in the biosynthesis of polyamines.

Interference with Glycolysis

As already seen (Chapter 9), a number of important parasites, including trypanosomes in their bloodstream form, malarial parasites, and schistosomes, obtain most of their energy via the glycolytic pathway. In the bloodstream trypanosomes, lactate dehydrogenase is absent and the regeneration of NAD from NADH depends on a dihydroxyacetone phosphate : glycerol-3-phosphate shuttle and a glycerol-3-phosphate oxidase. Under anaerobic conditions glycerol accumulates as an end product. Inhibition of the glycerol-3-phosphate oxidase by salicyl hydroxamic acid (SHAM) renders the metabolism like that under anaerobic conditions. If glycerol is then added to inhibit the glycerol kinase, the metabolism is stopped and the trypanosomes are killed. Mice infected with the cattle parasite *T. vivax* could be cured by a single dose of 430 mg SHAM/kg plus 3.6 g glycerol/kg. These dosages are too high for practical purposes.

The well-established antitrypanosomal drug suramin acts by inhibition of the glycerol-3-phosphate dehydrogenase, whereas melarsen oxide affects trypanosome glycolysis at the pyruvate kinase. Ernest Bueding showed long ago that antimonials affect the phosphofructokinase of schistosomes and that it is more sensitive to these compounds than is the mammalian phosphofructokinase. Such differences point to an important requirement for agents that affect systems as vital to the host as to the parasite.

Effects on Microtubules

Here we find a number of important anthelmintics, the benzimidazoles, that were developed in screening programs and later found to block transport of secretory granules and movement of subcellular organelles in intestinal cells of nematode parasites. These effects coincided with the disappearance of cytoplasmic microtubules. Mebendazole (Fig. 25.13a) and the related anthelmintic fenbendazole inhibited binding of colchicine to *Ascaris* embryonic tubulins with inhibition constants of 1.9×10^{-8} M and 6.5×10^{-8} M, respectively, as compared with values 250–400 times higher for their inhibition of binding of colchicine to bovine brain tubulin (7.3×10^{-6} M and 1.7×10^{-5} M, respectively). This differential affinity of the benzimidazoles for the nem-

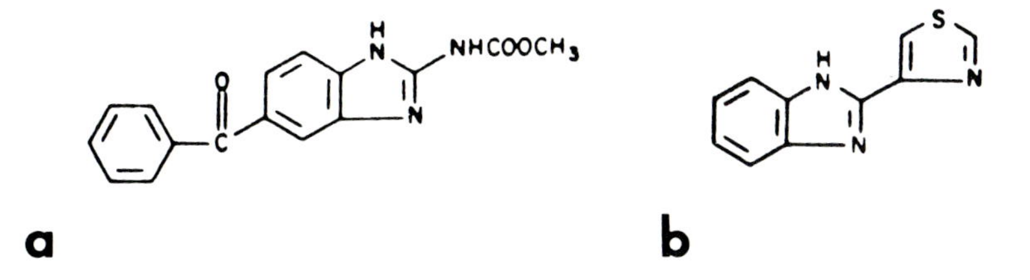

FIGURE 25.13. Two anthelmintics, the benzimidazoles mebendazole (a) and thiabendazole (b).

atode parasite and the host tubulins probably accounts for their anthelmintic efficacy.

Thiabendazole (Fig. 25.13b), another member of this group of compounds, has been found to also have antifungal action, again evidently based on its effect on the organism's tubulins. This suggests the possibility of antiprotozoal effects, but none have been reported.

Agents Affecting Nerve and Muscle and Motility

Obviously these can be effective mainly against metazoan parasites, such as helminths and arthropods, with a nervous system and neuromuscular junctions. This efficacy depends on certain important differences between the nervous systems of the parasites and those of their vertebrate hosts. Vertebrates have mainly nicotinic cholinergic receptors at the neuromuscular junctions whereas nerves with γ-aminobutyric acid (GABA) as transmitter are confined to the central nervous system and are protected from agents in blood by the blood–brain barrier. In insects the muscles have an excitatory synapse that uses L-glutamic acid and an inhibitory nerve with GABA as transmitter, the cholinergic nerves being mainly in the central nervous system. In nematodes cholinergic and GABAergic synapses are distributed throughout the organism. Levamisole (Fig. 25.14) evidently can penetrate the cuticle of nematodes and then act on the cholinergic receptors at the neuromuscular junctions to paralyze the worm. Once rendered immotile, a worm such as *Ascaris* is unable to maintain its position in the alimentary tract and is excreted.

Piperazine, a drug used for many years as a relatively ineffective remedy for gout, turned out to be highly active in a screening test using *Aspiculuris* infections in mice, and was then developed as one of the first effective anthelmintics. It also paralyzes nematodes, but does so by acting as a GABA agonist at the neuromuscular junctions. A particularly important group of agents of this second type has recently been developed. These are the avermectins (Fig. 25.15), agents that act as GABA agonists in arthropods and nematodes to cause paralysis. Since these compounds do not readily pass the blood–brain barrier and since GABA-mediated nerves in mammals are confined to the central nervous system, these drugs have a very wide margin of safety. For example, rats injected with tritium-labeled ivermectin at 0.3 mg/kg showed a maximum concentration in the brain of only 20 parts per billion whereas muscle (the tissue with the next lowest concentration) had 300 ppb. Correspondingly, heartworm could be suppressed in dogs with a dose of only 1 μg/kg whereas toxic effects first appeared at 5 mg/kg—5000 times more. It

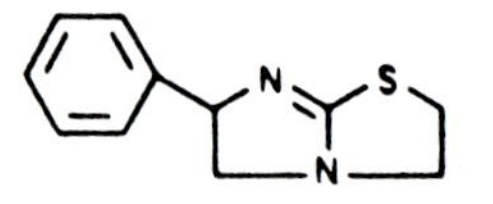

FIGURE 25.14. Levamisole.

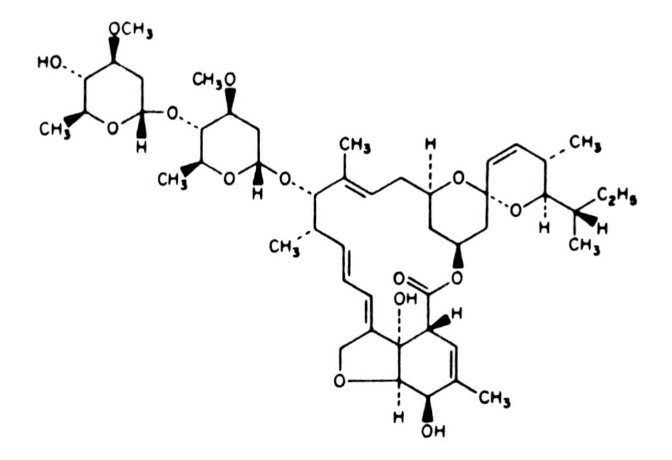

FIGURE 25.15. Ivermectin (avermectin B_1a).

is interesting to note that the avermectins were first discovered as antibiotics in the fermentation broth from a single actinomycete culture with action in a screening test using mice infected with the nematode *Nematospiroides dubius*. They have turned out to be also very effective insecticides, especially against ectoparasitic insects. Already there are strong indications that they may be effective in the treatment of onchocerciasis, a major disease for which there has not been any good therapeutic agent. Meanwhile, ivermectin is in wide use against nematode parasites of domestic animals.

The lack of effect of avermectins against trematodes and cestodes suggests that these helminths have nervous systems quite different from those of nematodes. Here again, however, the antischistosomal agents metrifonate and hycanthone cause paralysis probably by affecting cholinergic nerves. Of special interest and importance is the recently developed drug praziquantel (Fig. 25.16a), effective against the three major species of human schistosomes. This compound affects muscle contraction in these blood flukes by enhancing Ca^{2+} influx. In schistosomes muscle contraction is dependent on uptake of external Ca^{2+} but in cestodes it depends on endogenous Ca^{2+}. Thus, further detailed study of nerve and muscle physiology in cestodes and trematodes may lead to additional useful chemotherapeutic agents.

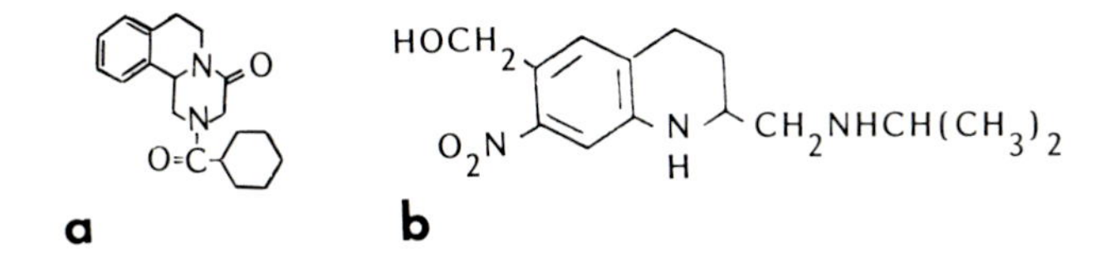

FIGURE 25.16. Two antischistosomal drugs: praziquantel (a) effective against all three main human species, and oxamniquine (b) effective only against *S. haematobium*.

Motility of these worms can also be affected via their metabolism. We have seen (Chapter 9) that schistosomes have mainly a glycolytic metabolism but that oxygen is required for egg formation. Aerobic mitochondrial metabolism apparently also is essential for their motility, as shown by the effects of closantel, a salicyl anilide that has found practical use against liver flukes of cattle and sheep as well as against the sheep nematode *Haemonchus* and against the dipteran larvae (*Oestrus ovis, Dermatobia* spp., and *Hypoderma* spp.) parasitic in domestic animals. Closantel inhibits ATP synthesis in mitochondria of *Fasciola hepatica;* in *Schistosoma mansoni* it completely inhibits motility at 10^{-6} M and reduces the energy charge to the same level that occurs under anaerobic conditions. The importance of motility to the life of the schistosomes is well illustrated by the differential effect of metrifonate (0,0-dimethyl-2,2,2-trichloro-1-hydroxyethylphosphonate) on two species of schistosomes. This compound is effective against *S. hematobium,* the bladder fluke, but not against *S. mansoni.* The drug, an organophosphorus derivative, inhibits cholinesterase and acetylcholinesterase in both species of worms to produce a reversible paralysis. The paralyzed *S. haemotobium* release their hold in the bladder veins and are carried eventually to the lungs, where they are unable to relocate even though they recover from the paralysis. The paralyzed *S. mansoni,* located in the mesenteric veins, are carried only to the liver whence they can return to the mesenteric veins once the effect of the drug wears off.

Miscellaneous Other Targets

The types of action discussed above by no means exhaust the possibilities. Certain antibiotics, such as tetracycline, act as antimalarials perhaps by inhibiting mitochondrial protein synthesis. The glaucorubinones, which include effective anticancer drugs thought to act as general inhibitors of protein synthesis, are also very active as antimalarials *in vitro.* Certain metal chelators are effective antiprotozoal drugs, particularly if they can readily be taken up by the cells. These may act by depriving the parasite of essential iron. This is the mode of action of desferrioxamine against *Plasmodium falciparum in vitro;* the effect is reversed by adding iron to the medium. This is also the probable mode of action of a series of chelators selected by screening with cultures of the insect hemoflagellate *Crithidia fasciculata,* some of which were then shown to be active *in vivo* against *Trypanosoma brucei.* Metal chelators may also act by interfering with metallo protein oxidases, enzymes that play special roles in metabolism. For example, a phenol oxidase is essential for egg formation in schistosomes and other fluke (see Chapter 9) and the drug disulfuram, known to inhibit such oxidases, causes abnormal egg production. Disulfuram (diethyldithiocarbamate) and a number of 8-hydroxyquinolines have been found to inhibit *P. falciparum in vitro,* an effect that may result from inhibition of an essential oxidase of as yet unknown nature.

Malarial parasites depend on the digestion of red cell hemoglobin for most of their amino acids. It is not surprising, therefore, that some protease inhibitors also actively inhibit growth of the parasite. Of special interest is cyclosporin A, originally used as an immunosuppressive agent and then found to have antimalarial effects *in vivo* as well as *in vitro*. Work with derivatives of cyclosporin A has dissociated the antimalarial effect from the immunosuppressive one and suggests that the antimalarial effect may result from inhibition of the parasites' hemoglobinase.

A relatively high requirement of intraerythrocytic malarial parasites for pantothenic acid was first noted in *in vitro* experiments with the avian parasite *P. lophurae* and has recently been demonstrated for *P. falciparum* in culture. In keeping with this, certain analogues of pantothenic acid have antimalarial effects *in vivo* as well as *in vitro*. None, however, has been developed as a practical drug. The effectiveness of these compounds may be limited in part because they must act on host rather than parasite enzymes, since it has been shown that erythrocytic stages of *P. lophurae* lack the CoA biosynthetic pathway and obtain this essential cofactor from the host erythrocyte.

The complex life cycles of many parasites suggest a number of possible targets for chemotherapeutic intervention. These have been little exploited because sufficiently detailed information is not yet available. Among helminthic parasites having separate males and females, there must be attractants and receptors that function to bring the sexes together and effect mating. Interference with these would prevent reproduction. In the special case of the schistosomes, preventing egg production means eliminating most of the pathology. Furthermore, as noted earlier (Chapter 8), the female schistosome lies permanently in the gynecophoric canal of the male and depends for its maturation on some essential factor or factors supplied to it by the male. At least one useful antischistosomal drug, oxamniquine (Fig. 25.16b), differentially kills the male schistosomes. The surviving females show regressive changes in the reproductive system that are the same as those seen in females deprived of their male partner without the use of any drug. These changes, furthermore, are reversed if the females are placed in new hosts with normal males.

With intracellular parasites, especially those living in nonphagocytic cells, it is likely that specific products of the parasite induce internalization by the host cell (see Chapters 3, 5). These again offer a target for chemical intervention, once their nature is known.

Important Antiparasitic Drugs of Undetermined Mode of Action

Among these are the major antimalarials quinine, chloroquine, and the recently developed mefloquine (Figs. 25.2, 25.17) and qinghaosu or artemisinin (Fig. 25.1). There are indications for cross-resistance of malarial parasites between quinine and mefloquine, suggesting that they may affect the same or related targets. Both drugs are 4-quinolinemethanols and both produce somewhat similar changes in ultrastructure of the malarial pigment. Meflo-

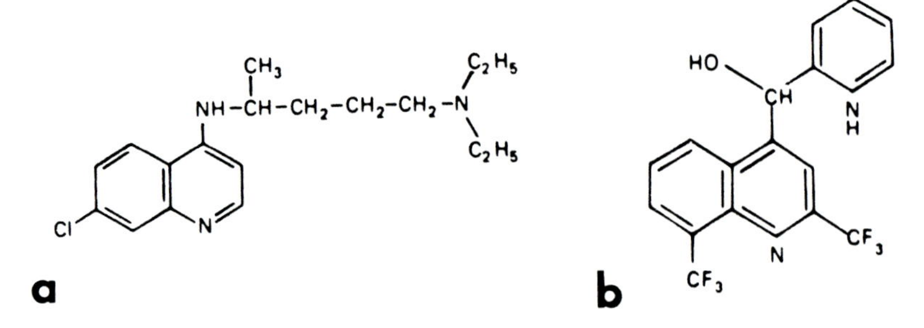

FIGURE 25.17. Chloroquine (a), a 4-aminoquinoline, and the recently developed antimalarial mefloquine (b), a 4-quinolinemethanol.

quine is much more active than quinine on a weight basis and its pharmacodynamics are quite different. It accumulates in uninfected as well as infected erythrocytes. The red cell membrane apparently has high-affinity binding sites for mefloquine. The drug binds weakly to hemoglobin, but since the hemoglobin concentration is so high this binding is probably of major importance. Mefloquine and quinine do not intercalate into DNA. Chloroquine, for 40 years the outstanding antimalarial, does intercalate into DNA but this is probably not its primary mode of action. It has a high affinity for the hemin which, as a product of digestion of hemoglobin by the malarial parasites, accumulates in infected red cells. These cells have a much higher chloroquine accumulation than do uninfected erythrocytes. Chloroquine–hemin adducts are toxic, but again this is probably not the principal mode of antimalarial action. Chloroquine accumulates in the acidic food vacuoles of the parasite (see Chapter 6) and it seems likely that here it interferes with the digestion of hemoglobin, perhaps in part by an effect on an acid protease. Qinghaosu, like chloroquine, accumulates in infected cells. It probably affects protein metabolism, perhaps via the formation of oxidative free radicals (see earlier section in this chapter). In *Plasmodium inui* parasites in rhesus monkeys treated with qinghaosu at the relatively high dose of 50 mg/kg, the mitochondria as seen by transmission electron microscopy appeared swollen by 2½ hr after administration of the drug. All four of these antimalarials act only against the erythrocytic stages of malaria. None is effective against the preerythrocytic hepatic stages or against the mature gametocytes of *P. falciparum* (see later in this chapter for further discussion of this). Chloroquine and mefloquine have long half lives and accumulate in the tissues. Both are therefore suitable as suppressive agents. Qinghaosu is quite rapidly excreted, but also acts very quickly, even more quickly than chloroquine. It has therefore been found particularly useful in restoring consciousness in patients comatose with cerebral malaria.

Among the drugs used against the African trypanosomes, suramin (Fig. 25.18), as noted earlier, has recently been found to inhibit the α-glycerophosphate oxidase, and the α-glycerophosphate dehydrogenase of blood-

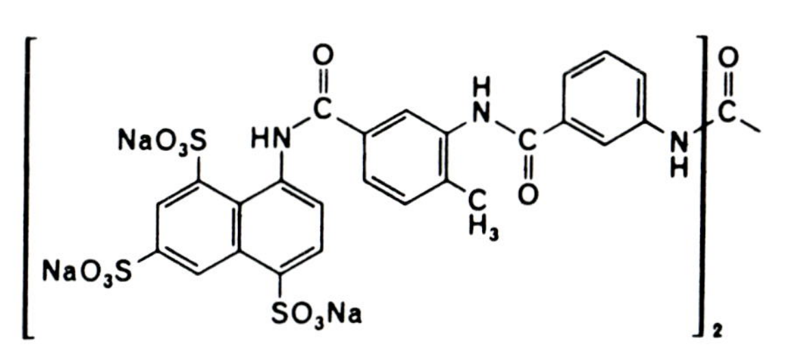

FIGURE 25.18. Suramin, an anionic trypanocidal drug.

stream trypanosomes. This is probably its principal mode of action. Tryparsamide (Fig. 25.19A), developed over 50 years ago and for many years the most dependable drug for treatment of early human trypanosomiasis, is reduced to a trivalent arsenical and probably acts on essential thiol groups, perhaps on the recently discovered trypanothione. Another arsenical, Mel B (Fig. 25.19B), developed by E. Friedheim, has been the only drug available for the treatment of central nervous system trypanosomiasis; it passes the blood–brain barrier. This was prepared by coupling melarsine, a melaminyl arsenical, with the disulfide originally known as British anti-lewisite (developed as a detoxifying agent against poison gas). Mel B has also been of tremendous value, despite its toxicity, because it is effective against tryparsamide-resistant strains which, by 1950, had become very widespread. This suggests that its mode of action is different from that of tryparsamide. An analogue with antimony replacing the arsenic, designated MSb, appears to be much less toxic and is also effective against filarial infections, but has not come into general use.

All of the important trypanocidal agents other than the arsenicals (or the newly developed polyamine antagonists already discussed) have notable effects on the kinetoplast DNA (see Chapter 11). They may be divided into two groups: (1) compounds that intercalate into DNA including ethidium (Fig. 25.20A); (2) compounds not intercalating, but binding across the small groove of the DNA helix, including the diamidines, pentamidine and berenil (Fig. 25.20B,C) and probably antrycide (Fig. 25.20D). Of these four major trypan-

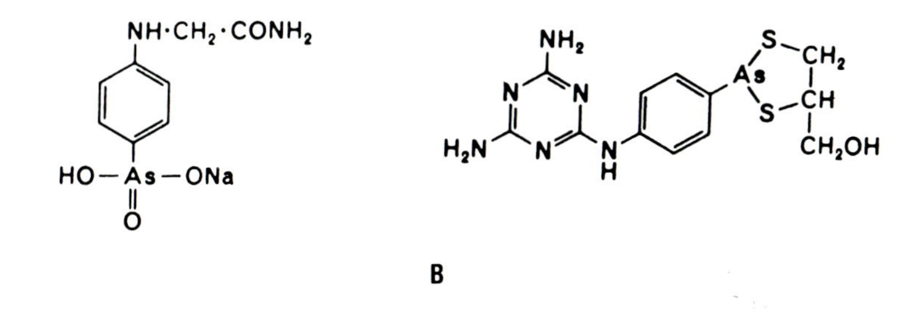

FIGURE 25.19. Trypanocidal arsenical drugs: tryparsamide (A) and Mel B (B). Only the latter is effective against central nervous system trypanosomiasis (African sleeping sickness).

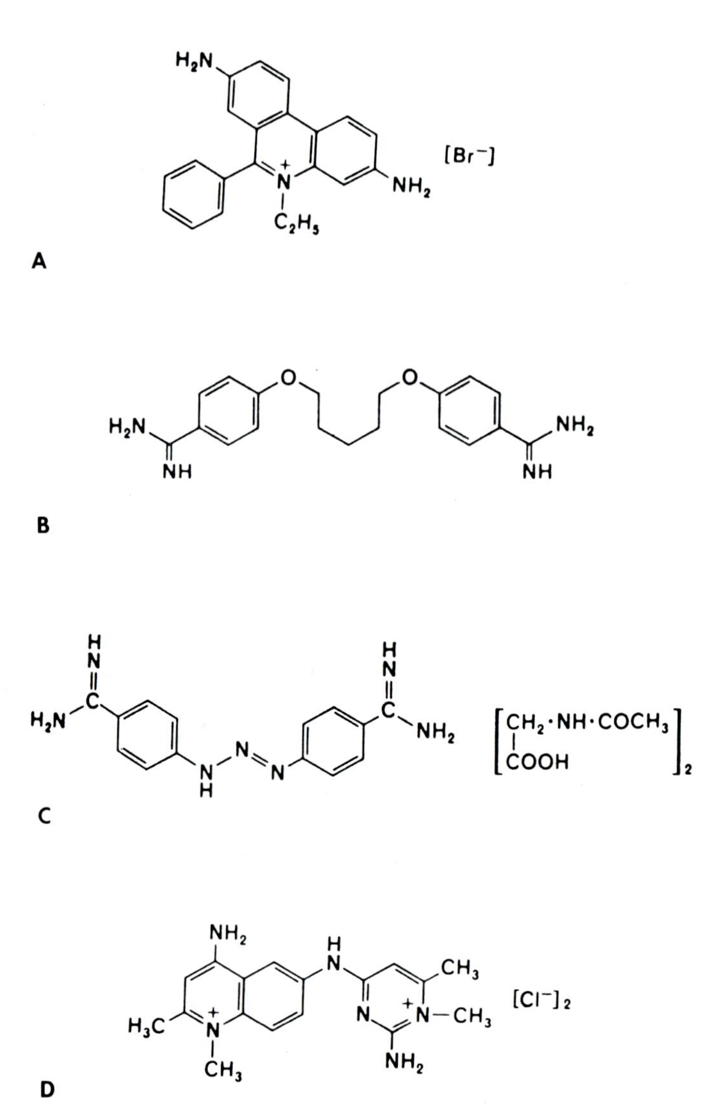

FIGURE 25.20. Four major trypanocidal drugs: (A) ethidium, (B) pentamidine, (C) berenil, and (D) antrycide. Only pentamidine is used against human trypanosomiasis.

ocidal drugs, only pentamidine has been widely used in human medicine. It was of great importance in the control of sleeping sickness outbreaks and was effectively used as a prophylactic agent in efforts at local eradication of the disease. Berenil is now one of the most widely used drugs for treatment of animal trypanosomiasis because of its high therapeutic index and the low incidence of drug-resistant strains induced by it. Berenil is rapidly degraded metabolically and excreted; this is probably responsible both for its lack of prophylactic activity and for the low incidence of drug-resistant strains. Ethidium and especially antrycide, on the other hand, have been useful for prophylaxis in cattle but also readily induce drug resistance, so that they are no longer effective. All of these compounds rapidly produce effects on the kinetoplast DNA of trypanosomes. For example, in mice infected with *T. rhodesiense,* the trypanosomes showed, within 5 hr after an intraperitoneal curative dose of berenil at 5 mg/kg, selective damage to the kDNA (Fig. 25.21). The mitochondrion was enlarged with a clear matrix. These effects on kDNA

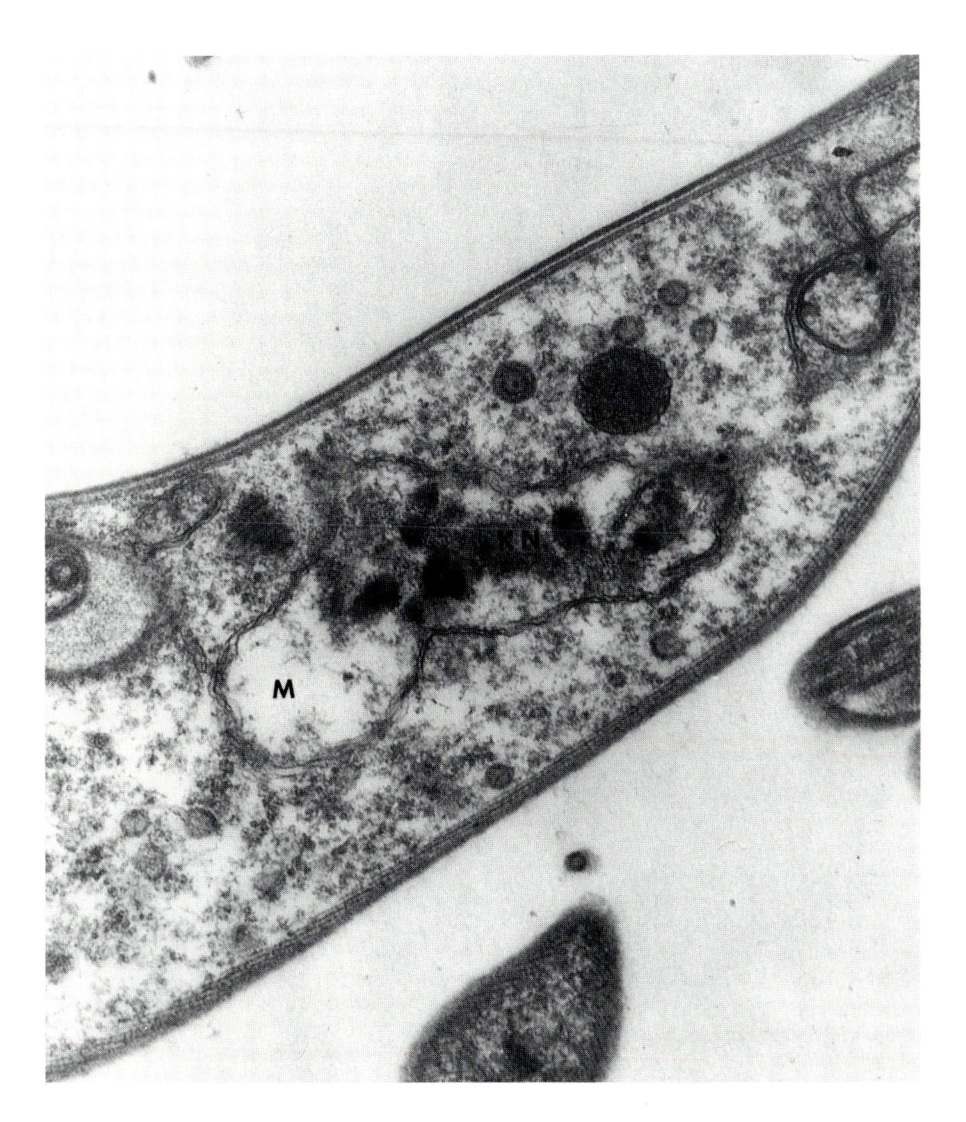

FIGURE 25.21. Effects of berenil on *Trypanosoma rhodesiense*. Note the discoid condensation of the kinetoplast DNA (KN) and the dilatation of the mitochondrion (M). TEM. × 13.000. (From Williamson, 1979. Original print courtesy of Dr. J. Williamson.)

resemble those produced by acriflavine (see Fig. 11.3), but it is not certain that they are brought about in the same way. In *T. cruzi,* for which these compounds are not effective, perhaps because development is intracellular, berenil was found to bind to isolated DNA circles at the same sites that bound RNA polymerase. Again like acriflavine, ethidium and antrycide are very effective in inducing the "petite" mutant of yeast, and all four drugs readily produce dyskinetoplastic trypanosomes (see Chapter 11). In a strain of *T. rhodesiense* maintained in mice and made resistant to stilbamidine by treatment for over a year, total loss of the kDNA occurred. There was no fluorescence whatever with the fluorochrome DAPI (see Chapter 11) and no kDNA band in a cesium chloride gradient. This strain was cross-resistant to almost all trypanocidal agents. Of course, such a strain would have little significance in nature for trypanosomes that have to be cyclically transmitted, since dyskinetoplastic trypanosomes cannot develop in the tsetse fly (see Chapter 11).

It is of interest that these drugs, and especially berenil, are also highly effective against *Babesia* in cattle. Treatment of cattle with berenil cures both trypanosomal and babesial infections. *Babesia,* an intraerythrocytic parasite (see Chapter 19), has no kinetoplast but its nuclear DNA is AT-rich, like kDNA.

The main anthelmintic drugs mebendazole and its analogues, piperazine, levamisole, the avermectins, metrifonate, praziquantel, and oxamniquine, have already been discussed in relation to their mode of action. Here I must call attention to hycanthone, a drug that was for some time used for mass treatment for *Schistosoma mansoni* and *S. haematobium.* It is not effective against *S. japonicum.* It is mutagenic and schistosomes develop resistance to it (see below) so that it has largely been replaced by praziquantel effective against all three species of schistosomes. A newer member of the benzimidazoles, albendazole, has been found active against the tapeworm *Echinococcus.* It is not known whether the effect is mediated via tubulins, as it is in nematodes (see previous section).

Drug Resistance

Drug resistance may be expected to occur with any parasitic organism exposed for any length of time to sublethal doses of a cytocidal or cytostatic agent. For reasons not fully understood, it occurs more readily with some kinds of organisms than with others, and against some kinds of drugs than against other kinds. Mainly for this reason there is continued need for the development of new drugs. One can never rest on one's laurels, especially in view of the long time required to bring a new agent into practical use. A practical drug must not only be active against the parasite at doses far below those producing harmful effects in the host. It must also be reasonably stable even under conditions of high temperature and high humidity. It should preferably be effective when taken orally as well as if injected and it must not be too difficult to manufacture and not too expensive. Even cosmetic factors have sometimes to be taken into account. For example, suppressive Atabrine, which effectively protected the U.S. Army from malaria during World War II, could not have been used in a civilian population because of the yellow color, with a green fluorescence, that it gave to the skin. The new antimalarial BW 58C (see above) imparts a bright red color to the urine of people taking it, making it unacceptable. Analogues must be sought that are as active but do not give this colored metabolite, prolonging still further the time to a practical new drug. And meanwhile the parasites become resistant to the few drugs on hand.

Drug resistance occurs among helminths as well as protozoa. Resistance of vector insects to insecticides is also of major importance but will be treated in the next chapter. Among the nematode parasites of domestic animals, resistance may appear to the three groups of available anthelmintics: benzimidazoles, levamisole, and avermectins. It is important to try to reduce the selection pressure for resistance by treating at a time when the worms have already produced most of their offspring.

When mice infected with *S. mansoni* were given one dose of hycanthone at 30 or 60 mg/kg, the worms all moved from the mesenteric veins into the liver and stopped laying eggs. But about 5 to 12 months later the 10 to 20% that survived were back in the mesenteric veins and laying viable eggs. Most interesting, the offspring hatching from these eggs were resistant to hycanthone and related compounds. The resistance has been analyzed genetically and found to behave as an autosomal recessive trait.

Since protozoa, unlike most helminthic parasites, multiply extensively within a single host, one might expect them to exhibit mutant drug-resistant strains more readily. Drug resistance does occur with all the parasitic protozoa. It has been a problem of tremendous practical importance in the poultry industry in relation to the control of coccidiosis, in trypanosomiasis of humans and domestic animals in Africa, and in malaria. Resistance has readily appeared to all classes of antitrypanosomal drugs with the exception of berenil. It has been especially apparent in cattle where the attempt has been made to use the drugs for prophylaxis. Such use could well provide particularly good conditions for the selection of resistant mutants and their propagation in tsetse. As we have seen, the precise mode of action of most trypanocidal drugs is not understood. Accordingly, even less is known about the metabolic basis of resistance. The production of dyskinetoplasty by a number of different drugs could well be responsible for cross-resistance among them under laboratory conditions of syringe transmission. But this is hardly relevant to the situation in nature, since dyskinetoplastic trypanosomes cannot infect tsetse flies. This is a subject which may now be ready for fruitful new approaches combining the recently available culture methods with the still more recent methods of molecular genetics.

Just these methods are now beginning to be applied to the study of drug resistance in malaria. Here an alarming situation has appeared. Drug resistance in malaria is not altogether recent. As far back as 1910 it was noted that European laborers in some areas of South American were not protected from malaria by the usual daily dose of quinine which was effective for natives in these regions and for Europeans in other regions. This pointed to two factors whose vital significance was not at first appreciated: (1) the existence of local strains relatively resistant to quinine, then the only antimalarial drug; (2) the role of partial acquired immunity in relation to drug treatment. The importance of the latter factor is now much more appreciated than it was formerly. It has been beautifully demonstrated experimentally in mice infected with *P. chabaudi* (Fig. 25.22). Despite the observed variation in efficacy of quinine treatment in nonimmune people, and despite the rapid emergence in the field of strains of both *P. falciparum* and *P. vivax* solidly resistant to antifolates such as proguanil and pyrimethamine, malariologists in the 1950s were remarkably complacent with regard to antimalarial chemotherapy. They had chloroquine, and it seemed to be the ideal drug: cheap, stable, nontoxic at the effective dosage, and persisting for a long time in the blood. It was not only curative of all species of malaria but could also be used at low dosage for suppressive prophylaxis of *P. falciparum*. It was widely used in this way by nonimmune people, civilians as well as military, having to spend short or even fairly prolonged periods of time in a malarious region. Furthermore, whereas various species of malarial parasites of experimental animals could easily be made resistant to sulfonamides and other antifolates, this was not true with chloroquine. Most attempts to induce resistance to chloroquine failed. There was one report by Ramakrishnan of its induction in *P. berghei* in mice, but this was ignored. Indeed, not too much attention was paid to the first reports of failure of chloroquine treatment in people, which appeared almost simultaneously from South America and from Southeast Asia. Only with the prolonged presence of U.S. military personnel in Vietnam did the potential threat of chloroquine resistance become apparent, and the urgent need for new drugs. Even as late as 1977, however, there were some who maintained that chloroquine resistance of *P. falciparum* would not appear in Africa, where falciparum malaria remains a prominent cause of mortality and morbidity.

The strength of such prejudice is well illustrated by the story of an American zoologist. Dr. Van Gelder has himself told it engagingly, and it should be read in the original. Here I will briefly summarize it. As mammalogist for the American Museum of Natural History, Van Gelder regularly visited East Africa. He always faithfully took the prescribed weekly dosage of 500 mg chloroquine begun 2 weeks before arrival and continued for a month after leaving the malarious region. He had no trouble until his 1977 trip. On his return he was ill with a fever. Malaria was diagnosed and he was treated with chloroquine. He recovered, only to have a recurrence some weeks later. At the third such episode he suggested to his physicians that possibly he had chloroquine-resistant malaria. Impossible, they said, in Africa. But he pointed out to them a recent paper in *Science* (by then it was 1978 and he was still struggling with his malaria). In this paper an *in vitro* culture of *P. falciparum*

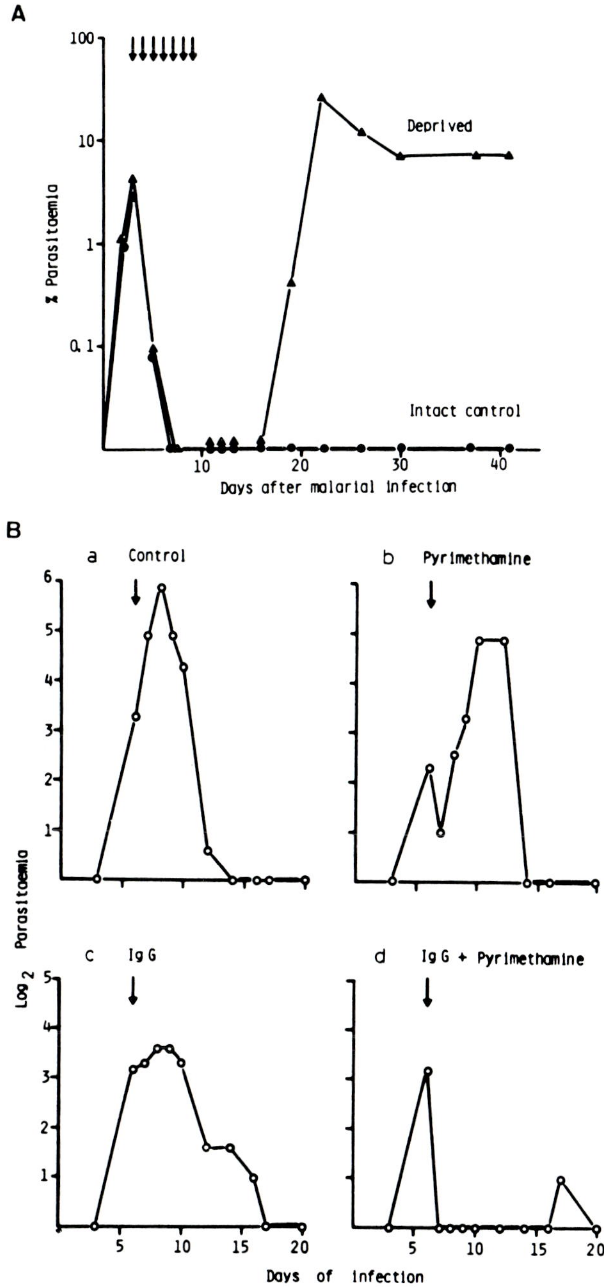

FIGURE 25.22. Interaction of chemotherapy and immune response in mice infected with *Plasmodium chabaudi.* (A) T-cell-deprived (▲) and intact controls were inoculated with 10^7 infected red blood cells and treated with chloroquine (arrows) at 4 mg/kg on days 3–9. (B) Inoculated mice in four groups received intraperitoneally on the day indicated by the arrow either: (a) 0.1 ml saline; (b) pyrimethamine at 1 mg/kg (a subcurative dose); (c) 1 mg IgG from mice hyperimmunized with *P. chabaudi;* (d) pyrimethamine at 1 mg/kg plus 1 mg of the hyperimmune IgG. (From Targett, 1984.)

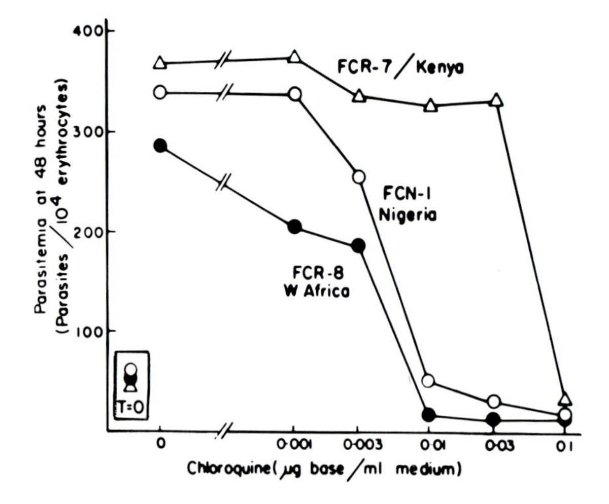

FIGURE 25.23. Chloroquine sensitivity of three African isolates of *Plasmodium falciparum* as measured by a 48-hr *in vitro* test. FCN-1 and especially FCR-8 were sensitive, whereas FCR-7 (from a patient showing clinical resistance) was resistant, not being affected at 0.03 μg base/ml (about 0.1 μM). (From Nguyen-Dinh and Trager, 1980.)

derived from the Gambia in West Africa had been made resistant to chloroquine with surprising rapidity. Van Gelder was then treated with Fansidar, the sulfadoxine–pyrimethamine combination, and was finally cured of his falciparum malaria. Before his cure, however, parasites from his blood were established in culture (as FCR-7/Kenya) and shown to be chloroquine-resistant (Fig. 25.23). It will be noted that FCR-7 was not affected *in vitro* by a chloroquine concentration of 0.1 μM, whereas *in vivo* plasma levels of 0.03 to 0.1 μM are ordinarily effective against sensitive strains.

With the advent of several tests to monitor chloroquine resistance *in vitro*, it has been possible to follow its spread. The rapidity of spread is astounding. In Southeast Asia and South America, virtually all isolates of *P. falciparum* are now chloroquine-resistant. East and Central Africa show many regions with chloroquine resistance. Especially alarming is the westward spread of chloroquine-resistant *P. falciparum* in India, where *P. vivax* had been the principal cause of malaria. Cases of *vivax* malaria in young children that were apparently resistant to chloroquine have been reported but without information as to whether adequate blood levels of chloroquine were attained. Most interesting, but also very alarming, was the finding in Thailand that increased levels of chloroquine resistance were correlated with increased infectivity to mosquitoes, both *Anopheles stephensi* and *A. balabacensis*.

For a time Fansidar was a useful drug against chloroquine-resistant falciparum malaria. But resistance to Fansidar is now prevalent in Southeast Asia and parts of South America, and this too is rapidly spreading elsewhere. Multi-drug-resistant strains have appeared that are also relatively resistant to quinine. This is alarming with regard to the efficacy of the new drug mefloquine since in laboratory experiments there is some cross-resistance between mefloquine and quinine. Mefloquine resistance and artemisinin resistance have already been observed in the field.

The molecular basis for chloroquine resistance is unknown. Beginnings are being made with the growth *in vitro* of sensitive and resistant clones of *P. falciparum* and with a veritable explosion of work on the molecular biology of this organism (see also Chapter 19). More is known about resistance to antifolates, in keeping with our better understanding of their mode of action. Early work with *P. berghei* made resistant to pyrimethamine established that the dihydrofolate reductase of the resistant parasites had a much lower affinity for pyrimethamine than did the dihydrofolate reductase of sensitive strains. A somewhat similar situation has been found for clones from two naturally resistant isolates of *P. falciparum*. Their dihydrofolate reductase activity had a 15-fold higher inhibition constant for pyrimethamine than did that of a clone from a sensitive isolate. Moreover, their dihydrofolate reductase activity was about 4-fold higher and considerably more stable than that of the sensitive parasites. These factors together probably could account for the 1000-fold difference in sensitivity to pyrimethamine *in vitro* (50% inhibitory level 2.5 μg/ml for the resistant parasites as compared to 3 ng/ml for the sensitive parasites). No evidence for gene amplification could be found. Such gene amplification with a corresponding increase in dihydrofolate reductase activity was shown to account for the resistance to methotrexate (another dihydrofolate reductase inhibitor) of a line of *Leishmania tropica* artifically made resistant to this drug. In one other study of dihydrofolate reductase of *P. falciparum*, no difference in susceptiblity of the enzyme could be found between pyrimethamine-resistant and susceptible strains. With one pyrimethamine-resistant isolate (Honduras-I CDC) and clones derived from it, characteristic intranuclear structures and a nucleolus have been observed in the erythrocytic stages; these have not been seen in other isolates examined by electron microscopy. Further study in depth of the genome of *P. falciparum* and the application of methods such as pulsed field gradient electrophoresis in relation to drug-resistant mutants should be of great interest. Resistant mutants of *P. falciparum* have not only been selected from culture but have also been induced with mutagenic agents. This too should facilitate the further study of this important problem.

Agents Active against Stages of Malaria Other Than the Erythrocytic Stage

Most antimalarials affect the erythrocytic parasites. This is where the action has to be for cure of the malaria attack. Although drugs such as chloroquine can effect permanent cure of *P. falciparum* infection (provided of course that the parasites are not resistant to the drug), they generally will not effect a permanent cure of *P. vivax* (or *P. ovale*) infection. In these relapsing malarias there exist in the liver resting stages, called hypnozoites, that, after a longer or shorter dormant period, develop into the preerythrocytic liver schizont, the merozoites from which then again initiate a blood infection. The preerythrocytic tissue schizonts of all species of malaria are not affected by chloroquine or most of the other antimalarials that act on the erythrocytic parasites.

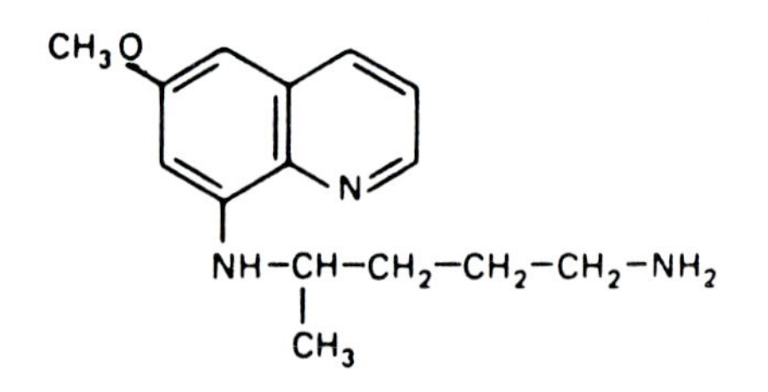

FIGURE 25.24. Primaquine, an 8-aminoquinoline.

This is also true for *P. falciparum* but here there are no hypnozoites; only the initial preerythrocytic cycle develops in the liver from sporozoites injected by the mosquito. This is why suppressive treatment with chloroquine is effective against chloroquine-sensitive *P. falciparum.* If a person, visiting a malarious region and bitten by an infectious mosquito even on the last day of his/her stay, continues the regime of suppressive chloroquine for another 3–4 weeks, blood infection will be prevented despite the fact that the hepatic preerythrocytic stages will have developed. With *P. vivax,* however, additional later crops of hepatic schizonts will develop, and, as the chloroquine level falls, will succeed in initiating a new blood infection. Such initial attacks after cessation of suppressive chloroquine have occurred as long as 2 years later with *P. vivax* and up to 3½ years later with *P. ovale.*

Hence, there has been need for an agent to give radical cure of these relapsing malarias. Only one drug, primaquine (Fig. 25.24), is available for this purpose. It seems to be more effective against temperate zone strains of *P. vivax* than against those from Southeast Asia or New Guinea. It also has a rather narrow therapeutic index and may be dangerously toxic in individuals having erythrocytes deficient in glucose-6-phosphate dehydrogenase, a mutant common especially in malarious regions (see Chapter 14).

If a better drug than primaquine could be found effective against the preerythrocytic stage, it could serve as a true causal prophylactic agent, preventing the initial development of sporozoites in the liver.

Primaquine is also of special interest in that it is the only drug affecting the later developmental stages of the gametocytes of *P. falciparum.* Such a gametocytocidal drug could be used to try to prevent the spread of chloroquine-resistant malaria in a malaria control program (see Chapter 26). Drugs that inhibit development of the parasites in the mosquito could be used in a similar way, but none of practical value is at present available.

General Discussion

As is evident from Table 25.2 of principal antiparasitic drugs, the situation with regard to adequate treatment of many of the parasitic diseases is far from satisfactory. For the treatment of African sleeping sickness, Mel B is the only

TABLE 25.2. Some of the Principal Antiparasitic Drugs: Their Uses and Limitations

Parasite	Drug	Uses	Limitations
Entamoeba histolytica	Metronidazole	Treatment of intestinal amebiasis and amebic liver abscess	May be carcinogenic. Drug resistance occurs
Trichomonas vaginalis	Metronidazole	Treatment of trichomonal vaginitis	As above
Giardia lamblia	Metronidazole	Treatment of giardiasis	As above
	Atabrine	"	Yellow discoloration of skin
Trypanosoma rhodesiense, T. gambiense	Pentamidine	Treatment of early human trypanosomiasis	Drug resistance now common
	Suramin	"	May be nephrotoxic
	Mel B	Treatment of central nervous system trypanosomiasis (African sleeping sickness)	Frequently produces encephalopathy
	Difluoromethyl ornithine	Treatment of central nervous system as well as early trypanosomiasis	New, limitations not yet apparent
T. brucei, T. vivax, T. congolense	Ethidium	Treatment and prophylaxis of trypanosomiasis of cattle and other domestic animals	Widespread drug resistance
	Berenil	Treatment of trypanosomiasis of domestic animals	
Plasmodium falciparum	Chloroquine	For cure and suppressive prophylaxis of falciparum malaria	Drug-resistant strains now common and spreading rapidly. Accumulates in tissues and may damage retina after prolonged use

(*continued*)

TABLE 25.2. *(continued)*

Parasite	Drug	Uses	Limitations
	Pyrimethamine	"	Drug resistance appears quickly and is widespread
	Pyrimethamine–sulfadoxine (Fansidar)	"	Drug-resistant strains now appearing. Allergic reactions relatively common
	Quinine	For cure and suppressive prophylaxis of falciparum malaria	Unpleasant side effects common. Drug resistance occurs
	Mefloquine	For cure of falciparum malaria	New drug. Resistance already noted in field
	Qinghaosu	For cure of falciparum malaria. Especially good for cerebral form	New drug
P. malariae	All of the drugs listed above can be used	Treatment of acute attack	
P. vivax, P. ovale	All of the drugs listed above can be used	Treatment and suppression of acute attack	Do not prevent relapses. Resistance to pyrimethamine common
	Primaquine	For radical cure of vivax malaria	Therapeutic index relatively low. May cause hemolytic anemia in G6PD-deficient people
Leishmania spp.	Pentamidine	For treatment of leishmaniasis	Must be given intravenously
	Pentostam	"	Must be given intravenously. Long course required
	Allopurinol	"	New drug
Eimeria tenella and other coccidia of domestic animals	Amprolium	For prevention and treatment of coccidiosis of poultry, and of other domestic animals	Drug resistance occurs
	Monensin	"	"
Onchocerca volvulus	Diethylcarbamazine	Treatment of onchocerciasis	Mazzotti reaction in skin common

	Suramin	″	May be nephrotoxic
	Ivermectin	″	New drug
Wuchereria bancrofti	As for *Onchocerca*	Treatment of human filariasis	As above
		Treatment of heartworm of dogs	
Other nematode parasites of humans and domestic animals	Piperazine	Treatment of ascariasis, hookworm and other intestinal nematodes of people	
	Mebendazole	Treatment and prophylaxis of alimentary tract and lung nematodes of ruminants	Resistance occurs
	Thiabendazole	″	″
	Levamisole	″	″
	Ivermectin	″	New drug
Schistosoma mansoni	Oxamniquine	Treatment of schistosomiasis caused by *S. mansoni*	Not effective against other species of schistosome
	Praziquantel	Treatment of all forms of schistosomiasis	
S. haematobium	Praziquantel	″	
	Metrifonate	Treatment of haematobium schistosomiasis only	″
S. japonicum	Praziquantel		
Taenia solium	Praziqantel	Treatment of cysticercosis	
Echinococcus	Albendazole	Treatment of echinococcosis	

proven drug and it often is found to be toxic. For trypanosomiasis of cattle, the long-acting drugs such as ethidium have readily induced resistance. For leishmaniasis, pentostam, an antimony compound, still has to be mainly relied on. For malaria, there are several excellent new as well as old drugs for treatment of the erythrocytic stages, but *P. falciparum* in particular seems able to develop resistance readily and there are no grounds for complacency. Primaquine, not a very satisfactory agent, remains as the only drug able to effect radical cure of *P. vivax* infection, and this not always. Metronidazole and related nitroimidazoles are highly effective against amebiasis and trichomoniasis and resistance has not yet been a serious problem, but these compounds may be carcinogenic and new agents would be desirable. For schistosomiasis, praziquantel, with oxamniquine and metrifonate, seem to provide for effective and safe therapy at least for the near future. For onchocerciasis and other forms of filarial infection, there has been no safe treatment. It may be that this will now be supplied by the avermectins. These will also be useful, in addition to mebandazole and related compounds and levamisole, for the treatment of other nematode infections and of some cestode infections. For cysticercosis and echinococcosis, there has been no satisfactory chemotherapy, but praziquantel for the former and albendazole for the latter look promising.

This situation emphasizes the need for continued and more intensive research on the biochemistry of parasites. With the development of culture methods, with the improved understanding of mode of action of some antiparasitic and anticancer drugs, with better knowledge of the metabolism of parasites and of specific receptors playing key roles at different stages of the life cycle, and with the application of methods of molecular biology, the rational development of antiparasitic drugs should become possible.

Bibliography

Amole, B. O., and Clarkson, A. B., Jr., 1981, *Trypanosoma brucei:* Host parasite interaction in parasite destruction by salicylhydroxamic acid and glycerol in mice, *Exp. Parasitol.* **51:**133–140.

Avila, J. L., 1983, New rational approaches to Chagas' disease chemotherapy, *Interciencia* **8:** 405–417.

Bacchi, C. J., 1981, Content, synthesis, and function of polyamines in trypanosomatids: Relationship to chemotherapy, *J. Protozool.* **28:**20–27.

Berens, R. L., Marr, J. J., Looker, D. L., Nelson, D. J., and LaFon, S. W., 1984, Efficacy of pyrazolopyrimidine ribonucleosides against *Trypanosoma cruzi:* Studies *in vitro* and *in vivo* with sensitive and resistant strains, *J. Infect. Dis.* **150:**602–608.

Borchardt, R. T., 1980, S-Adenosyl-L-methionine-dependent macromolecule methyltransferases: Potential targets for the design of chemotherapeutic agents, *J. Med. Chem.* **23:**347–357.

Brack, C., and Delain, E., 1975, Electron microscopic mapping of AT-rich regions and of *E. coli* RNA polymerase-binding sites on the circular kinetoplast DNA of *Trypanosoma cruzi, J. Cell Sci.* **17:**287–306.

Bray, D. H., O'Neill, M. J., Boardman, P., Peters, W., Phillipson, J. D., and Warhurst, D. C., 1985, Plants of the family Simaroubaceae as potential antimalarial agents, *Trans. R. Soc. Trop. Med. Hyg.* **79:**426.

Bruce-Chwatt, L. J. (ed.), 1981, with Black, R. H., Canfield, C. J., Clyde, D. F., Peters, W., and Wernsdorfer, W. H., *Chemotherapy of Malaria* (2nd edition), World Health Organization Monograph Series No. 27, Geneva.

Campbell, W. C., Fisher, M. H., Stapley, E. O., Albers-Schönberg, G., and Jacob, T. A., 1983, Ivermectin: A potent new antiparasitic agent, *Science* **221**:823–828.

Chabala, J. C., Mrozik, H., Tolman, R. L., Eskola, P., Lusi, A., Peterson, L. H., Campbell, W. C., Egerton, J. R., and Ostlind, D. A., 1980, Ivermectin, a new broad-spectrum antiparasitic agent, *J. Med. Chem.* **23**:1134–1136.

Cioli, D., and Mattoccia, L. P., 1984, Genetic analysis of hycanthone resistance in *Schistosoma mansoni, Am. J. Trop. Med. Hyg.* **33**:80–88.

Clarkson, A. H., and Brohn, F. H., 1976, Trypanosomiasis: An approach to chemotherapy by the inhibition of carbohydrate catabolism, *Science* **194**:204–206.

Denham, D. A. (ed.), 1985, Chemotherapy of Parasites, Symposium of the British Society for Parasitology 22, *Parasitology* **90**:615–721.

Desjardins, R. E., Canfield, C. J., Haynes, J. D., and Chulay, J. D., 1979, Quantitative assessment of antimalarial activity *in vitro* by a semiautomated microdilution technique, *Antimicrob. Agents Chemother.* **16**:710–718.

Docampo, R., and Moreno, S. N. J., 1984, Free-radical intermediates in the antiparasitic action of drugs and phagocytic cells, in: *Free Radicals in Biology*, Volume VI (W. A. Pryor, ed.), Academic Press, New York, pp. 243–288.

Docampo, R., and Moreno, S. N. J., 1984, Free radical metabolites in the mode of action of chemotherapeutic agents and phagocytic cells on *Trypanosoma cruzi, Rev. Infect. Dis.* **6**:223–237.

Draper, C. C., Brubaker, G., Geser, A., Kilimali, V. A. E. B., and Wernsdorfer, W. H., 1985, Serial studies on the evolution of chloroquine resistance in an area of East Africa receiving intermittent malaria chemosuppression, *Bull. WHO* **63**:109–118.

Evans, D. A., and Holland, M. F., 1978, Effective treatment of *Trypanosoma vivax* infections with salicylhydroxamic acid (SHAM), *Trans. R. Soc. Trop. Med. Hyg.* **72**:203–204.

Fairlamb, A. H., and Bowman, I. B. R., 1977, *Trypanosoma brucei:* Suramin and other trypanocidal compounds' effects on sn-glycerol-3-phosphate oxidase, *Exp. Parasitol.* **43**:353–361.

Ferone, R., Burchall, J. J., and Hitchings, G. H., 1969, *Plasmodium berghei* dihydrofolate reductase: Isolation, properties and inhibition of antifolates, *Mol. Pharmacol.* **5**:49–59.

Fish, W. R., Marr, J. J., Berens, R. L., Looker, D. L., Nelson, D. J., LaFon, S. W., and Balber, A. E., 1985, Inosine analogs as chemotherapeutic agents for African trypanosomes: Metabolism in trypanosomes and efficacy in tissue culture, *Antimicrob. Agents Chemother.* **27**:33–36.

Friedman, P. A., and Platzer, E. G., 1980, Interaction of anthelmintic benzimidazoles with *Ascaris suum* embryonic tubulin, *Biochim. Biophys. Acta* **630**:271–278.

Gilles, H. M., 1981, The treatment of schistosomiasis, *J. Antimicrob. Chemother.* **7**:113–115.

Gönnert, R., and Andrews, P., 1977, Praziquantel, a new broad-spectrum antischistosomal agent, *Z. Parasitenkd.* **52**:129–150.

Goodwin, L. G., 1980, New drugs for old diseases, *Trans. R. Soc. Trop. Med. Hyg.* **74**:1–7.

Guru, P. Y., Warhurst, D. C., Harris, A., and Phillipson, J. D., 1983, Antimalarial activity of bruceantin *in vitro, Ann. Trop. Med. Parasitol.* **77**:433–435.

Hammond, D. J., Burchell, J. R., and Pudney, M., 1985, Inhibition of pyrimidine biosynthesis de novo in *Plasmodium falciparum* by 2-(4-t-butylcyclohexyl)-3-hydroxy-1,4-naphthoquinone *in vitro, Mol. Biochem. Parasitol.* **14**:97–109.

Hudson, A. T., Randall, A. W., Fry, M., Ginger, C. D., Hill, B., Latter, V. S., McHardy, N., and Williams, R. B., 1985, Novel anti-malarial hydroxynaphthoquinones with potent broad spectrum anti-protozoal activity, *Parasitology* **90**:45–55.

Inselburg, J., 1984, Induction and selection of drug resistant mutants of *Plasmodium falciparum, Mol. Biochem. Parasitol.* **10**:89–98.

Inselburg, J., 1985, Induction and isolation of artemisinine-resistant mutants of *Plasmodium falciparum, Am. J. Trop. Med. Hyg.* **34**:417–418.

Jiang, J.-B., Jacobs, G., Liang, D.-S., and Aikawa, M., 1985, Qinghaosu-induced changes in the morphology of *Plasmodium inui, Am. J. Trop. Med. Hyg.* **34**:424–428.

Kan, S. C., and Siddiqui, W. A., 1979, Comparative studies of dihydrofolate reductases from *Plasmodium falciparum* and *Aotus trivirgatus*, *J. Protozool.* **26:**660–664.

Kass, I. S., Wang, C. C., Walrond, J. P., and Stretton, A. O. W., 1980, Avermectin B_{1a}, a paralyzing anthelmintic that affects interneurons and inhibitory motoneurons in *Ascaris*, *Proc. Natl. Acad. Sci. USA* **77:**6211–6215.

Klayman, D. L., 1985, Qinghaosu (artemisinin): An antimalarial drug from China, *Science* **228:**1049–1055.

Königk, E., and Putfarken, B., 1983, Inhibition of ornithine decarboxylase of *in vitro* cultured *Plasmodium falciparum* by chloroquine, *Tropenmed. Parasitol.* **34:**1–3.

Krotoski, W. A., 1980, Frequency of relapse and primaquine resistance in Southeast Asian vivax malaria, *N. Engl. J. Med.* **303:**587.

Lanners, H. N., and Trager, W., 1984, Intranuclear structures in pyrimethamine-resistant isolates of the malaria parasite *Plasmodium falciparum*, *Cell Biol. Int. Rep.* **8:**221–225.

McCutchan, T. F., Welsh, J. A., Dame, J. B., Quakyi, I. A., Graves, P. M., Drake, J. C., and Allegra, C. J., 1984, Mechanism of pyrimethamine resistance in recent isolates of *Plasmodium falciparum*, *Antimicrob. Agents Chemother.* **26:**656–659.

Mansour, T. E., 1979, Chemotherapy of parasitic worms: New biochemical strategies, *Science* **205:**462–469.

Marr, J. J., Berens, R. L., Nelson, D. J., Krenitsky, T. A., Spector, T., LaFon, S. W., and Elion, G. B., 1982, Antileishmanial action of 4-thiapyrazolo(3,4-*d*)pyrimidine and its ribonucleoside: Biological effects and metabolism, *Biochem. Pharmacol.* **31:**143–148.

Moreno, S. N. J., Mason, R. P., and Docampo, R., 1984, Distinct reduction of nitrofurans and metronidazole to free radical metabolites by *Tritrichomonas feotus* hydrogenosomal and cytosolic enzymes, *J. Biol. Chem.* **259:**8252–8259.

Moreno, S. N. J., Mason, R. P., and Docampo, R., 1984, Reduction of nifurtimox and nitrofurantoin to free radical metabolites by rat liver mitochondria, *J. Biol. Chem.* **259:**6298–6305.

Müller, M., 1983, Mode of action of metonidazole on anaerobic bacteria and protozoa, *Surgery* **93:**165–171.

Nathan, H. C., Bacchi, C. J., Hutner, S. H., Rescigno, D., McCann, P. P., and Sjoerdsma, A., 1981, Antagonism by polyamines of the curative effects of alpha-difluoromethylornithine in *Trypanosoma brucei brucei* infections, *Biochem. Pharmacol.* **30:**3010–3013.

Neal, R. A., 1983, Experimental amoebiasis and the development of anti-amoebic compounds, *Parasitology* **86:**175–191.

Nguyen-Dinh, P., and Trager, W., 1980, *Plasmodium falciparum in vitro:* Determination of chloroquine sensitivity of three new strains by a modified 48-hour test, *Am. J. Trop. Med. Hyg.* **29:**339–342.

Peters, W., 1970, *Chemotherapy and Drug Resistance in Malaria,* Academic Press, New York.

Peters, W., and Richards, W. H. G. (eds.), 1984, *Antimalarial Drugs I. Biological Background, Experimental Methods, and Drug Resistance,* Springer-Verlag, Berlin.

Peters, W., and Richards, W. H. G. (eds.), 1984, *Antimalarial Drugs II. Current Antimalarials and New Drug Developments,* Springer-Verlag, Berlin.

Popiel, I., and Erasmus, D. A., 1984, *Schistosoma mansoni:* Ultrastructure of adults from mice treated with oxamniquine, *Exp. Parasitol.* **58:**254–262.

Qinghaosu Antimalaria Coordinating Research Group, 1984, Antimalaria studies on qinghaosu, *Chin. Med. J.* **92:**811–816.

Ramakrishnan, S. P., Prakash, S. and Choudury, D. S., 1957, Studies on *Plasmodium berghei,* Vincke and Lips 1948. XXIV. Selection of a chloroquine-resistant strain, *Indian J. Malariol.* **11:**213–220.

Rathod, P. K., and Reyes, P., 1983, Orotidylate-metabolizing enzymes of the human malarial parasite, *Plasmodium falciparum,* differ from host cell enzymes, *J. Biol. Chem.* **258:**2852–2855.

Rogers, S. H., and Bueding, E., 1971, Hycanthone resistance: Development in *Schistosoma mansoni, Science* **172:**1057–1058.

Scheibel, L. W., 1984, *In vitro* inhibition of the human malaria parasite by selected lipophilic chelators, in: *The Red Cell,* Sixth Ann Arbor Conference, Liss, New York, pp. 377–394.

Scheibel, L. W., Bueding, E., Fish, W. R., and Hawkins, J. T., 1984, Protease inhibitors and antimalarial effects, in: *Malaria and the Red Cell* (J. W. Eaton and G. J. Brewer, eds.), Liss, New York, pp. 131–142.

Schmidt, L. H., Crosby, R., Rasco, J., and Vaughan, D., 1978, Antimalarial activities of various 4-quinoline-methanols with special attention to WR-142,490 (mefloquine), *Antimicrob. Agents Chemother.* **13:**1011–1030.

Shapiro, A., Hutner, S. H., Katz, L., and Bacchi, C. J., 1981, Rapid *in vitro* prescreen for chelators as potential trypanocides based on growth of *Crithidia fasciculata, J. Protozool.* **28:**370–377.

Sjoerdsma, A., and Schechter, P. J., 1984, Chemotherapeutic implications of polyamine biosynthesis inhibition, *Clin. Pharmacol. Ther.* **35:**287–300.

Sucharit, S., Surathin, K., Tumrasvin, W., and Sucharit, P., 1977, Chloroquine-resistant *Plasmodium falciparum* in Thailand—susceptiblity of Anopheles, *J. Med. Assoc. Thailand* **60:**648–654.

Talib, V. H., Kiran, P. C., Talib, N. J., and Choudhury, M., 1979, Chloroquine-resistant *Plasmodium vivax* malaria in infancy and childhood, *Indian J. Pediatr.* **46:**158–162.

Targett, G. A. T., 1984, Interactions between chemotherapy and immunity, in: *Antimalarial Drugs I. Biological Background, Experimental Methods, and Drug Resistance* (W. Peters and W. H. G. Richards, eds.), Springer-Verlag, Berlin, pp. 331–348.

Trager, W., and Polonsky, J., 1981, Antimalarial activity of quassinoids against chloroquine-resistant *Plasmodium falciparum in vitro, Am. J. Trop. Med. Hyg.* **30:**531–537.

Trager, W., Tershakovec, M., Chiang, P. K., and Cantoni, G. L., 1980, *Plasmodium falciparum:* Antimalarial activity in culture of sinefungin and other methylation inhibitors, *Exp. Parasitol.* **50:**83–89.

Van Gelder, R., 1980, Malaria safari, *Nat. Hist.* **May:**10–18.

Wang, C. C., 1984, Parasite enzymes as potential targets for antiparasitic chemotherapy, *J. Med. Chem.* **27:**1–9.

Webster, H. K., and Whaun, J. M., 1982, Antimalarial properties of bredinin: Prediction based on identification of differences in human host–parasite purine metabolism, *J. Clin. Invest.* **70:**461–469.

Williamson, J., 1962, Chemotherapy and chemoprophylaxis of African trypanosomiasis, *Exp. Parasitol.* **12:**274–322.

Williamson, J., 1979, Effects of trypanocides on the fine structure of target organisms, *Pharmacol. Ther.* **7:**445–512.

Woolley, D. W., 1952, *A Study of Antimetabolites,* Wiley, New York.

World Health Organization Scientific Group, 1984, *Advances in Malaria Chemotherapy,* Technical Report Series 711, Geneva.

Yayon, A., Cabantchik, Z. I., and Ginsburg, H., 1985, Susceptibility of human malaria parasites to chloroquine is pH dependent, *Proc. Natl. Acad. Sci. USA* **82:**2784–2788.

CHAPTER 26

Ecology and Population Biology of Parasites. Sanitation and Vector Control

Good plumbing has done more for good health than has good medicine. Appropriate methods of sewage disposal and a clean water supply protect large populations from intestinal parasites transmitted by fecal contamination. Similarly, good housing tends to reduce the transmission of insect vector-borne diseases. This would be particularly true of Chagas' disease in South and Central America. The rather large bugs that are essential to the transmission of *Trypanosoma cruzi* live in thatched roofs and the cracks of mud houses (Fig. 26.1). They would find fewer dwelling places in a well-built house. The provision of such housing for the large populations concerned is, however, economically not feasible under present conditions, so that other methods for control must be used. With a disease like malaria, while good houses with screens do help to reduce transmission, they are not sufficient by themselves to interrupt it, especially under tropical conditions. Some anopheline mosquitoes bite people while they are outdoors in the early evening. Knowledge of the habits of the vector is essential. This becomes even more obvious with vectors that bite by day, such as tsetse flies or *Simulium*. People are unavoidably exposed to these in their ordinary daily activities. Workers clearing jungle are at special risk to infection with *Leishmania* spp. borne by sand flies infected from various animal reservoir hosts. People out camping or simply enjoying woods and fields may be bitten by ticks, again transmitting infectious agents they have acquired from various animal hosts. Such diseases are zoonoses; humans are only an incidental host in a cycle depending on wild animals. The number of these animals as well as the abundance of the vectors will both play a role in determining the incidence of the disease. With diseases of domestic animals, all of these essentially ecological factors become even more important. They are sometimes of such overriding significance that an experienced observer can tell simply by looking at the vegetation whether a particular region is likely to harbor a particular disease agent. Scrub typhus provides a good example. Four requirements must be fulfilled: the disease agent, *Rickettsia tsutsugamushi;* the vector, trombiculid mites; the small

FIGURE 26.1. Breeding place of *Triatoma infestans*, a vector of *Trypanosoma cruzi*, the cause of Chagas' disease. All stages of development of the bug are shown in this crack in the wall of a mud hut. (From Hoare, 1972, after Geigy and Herbig, 1955.)

mammalian hosts of the mites, chiefly of the genus *Rattus;* all three of these being found together in a stage in ecological succession of vegetation characterized by scrubby conditions on recently disturbed land. Appropriate modification of any one or several of these ecological factors could interfere with transmission of the disease. In fact, some of the greatest triumphs of preventive medicine have been in sanitation and in vector control. In each case an ecological modification breaks the chain of transmission. The sanitary disposal of human wastes prevents infection from cysts or eggs excreted in feces or urine. Eradication or merely a sufficient reduction in the population size of an essential vector species stops the transmission of a vector-borne disease.

Malaria, though still the world's leading infectious disease of people, at the same time provides the most outstanding examples of successful control and even local eradication by reduction in numbers of the vector mosquitoes. In the early work, soon after Ross and Grassi had shown that only mosquitoes of the genus *Anopheles* could transmit human malaria, emphasis was placed on elimination of the larval breeding places and killing of larvae, as by drainage of swamps and application of larvicides. These methods were successful but expensive. These were the methods used during the building of the Panama

Canal, and later in the control of malaria in Italy, Greece, and elsewhere in Europe, in Israel, and in the southeastern United States. In Europe it was observed that there were regions with *Anopheles maculipennis,* one of the major local vectors, but without malaria. It developed that there were races or subspecies of *A. maculipennis* that were zoophilic, preferring to feed on animals rather than people and accordingly not involved in the transmission of malaria. From this observation grew a whole study of species complexes, consisting of species or subspecies morphologically very similar (though sometimes distinguishable, as by egg characteristics) but differing markedly in behavior. Not only were some zoophilic and others anthropophilic, they differed also in choice of resting places and of breeding places. Detailed study of the bionomics of the vector species sometimes revealed simple ecological ways of control of their breeding, as by a small change in salinity of a larval habitat, or by introduction of larvivorous fish. Fascinating stories of this type are to be read in the book *Malaria in Europe* by L. W. Hackett. This book summarizes well the malaria situation in temperate zone regions before the advent of DDT.

DDT [1,1,1-trichloro-2,2-di(*p*-chlorophenyl) ethane] and the other residual insecticides depend for their efficacy on the behavior of the mosquitoes. Anopheline females that feed indoors generally rest after engorgement on the interior walls of the house. If the walls have been sprayed with a residual insecticide, this contact often suffices for the uptake of a lethal dose by the resting mosquito. If the daily mortality of female anophelines effected in this way is only 25%, it will suffice to interrupt transmission in many malarious regions. This cheap and relatively easy method, relying on periodic spraying of the interior of dwellings, greatly reduced malaria in many areas and eradicated it from some. With the most potent vectors in holoendemic regions, a daily mosquito mortality of 40 to 50% would be required, and this can only be achieved with very careful application of the insecticide. Furthermore, as already noted, the mosquitoes become resistant. The resistance is of at least two types: some with a true metabolic resistance, others with a change in behavior so that they no longer rest on the sprayed walls, in this way avoiding contact with the insecticide.

Even in the absence of resistance, house spraying with residual insecticide has given poor results in holoendemic regions such as the Sudan savanna in Africa. This was especially well demonstrated in the Garki Project in northern Nigeria. This trial was carried out with utmost care and attention to detail and without regard to cost. It showed that regular spraying with propoxur (*O*-isopropoxyphenyl methylcarbamate) over a 2-year period did indeed reduce the vectorial capacity (the risk of transmission) by 90%, but the prevalence of *P. falciparum* was reduced by only about 25%. Furthermore, in the villages with the highest transmission before treatment, a new equilibrium had been reached after 2 years, indicating that further spraying would not improve the result. It was concluded that this failure resulted from the very high baseline level of transmission and from the fact that the two vector species of the *A. gambiae* complex present in the region rested *outdoors* to a significant degree. This exophilic behavior is presumably genetically determined; hence, a significant proportion of the anophelines were never exposed to the insecticide. The vectorial capacity, though greatly reduced by propoxur, nevertheless remained well above its critical level, i.e., the minimum level required to maintain endemic falciparum malaria. With a baseline cumulative entomological inoculation rate reaching a maximum of 145 sporozoite-positive bites in 1 year (of which 132 were in the wet season), it can be seen that even a large reduction still leaves an ample margin of transmission. Not surprising under these conditions was the rapid resurgence of falciparum malaria after the cessation of spraying (Fig. 26.2).

Entomological and parasitological data of the kind gathered in the Garki Project can be used to test the adequacy of mathematical models for disease transmission and to improve these models. The models in turn can be used in attempts to devise more effective and better integrated methods of control. Ronald Ross was among the first to appreciate the possible utility of such models, and a number have been devised especially for malaria. The number and variety of the factors that have to be entered into such a mathematical model are illustrated in Table 26.1. It is clear that this attempts to take into consideration both the human and the vector populations, their extent of infection and their interactions over time. If more than one vector is present,

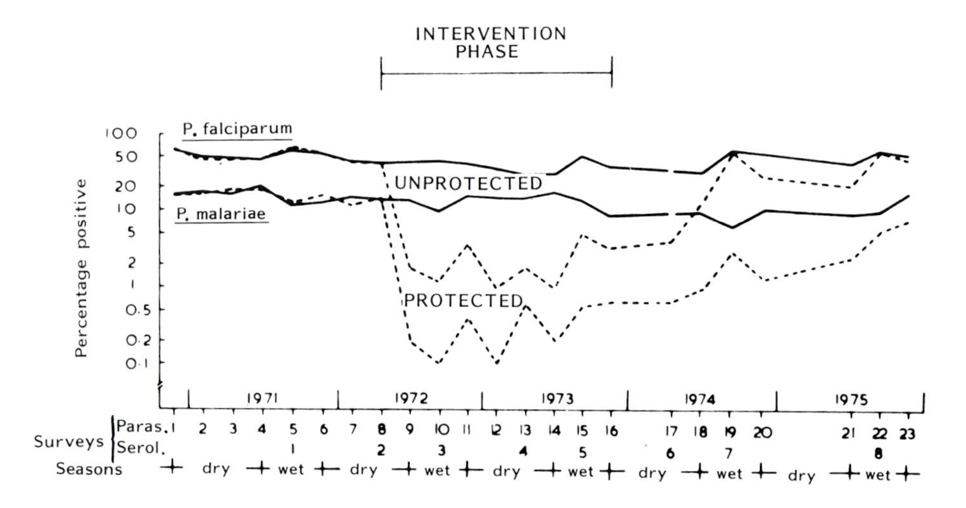

FIGURE 26.2. The resurgence of malaria after a nearly 2-year period of spraying with propoxur and mass drug administration. It will be seen that the percentage positive, which remained essentially constant for a 5-year period in the unprotected control area, decreased significantly during the period of intervention but returned to control level within 2 years. (From Molineaux and Gramiccia, 1980.)

TABLE 26.1. Symbols and Definitions Used in the Mathematical Model for Transmission in the Garki Project[a]

Symbol	Name	Definition and/or comments
a	Human-biting habit	No. of bloodmeals taken on humans by one vector in 1 day.
m	Relative density of vectors	No. of vectors per human, i.e., ratio between the size of the vector population and the size of the human population.
n	Extrinsic incubation period	Incubation in the vector.
P	Probability of surviving 1 day	Probability that the vector survives 1 day.
C	Vectoral capacity	$C = ma^2p^n/(-\ln p)$. This defines the number of infective contacts one person makes, through the vector population, per unit time.
g	Susceptibility	Probability that an infection results, given that at least one contact has occurred, in one time-unit.
h	Effective inoculation rate	Probability that a negative acquires the infection (becomes incubating), in one time-unit; $h = g\,[1-\exp(-Cy_1)]$.
N	Intrinsic incubation period	Incubation period in humans.
q_1, q_2, q_3	Probability of detection	Probabilities that the three kinds of positives (y_1, y_2, y_3) are detected by a standard parasitological examination. q_3 for "immune positives" (y_3) always smaller than q_1 and q_2.
r_1, r_2	Basic recovery rates of nonimmunes, immunes	Recovery rate from one infection of the y_2, y_3, respectively.
$R_1(h), R_2(h)$	Actual recovery rates of nonimmunes, immunes	Actual rate at which the y_2, y_3, recover, taking into account the superinfections resulting from a given h; $R_i(h) = h/[\exp(h/r_i)-1]$, $i = 1, 2$.
t	Time	$x_1(t)$ designates x_1 at time t, $x_1(t-N)$ designates x_1 at time $(t-N)$, and so on.
T_1, T_2	Expected duration of states y_2, y_3	$T_i = 1/R_i(h)$, $i = 1, 2$.
x_1, x_3	Nonimmune, immune negatives	Newborn are in nonimmune negative state x_1.
x_2, x_4	Nonimmune, immune incubating	These are inoculated at a rate of h to enter incubating class x_2 where they stay N days.
y_1, y_2, y_3	Three kinds of positives	These become positive and infectious to enter class y_1. When still positive but noninfective, they enter class y_2. May then recover and return to nonimmune negative x_1 or become an "immune

(*continued*)

TABLE 26.1. (*continued*)

Symbol	Name	Definition and/or comments
		through several cycles $x_1 \rightarrow x_2 \rightarrow y_1 \rightarrow y_2 \rightarrow x_1$ and so on or $y_3 \rightarrow x_3 \rightarrow x_4 \rightarrow y_3$. With increased inoculation rate, R_1 and R_2 decrease and more persons go the route $x_1 \rightarrow x_2 \rightarrow y_1 \rightarrow y_2 \rightarrow y_3$ and stay there or, if they recover, are reinoculated and return through x_4 to y_3.
z	Observed proportion positive	$q_1y_1 + q_2y_2 + q_3y_3$, or $\sum_{i=1}^{3} q_iy_i$
α_1	Rate of loss of infectivity	
α_2	Rate of acquisition of a high recovery rate	
δ	Death rate	For simplicity a single death rate adopted and birth rate made equal to this, giving a stationary population with an exponential age distribution.

[a] Modified from Molineaux and Gramiccia (1980).

the total vectorial capacity must be calculated from the sum of the individual vectorial capacities. The equation constituting the model can be tested by calculating the expected proportion of persons found positive for *P. falciparum* by examination of 200 fields of a standard thick blood film, as a function of age and time, given the vectorial capacity and the birth and death rates of the human population, and comparing them with the observed data. In the model used for the Garki Project the fit was fairly good, in part as a result of the fact that preparation of the model went along hand in hand with gathering and analysis of data from the field. Such models are helpful in attempts to decide such questions as: to what extent can available measures suffice for control; what additional information, as from a pilot trial, is needed; what could be expected from a new tool, such as a vaccine, or a better insecticide?

A close look at Table 26.1 reveals that most of the quantities used for the model cannot be measured very precisely. Thus, the parasite rate in people is determined from a standard examination of 200 fields of a thick blood film by a trained microscopist. It is easy to see how this would be subject to considerable error depending on the competence of the microscopist and even on whether he or she is fresh or tired. Even less precise has been the determination of the sporozoite rate among the mosquitoes. This depends on dissection of a large number of mosquitoes and microscopic examination of the salivary glands for sporozoites, a very laborious procedure and one which cannot determine the species of the sporozoites found. Under development are serological tests for parasite antigens or for parasite DNA that hold some promise of replacing these microscopic procedures and of providing more and better information. Especially likely to be useful is an immunological method based on the surface antigen of sporozoites. With this assay, whole mosquitoes can be used to determine their infection with a particular species of malaria. A two-site enzyme-linked immunosorbent assay (ELISA) has been devised that uses an extract of dried mosquitoes as antigen. The specificity and sensitivity of this assay are illustrated in experiments with mosquitoes infected with *P. falciparum* (Fig. 26.3).

In view of the environmental hazards that may accompany use of chemical insecticides against mosquitoes, and in view of the increasing prevalence of genetic resistance to these insecticides, more and more attention is being given to biological control methods. Some of these are old and of proven effectiveness under appropriate conditions, such as the use of larvivorous fish, notably species of *Gambusia* and *Poecilia.* Others are new and not yet widely tried. Especially promising is a material prepared from bacteria of the strain *Bacillus thuringiensis israelensis.* This is essentially an insecticide of microbial origin very active against mosquito larvae. Also highly effective against mosquito larvae is the juvenile hormone mimic methoprene. These two agents were successfully used against the yellow fever mosquito *Aedes aegypti,* together with spraying of interiors of houses and the elimination of breeding places, the last largely by community participation, in an integrated control program carried out by Marshall Laird on the Tuvalu islands in Polynesia. Whereas the toxic material from *B. t. israelensis,* like a chemical insecticide, must be applied at frequent intervals, other biological control agents under development against mosquitoes would be used as the living self-replicating organisms. Among these are the fungus *Culicinomyces clavisporus* and the nematode *Romanomermis culicivorax.* Despite their self-replication the indications are that both of these parasites of mosquito larvae have to be applied intermittently in overwhelming numbers. For this purpose it would be very advantageous if they could be cultivated apart from their living hosts, something that has not yet been achieved. The same is true for another potential agent for the biological control of mosquitoes, the microsporidian parasite *Nosema algerae.* Nematodes and microsporidia have already been used for control of agricultural pest insects. Work of this nature incidentally points to another broad field of application of studies in parasitism, one which I unfortunately cannot enlarge upon here. Biological control would also include genetic manipulation, such as the release of large numbers of sterile males or males carrying lethal genes, or attempts to overwhelm the natural vector population with a population resistant to the parasite concerned. None of these methods has yet proved practical against mosquitoes or other insect vectors of protozoan or helminthic parasites. The concept of integrated control has been applied mainly to control of insects, either agricultural or domestic

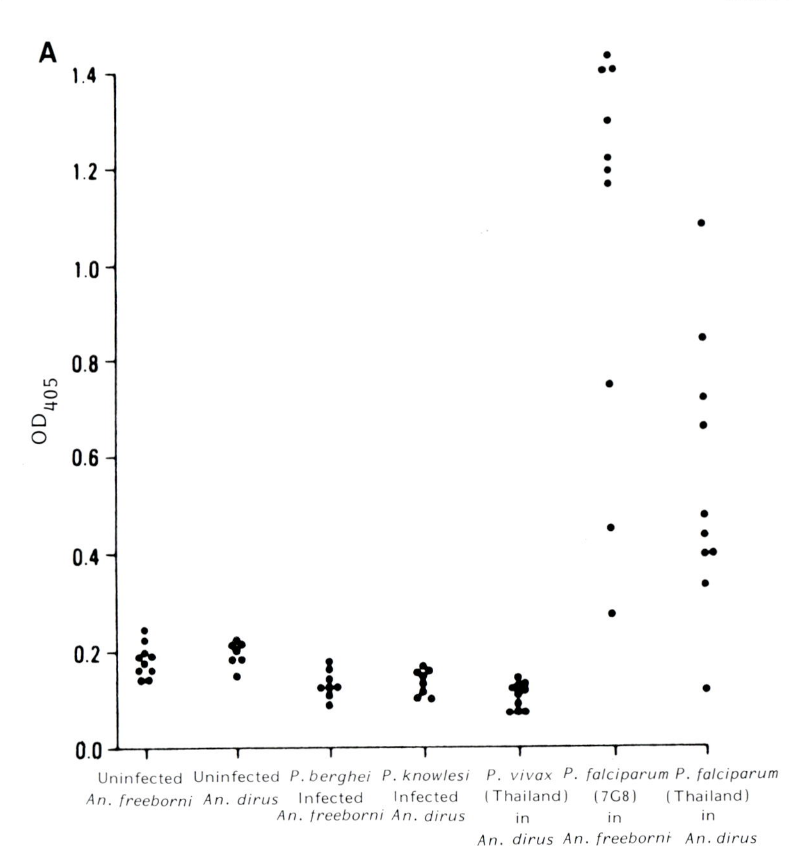

FIGURE 26.3. An ELISA test for identification of mosquitoes infected with *Plasmodium falciparum* sporozoites. Results with individual mosquitoes. (A) Specificity of the assay. Note that a positive reaction was obtained only with the mosquitoes harboring *P. falciparum* sporozoites.

pests or disease vectors. It implies the judicious and balanced use of chemical, environmental, and biological agents. This same concept, however, can equally well be applied to the control of a parasitic organism. With reference to malaria it would entail the combined use of all or a number of available methods, their choice depending on a knowledge of the local situation and resources. For example, one might combine antimosquito measures, such as larvivorous fish and indoor house spraying, with mass drug administration and the use of vaccines, if these are shown to be effective. Especially where drug-resistant *P. falciparum* is present, it would be important to attempt to reduce its transmission, as with a gametocytocidal drug or a gamete vaccine. There is here a need for a better gametocytocidal drug than primaquine (see Chapter 25). Vaccines are under experimental study that induce in the vertebrate host antibodies to the gametes. When these antibodies are then present with ingested blood in the midgut of a recently fed mosquito, they will inactivate

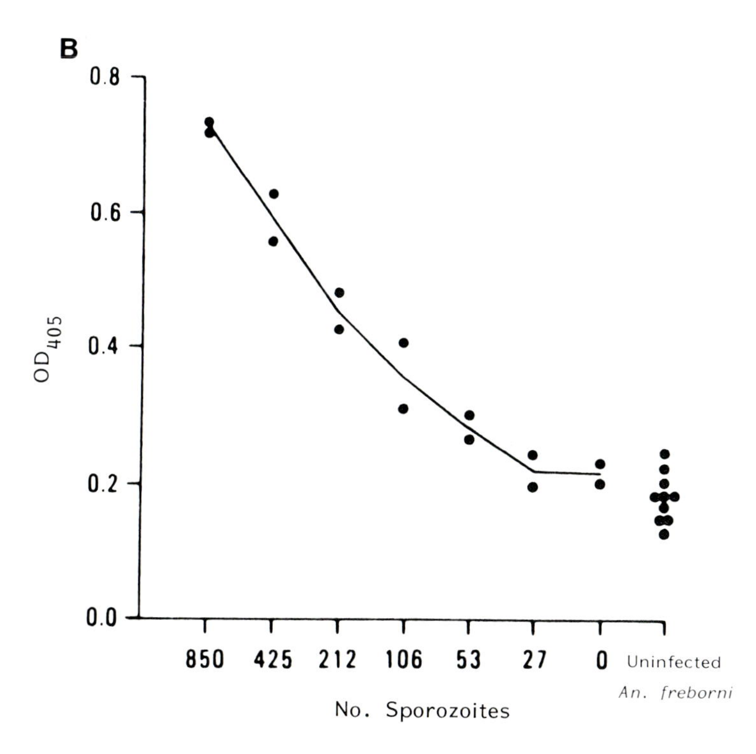

FIGURE 26.3. *(continued)* (B) Sensitivity of the assay. As few as 100 sporozoites could be detected. (From Burkot *et al.*, 1984.)

any malarial gametes that form from gametocytes taken up with the blood meal. In this way infection of the mosquito and subsequent transmission are blocked. Such a vaccine might be useful as an additional ingredient in a vaccine directed against the sporozoites and the erythrocytic stages (see Chapter 19).

Onchocerciasis, like malaria, is caused by a parasite restricted to humans as the definitive host and transmitted by an insect vector in which an obligate developmental cycle occurs (see Diagram XXII). The same kinds of factors enter into the population dynamics of this disease, but of course the details differ. Instead of the parasitemia determined from thick blood films, one has here the average microfilarial density determined from standard snips of skin. These skin microfilariae are the source of infection of the blackflies (*Simulium* spp.). The man–fly contact is here expressed as the annual biting rate (ABR). The ABR is the annual total of the monthly biting rates calculated by dividing the number of flies caught biting (under standardized conditions on the legs and feet of a collector) by the number of catching days and multiplying the quotient by the number of days in the month. Instead of the sporozoite rate, one has now the average number of infectious larvae per fly. Like the sporozoites of malaria, these are the stages that initiate the infection in people.

Infective larvae are those found in the head of the fly. The numbers of these found in the caught flies are used to define the annual transmission potential. This is the annual total of the monthly transmission potentials (MTP) obtained from the following formula:

$$\text{MTP} = \frac{\text{No. of days in month} \times \text{No. of infective larvae}}{\text{No. of days worked}} \times \frac{\text{No. of flies caught}}{\text{No. of flies dissected}}$$

With onchocerciasis the average density of microfilariae in the skin of an infected person decreases only slightly with age, and this may be in part a result of the premature death of individuals made blind by the disease. Eye lesions and subsequent blindness are the most serious consequences of onchocerciasis. In some West African villages afflicted with this disease, blind people are conspicuously numerous. The prevalence of eye lesions and of blindness increase with the average microfilarial density (Table 26.2). This in turn is directly related to the ABR and to the annual transmission potential. Dietz has used a mathematical model based on data from West African villages to predict the effects of three different vector control strategies: (1) 100% vector control for 9 years, starting in year 2, followed by return of vector density to its original level; (2) 100% vector control for 20 years, starting in year 2, followed again by return of vector density to its original level; (3) reduction of the original ABR of about 21,000 to 2000 in year 2, and maintenance of this level for 29 years. With either of the first two strategies, microfilarial density would return quickly to the precontrol level after cessation of control measures. Incidence of eye lesions and blindness at 30 years would be as low

TABLE 26.2. Comparison of Observed Density of Microfilariae of *Onchocerca volvulus* and of Onchocercal Eye Lesions and Blindness in Four West African Villages with the Predictions from a Mathematical Model[a,b]

Average microfilarial density		Prevalence of eye lesions		Prevalence of blindness	
Observed	Expected	Observed (%)	Expected (%)	Observed (%)	Expected (%)
12.3	16.1	2.4	0.5	0.0	0.0
23.3	17.2	4.8	0.6	0.0	0.0
28.0	28.5	8.3	3.1	0.2	0.0
41.5	51.3	14.7	17.5	2.1	0.4
46.3	59.4	15.2	21.3	2.6	2.4
54.3	64.0	20.9	23.3	4.5	4.8
66.0	69.9	25.0	25.4	7.6	7.0

[a] Modified from Dietz (1982).
[b] Note that the incidence of eye lesions and blindness rises with increasing microfilarial density. At lower densities the observed incidence of both was higher than that predicted by the model.

with strategy 3 of about 90% effective vector control for 29 years as with the much more difficult to implement strategy 2 of 100% vector control for 20 years (and considerably lower with either strategy than with strategy 1). Unlike the situation with malaria, for onchocerciasis there has been no simple and relatively safe method of mass chemotherapy, though this situation may change with further trials of the avermectins (see Chapter 25). No vaccine is in sight, and control of the disease must rest on control of *Simulium.* Here again an integrated approach, based on thorough knowledge of the bionomics of the vectors, is desirable.

What can be accomplished has been demonstrated by the Onchocerciasis Control Programme (OCP) of the World Health Organization in West Africa. This was organized in 1974 and covers an area of about 760,000 km^2 mostly in the countries of Mali and Burkina Faso (formerly Upper Volta). At the start of the program this area had close to 13 million people of whom 10% had onchocerciasis and 1% were blind. At most sites in the area, ABRs ranged from 4000 to over 8000. The control program relied on insecticide treatment of breeding sites in rivers and streams at appropriate time intervals with Temephos (also known as Abate), a material relatively nontoxic to fish and other nontarget fauna. This was mostly applied from airplanes or helicopters. By 1978 the ABRs at most sites in the treated region were less than 500, values over 1000 being found only at the periphery (and, as before, outside the treated zone) (Fig. 26.4). A reduction of the ABR to less than 1000 was considered satisfactory. Similarly, the annual transmission potential was reduced to less than 100 infective larvae per man-year. In keeping with these effects the prevalence of ocular onchocerciasis fell by about 28%. Most striking was the finding that in at least some regions, all of the children under 5 years old, and hence born after the start of vector control, were free from onchocerciasis. The overall diminution in mean microfilarial load, however, has been slow, presumably because of the longevity of the adult worm. After 5 years of the program, 40% of the villages were still hyperendemic, as against 67% before control, 39% were mesoendemic and 20% were hypoendemic, as compared to 24 and 9%, respectively, before control. Such results, though encouraging, nevertheless underline the great difficulty of adequately controlling such vector-borne infections by a method that relies on continued repeated application of insecticides. It is an expensive proposition. But it has already saved many from blindness and it is too early to evaluate its long-term impact.

Unlike malaria and onchocerciasis, some other vector-borne parasitic infections of humans also occur in animal reservoir hosts. Hence, these hosts have to be considered in relation to the ecology and epidemiology of these parasitoses. This is the situation with both African and South American trypanosomiasis and with the various kinds of leishmaniases. The leishmaniases caused by different species of *Leishmania* (Diagram III) provide examples covering the whole spectrum from zoonoses with only occasional human infections through those with frequent human cases to others where humans are the principal host and the disease is therefore anthroponotic. The epide-

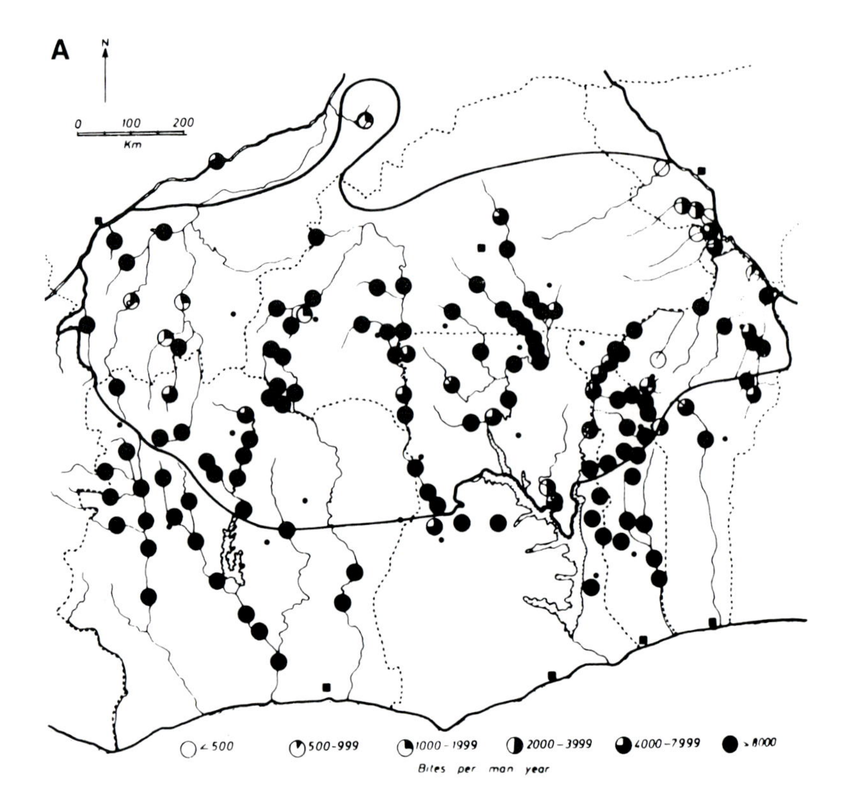

FIGURE 26.4. The Onchocerciasis Control Project in West Africa (region outlined by the heavy line). Annual biting rates before control (A) and after 3 years (in 1977–1978) (B). (From Walsh *et al.*, 1979.)

miology is complex, depending on the kinds of reservoir hosts and the species of vector sand flies and their behavior. There are regions where the infection is so intensely enzootic, and where the sand flies will readily feed on people, that anyone entering even for a short stay is highly likely to become infected. In such areas, workers on development projects are especially at risk; 50 to 100% may become infected. This is the case in some regions of the Irano-Turanian vegetational zone where *Leishmania major*, a cause of cutaneous leishmaniasis, is a parasite of rodents in whose burrows sand flies (*Phlebotomus* spp.) live and breed. This is also the case in some jungle regions of Brazil where parasites of the *L. braziliense* complex are maintained among the sloths and anteaters and various forest rodents by *Lutzomyia* spp. of sand flies. In the open steppes of the first environment, spraying with insecticides and destruction of the gerbil hosts can greatly reduce transmission. In the second environment, as the forest is cleared, the disease may acquire a peridomestic epidemiology involving dogs as the reservoir hosts and species of sand flies able to adapt to the new conditions. With the classical kala-azar or visceral

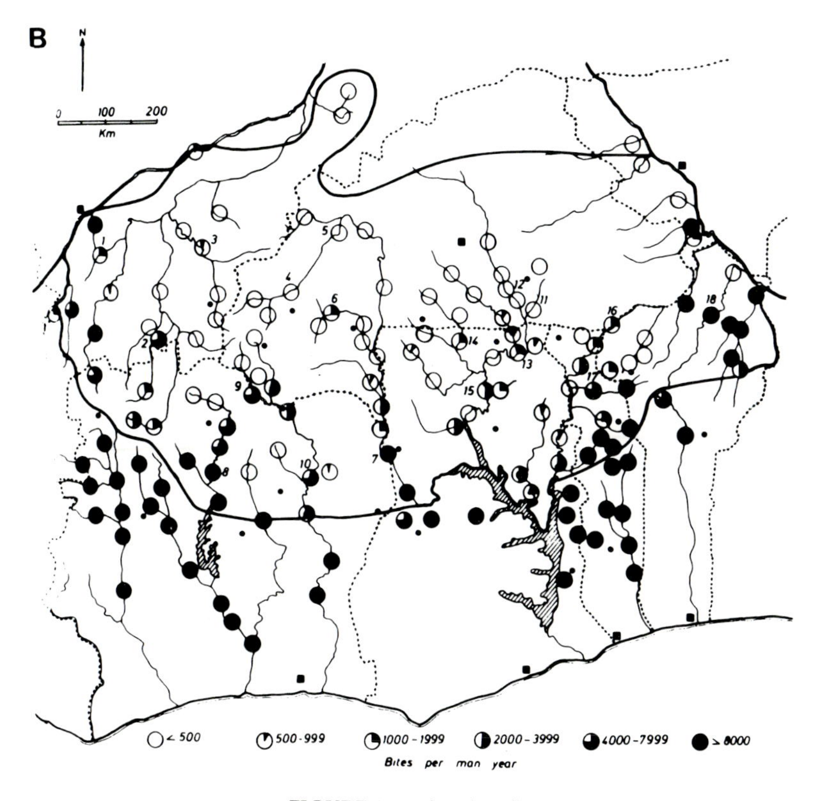

FIGURE 26.4. *(continued)*

leishmaniasis of China and India, caused by *L. donovani,* dogs were important reservoir hosts and this is also true with the Mediterranean form of this infection caused by *L. d. infantum.* With a drastic reduction in the dog population and with use of insecticides, the disease has been almost eliminated in China. It has also been greatly reduced in India, largely as an incidental result of insecticide application directed against mosquitoes. In many regions of Africa, the Middle East, Central Asia, South and Central America, both visceral and cutaneous leishmaniasis continue to be serious problems. The leishmaniases of the New World are particularly complex and only now beginning to be sorted out as to species, reservoir hosts, and vectors involved.

In Africa many different kinds of wild animals serve as reservoir hosts for the three species of trypanosomes that are pathogenic to cattle and other domestic animals: *Trypanosoma brucei* (Diagram V), *T. congolense,* and *T. vivax.* Transmission to cattle is dependent especially on the tsetse flies of the *Glossina morsitans* group. The abundance and distribution of these tsetses in turn depend on complex ecological factors including the abundance and the species of wild animals, the density of human populations, and human agricultural practices. The fascinating story of the impact of human activities and of other ecological factors, such as the rinderpest epidemics of the late 19th and early 20th century, on the rise and fall and rise again of tsetse populations, has been well told by John Ford in his book.

Despite many years of work and extensive attempts at control by insecticides and drugs, tsetse still infest about 10 million km^2 of Africa and about 30% of the 147 million cattle in this area are exposed to trypanosomiasis. As we have seen (Chapter 18), antigenic variation makes the development of a vaccine difficult if not unlikely. Adequate control by drugs is hampered by drug resistance. Control by insecticides has not been effective on a large scale. Extermination of the reservoir hosts does not seem desirable, even if it were practicable. Actually the most promising approach is the utilization of breeds of cattle that resemble the wild animal hosts in being able to survive and be productive despite presence of the trypanosomes and without the need for treatment. This is the case with the small cattle of the N'Dama and Muturu breeds. These animals of the species *Bos taurus* have long been known to survive in tsetse-infected regions. Their resistance to trypanosomes seems to involve both innate and acquired immunity and has been experimentally demonstrated under field as well as laboratory conditions. This is well shown by the field experiment summarized in Fig. 26.5. Ten N'Dama and ten Zebu (*B. indicus*) cattle, all 3-year-old females never previously exposed to trypanosomiasis, were exposed to natural challenge by tsetse flies (*Glossina morsitans submorsitans*). Within a few weeks all showed trypanosomes, but the levels of parasitemia were much higher in the Zebus and the anemia much more severe. Only one N'Dama died (from anthrax) as against all of the Zebus, seven of them within the first 14 weeks. Furthermore, the N'Dama had no abortions, produced five calves, and three more were pregnant at the termination of the experiment. The Zebu aborted in early and late pregnancy and produced no calves. Comparable results have been repeatedly obtained in both East and West Africa and one wonders why there are not more N'Dama and fewer Zebu in African herds; N'Dama and Muturu together represent only about 5% of the cattle population in the countries where tsetse occur. Apparently it has been generally assumed that these breeds were unproductive because they were smaller. More recent work refutes this view; with no tsetse challenge the productivity of N'Dama was found to be only 4% lower than that of Zebu. Even N'Dama can be severely affected by trypanosomiasis if they are exposed to a very high tsetse challenge. Hence, some measures for fly reduction would still be essential even as use of the N'Dama becomes more widespread. The physiological bases for the resistance of N'Dama are not known in detail. It seems likely that their immune response is triggered sooner and that differentiated nonmultiplying forms of the trypanosomes appear sooner. In any case, more extensive use of N'Dama cattle and further breeding programs for resistance seem to offer at present the best practical approach against trypanosomiasis of food animals in Africa.

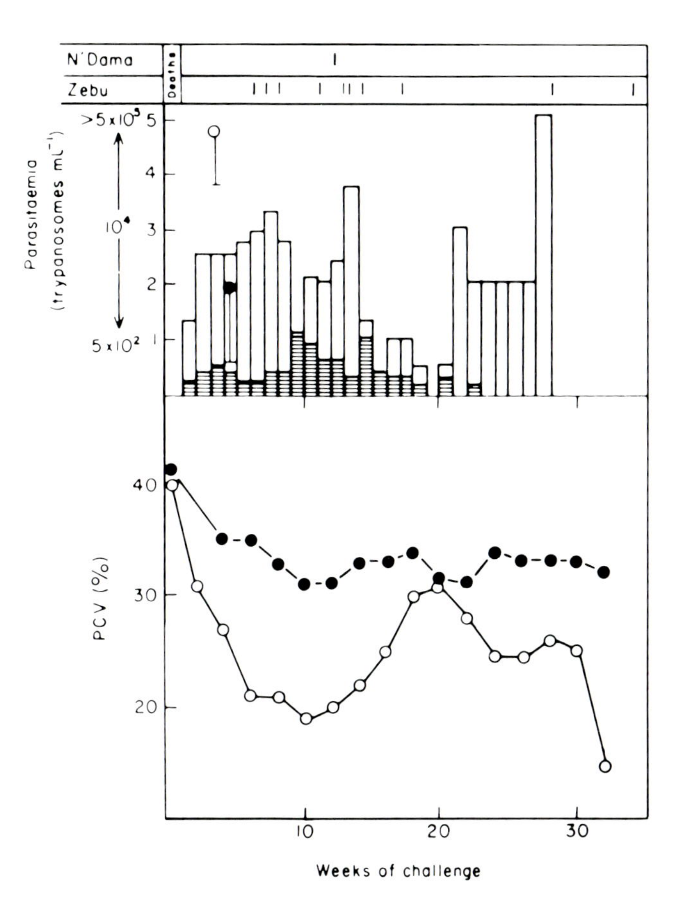

FIGURE 26.5. Comparison of trypanosomiasis in N'Dama and Zebu cattle exposed to natural tsetse challenge. At top, mortality. Middle shows average weekly parasitemia; hatched areas represent N'Dama. The level of the first peak of parasitemia plus one standard deviation is shown for the N'Dama (●) and the Zebu (○). The bottom part shows the packed cell volumes (PCV). (From Murray *et al.*, 1982.)

This obviously is not an approach to control of human trypanosomiasis. African sleeping sickness smolders on in a tsetse-infested belt of about 4 million km^2 with about 45 million people at risk. Outbreaks occur periodically, and the disease is still a major cause of depopulation, though not to the same extent as onchocerciasis. Two forms of human trypanosomiasis are found in Africa, both caused by subspecies of *T. brucei*. One form, that caused by *T. b. rhodesiense*, is a typical zoonosis. Like its sister subspecies *T. b. brucei*, this organism is maintained among the wild ruminants and other African mammals mainly by tsetse flies of the *G. morsitans* group. I have already discussed in Chapter 14 the remarkable difference that permits *T. rhodesiense* to survive in human serum whereas *T. brucei* is killed by human high-density lipoprotein. People run the risk of infection when they travel or work in regions where the animal reservoirs and tsetse flies are abundant. This would include visitors in game parks, as well as native inhabitants whose work or activities bring them into such regions for any length of time. The infection is also sometimes brought close to human habitation via infected bushbuck and even more by infections of domestic cattle. Subsequent human–fly–human transmission may then occur, possibly involving species of tsetse other than the *G. morsitans* group, with the sudden development of a localized epidemic.

It is just such human–fly–human transmission that accounts for the main features of the epidemiology of the second form of human trypanosomiasis in Africa, that caused by *T. b. gambiense.* This is associated especially with the so-called riverine tsetse of the *Glossina palpalis* group, *G. palpalis* and *G. tachinoides.* There is here a particularly interesting ecological situation responsible for human–fly contact that was demonstrated largely through the work of T. A. M. Nash in Nigeria. *G. palpalis* extends from the coastal forest northward and inland to progressively drier regions where it is closely associated with streams and moist areas. *G. tachinoides* is absent from the coast but extends farther into drier regions and is again associated with water. Especially during the dry season when water sources become fewer and more localized, people and tsetse must come to the same places. Intensive and so-called "personal" human–fly contact occurs. This situation is accentuated if, as is the case in much of northern Nigeria, wild animals have been driven out and domestic animals may be few, leaving humans as the chief source of blood for the tsetse. Both *G. palpalis* and *G. tachinoides* are highly adaptable flies able to live under such conditions. A typical place for such close contact with tsetse is provided by the "sacred groves," common in northern Nigeria and corresponding parts of other West African countries. These small islands of forest, about an acre or less, situated in the midst of fully cleared cultivated land, often surround the village water supply. They provide a place of worship, sometimes a cemetery, and a perfect habitat for tsetse. If one person infected with *T. gambiense* visits such a place, he or she may infect a number of flies that will subsequently transmit the infection. Nash provided an extreme example where a few *G. palpalis,* isolated during the dry season at a watering point in northern Nigeria, were responsible for an infection rate of 70% in a population of 43. On the other hand, in moist coastal forest regions the tsetse are not so restricted, have a wider range of hosts, and their contact with humans is impersonal. Human activities and increasing population density up to a certain level tend to restrict the tsetse to the same places as people and increase the human–fly contact. But with still higher population density, as in cities, tsetse habitats are destroyed, a situation quite different from that with anopheline mosquitoes. Sleeping sickness in West Africa is thus a disease of the villages in the savanna region.

Whether any animal reservoir is involved in the gambiense form of African sleeping sickness is a question that has been long debated and generally answered in the negative. This trypanosomiasis appears relatively well adapted to the human host. It produces a chronic disease of long duration providing ample opportunity for fly infection; the rhodesiense form on the other hand tends to be acute, in keeping with the fact that its natural hosts are the wild ruminants. Just the same, recent work has shown that domestic animals, pigs in particular, are suitable hosts for *T. gambiense*. Some animal reservoirs might or might not be associated with the continuing endemic foci of the disease that appeared after the great West African epidemic of the 1920s to 1940s receded. These foci persist even now, despite considerable efforts to eliminate them. They can serve as sources of epidemics, particularly at times of civil disturbances. Until the basis for these foci is understood and they can be eradicated, they must be kept under continual surveillance.

South American trypanosomiasis, or Chagas' disease, caused by *T. cruzi* (see Diagram IV and Chapter 20), provides another example of a zoonosis, but quite different from that provided by the rhodesiense form of African trypanosomiasis. Many kinds of domestic as well as wild animals are susceptible to *T. cruzi*, and this reservoir of infection is maintained by transmission by various reduviid bugs. Humans, however, are not an infrequent accidental host; on the contrary, so many people are infected that this is the most widespread and dangerous disease of Latin America. Both mammals and triatomine bugs infected with *T. cruzi* have been found in the United States but autochthonous human cases have not been reported north of Mexico. This probably results from the presence in northern regions of different species of bugs less likely to live in houses, and from the different type of housing, both factors greatly reducing the likelihood of human contact with the bugs (see Fig. 26.1). Unlike tsetse flies or mosquitoes, these reduviid bugs live and breed in houses where, like the common bedbug, they feed on the inhabitants at night while they are asleep. South American bugs typically defecate while they are feeding, depositing infective metacyclic trypanosomes on skin or mucous membranes, where they can enter the host. North American species tend to finish feeding and move away from the host before defecating, making transmission much less likely. The main epidemiological factor in the maintenance of Chagas' disease among people is the presence of species of triatomine bugs that have become domiciliary. Thus, the disease may be brought in originally from outside, as in infected rats present in palm leaves or other material used for roofing. But it then becomes established in the domiciliated bugs. Species such as *Triatoma infestans, Panstrongylus megistus,* and *Rhodnius prolixus* are especially adapted to living in houses. Such insects may live over a year, and once infected with *T. cruzi* continue to transmit the parasites to people and domestic animals. In this situation dogs and cats are especially important as reservoir hosts; in some areas about 20% have been found infected. As indicated at the beginning of this chapter, Chagas' disease could be eliminated by the provision of appropriate housing. For the very large populations concerned and with the economic conditions obtaining in Latin America, this is not likely to be accomplished in the foreseeable future. Control of the vectors by insecticide application is difficult; it requires very thorough application but it can be effective. The γ isomer of hexachlorocyclohexane is considered best; it seems to be the only residual chlorinated insecticide not inactivated by the type of mud used for houses in some parts of Brazil. This, together with education of the people as to the great importance of eliminating the bugs from their houses, is the only practical approach to control at present available.

Education and human behavior are of prime importance in relation to other major parasitic diseases having varied ecology and epidemiology. One of these is schistosomiasis. In this infection, as in those I have been considering, transmission is effected via an invertebrate vector in which the parasite must undergo a developmental cycle in order to become infective to its definitive host. The details are, however, very different (see Diagram XIII). The vector, a snail, becomes infected by a free-swimming form that hatches from eggs excreted by the human host with either feces (for *Schistosoma mansoni* and *S. japonicum*) or urine (for *S. haematobium*). After development in the snail, it is again a free-swimming form that will penetrate through the skin of a person entering the water. Hence, the important factor in transmission is not human–vector contact but rather human–water contact. The likelihood of acquiring infection will additionally depend on the numbers of infected snails, and this in turn will depend on the number of eggs getting into the water. This last number is determined by fecal contamination and by the population and the number of worms it harbors. As with insect-transmitted diseases, all of these parameters can potentially be measured and incorporated into a mathematical model. Several such models for schistosomiasis have been constructed. Although none is of high precision, they are useful in clarifying relationships between components of this complex system. One relatively simple model, applied to data on *S. haematobium* in a region in Tanzania and also to data on *S. japonicum* in a region in the Philippines, indicated that in both, human immunity was the most important natural factor limiting transmission (see also Chapter 21). Furthermore, the analysis indicated that with *S. haematobium*, immunity to further cercarial penetration became effective after two infections, not after only one. The models are also useful in attempts to predict the relative impacts of different control strategies.

It is clear that the chain of transmission of schistosomiasis can be broken by avoidance of contact with infected water, as by use of a safe water supply, or by proper disposal of human waste so that the snails do not become infected. Both of these methods involve human behavior. Not only must the proper facilities be provided, but in addition the population must be educated in their use. In practice, this is not easy. Even in China, with its excellent system of both education and propaganda at the village level, schistosomiasis has been eradicated only north of the Yangtze river. Elsewhere it has been reduced but is still a problem. The species present in China, *S. japonicum*, infects domestic animals as well as people, making its control even more difficult than that of the other two species, for which humans are the only host. The chain of transmission can also be broken by chemotherapy of infected individuals. This requires case-finding and again the educated cooperation of the local populace. With the several excellent new drugs now available (see Chapter 25), this is a cost-effective approach. In particular, there are indications that one need only treat those individuals who (for reasons not understood) have heavy infection and produce most of the eggs. Finally, transmission can be stopped by elimination of the snails, as with a molluscicide. One or two useful synthetic molluscicides are available but are expensive. A strong case can be made for the use of endod, a natural product of the soapberry plant *Phytolacca dodecandra*. This plant, which grows throughout Africa, produces a type of saponin that is most concentrated in the pericarp of the immature fruit. In trials in Ethiopia, application of the crude ground berries reduced the snail populations enough so that the incidence of schistosomiasis in children 1–5 years old declined by 85%. This compares very favorably with results obtained with much more expensive synthetic molluscicides such as Bayluscide. Strains of *Phytolacca* showing rapid growth, resistance to insect pests and drought, high yield of berries, and high potency are available and constitute a promising means of snail control. A few studies that have been done to compare snail control with chemotherapy and with a safe water supply show that each of these gives about the same extent of control (Fig. 26.6). Once again, an integrated approach would be most effective using all available methods in relation to the local situation. Such an

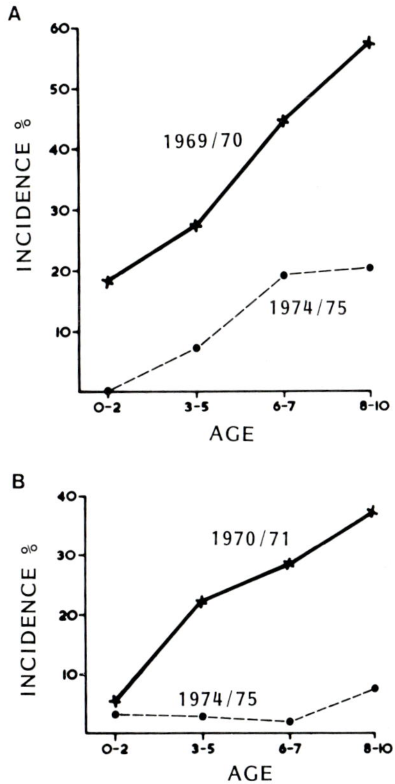

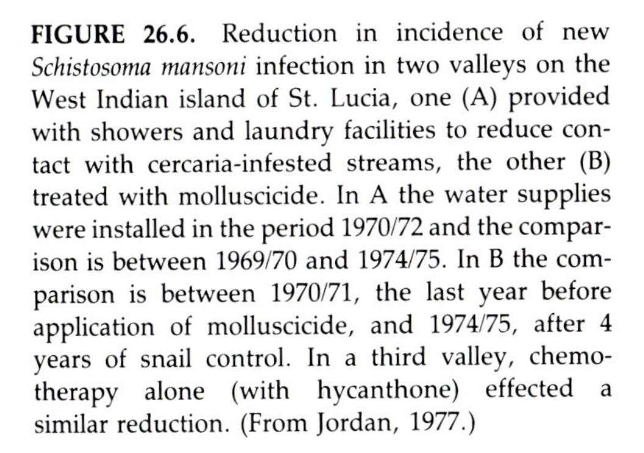

FIGURE 26.6. Reduction in incidence of new *Schistosoma mansoni* infection in two valleys on the West Indian island of St. Lucia, one (A) provided with showers and laundry facilities to reduce contact with cercaria-infested streams, the other (B) treated with molluscicide. In A the water supplies were installed in the period 1970/72 and the comparison is between 1969/70 and 1974/75. In B the comparison is between 1970/71, the last year before application of molluscicide, and 1974/75, after 4 years of snail control. In a third valley, chemotherapy alone (with hycanthone) effected a similar reduction. (From Jordan, 1977.)

integrated approach must include education in order to obtain the cooperation of the people concerned.

A safe water supply, for bathing and laundry as well as for drinking, and sanitation, i.e., disposal of feces in such a way as to minimize contamination of food, water, and the environment, are most effective with a population that understands their significance to public and individual well-being. Together they reduce the incidence of all infections involving transmission via a stage excreted with feces or urine. Among the common parasitic diseases of humans one finds here not only schistosomiasis but also amebic dysentery, giardiasis, hookworm disease, ascariasis and other intestinal helminthic infections. In all of these an egg or a cyst is passed with the feces and is either resistant and directly infective (as for *Entamoeba*, *Giardia*, and *Ascaris*) or gives rise to subsequent stages that are infective (as for *Schistosoma* and the hookworms *Necator* and *Ancylostoma*). Hookworm remains today as a widespread infection and a significant cause of morbidity in most developing countries. In 1975 it was estimated that 900 million people were infected out of the world population of 3900 million. In these regions, adequate sanitation does not exist, especially in rural areas, and soil is contaminated with feces which decay rapidly but leave the soil infested with the third-stage infective larvae ready to penetrate through the skin of a person walking by (Diagram XX). Transmission is of course facilitated if, as is often the case, a particular area is used by a number of people for defecation purposes. The survival of the infective larvae depends especially on adequate humidity and this is enhanced by shading from direct sunlight, as is usually the situation in such an area.

The life expectancy of the infective larvae (usually estimated at about 5 days for *Necator americanus*) is only one of a whole series of parameters that affect worm populations. Others are human population density, human life expectancy, life expectancy of the mature worms (about 3–4 years), maturation time in the human host (6 weeks), proportion of infective larvae that penetrate the host and survive to maturity (estimated at 0.1), egg production per female worm (15,000 per day), time from egg release to development of the third-stage infective larvae (5 days), proportion of female worms in the population (0.5), and basic reproductive rate (which must be over 1 and is usually estimated at 2–3). There is an additional less obvious but very interesting parameter—the degree of worm aggregation within the host population. This may be very marked, the majority of worms occurring in a relatively small proportion of the host population; e.g., in one survey 60% of the worms in less than 10% of the people. This predisposition of certain individuals to heavy infection is well shown by following fecal egg counts in the same individuals before and after anthelmintic treatment. As they become reinfected, the same people who had exceptionally heavy infections before treatment tend to again develop such infections (Fig. 26.7). This situation must be the result of a number of factors such as the habits of the individuals, their personal hygiene, their innate and acquired immunity, and the fecundity of the worms they harbor (since intensity of infection is measured by egg output). Such patterns

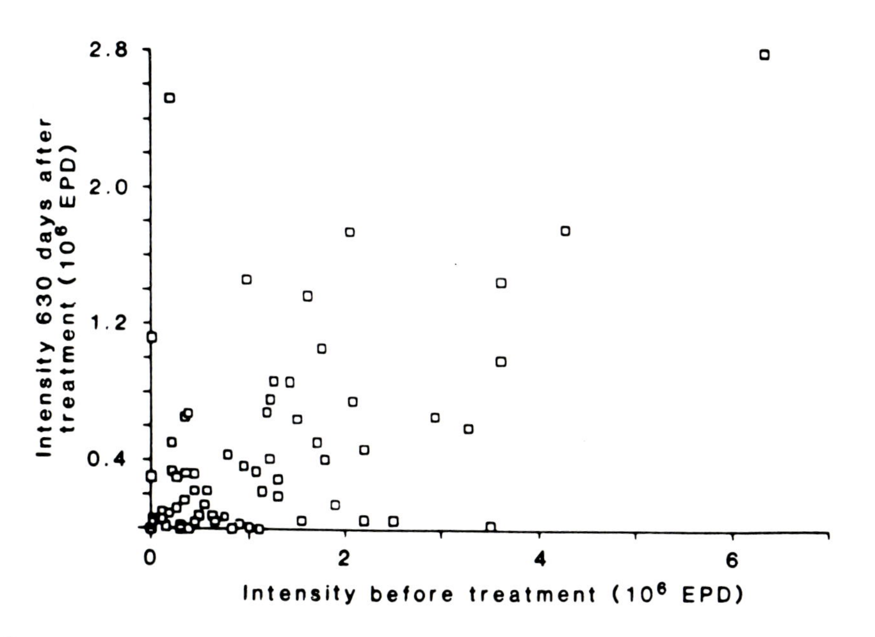

FIGURE 26.7. The association between fecal egg counts (EPD) before treatment and those 630 days after treatment for a sample of patients in a West Bengal village. Kendall's rank correlation coefficient $\tau = 0.33$, $n = 65$, $p < 0.005$. (From Schad and Anderson, 1985.)

of worm numbers per person are well described empirically by a negative binomial probability distribution:

$$p(a) = 1 - \left[1 + \frac{M(a)}{k}\right]^{-k} \tag{1}$$

where $p(a)$ represents prevalence at age a, $M(a)$ is the mean worm burden, which rises rapidly with host age (a), and k is the inverse measure of the degree of worm aggregation. For hookworm this ranges from 0.01 to 0.6 with an average value of 0.34. This uneven distribution of the worm burden, which is true also of schistosomiasis and ascariasis, suggests that it might be both economic and effective to use chemotherapy only in this small proportion of individuals harboring the most worms, rather than on a mass basis.

Certainly mass chemotherapy, even though it may temporarily reduce worm numbers and prevalence apparently to 0, has little lasting effect unless it is continued with high frequency and intensity for a number of years. This would be predicted from both equation (1) above for worm prevalence (usually expressed as % infection) and from the following equation for the mean worm burden:

$$M(a) = \Lambda\ell(1 - e^{-a/\ell}) \tag{2}$$

where Λ represents the "force of infection" and ℓ denotes the life expectancy of the adult parasites. This is also what has been repeatedly observed (Fig. 26.8).

We are led once more to the conclusion that a clean water supply and sanitation are in the long run the most economical and most effective methods for control of these feces-borne parasites. But this is true only if, and this bears repeating, the people at risk have been educated to the point where they understand how to use the facilities provided and why they are important. The difficulty of effecting such cultural changes should not be underestimated. It is as difficult in developed as in developing countries; one need only note how hard it is to get people to stop smoking cigarettes despite clear evidence of its harmfulness. Chemotherapy on either a mass or a selective basis will therefore continue to have utility for short-term control of helminthic infections of people.

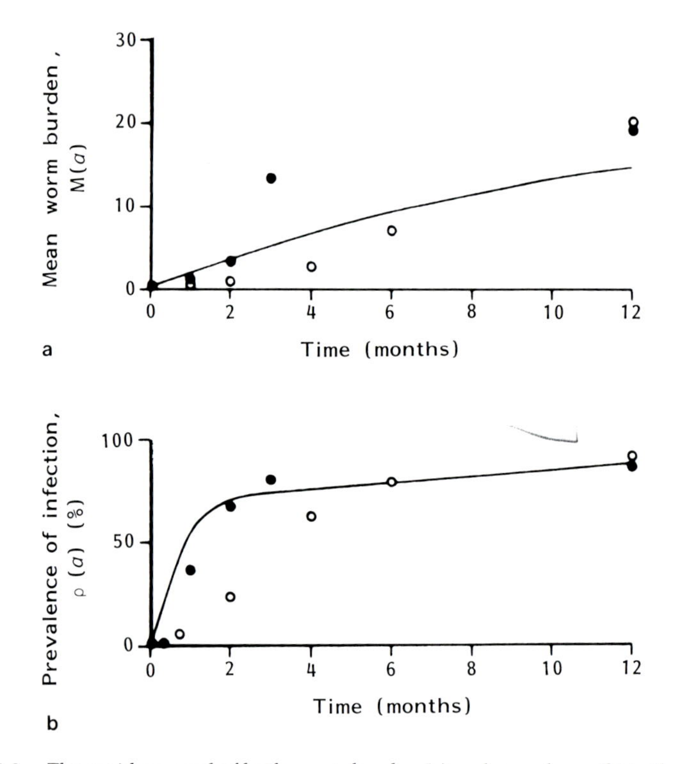

FIGURE 26.8. The rapid reversal of both worm burden (a) and prevalence (b) to the preexisting levels within a year after mass chemotherapy. The open and solid circles are the data for *Ascaris* infections from two separate studies. The solid lines are the predictions from equations (1) and (2) in the text, with parameter values as follows: basic reproductive rate (R) = 4.3, ℓ = 1 year, M = 22, k = 0.57, and z (inverse measure of the magnitude of density-dependent constraints on worm fecundity, a factor entering into the determination of Λ) = 0.96. (From Anderson, 1982b.)

Chemotherapy is the principal method for control of feces-borne helminthic infections of domestic animals. For its most effective use it is important to understand the population biology of the parasite. *Ostertagia ostertagi,* one of the most harmful nematode parasites of cattle in the temperate zones, provides an interesting example. Like *Haemonchus contortus* of sheep (Diagram XXI), this parasite of the abomasum has a direct life cycle. Eggs excreted in the feces produce free-living larvae that feed on bacteria and molt twice to give third-stage infective larvae. These nonfeeding forms retain the sheath of the previous second instar and are very long-lived and resistant to external conditions. They migrate away from the fecal pat into the pasture; when ingested by a grazing animal they initiate the parasitic development. After two more molts and maturation of the males and females, egg production begins about 3 weeks after infection. The population dynamics of the free-living stages are determined by density-independent factors, such as soil surface moisture and temperature, but those of the parasitic stages are dominated by the immunological status of the host and are therefore density-dependent. Whether calves are infected once or subjected to continual daily infection, the same pattern of daily fecal egg output is seen. It rises rapidly at first to reach a peak 2–3 weeks after patency; then it falls in an exponential manner. This pattern results from immunological responses that effect a reduction in fecundity of the female worms, an increase in the death rate of mature worms, and a reduction in the proportion of newly entering third-stage larvae able to establish themselves in the abomasal mucosa. Populations of *O. ostertagi* are furthermore influenced by a phenomenon I have not previously discussed: arrested development or diapause. Whereas the newly molted fourth stage (first parasitic stage) larvae ordinarily grow rapidly from about 1 mm in length to 3 to 4 mm, under certain circumstances their growth is arrested; they remain about 1.1 mm long. Such arrested larvae accumulate in large numbers in infected calves. Arrestment is at a maximum just before periods of unfavorable environmental conditions, such as winter cold, summer drought, or a housing period. In northern Europe and North America this occurs in autumn and winter; in the southern United States, Australia, and New Zealand, in the spring and summer. This arrestment is clearly of adaptive value to the parasite. Like diapause in free-living animals, it helps to carry the population through periods of unfavorable conditions. Furthermore, since development is arrested very early in the parasitic cycle, the arrested larvae induce no effective immunity and very few die. They then resume development and reach the adult stage when the animals are again grazing and conditions for transmission are again favorable. Nothing is known as to the physiological mechanism determining arrest and resumption of development; this would be part of a study of the whole problem of diapause in plants and animals. It is known that strains of the nematode differ in their propensity to arrest. In any case the number of arrested larvae is a result of the extent of exposure to infective larvae in the previous season and in turn determines the severity of the effects of the worms when they mature. This also determines the extent of the infestation of the pasture, especially if the intervening conditions (winter cold or summer drought) have been sufficiently severe so as to eliminate most of the free-living infective larvae that were present.

A mathematical model has been formulated that takes into account: duration of infection in calves, development time of the free-living noninfective stages (first two larval instars) (a function of temperature), period of arrestment, average maturation time, death rates of all the stages, proportion of infective larvae that become established, probability of larval arrestment, fecundity of the female worms. The model can be used, for example, to predict the extent of contamination of a pasture with third-stage infective larvae after different schedules of application of chemotherapy. The model predicts, and experience bears out, that therapeutic administration of anthelmintic at the time of the midsummer rise in infestation (in the climate of northern Europe) has no effect, whereas prophylactic administration at 3 weeks after the animals have been turned out and again a few weeks later greatly reduces the infestation of the pasture.

Bibliography

Anderson, R. M. (ed.), 1982a, *Population Dynamics of Infectious Diseases: Theory and Applications*, Chapman & Hall, London.

Anderson, R. M., 1982b, The population dynamics and control of hookworm and roundworm infections, in: *Population Dynamics of Infectious Diseases: Theory and Applications* (R. M. Anderson, ed.), Chapman & Hall, London, pp. 67–108.

Aron, J. L., and May, R. M., 1982, The population dynamics of malaria, in: *Population Dynamics of Infectious Diseases: Theory and Applications* (R. M. Anderson, ed.), Chapman & Hall, London, pp. 139–179.

Barbour, A. D., 1982, Schistosomiasis, in: *Population Dynamics of Infectious Diseases: Theory and Applications* (R. M. Anderson, ed.), Chapman & Hall, London, pp. 180–208.

Beausoleil, E. G., 1984, A review of present antimalaria activities in Africa, *Bull. WHO* **62:**13–17.

Bruce-Chwatt, L. J., 1984, Lessons learned from applied field research activities in Africa during the malaria eradication era, *Bull. WHO* **62**(Suppl.):19–29.

Bruce-Chwatt, L. J., 1985, *Essential Malariology* (2nd edition), Heinemann, London.

Burkot, T. R., Williams, J. L., and Schneider, I., 1984, Identification of *Plasmodium falciparum*-infected mosquitoes by a double antibody enzyme-linked immunosorbent assay, *Am. J. Trop. Med. Hyg.* **33:**783–788.

Cohen, J. E., 1976, Schistosomiasis: A human host–parasite system, in: *Theoretical Ecology: Principles and Applications* (R. M. May, ed.), Saunders, Philadelphia, pp. 237–256.

Dietz, K., 1982, The population dynamics of onchocerciasis, in: *Population Dynamics of Infectious Diseases: Theory and Applications* (R. M. Anderson, ed.), Chapman & Hall, London, pp. 209–241.

Ford, J., 1971, *The Role of the Trypanosomiases in African Ecology: A Study of the Tsetse Fly Problem*, Clarendon Press, Oxford.

Geigy, R., and Herbig, A., 1955, Erreger and Überträger tropischer Krankheiten, *Acta Trop. Suppl.* **6.**

Georgis, R., and Poinar, G., Jr., 1982, Field control of the strawberry root weevil, *Nemocestes incomptus*, by neoaplectanid nematodes (Steinernematidae: Nematode), *J. Invert. Path.* **43:**130–131.

Gillett, J. D., 1985, The behaviour of *Homo sapiens*, the forgotten factor in the transmission of tropical disease, *Trans. R. Soc. Trop. Med. Hyg.* **79:**12–20.

Glasgow, J. P., 1963, *The Distribution and Abundance of Tsetse*, Macmillan Co., New York.

Greany, P. D., Vinson, S. B., and Lewis, W. J., 1984, Insect parasitoids: Finding new opportunities for biological control, *BioScience* **34:**690–696.

Hackett, L. W., 1937, *Malaria in Europe*, Oxford University Press, London.

Harte, P. G., Rogers, N., and Targett, G. A. T., 1985, Vaccination with purified microgamete antigens prevents transmission of rodent malaria, *Nature* **316:**258–259.

Hazard, E. I., and Chapman, H. C., 1977, Microsporidian pathogens of Culicidae (mosquitoes), *Bull. WHO* **55**(Suppl. 1):63–67.

Henry, J. E., 1978, Microbial control of grasshoppers with *Nosema locustae* Canning, *Misc. Publ. Entomol. Soc. Am.* **11**(1):85–95.

Hill, D. H., and Esuruoso, G. O., 1976, Trypanosomiasis in N'Dama and white Fulani heifers exposed to natural infection on a ranch in western Nigeria, *Bull. Anim. Health Prod. Afr.* **24:**117–124.

Hoare, C. A., 1972, *The Trypanosomes of Mammals: A Zoological Monograph*, Blackwell, Oxford.

Jordan, P., 1977, Schistosomiasis—Research to control, *Am. J. Trop. Med. Hyg.* **26:**877–886.

Kennedy, C. R. (ed.), 1976, *Ecological Aspects of Parasitology*, North-Holland, Amsterdam.

Laird, M., 1985, New answers to malaria problems through vector control?, *Experientia* **41:**446–456.

Lambert, J. D. H., Wolde-Yohannas, L., and Makhubu, L., 1985, Endod: Potential for controlling schistosomiasis, *BioScience* **35:**364–366.

McKelvey, J. J., Jr., 1973, *Man Against Tsetse—Struggle for Africa*, Cornell University Press, Ithaca, N.Y.

Molineaux, L., and Gramiccia, G., 1980, The Garki Project: Research on the Epidemiology and Control of Malaria in the Sudan Savanna of West Africa, World Health Organization, Geneva.

Mulligan, H. W. (ed.), 1970, *The African Trypanosomiases*, Allen & Unwin, London.

Murray, M., Morrison, W. I., and Whitelaw, D. D., 1982, Host susceptibility to African trypanosomiasis: Trypanotolerance, *Adv. Parasitol.* **21:**1–68.

Murray, M., Trail, J. C. M., Davis, C. E., and Black, S. J., 1984, Genetic resistance to African trypanosomiasis, *J. Infect. Dis.* **149:**311–319.

Price, P. W., 1980, Evolutionary biology of parasites, *Monogr. Popul. Biol.* **15.**

Rener, J., Graves, P. M., Carter, R., Williams, J. L., and Burkot, T. R., 1983, Target antigens of transmission-blocking immunity on gametes of *Plasmodium falciparum*, *J. Exp. Med.* **158:**976–981.

Schad, G. A., and Anderson, R. M., 1985, Predisposition to hookworm infection in humans, *Science* **228:**1537–1539.

Scott, C. M., Frézil, J.-L., Toudic, A., and Godfrey, D. G., 1983, The sheep as a potential reservoir of human trypanosomiasis in the Republic of the Congo, *Trans. R. Soc. Trop. Med. Hyg.* **77:**397–401.

Smith, G., and Grenfell, B. T., 1985, The population biology of *Ostertagia ostertagi*, *Parasitol. Today* **1**(3):76–81.

Traub, R., and Wisseman, C. L., Jr., 1974, The ecology of chigger-borne rickettsiosis (scrub typhus), *J. Med. Entomol.* **11:**237–303.

Walsh, J. F., Davies, J. B., and LeBerre, R., 1978, Standardization of criteria for assessing the effect of *Simulium* control in onchocerciasis control programmes, *Trans. R. Soc. Trop. Med. Hyg.* **72:**675.

Walsh, J. F., Davies, J. B., and LeBerre, R., 1979, Entomological aspects of the first five years of the onchocerciasis control programme in the Volta River basin, *Tropenmed. Parasitol.* **30:**328–344.

Wilson, G. G., and Kaupp, W. J., 1976, Incidence of *Nosema fumiferanae* in spruce budworm, *Choristoneura fumiferana*, in the year following application, *Rev. Appl. Entomol. Ser. A* **65:**5064.

World Health Organization, 1982, Scientific Activities WHO–OMS—Activitès scientifiques: Evaluation of the onchocerciasis control programme, *Bull. WHO* **60:**185–188.

WHO Expert Committee, 1984, *The Leishmaniases*, WHO Technical Report Series, No. 701, Geneva.

Epilogue

In the preceding discussion (Chapter 26), centering largely on the epidemiology of a number of parasitic infections and how this is affected by various ecological conditions, I have considered parasite populations in terms of their average distribution in a given population of hosts. I have ignored the salient fact that parasite ecology involves exploitation of a small patch of resources, the individual host. Each host is like an island, separated by a smaller or greater expanse of sea, the hostile external environment. The patchy resources and their ephemeral nature all contribute to the nonequilibrium conditions in which parasites exist. They typically have high rates of reproduction with strong tendencies to parthenogenesis and hermaphrodism. They show rapid rates of evolution with often polytypic species. Much of the evolutionary history of parasites has been determined by the development of different ways to get from island to island, i.e., from host to host. This brings us back full circle to the beginning of this book where I discussed the establishment of infection. We are just beginning to understand the physiological and develomental mechanisms, the cellular and molecular interactions involved in getting out from one host and into another. This remains as one of the most fascinating and most important fields for future inquiry. With intracellular protozoa, receptors on their surface evidently interact with molecules on the surface of the host cell. Here the malarial parasites in their erythrocytic stage have provided a favorable material for study, but many other host–parasite combinations are available and should be investigated. The responses and tropisms involved in the finding of special sites within a host by parasitic helminths have been little studied. Here there is need for the development of appropriate systems amenable to experimental work. Perhaps the cercaria of *Diplostomum flexicaudum,* that seeks out the lens of its fish host, would be such a material (see Chapter 3).

Equally interesting and important are the problems concerned with developmental cues, what directs development along one of several possible lines. Why, for example, do many of the newly molted fourth-stage larvae of the nematode *Ostertagia,* at a certain time of the year, enter into diapause instead of developing rapidly to the adult stage? Why do some sporozoites of the malarial parasite *Plasmodium vivax* remain in a liver cell for months as dormant stages (hypnozoites) only to resume development (in response to what?) and produce a relapse? What determines the differential development of presumably genetically identical merozoites of *P. falciparum* so that some develop into male gametocytes, some into female gametocytes, but most into schizonts that reproduce and continue the asexual cycle?

As already noted several times, parasites must be especially adept at switching genes on and off as they go from one host to another, or from the external environment into a new host. Extension of this same facility presumably enables them to respond to the more subtle environmental differences that direct development to a gametocyte rather than an asexual schizont, or to a diapause larva or a resting hypnozoite instead of an actively developing form. An understanding of antigenic variation in trypanosomes might lead to better understanding of the many other biological phenomena that seem to depend on a turning on and off of particular genes at appropriate times.

Antigenic variation is only one of the ways parasites have to evade the immune system of a vertebrate host. Some of the helminthic parasites in particular have developed the ability to live on together with their host for many years despite the presence of a measure of acquired immunity that limits the extent of superinfection, thereby largely preventing excessive injury to the host. Immunologists have already been attracted to these problems, which promise to throw light as much on immunological mechanisms as on parasite physiology. Here there are a number of medically important parasites already studied in depth and waiting to be further explored (see Chapters 18–23).

For some other kinds of basic problems in parasitism, however, medically important species may not provide the best material. Some effects on host behavior and sexual characters are especially prominent in certain invertebrate hosts (see Chapter 12). The *Sacculina*–crab system would be particularly interesting if it could be brought under controlled laboratory conditions. Could *Sacculina* be induced to metamorphose if provided with appropriate stimuli and a nonliving nutrient medium? With its complex branching root system within its host, *Sacculina* seems as intimately host-dependent as the intracellular parasites.

The extent and manner of integration of intracellular parasites into the economy of their host cells is again of as much interest to cell biologists as to parasitologists (see J. W. Moulder, 1985, cited in Chapter 5). There are here two general sets of problems. First, what special environmental factors does the host cell supply to the parasite? Can these be furnished in a nonliving culture medium to permit axenic growth of the parasite? Second, how do products of the parasite effect appropriate modifications of the host cell? These changes can include a whole range from increased permeability to special surface structures to hyperplasia and hypertrophy. Studies with intracellular protozoa like *Plasmodium* and *Leishmania* are beginning to provide answers.

Much more should be done with the microsporidia, some of which have now been grown in tissue culture. As in so many other kinds of biological problems, application of techniques of cell culture and molecular biology should lead to significant progress.

Those who study parasitism are by definition parasitologists. However esoteric their methods, however enamored they may be of a pet host–parasite system, they should not forget the ultimate practical importance of their subject. Much of humanity still suffers dearly from parasitic diseases. The parasites, of course, have been around for a long time. We are just beginning to learn.

Taxonomic Index to the Parasites and Vectors Considered

Subject Index